Adult Dental Health Survey

Oral Health in the United Kingdom 1998

A survey commissioned by the United Kingdom Health Departments carried out by the Social Survey Division of the Office for National Statistics in collaboration with the Dental Schools of Birmingham, Dundee, Newcastle and Wales, and the Central Survey Unit of the Northern Ireland Statistics and Research Agency

Maureen Kelly, Jimmy Steele, Nigel Nuttall, Gillian Bradnock, John Morris, June Nunn, Cynthia Pine, Nigel Pitts, Elizabeth Treasure, Deborah White.

Edited by Alison Walker and Ian Cooper.

London: The Stationery Office

About the Office for National Statistics

The Office for National Statistics (ONS) is the Government Agency responsible for compiling, analysing and disseminating many of the United Kingdom's economic, social and demographic statistics, including the retail prices index, trade figures and labour market data, as well as the periodic census of the population and health statistics. The Director of ONS is also Head of the Government Statistical Service (GSS) and Registrar-General in England and Wales and the agency carries out all statutory registration of births, marriages and deaths there.

Editorial policy statement

The Office for National Statistics works in partnership with others in the Government Statistical Service to provide Parliament, government and the wider community with the statistical information, analysis and advice needed to improve decision-making, stimulate research and inform debate. It also registers key life events. It aims to provide an authoritative and impartial picture of society and a window on the work and performance of government, allowing the impact of government policies and actions to be assessed.

Information services

For general enquiries about official statistics, please contact the National Statistics Public Enquiry Service:

tel: 020-7533 5888
textphone (minicom): 01633 812399
fax: 020-7533 6261
e-mail: info@ons.gov.uk.
written enquiries: room DG/18, 1 Drummond Gate, London, SW1V 2QQ.

ONS can also be contacted on the Internet: www.ons.gov.uk

Most National Statistics publications are published by The Stationery Office and can be obtained from PO Box 29, Norwich, NR3 1GN, tel 0870 600 5522 or fax 0870 600 5533.

Contents

Contents

Contents

Foreword

The survey of Adult Dental Health in England and Wales in 1968 brought together, for the first time in this country at a national level, dental epidemiologists and social researchers. The authors were Gray, Todd, Slack and Bulman. Professor Slack called on senior dental colleagues for advice and guidance and was assisted by a young research fellow Roger (Andy) Anderson, who carried out one of the pilot studies for the Adult Dental Health Survey. A research assistant, Jean Todd, in the Government Social Survey (now ONS), contributed considerably to the writing of the text. Jean Todd and Andy Anderson became the backbone or foundation on which the national surveys of child and adult dental health developed over the next twenty-five years. The intertwining of their ideas and their total commitment to working together ensured that these surveys became established nationally and were recognised internationally.

In one sense the 1968 survey was only a partial success from a dental epidemiological point of view. Because of the large number of edentate adults, the data on dentate adults could only be divided into two age groups, 16 to 34 year olds and those aged 35 years and over. Over the life of the surveys, various changes have been introduced. In 1978, edentate adults were not examined in order to concentrate dental manpower on the dentate population. By 1988, the calibration and training sessions benefited from the fact that there was a core of dental examiners who had taken part in previous surveys. Parts of the dental examination were expanded so that more attention could be paid to periodontal disease in adults and to the increasing amount of complex restorative dentistry (crowns and bridges) that was now evident in the adult population.

Further developments have been incorporated into the 1998 Adult Dental Health Survey. The diagnosis of caries has been expanded to include visual evidence, to ensure a more complete assessment of this disease; cavitated lesions were recorded separately to allow comparison with earlier surveys. For the first time an assessment was made of adhesive bridges and veneers, so that there is scope to record new technologies in a meaningful way. The assessment of periodontal disease has been expanded, because this is an important parameter in an ageing dentate population. A measurement of plaque, a matter of contention in 1968, has now been included in the present survey. This is important if improvements in oral hygiene are to be measured, particularly in ageing dentate populations at risk to root caries as well as periodontal disease. The state of the occlusion is recorded by determining the proportion of the dentate population with 21 or more teeth and the distribution of functioning teeth, emphasising the trend towards the concept of a functional dentition. More comprehensive sociological measures have been introduced, including the Oral Health Impact Profile, to ensure a patient based measure of health, rather than focusing merely on disease.

The improvements that have been made to the 1998 Adult Dental Health Survey reflect the changing nature of oral health in the population and enable new benchmarks to be set so that variations in the oral health of the population in the future can be monitored. At the same time the results of the present study can be set in context and compared with data collected over the last thirty years. Thirty years ago 37% of adults in England and Wales aged 16 years and over were edentate, now the proportion has fallen to 12%.

All those responsible for the 1998 survey are to be congratulated for producing an excellent report, which should be of great interest to the dental profession and to all those who believe that good oral health is an important component of the general health of the population.

J J Murray C B E
The Dental School, University of Newcastle upon Tyne

Acknowledgements

Surveys of dental health depend on the efforts of many different people and we would like to thank everybody who contributed to the survey and to the production of this report.

We would like to thank all the specialist colleagues at ONS who carried out the sampling and data preparation and in particular the interviewers who, as well as conducting interviews, took on the task of dental recorders for the home dental examination.

We would also like to congratulate the survey dental examiners on their energy and enthusiasm, and for being willing to adapt to the unfamiliar circumstances of carrying out survey dental examinations in people's own homes. Their names, along with those who trained them, will be found at the end of this report.

As well as thanking the dental examiners and trainers we would also like to thank those who volunteered to be examined during the training of the survey dental team. This vital contribution to the survey involves being examined as many as ten times in an hour. Staff working at the ONS Offices and the Patent Office in Newport undertook this somewhat arduous role, many of them volunteering for several sessions.

Finally, we would like to thank the people who agreed to take part in the survey. Without their willingness to help, the survey could not have happened.

This report is dedicated to the memory of
Jean Todd and Andy Anderson

Notes on the tables

All proportions and means presented in the tables are taken from data weighted to compensate for the differential probabilities of selection and non-response. Base numbers for the 1998 data are presented in the unweighted state and represent the actual number of people in any specified group. All base numbers have been given in italics. Base numbers for data in the trend tables have been taken from the published reports of the earlier surveys and are presented in Appendix G.

Symbols used in the tables:
0   less than 0.5% or a mean of less than 0.05 or a zero value
-       base number less than 30, statistic not given
..   data not available or applicable.

The varying positions of percentage signs and labels and of bases in the tables denote the presentation of different types of information. Where the percentage sign is at the head of a column and the base is at its foot the whole distribution is presented and the unrounded figures add to 100% unless otherwise stated. A percentage label in italics across the top of the table and bases at the side or foot of the table signify that the figures show the percentage of people with the attribute being discussed. The complementary percentage (those without the attribute) is not shown in the table.

The total column may include cases from small sub-groups not shown separately elsewhere on the table, therefore the individual column bases may not add to the base in the total column. Rounding may mean that percentages may add to between 99% and 101%.

Footnotes identify the equivalent table from the 1988 report where appropriate.

All differences commented on in the text have been found to be significant at the 95% confidence level (p<0.05). Throughout the report, the terms 'significant' and 'statistically significant' are used interchangeably. Where differences are described as 'not (statistically) significant' this indicates that p<0.05. The formulae used to test for significant differences are shown in Appendix F.

Corrections to report of the 1988 survey (see Chapter 1.1 for reference)
Table 2.4 UK 1978 aged 55-64;   for 50% *read 56% and for +7% read +1%.*
Figure 10.4 heading: *for 16-34 read 16-24.*
Appendix 2 Tables on total tooth loss. *For headings* all dentate adults *read* all adults.

Part I

Background and methodology

1.1   Background and methodology

Summary

- This report presents the results from the 1998 Adult Dental Health Survey, the fourth in a series of national dental surveys that have been carried out every ten years in England and Wales since 1968 and across the United Kingdom since 1978.
- The survey was based on a representative sample of adults aged 16 and over living in private households in the United Kingdom.
- All adults living in the selected households were asked to take part in an interview; followed at a later date by a home dental examination for adults with some natural teeth, carried out by a dentist specially trained for the survey.
- The fieldwork for the survey was carried out between September and December 1998, following a pilot study carried out in February 1998.
- The interview covered dental attitudes, experiences and behaviour.
- The examination included the existence and condition of natural teeth, the condition of the root surfaces, the nature of contacts between the upper and lower teeth, whether there were spaces between the teeth and if these were filled by dentures or bridges, the type and condition of any dentures, and the condition of the gums.
- The purpose of the survey is to provide information on the current state of adults' teeth and oral health in the four countries of the United Kingdom and to measure changes in oral health since 1988.

1.1.1 Aims of the 1998 Adult Dental Health Survey

The 1998 Adult Dental Health Survey, commissioned by the four United Kingdom (UK) Health Departments, is the fourth in a series of national dental surveys that have been carried out every ten years since 1968[1]. The purpose of the survey is to provide information on the current state of adults' teeth and oral health in the four countries of the United Kingdom and to measure changes in oral health since 1988.

The specific aims of the survey are:
- *to establish the condition of the natural teeth and supporting tissues*
- *to investigate dental experiences, attitudes and knowledge, dental care and oral hygiene*
- *to establish the state and use made of dentures worn in conjunction with natural teeth*
- *to identify those who have lost all their natural teeth and investigate their use of dentures*
- *to examine the change over time in dental health, attitudes and behaviour*
- *to monitor the extent to which oral health targets set by government are being met.*

1.1.2 Survey overview

The survey was based on a representative sample of adults aged 16 and over living in private households in the UK. All selected adults were asked to take part in an interview; followed at a later date by a home dental examination for adults with some natural teeth, carried out by a dentist specially trained for the survey. The fieldwork for the survey was carried out between September and December 1998, following a pilot study carried out in February 1998.

1.1.3 Sample design

The sample size and design were determined by the need to provide data for each of the constituent countries of the UK and to measure changes in oral health since 1988. In order to achieve this, proportionately larger samples were selected in Northern Ireland, Scotland and Wales than in England. The sampling proportions in those three countries were increased by factors, in rounded terms, of six, four and three and a half respectively. This design produced a self-weighting sample for each of the constituent countries of the UK but necessitates re-weighting to produce representative figures for the UK as a whole (see Section 1.9).

A multi-stage sampling procedure was used, with the sample size at the final stage determined partly by consideration of the number of dental examinations that each examiner could reasonably be expected to cover both in terms of time and cost. In Great Britain, a random sample of addresses was selected from the Postcode Address File (PAF) using a two-stage sample design. At the first stage, 76 postcode sectors in England, 32 postcode sectors in Scotland and 16 postcode sectors in Wales were selected from a list stratified by region (Government Office Regions), socio-economic group, and car ownership. Forty addresses within each sector were then selected. This gave a total of 4,960 sampled addresses in Great Britain. In Northern Ireland a simple random sample of 580 private addresses was selected from the Rating and Valuation list. In total, 5,540 addresses were sampled for the survey. Interviewers were instructed to seek interviews with all adults aged 16 and over living at each address.

1.1.4 The interview

An interviewer visited each household selected in the sample and interviewed every adult aged 16 and over who agreed to participate in the survey. For the first time in this series of Adult Dental Health Surveys, the interviews were conducted using computer-assisted interviewing (CAI). Laptop computers held the questionnaire in the form of a computer program and the respondent's answers were entered into the laptop by the interviewer (or by the respondent in the case of the self completion sections of the interview).

The interview covered the following topics:

- *a self assessment of the presence of natural teeth, fillings and dentures;*
- *satisfaction with their teeth and mouth, including appearance and the ability to speak, chew and swallow;*
- *opinions on the need for dental treatment;*
- *past dental experience and care received;*
- *patterns of past, present and future dental attendance including the most recent dental visit;*
- *attitudes to dental treatment;*
- *oral hygiene habits and advice received;*
- *patterns and reasons for tooth loss;*
- *pattern of denture wearing and attitudes to dentures.*

Most of the questions were the same as in the 1988 survey in order to measure any differences in experiences, attitudes and opinions since 1988. A paper version of the questionnaire can be found in Appendix H.

1.1.5 The examination

At the end of the interview, adults with some natural teeth were asked whether they would take part in a home dental examination. If they agreed, the interviewer made an appointment to return at a later date with a dentist to conduct the dental examination.

Each respondent signed a consent form before the dental examination took place. If they were aged 16 or 17 and living with their parents, the parents' consent was also required. The dentist also asked if the respondent had a history of certain medical conditions that could have related to any risk the dental examination may pose. (see Appendix C for a list of the conditions). If the respondent had any of the conditions listed then the part of the examination relating to the condition of the gums (the periodontal examination) was omitted.

During the dental examination the dentist examined for:

- *the existence and condition of natural teeth;*
- *the condition of the root surfaces;*
- *wear of the tooth surfaces;*
- *the nature of contacts between the upper and lower teeth;*
- *whether there were spaces between the teeth and if these were filled by dentures or bridges;*
- *the type and condition of any dentures;*
- *the condition of the gums (the periodontal examination).*

Interviewers recorded the details of each dental examination on their laptop computers using an electronic dental chart specifically designed for this survey.

The criteria for the examination were designed to take into account changes in the relative importance of the different aspects of oral health while maintaining comparability with previous Adult Dental Health Surveys. One of the most significant changes with respect to the criteria for the survey dental examination was that in the assessment of decay. In the 1998 survey the criteria for measuring decay were changed from those used in earlier surveys in this series. Previously, a tooth was recorded as decayed only if cavitated caries was present, but for the first time in 1998, an assessment of visual caries (that is early caries that has caused demineralisation of the tooth but without cavitation) was added to the assessment of decay. Bearing in mind the need for comparability, visual caries was identified separately from cavitated caries so that summaries of the data could be derived with or without visual caries (see Chapter 3.1).

The examination also included new sections on tooth wear, and more detail on tooth contacts. Parts of the criteria, such as tooth wear, condition of root surfaces and tooth contacts, were also designed to be comparable to previous national surveys of children's teeth or the oral health surveys undertaken within the series of National Diet and Nutrition

(NDNS) surveys in particular that of older adults[2]. The criteria and reporting for periodontal disease has caused problems in previous Adult Dental Health Surveys. The criteria for this part of the examination were designed to provide data that could be compared with some of the data collected in the 1988 survey. Full details of the examination criteria and procedures are given in Appendix C.

1.1.6 Recruitment and training

In Great Britain, interviewers from the permanent field force employed and trained by the Social Survey Division of the Office for National Statistics (ONS) carried out the interviews and recorded the dental examinations. In Northern Ireland, the Central Survey Unit (which is part of the Northern Ireland Statistics and Research Agency) carried out the fieldwork. All the interviewers were briefed on the dental questionnaire and specially trained to record the details of the dental examinations.

The home dental examinations were conducted by 75 dentists from the community dental services and health authorities of the National Health Service. The majority of examiners were recruited via the Regional Offices of the NHS Executive and British Association for the Study of Community Dentistry (BASCD) epidemiology advisors. Many had previous experience in collecting data using similar clinical criteria and procedures on other surveys, such as the BASCD surveys of children. The examiners who took part in the survey are listed in Appendix D.

The dentists each attended one of two training courses held at the ONS office in Newport, Wales. The training consisted of an initial briefing on the criteria used in the dental examination, two days of practice dental examinations, and a calibration exercise on the final morning. The practice dental examinations were conducted on volunteers from ONS Newport and the nearby Patent Office. Part of the training included working with the interviewers who were to record the details of the dental examinations in the field.

As there were many dentists working on the survey there was scope for variation in how individual dentists interpreted what they saw during an examination. The training was designed to ensure that the dentists conducted the examinations in the most consistent way possible. During the training, three or four different dentists examined each volunteer and any inconsistencies between the dentists were identified and resolved. After the training was completed, a calibration exercise was conducted to measure the variation between the dentists. Calibration data were calculated for conditions of tooth and root surfaces, tooth wear, contacts and spacing. The results of this exercise are presented in Appendix B.

Each dentist worked with between one and three interviewers. One of these interviewers attended the training course in Newport. Interviewers who had not met the dentist at the training course arranged a short practice session before starting fieldwork. The dentists had only a limited amount of time set aside to help with the survey. To make the most efficient use of this time, the dentists arranged with the interviewers when they would be able to conduct dental examinations. This enabled interviewers to make appointments with members of the public knowing that the dentist was available.

1.1.7 Ethical clearance

Ethical approval for the survey as a whole was obtained from the North Thames Multi-Centre Research Ethics Committee and by the Local Research Ethics Committees (LRECs) in all the areas covered by the survey. The LREC for Gwent Health Authority gave ethical approval for the training that took place in Newport.

As part of the survey procedures that gained ethical approval, emphasis was placed on obtaining the informed consent of the respondents at each stage of the survey. Those who agreed to co-operate with the interview and had some natural teeth were given the opportunity to opt out of the dental examination. Respondents were asked to have a dental examination only if they had completed the interview. The dental examination never took place on the same day as the interview.

1.1.8 Response

In total, 6,204 adults were interviewed and 3,817 dental examinations were carried out. Table 1.1.1 provides further details.

Of the 5,540 selected addresses in the sample, 599 (11%) did not contain a private household and were therefore ineligible for the survey; these included institutions, demolished or vacant premises and addresses used solely for business purposes. Some addresses contained more than one private household. Where this occurred, set procedures were followed to ensure each household had an equal chance of selection (see Appendix F). A total of 43 extra households were identified, yielding a total sample of 4984 eligible households.

Interviewers were instructed to seek interviews with all adults aged 16 and over living in the selected households. At least one person was interviewed in 74% of all the households that were visited, 21% of the households declined to co-operate at all and at 5% no contact was made with any member of the household. The interviewers were instructed to carry out an interview with each adult if at all possible. However, proxy interviews were allowed for respondents who had no natural teeth, but only if the person was difficult to contact or difficult to interview due to their age or an illness. During the survey 50 such interviews were carried out.

As previously mentioned, respondents were eligible for the dental examination only if they had some natural teeth and if they had completed the interview; 72% of the dentate adults interviewed had a dental examination.

Table 1.1.1

Any survey that is based on several stages of data collection is liable to non-response at each stage. In this survey non-response occurred at the household level, at the interview and at the dental examination. The level of non-response at each of these stages was higher than it was in 1988. One of the possible reasons for this is that over the last decade there has been a general decrease in the levels of co-operation in surveys. Although response was not at the level it was in 1988 it was still relatively high and the levels achieved reflect the effort and resource that were put into the fieldwork by the interviewers and dentists. The effect of non-response on the data and any bias it may have caused is discussed below.

1.1.9 Weighting

As mentioned earlier, proportionately larger samples were drawn in Wales, Scotland and Northern Ireland than in England. In order to produce representative figures for the UK as a whole, the sample has to be weighted to restore the correct balance between the constituent countries.

In addition to the weighting required by the sample design, weighting was also used to reduce the risk of possible bias caused by non-response at the interview and dental examination stages (see Section 1.8). Non-response may bias the results of a survey if the characteristics of people who did not participate are different from those of people who did take part; the results based on data from the responding adults would not be representative of the population as a whole. For example, if adults with natural teeth were more likely to respond to the survey than those without natural teeth, people with no natural teeth would be under-represented in the sample. If this occurred the results from the survey would give a higher proportion of dentate adults than existed in the population.

Weighting is used to try and reduce the risk of any non-response bias. One of the difficulties in trying to correct for non-response bias is identifying where and in what direction the bias may be found.

1.1.10 Weighting the interview data

Little or no information was known about the adults who were not contacted or refused to take part in the survey at the interview stage. Previous dental surveys have shown that several factors that relate to dental health vary according to age, sex and region. It was therefore considered important that the age, sex and regional distributions of the responding sample reflected those of the population as a whole. The distributions of the interviewed sample were compared with the population data provided by the Labour Force Survey (LFS) for September to November 1998. The LFS is a large sample survey of the general population carried out in the UK on a quarterly basis; the data are weighted and grossed to give estimates of the total UK population living in private households. The LFS estimates were taken as the most appropriate for comparison with the survey sample as they provide estimates of the age, sex and regional distribution of people living in private households in the UK.

Taken as a whole, the sample of responding adults under-represented men. The sample also under-represented people aged 16 to 24 and over-represented people aged 45 to 54 and 65 to 74. Within each region, the age and sex distributions of the sample also varied when compared with the LFS estimates, although the sub-groups that were under and over-represented differed in each region. The data were weighted to reflect the age and sex distributions within each region given by the LFS estimates. This weighting incorporated both the adjustment for possible non-response bias and the adjustment of the over-sampling in Wales, Scotland and Northern Ireland.

1.1.11 Weighting the examination data

All dentate adults who were interviewed were asked if they would have a home dental examination; 72% of the dentate adults interviewed also had a dental examination. To reduce the risk of bias it was possible to weight the data for the sample of people who had dental examinations to reflect the characteristics of the full sample of dentate adults who were interviewed.

During the interview a large amount of data were gathered from dentate adults on a variety of topics including demographic information; social background; and dental behaviour, attitudes and opinions. These data were analysed,

using a program called CHAID[3] (Chi-squared automatic interaction detector) to find which characteristics were most significant in distinguishing between those who took part in the dental examination and those who did not. Nine characteristics were found to be significant which included characteristics relating to dental behaviour and opinions as well as socio-demographic information. The examination data were then weighted according to these characteristics, so that their distributions for the examined adults matched the distributions for the dentate adults as a whole.

A full description of the weighting procedures carried for both the interview and examination stages is given in Appendix F.

1.1.12 Sample sizes

The interview weights were calculated so that England retained the same sample size when weighted and unweighted. As a result, the weighted sample sizes for Wales, Scotland and Northern Ireland were much lower than the unweighted sample sizes. The examination weights were

---

Fig 1.1   **Geographical areas used in the 1998 Adult Dental Health Survey**

calculated to retain (approximately) the same sample size of dental examinations. Table 1.1.2 shows the unweighted and weighted sample sizes for the constituent countries at the various stages of the survey.

Table 1.1.2

The tables in the main body of the report present weighted proportions, using the appropriate weight, and unweighted bases.

1.1.13 Regions

The sample size was large enough to produce results on a regional basis in England. The English regions used in the analysis of the 1998 data and presented in this report were slightly different from those used on the 1988 survey. In 1988, England was subdivided into three regions, Northern, Midlands and Southern, based on amalgamations of the Regional Health Authorities (RHAs) that existed in 1988. Over the past ten years the administrative areas of the National Health Service have been reorganised. The RHAs were abolished and were replaced by Regional Health Offices, with different geographical boundaries. There have also been boundary changes to the constituent Health Authorities that form the larger Regional Health Offices (and that used to form the RHAs).

Using the 1998 administrative boundaries the regions used in 1988 could not be replicated exactly. However, for comparison of data over time, equivalent regional groupings were required. Since re-analysis of the 1988 data was not within the remit of this study, the best approximations to the 1988 regions were produced based on an amalgamation of the new Health Authorities and Regional Health Offices. In some instances, the new Health Authorities crossed the boundaries of two of the regions used in the 1988 survey. Where this occurred the Health Authority was assigned to the region that the major part of the authority would have been in, in 1988. Figure 1.1 shows the geographical areas used.

Figure 1.1

In addition, the 1998 data by individual Regional Health Office are presented in the chapter which looks at the results for England (Chapter 7.1).

1.1.14 Contents of the report

This report is divided into 7 main parts:
1. Background and methodology (with additional details in the appendices)
2. Total tooth loss, number of teeth and function

3. The condition of the natural teeth, restorative treatment and supporting structures
4. Social and behavioural characteristics and oral health
5. Trends in tooth loss and the condition of the natural teeth
6. Dental attitudes, experience and reported behaviour
7. Reports by country

Parts 2 to 4 examine the oral health of adults in the United Kingdom in 1998 and Part 5 presents comparisons of these results with those of the previous surveys in the series. Part 6 looks at dental attitudes, experience and behaviour as reported by respondents during the interview. In Part 7, findings from the rest of the report are drawn together to present a cohesive picture for each of the countries of the United Kingdom together with reference to oral health targets for these countries.

It should be noted that results presented in Parts 2 to 4 are based on the new criteria introduced for the 1998 survey which included visual caries in the definition of decay (see Section 1.1.5). Results in Part 5 and some of those in Part 7 are based on the examination criteria used in the previous surveys of adult dental health, which did not include visual caries.

Notes and references

1. The references for the surveys of adult dental health in this series are as follows:

    Gray PG, Todd JE, Slack GL and Bulman JS. *Adult dental health in England and Wales in 1968* HMSO London 1970
    Todd JE and Walker AM. *Adult dental health Volume 1 England and Wales 1968-78* HMSO London, 1980
    Todd JE, Walker AM and Dodd P. *Adult dental health Volume 2 United Kingdom 1978* HMSO London, 1982
    Todd JE and Lader D. *Adult dental health 1988 United Kingdom* HMSO London, 1991

    Two separate surveys have been carried out during this period, one for Northern Ireland and one for Scotland.
    Todd JE and Whitworth A. *Adult dental health in Scotland 1972* HMSO London, 1974
    Rhodes JR and Haire TH. *Adult dental health survey Northern Ireland 1979* HMSO Belfast, 1981

2.   Steele JG, Sheiham A, Marcenes W, Walls AWG.
      *National Diet and Nutrition Survey: people aged
      65 and over Volume 2: Report of the oral health
      survey* TSO London 1998

3.   CHAID is a package within the statistical data
      analysis system SPSS.

Table 1.1.1 **Response at different stages of the survey**

Total sample (unweighted)		%
Sample of addresses		
Number of selected addresses	5540	100
Addresses containing eligible households	4941	89
Ineligible addresses	599	11
Household response†		
All eligible households	4984	100
Responding households	3666	74
Refusals	1063	21
Non-contact	255	5
Individual response		
Adults in contacted households	6764	100
Adults interviewed	6204	92
No interview obtained	560	8
Interviewed adults with natural teeth	5281	100
Dental examination obtained	3817	72
No dental examination	1464	28

† some addresses contained more than one household.

Table 1.1.2 **Effects of weighting by country**

	Actual interviews	Actual examinations	Final interviews (reweighted)	Final examinations (reweighted)
United Kingdom				
Interviewed adults	6204	-	4110	-
Adults with teeth	5281	-	3592	-
Adults with no teeth	923	-	518	-
Examined adults (with teeth)	3817	3817	2599	2598
England				
Interviewed adults	3436	-	3436	-
Adults with teeth	3010	-	3029	-
Adults with no teeth	426	-	407	-
Examined adults (with teeth)	2186	2186	2197	2190
Wales				
Interviewed adults	830	-	205	-
Adults with teeth	682	-	171	-
Adults with no teeth	178	-	34	-
Examined adults (with teeth)	502	502	126	127
Scotland				
Interviewed adults	1204	-	358	-
Adults with teeth	953	-	295	-
Adults with no teeth	251	-	63	-
Examined adults (with teeth)	668	668	206	208
Northern Ireland				
Interviewed adults	734	-	111	-
Adults with teeth	636	-	97	-
Adults with no teeth	98	-	14	-
Examined adults (with teeth)	461	461	70	73

Part 2

Total tooth loss, number of teeth and function

2.1   The loss of all natural teeth

Summary

- In 1998, 87% of the population of the United Kingdom had some natural teeth, while 13% had lost all their natural teeth.

- Within the four countries of the UK the proportion of adults who had lost all their natural teeth varied; 12% of adults in England and Northern Ireland, 17% in Wales and 18% in Scotland had lost all their natural teeth.

- Adults with no natural teeth were concentrated in the older age groups and it was rare for adults to have lost all of their teeth before the age of 45. Among adults in all age groups under 45 years 1% or less of adults had lost all their natural teeth compared with 58% of adults aged 75 and over.

- Total tooth loss varied with other characteristics: gender, English region, and social class. Men from non-manual backgrounds living in the South had the highest proportion of dentate adults (97%) whereas women from unskilled manual backgrounds from Scotland had the lowest (64%).

- The proportion of edentate adults in 1998 mainly reflects the experience of dental disease, treatment and practice from many years ago; very few adults (1%) reported becoming edentate between 1988 and 1998. These adults were drawn from almost all age, social and geographical categories.

- By 1998 the loss of natural teeth was predominately a gradual process. The removal of large numbers of teeth at one time to render adults edentate, which was once common, was a rarity. Under a quarter of those who became edentate between 1988 and 1998 had lost 12 or more teeth when they became edentate.

- Two thirds of adults who became edentate between 1988 and 1998 felt that tooth decay was the major reason for the removal of their remaining teeth and half reported gum disease or loose teeth prior to their removal.

2.1.1 Introduction

Despite being a relatively crude indicator, the proportion of the population who have no natural teeth (usually referred to as edentate or edentulous) is a very important measure of oral health. Existing evidence suggests that where all teeth are lost there may be a significant impact on diet, nutrition and general well being[1], so the implications of the loss of all of the natural teeth may extend well beyond the mouth. Reducing the proportion of the population who have no remaining natural teeth is still a high priority in terms of improving oral health.

The report of the 1988 Adult Dental Health Survey stated that some tooth loss was considered inevitable, but that "by and large, loss of all natural teeth was avoidable". With this in mind, in the 1988 report a threshold of 95% of the population with some natural teeth (dentate) was identified as a desirable and achievable marker against which the changing oral health of different groups of the United Kingdom population could be measured. In the subsequent Department of Health publication *An oral health strategy for England,* published in 1994, a national target for tooth retention was set: that 33% of adults aged over 75 years old should have some natural teeth by 1998. In Scotland a target of 95% of 45 to 54 year olds by 2010 has been set. For those who had some natural teeth, the possibilities and opportunities for retaining them were broadly similar in 1998 to the situation in 1988. The 95% threshold will be used again in this report and reference will also be made to the Department of Health's target set in 1994 for England. Although the latter was set specifically for England, it is referred to here as a means of measuring the progress of the population of the United Kingdom and its constituent parts as well.

As the proportion of the population without teeth diminishes, other indicators will be necessary to measure oral health and function. The simple presence of teeth infers nothing about their condition or health, and the proportion without teeth is only one of a range of key measures that will be used to evaluate oral health and function in this report. It is the one that will be addressed first.

2.1.2 Variations in total tooth loss in the United Kingdom in 1998

In 1998, 87% of adults in the UK had one or more natural teeth, while 13% had lost them all, marking a continued improvement over the previous 3 decades. Table 2.1.1 shows the age distribution of those with and without natural teeth in 1998 and illustrates clearly that the prevalence of total tooth loss is very much higher in the oldest age groups than

in the middle aged and young. Although an age related increase in prevalence is to be expected, in the report of the 1988 survey it was observed that the proportion of the population who were edentate was largely an historical statistic, because at one time loss of all natural teeth by the age of 50 was not at all uncommon. While early onset total tooth loss is now very rare indeed, this legacy of the past was still with us in 1998, and the generation involved (by and large those who were late adolescents or adults at the inception of the National Health Service in 1948) account for a large proportion of the current population of edentate adults in the United Kingdom.

Figure 2.1.1, Table 2.1.1

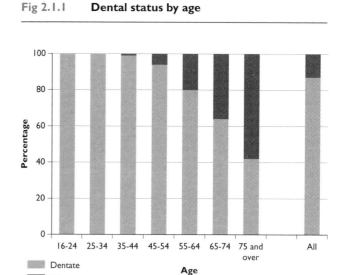

Fig 2.1.1   **Dental status by age**

If the data are looked at in a different way it can be seen how this, mainly residual, edentate population is distributed according to age and gender. Table 2.1.2 shows the percentage of the edentate population in 1998 in different age and gender groups. Over two thirds (69%) of all of the edentate people in the United Kingdom in 1998 were aged 65 and over and 39% were aged 75 and over. Although the total number of edentate people is now relatively low, a large part of the overall edentate population, around one in three, were relatively young, aged under 65 years of age (38% of all of the edentate men and 26% of the edentate women in the population are under 65 years of age). Gradually, over the next couple of decades, the legacy of widespread early loss of all natural teeth will disappear and the total number of people with no natural teeth is likely to diminish further. However, although fewer edentate people are expected, with less of the very old edentate adults, the proportion of this diminishing edentate population who are aged under 65 years of age may actually increase. Chapter 5.1 investigates in greater detail when the effect of early tooth loss will finally be lost.

Table 2.1.2

The prevalence by age of being dentate in 1998 gives a simple indication of the changes in the different groups of the United Kingdom population and of their progress towards existing targets. At the 95% threshold (which was also used in the report of the 1988 survey), all groups aged under 45 years in 1998 stand a very high chance of remaining dentate above this level for some time, with rates of total tooth loss being around 1%, and with 99% of the sample having at least one tooth. Table 2.1.1 shows a marked drop in the percentage with teeth at 45 years of age. Looking at the age distribution in a little more detail (see Table 2.1.3), the greatest change clearly occurs between the 40 to 44 year olds , among whom 1% are edentate (very similar to all of the other younger age groups), and those aged 45 to 49 years for whom 5% are already edentate. On current trends, for all age groups under the age of 45 in the United Kingdom in 1998, maintaining levels of tooth retention above the 95% threshold should be comfortably achievable over the next 20 years or more. The 45 to 49 year olds, by comparison, will now be most unlikely to stay above the 95% dentate target in the longer term; given that 5% of this age group have already been rendered edentate, it would seem likely that more of this age group will become edentate in the future.

The government target against which progress can be measured is the one published in the Department of Health document *An oral health strategy for England*. This aimed to achieve 33% of people aged 75 years and over being dentate by 1998. In terms of the United Kingdom as a whole, this target has been met comfortably, with 42% of people aged 75 years and over retaining at least some natural teeth.

Table 2.1.3

In 1998 there was considerable variation in the proportion of the population with and without natural teeth according to gender, region and social class, as well as according to age.

Overall, 90% of men had some natural teeth and men were slightly more likely to have some natural teeth than women (85%), although this difference was largely as a result of differences in the older age groups (55 and over). The sample has been separated into men and women for subsequent analyses of region and social class because of these gender differences, although data for all adults are presented at the end of this set of tables in Table 2.1.7. There were only subtle differences between the two gender groups in terms of the age distribution of the dentate. When comparing men with women in the older age groups, 69% of men aged 65 to 74 years had some remaining natural teeth compared with 60% of women at the same age and overall, men were more likely to retain some teeth beyond the age of 65. Amongst people

aged 16 to 44 years of age, both men and women exceeded the 95% threshold.

Tables 2.1.4 and 2.1.7

2.1.3 Variations in total tooth loss by country, region and social class

Table 2.1.4 compares the proportion of dentate men and dentate women across the constituent countries of the UK and three large regional groupings in England. Within England, the population was divided into 3 regions; the South, the Midlands, and the North, with the South of England including counties in both the South East and South West of England. These regions are similar to those used in the 1988 report (see Chapter 1.1). There were differences between countries, and between different regions of England. The tendency for women to be more likely to be edentate persisted in all four countries and in all three of the English regions, with the difference being between 2 and 5 percentage points in all areas except Scotland where the gender difference was 7 percentage points. The South had the highest proportion of the population who were dentate among both men and women, with 93% of men and 88% of women having at least some natural teeth. Across all ages, men in the South were the group who were closest to the threshold of 95% dentate. By contrast, at 79%, women in Scotland were the group with the lowest overall percentage with some natural teeth remaining. For men, the groups at the lower end of the range for the proportion dentate were those from Wales, at 85%, and Scotland at 86%.

Figure 2.1.2

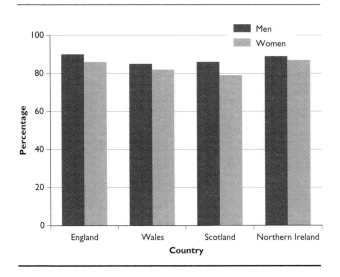

Fig 2.1.2 **Proportion of adults who were dentate by gender and country**

When comparing the four countries, for both men and women, it was in Scotland and Wales where there was the lowest proportion of the adult population with teeth in 1998, and in both cases the proportion was significantly lower than the rest of the United Kingdom. England and Northern Ireland were very similar to each other in terms of the proportion of the population with teeth; 90% and 89% respectively for men and 86% and 87% for women. Among men and women in each age group under the age of 45, and in all four countries, there was a chance, in many cases a very good chance, of the population staying above the 95% threshold for the proportion dentate. Among women in England (as a whole), the largest of all of these sub-groups, that is those up to 55 years of age, had a chance of meeting the 95% target.

When comparing the three regions of England, it was the South which stood out from the other two regions with 93% of men and 88% of women retaining some natural teeth. The Midlands and the North were closely matched, with the North having a slightly higher proportion of people with some natural teeth among both men and women (88% of men in the North, 87% in the Midlands, 85% of women in the North, 83% in the Midlands).

Table 2.1.4

When the sample was divided into three social class categories, as determined by the occupation of the head of household, further differences emerged. People from non-manual backgrounds were significantly more likely to be dentate than those from manual backgrounds among both men and women. Among women whose head of household was in the manual category there was a further significant difference between those from skilled (85%) and those from unskilled (73%) backgrounds. These differences held for both men and women, but were more marked among women.

Men from non-manual backgrounds had, as a group, already reached the 95% target for the proportion dentate. When broken down into different age groups, only those aged 65 and over were below the 95% threshold. Women from non-manual backgrounds also had a high proportion with natural teeth, with all age groups under the age of 55 exceeding the 95% threshold. The picture was rather different for people from manual backgrounds, among whom women from unskilled manual backgrounds stand out as particularly low, with only 73% having some natural teeth, compared with 83% of men from this social group and 85% of women from a skilled manual background.

Of all of the sub-groups examined, only women from unskilled manual backgrounds fell below the government target for England of 33% of those aged 75 and over retaining some natural teeth by 1998. Only 27% of women aged 75 and over from an unskilled manual background had some natural teeth. All other gender, regional and social class sub-groups had achieved this target in 1998.

Figure 2.1.3, Table 2.1.5

Where regions and social class groups were combined, some of the social and regional inequalities are drawn into even sharper focus. Men from non-manual households in the South reached 97% dentate, comfortably clear of the 95% threshold. This is the single regional or gender group with the highest percentage with some natural teeth, but in all regions men from non-manual backgrounds achieved 92% or more of the sample with natural teeth. For women, it was those from non-manual backgrounds in Northern Ireland and the South of England who experienced the highest proportion dentate, at 94% and 92% respectively. In other areas (except for England as a whole) women from non-manual households

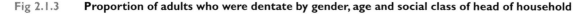

Fig 2.1.3   **Proportion of adults who were dentate by gender, age and social class of head of household**

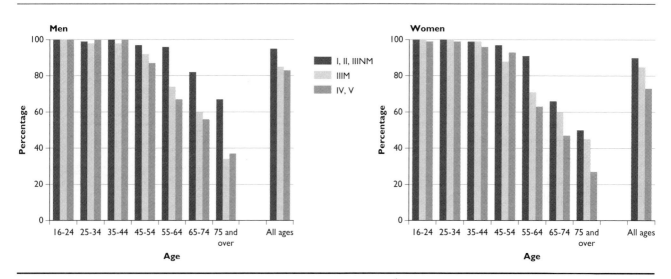

were just below 90%, but with Scottish women significantly lower at 85%. At the other extreme were women from unskilled manual backgrounds. In all areas the proportion of this group with natural teeth varied around 70%, with women from the South of England at the upper end of the range, at 76%, and those from Scotland significantly lower at 64%. Women from Scotland were significantly less likely, across all social groups, to have natural teeth than their counterparts elsewhere in the United Kingdom, with the exception of women in skilled manual households in Wales where the difference was not significant.

Among the detail of the various population sub-groups, the scale of the differences in oral health in the United Kingdom in 1998 is vividly illustrated by comparing men from non-manual backgrounds, particularly those from the South of England (97% dentate) with unskilled manual women, particularly those from Scotland (64% dentate).

Table 2.1.6

Where statistically significant differences occur between men and women, or between different countries and regions or between different social groups, there could be a number of explanations for why they exist. Gender differences resulting from a biological difference in the predisposition to dental diseases, is one possibility, but differences in predisposition to disease are unlikely to account for all of the regional or social inequalities reported here. These are more likely to relate to behavioural, cultural and attitudinal factors associated with oral health, which in turn affect the risk of disease or alter the decision making process about tooth removal. Further, more detailed analysis of the relationships of social factors to various key measures of oral health are explored in Part 4 of this report.

2.1.4 Adults who became edentate in the last decade

Although the number of people without natural teeth is now low, the continuing incidence of total tooth loss means that this is still an important group who require professional clinical skills and services which are different in some respects from the dentate population. The number of participants in the survey who reported losing their teeth in the last ten years is small (96 people) but the information they are able to provide is of considerable value. The small numbers mean that possible analyses are limited, but this section will give a profile of the types of individuals involved. Chapter 6.6 provides a broader picture of the history of tooth loss for all edentate people in the survey.

2.1.5 The demographic profile of adults who became edentate in the last decade

Ninety-six individuals from the sample (1%) said that they became edentate in the decade between 1988 and 1998. Table 2.1.8 shows the demographic profile of these individuals. Although they were widely spread through the age groups there was a strong bias towards the oldest groups. One individual of only 18 years, and a total of four under the age of 35, had lost the last of their natural teeth in the preceding decade, representing only a tiny proportion (less than 1%) of the sample at this age. Between 35 and 54 years of age there were 22 individuals (1% of that age group) who had recently become edentate, while the 70 individuals aged 55 years and over represented 3% of the sample at that age. Women (47 individuals) were as likely to have lost the last of their natural teeth in the last ten years as men (49 individuals).

With the exception of the age profile, which shows the expected tendency towards the highest incidence of total tooth loss in the last decade occurring in the oldest participants, the demographic distribution of this small group is very evenly spread, with people from all social groups and all areas of the country being affected. With such a small group it would be impossible to detect subtle trends, but neither is there any evidence of an obvious pattern emerging from these data.

Table 2.1.8

2.1.6 The process of total tooth loss in the last decade

The number of teeth lost on becoming edentate has reduced markedly in the last few decades. The normal complete natural dentition comprises 32 teeth, although for many people the number of erupted teeth is slightly fewer. Just under a quarter of those who reported losing the remainder of their natural teeth in the last ten years lost 12 or more teeth to become edentate, with the remainder losing less than 12. This shows a marked movement away from the removal of large numbers of natural teeth at one time, a pattern which was common in previous decades. In 1988 nearly half (47 of 99 participants) had lost 12 or more teeth to become edentate, while back in 1968 around two thirds had lost 12 or more teeth when they were rendered edentate and a third overall had had 21 or more teeth removed when they became edentate.

Although the dentist was the person most likely to suggest removal of the remaining natural teeth (55%), the respondent was almost as likely to make the suggestion (43%). Participants were asked why the remaining teeth were lost and were able to choose one or more options from a list of likely reasons. Two thirds (64%) said their teeth were decayed, and less than a third (28%) thought that they had had a problem with their gums. Other reasons included that they had been advised to have them removed, that the teeth could no longer be restored or that there had been problems with dentures. Participants were also asked to report the sort of problems which they had experienced in the period prior to the loss of their remaining teeth. There was scope for the participant to choose more than one response to this question, and a number of respondents had experienced two or more problems prior to the loss of their natural teeth . Despite the fact that only 28% had thought that gum problems were the

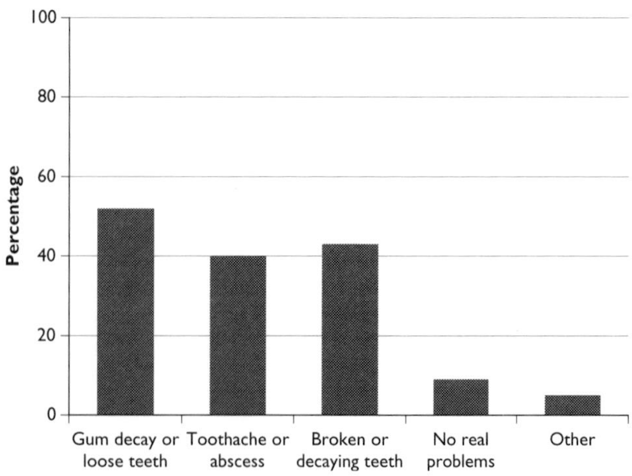

reason for the ultimate loss of their teeth, over half (52%) mentioned loose teeth as a problem prior to becoming edentate. Two out of five people reported toothache or an abscess, while a similar number had noticed broken or decaying teeth. Less than one in ten people reported having noticed no real problems prior to the removal of the last of their teeth.

Figures 2.1.4–5, Table 2.1.9

2.1.7 Behaviours and attitudes prior to the loss of the last natural teeth among people who became edentate in the last decade

Those who had lost their teeth in the previous decade were predominantly those who had attended the dentist only when they had trouble. Over half (59%) said that they had attended the dentist only when they had trouble, while 29% said that they had been regular attenders. These figures are almost the opposite to those found for adults who were still dentate (30% only went with trouble and 59% went for regular check-ups, see chapter 6.1). When asked whether or not they had expected to lose their teeth at the age they did, 65% said that they had expected it, and 27% that they were surprised to lose them at that age. All edentate participants who had lost the last of their teeth in the preceding decade were asked what advice they would offer to others who were about to lose their last teeth. The statements with which participants agreed were varied. In total, 32% agreed with the statement that they would advise people to keep their teeth as long as possible, while 22% said that if they were bad you should have them out. In total, 21% agreed with the statement that you should see or get advice from a good dentist while 12% agreed that it was nothing to worry about. Other statements of advice with which small numbers of individuals agreed related to when and how to cope with a new denture (data not in table). The majority (54%) of the participants who had lost the last of their natural teeth in the previous decade said that they were not at all upset by this loss, 28% said that they were a little upset and 18% that they were very upset.

Table 2.1.10

In terms of reducing further the incidence of total tooth loss, these data may give some indication of the challenges ahead. Participants who said they were upset represent a specific challenge as they are the group for whom the process of tooth loss is most traumatic, and who may make greater demands on dental services. On the other hand, those who did not feel at all upset may have recognised that the outcome was likely some time before it happened and so had some time to become accustomed to the idea. Alternatively, they may have been

Fig 2.1.4   **Reasons given for losing remaining teeth**

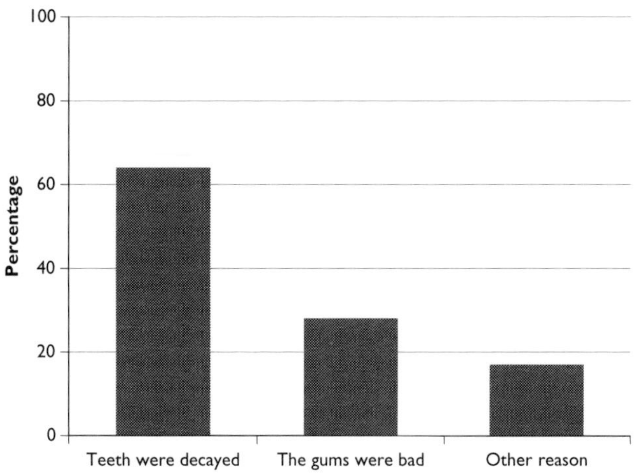

Fig 2.1.5   **Problems prior to losing remaining teeth**

apathetic all along, or have been so uncomfortable that the loss of the last teeth came as a relief. This group may perhaps be expected to be less motivated to maintaining their own oral health. Further analysis showed that two thirds of the group who were not upset expected to lose their teeth compared with a third of those who were upset.

The important point which emerges from this analysis is that the people who had lost all of their teeth in the last decade are a heterogeneous group. They came from a variety of age groups and backgrounds, held a range of attitudes and lost their teeth in a variety of circumstances, with little evidence for strong correlating factors. However, those who were the least upset at losing their teeth were likely to be the ones who most expected it.

Notes and references

1. Steele JG et al. *National Diet and Nutrition Survey: people aged 65 years and over Volume 2: Report of the oral health survey* London The Stationery Office 1998

2.1: The loss of all natural teeth

Table 2.1.1 Dental status by age

All adults   *United Kingdom*

Age		Dentate	Edentate	Base
16-24	%	100	0	702
25-34	%	100	0	1158
35-44	%	99	1	1114
45-54	%	94	6	1106
55-64	%	80	20	833
65-74	%	64	36	758
75 and over	%	42	58	533
All	%	87	13	6204

1988 Table 2.1

Table 2.1.2 Age and gender of edentate adults

Edentate adults   *United Kingdom*

Age	Males	Females	All
	%	%	%
Less than 44	3	1	2
45-54	12	7	8
55-64	23	18	20
65-74	30	30	30
75 and over	33	43	39
Base	360	563	923

Table 2.1.3 Dental status by age (in five year groups)

All adults   *United Kingdom*

Age		Dental status		Base
		Dentate	Edentate	
16-19	%	100	0	334
20-24	%	100	0	368
25-29	%	100	0	516
30-34	%	99	1	642
35-39	%	99	1	595
40-44	%	99	1	519
45-49	%	95	5	553
50-54	%	92	8	553
55-59	%	88	12	422
60-64	%	73	27	411
65-69	%	71	29	399
70-74	%	56	44	359
75-79	%	49	51	282
80 and over	%	35	65	251
All	%	87	13	6204

1988 Table 2.2

Table 2.1.4 **Dentate adults by age, gender, country and English region**

Dentate adults

Age and gender	English region			Country				
	North	Midlands	South	England	Wales	Scotland	Northern Ireland	United Kingdom
				percentage				
Men								
16-24	100	100	100	100	100	100	100	100
25-34	98	98	100	99	98	100	98	99
35-44	100	99	100	100	98	97	96	99
45-54	91	89	97	94	90	90	96	93
55-64	75	78	92	84	76	69	74	82
65-74	64	70	75	71	61	60	60	69
75 and over	46	49	49	49	36	44	47	48
All	88	87	93	90	85	86	89	90
Women								
16-24	100	100	100	100	100	99	100	100
25-34	100	100	100	100	100	99	99	100
35-44	100	99	99	99	100	96	100	99
45-54	98	89	98	96	83	85	93	94
55-64	76	71	87	80	74	65	84	79
65-74	51	61	68	62	61	42	52	60
75 and over	37	39	43	41	36	33	36	40
All	85	83	88	86	82	79	87	85
				Base				
Men								
16-24	48	40	84	172	52	68	39	331
25-34	81	65	140	286	60	79	52	477
35-44	78	90	147	315	49	99	57	520
45-54	78	87	133	298	82	89	51	520
55-64	59	45	94	198	59	80	42	379
65-74	63	49	81	193	44	65	38	340
75 and over	28	25	47	100	39	41	19	199
All	435	401	726	1562	385	521	298	2766
Women								
16-24	54	33	93	180	40	70	81	371
25-34	112	107	198	418	70	119	74	681
35-44	85	82	155	322	83	117	72	594
45-54	87	81	140	308	89	113	76	586
55-64	74	58	120	252	58	88	56	454
65-74	77	48	98	223	61	90	44	418
75 and over	51	46	74	171	44	86	33	334
All	540	455	879	1874	445	683	436	3438

1988 Table 2.10, 2.11

Table 2.1.5 **Dentate adults by age, gender, and social class of head of household**

Dentate adults *United Kingdom*

Age	Men			Women		
	I, II, IIINM	IIIM	IV, V	I, II, IIINM	IIIM	IV, V
			percentage			
16-24	100	100	100	100	100	99
25-34	99	98	100	100	100	99
35-44	100	98	100	99	99	96
45-54	97	92	87	97	88	93
55-64	96	74	67	91	71	63
65-74	82	60	56	66	60	47
75 and over	67	34	37	50	45	27
All	95	85	83	90	85	73
			Base			
16-24	130	94	63	149	82	70
25-34	216	154	89	305	185	176
35-44	245	174	87	285	131	94
45-54	232	183	95	267	139	97
55-64	157	126	93	188	123	103
65-74	144	129	63	194	91	107
75 and over	81	68	44	139	50	100
All	1205	928	534	1527	801	687

1988 Table 2.12

Table 2.1.6 **Dentate adults by country and English region, gender, and social class of head of household**

Dentate adults

Country and English region	Men			Women		
	I, II, IIINM	IIIM	IV, V	I, II, IIINM	IIIM	IV, V
			percentage			
All	95	85	83	90	85	73
Country						
England	95	86	84	90	86	74
Wales	92	80	79	89	80	74
Scotland	92	82	80	85	78	64
Northern Ireland	94	87	86	94	89	73
English region						
North	94	83	77	88	90	70
Midland	92	89	79	89	82	74
South	97	86	90	92	86	76
			Base			
All	1205	928	534	1527	801	687
Country						
England	723	519	274	888	429	354
Wales	161	139	76	182	127	89
Scotland	214	162	114	303	145	157
Northern Ireland	107	108	70	154	100	87
English region						
North	205	152	72	269	129	96
Midland	151	165	78	176	122	103
South	367	202	124	443	178	155

1988 Table 2.14

Table 2.1.7   **Dentate adults by gender, country, English region, social class of head of household and age**

Dentate adults   *United Kingdom*

	Age							All
	16-24	25-34	35-44	45-54	55-64	65-74	75 and over	
				Percentage				
Gender								
Male	100	99	99	93	82	69	48	90
Female	100	100	99	94	79	60	40	85
Country								
England	100	100	99	95	82	66	44	88
Wales	100	100	100	86	75	62	37	84
Scotland	100	100	97	88	66	51	36	83
Northern Ireland	100	100	100	94	79	54	43	88
English Region								
North	100	99	100	95	75	56	41	86
Midlands	100	99	99	89	74	65	43	85
South	100	100	99	98	89	71	45	90
Social Class								
I, II, IIINM	100	100	100	97	93	73	56	92
IIIM	100	99	98	90	72	60	38	85
IV,V	100	100	98	90	65	50	30	78

For bases see Tables Appendix

Table 2.1.8 **Demographic characteristics of adults who became edentate between 1988 and 1998**

All adults *United Kingdom*

	Percentage[†] who became edentate	Number who became edentate	Base
All	1	96	6204
Age			
under 35	0	4	1860
35-54	1	22	2220
55 and over	3	70	2124
Gender			
Men	1	49	2766
Women	1	47	3438
Social class			
of head of household			
I, II, IIINM	1	22	2732
IIIM	2	42	1729
IV,V	2	27	1221
Country / English region			
North	2	17	975
Midlands	1	8	856
South	1	19	1605
Scotland	2	14	1204
Wales	2	22	830
Northern Ireland	2	16	734

† Percentage calculated on weighted data, number and bases are unweighted, see Appendix F

Table 2.1.9 **Process of and reasons for loss of remaining teeth among adults who became edentate between 1988 and 1998**

Adults who became edentate between 1988 and 1998 *United Kingdom*

Process of and reasons for loss of remaining teeth	
Number of teeth lost	%
1-11	76
12-20	14
21 or more	10
Who suggested it	%
Respondent	43
Dentist	55
Reasons given for loss of remaining natural teeth[†]	%
Teeth were decayed	64
The gums were bad	28
Other reason	17
Reported problems prior to loss of remaining teeth[†]	%
Gum disease or loose teeth	52
Toothache or abscess	40
Broken or decaying teeth	43
No real problems	9
Other	5
Base	96

† Percentages add to more than 100% because respondents may have given more than one answer

1988 Table 2.17 (part)

Table 2.1.10 **Attitudes to tooth loss and advice offered among adults who became edentate between 1988 and 1998**

Adults who became edentate
between 1988 and 1998   *United Kingdom*

Attitudes to tooth loss and advice offered

Previous reason for dental attendance	%
Regular check-ups	29
Occasional check-ups	12
Only with trouble	59
Expectations of losing teeth at that age	%
Expected to lose them	65
Surprised to lose them	27
Other	8
Whether respondent found loss of	
remaining teeth upsetting	%
Very upsetting	18
A little upsetting	28
Not at all upsetting	54
Advice offered by respondent to others[†]	%
If they are bad, have them out	22
Keep natural teeth as long as possible	32
Get advice from a good dentist	21
Nothing to worry about	12
Would not presume to give advice	9
Base	96

† *Percentages may add to more than 100% because respondents may have given more than one answer*
1988 Table 2.17 (part), 2.18

2.2 The number and distribution of teeth, and the functional dentition

Summary

- Various measures have been used to describe the number and distribution of teeth, with particular reference to the ability to function.
- The presence of teeth in one jaw but not the other was found in 5% of dentate adults, but was much more common in older adults, representing 16% to 17% of dentate people aged 65 and over. The pattern of tooth retention in one jaw suggests that in many cases this is a transitional condition on the way to total tooth loss.
- The mean number of teeth decreased with age from a maximum of 28 teeth among 25 to 34 year olds.
- The largest differences in the mean number of teeth according to social and demographic variables occurred in the age groups of 55 years or more, and the most marked differences were associated with the reported reason for dental attendance. In the 55 to 64 year old group, people who said they attended the dentist for regular or occasional check-ups had, on average, 6 more natural teeth than those who attended only when they had trouble.
- The presence of 21 or more teeth (which was used as indicator of a functional dentition) varied with age and almost everyone up to the age of 34 had 21 or more teeth. The proportion then reduced to less than half of the dentate population age 55 and over.
- Even where there were 21 or more teeth, some people used dentures to restore spaces at the front of the mouth where teeth had been lost, but people with 21 or more teeth were as likely to have fixed bridges as dentures in these spaces (3% in each case), and were much more likely to have the space left unrestored (17%).
- The social and demographic variation for the retention of 21 or more natural teeth was greatest in the oldest age groups with people from non-manual backgrounds and from England, particularly in the South, most likely to have 21 or more natural teeth (47%, 40% and 47% of those aged 65 and over in each of these groups had 21 or more natural teeth).
- There was a large difference in the proportions of those aged 65 or more who had 21 or more teeth according to the reported usual reason for dental attendance, with people who attend for check-ups more than twice as likely to have 21 or more teeth than those who said they attended only with trouble (44% and 22% respectively).

2.2.1 Introduction

The previous chapter reported the prevalence of total tooth loss in the population and, conversely, tooth retention. However, having some natural teeth provides no guarantee that oral function will be adequate. Oral health was defined in terms of function in the 1994 Department of Health publication *An oral health strategy for England*. The definition of oral health referred to the ability to "eat, speak and socialise without active disease, discomfort or embarrassment". Such attributes as eating comfortably and socialising without embarrassment can be related directly to the number and distribution of natural teeth. With retention of some natural teeth into older age now such a common occurrence, additional simple ways of reporting the oral health within this growing group of dentate people are increasingly required. If the population is to benefit from the widespread retention of natural teeth into old age, it is important that as many people as possible have enough natural teeth in the right positions in the mouth to provide adequate function, and to maintain that function throughout a lifetime.

This chapter looks in more detail at dentate adults and considers the number and distribution of their natural teeth. Several measures are used which represent different numbers and distributions of teeth, and different functional conditions. These include the proportion of people who have teeth in only one jaw (representing a transitional condition between being dentate and being edentate), the number of natural teeth and the proportion of the population with 21 or more teeth. The latter was used in the report of the 1988 Survey of Adult Dental Health[1] as an indicator of the proportion of the population with a "functional" dentition and is reported again here, in combination with other indicators with known relationships to function, such as spaces at the front of the mouth and contacting pairs of natural teeth at the back.

2.2.2 Teeth in only one jaw

In the previous chapter it became clear that the process of becoming edentate is now usually a rather gradual one, and losing large numbers of teeth at one time is now rare. It is not unusual for dentists to encounter people who have natural teeth in one jaw and a full denture in the other. Although some people may lose teeth from both jaws when they finally become edentate, the retention of some teeth in one jaw only is likely to represent a late stage in the progression towards total tooth loss, and can be considered as an indicator, perhaps, of a likely progression to total tooth loss in the future. Loss of teeth in one jaw also has implications for the provision of dental services. From a technical dental standpoint providing

successful dentures may be more difficult for this group of people than for those with no natural teeth, while from a patient's point of view it can sometimes be more difficult to function where there are a few remaining teeth in just one jaw. Individuals who are edentate in one jaw may well also require routine treatment such as fillings and periodontal (gum) treatment.

Overall, 5% of dentate adults were dentate in only one jaw. People were far more likely to have lost all their upper teeth and retained some lower teeth than the reverse, 4% were dentate in the lower jaw only while less than 1% were dentate in the upper arch. Thus, among the 5% of dentate adults who had teeth in only one jaw, very few indeed (less than one in ten) had retained some upper teeth.

The general pattern of this condition according to age, gender, social class and geography mirrored that for total tooth loss. Up to the age of 54, only very small numbers of people were affected (0 to 4%), but the proportion with natural teeth in only one arch increased sharply for the 55 to 64 year age group to 14% and then showed a slight but non-significant increase to 17% of those aged 65 to 74 and 19% of those aged 75 and over. If these proportions are recalculated as proportions of the whole population (see Table 2.2.2), including the edentate, this pattern is further emphasised showing the increase at age 55 to 64 (to 11%) and then remaining at around this level for the two oldest age groups (11% and 8% respectively). This pattern, of a low proportion of people losing teeth from one jaw up until the age of 54 before increasing sharply and then leveling off, was reflected to some degree in many of the social and demographic sub-groups (see Tables 2.2.3 to 2.2.7), suggesting that this condition does indicate a transitional phase on the route to total tooth loss for many people.

Tables 2.2.1-2

There was little gender difference in the pattern, but there were differences according to the social class of the head of household, with those from non-manual households less likely to be dentate in only one jaw than those from either skilled or unskilled manual households. For example, among those aged 65 and over 11% of those from non-manual backgrounds were dentate in only one jaw compared with 25% and 20%, respectively, of those from skilled and unskilled manual groups.

Tables 2.2.3-4

When analysed by country, the overall pattern according to age was similar for the four countries surveyed, although Scotland showed signs of an earlier increase than the other countries, with 9% of 45 to 54 year olds affected compared with 7% for Northern Ireland, 6% for Wales and 3% for England at the same age. The proportion of dentate people aged 65 years and older with teeth in one arch appeared to be much lower in England than the other countries (16% compared with 27% to 30%) although only the Scottish figure is significantly different. When different areas of England were analysed, the pattern by age was similar for all areas. The North showed a particularly marked step between the 45 to 54 year old dentate adults (4% affected) and those in the next age group, the 55 to 64 year olds (20%). Although the differences were not significant, the South demonstrated lower values than the other two areas, particularly in those aged 65 years and over (10% compared with 20% to 26%), mirroring the findings for total tooth loss.

Tables 2.2.5-6

The largest differences in the proportions of the population with teeth in only one jaw were found between groups with different reported dental attendance patterns. People were asked about their usual reason for dental attendance; whether they went for regular or occasional check-ups or whether they only attended the dentist with trouble. Few people attended only occasionally for check-ups and, as their dental status was generally closer to the regular attenders, these two groups have been merged. When "check-up attenders" are compared with those who said they attended only when they had trouble, there were substantial differences between the groups particularly at older ages. In total, 3% of dentate people who attend for check-ups, compared with 9% of dentate people who said they only attend with trouble had teeth in one jaw. Around a third (31% to 36%) of the latter group aged 55 years and over had teeth in only one jaw. In contrast, only 7% of check-up attenders aged 55 to 64 years, and 11% of those aged 65 years and over were similarly affected.

Figure 2.2.1, Table 2.2.7

Among those with teeth in just one jaw, no-one had more than 13 teeth. Although people with 1 to 8 teeth accounted for 70% of those with teeth in one jaw, the greatest concentration had 6,7 or 8 teeth (42% of all people with teeth in one jaw). Having just a few teeth (less than 6) was much less common, accounting for 27% of those with teeth in one jaw.

Table not shown

2.2.3 The variation in numbers of natural teeth

In the 1988 Adult Dental Health Survey, data on the number of teeth were presented in the form of the number of missing teeth. The number of teeth missing starts from a nominal maximum of 32, but teeth can be lost for a variety of reasons, indeed some teeth may be missing congenitally or not erupt at all. As a consequence, little can be inferred by considering only the number of teeth missing and there is also uncertainty about the history of the teeth which are absent. The number of teeth retained, however, should give an indication of the adequacy of oral function and the prospects for tooth retention. In this report, data are presented as the number of retained natural teeth, but using the same groups as in 1988. The equivalent number of missing natural teeth is also given in Table 2.2.8.

The number of natural teeth retained is strongly associated with age. At all ages a complete dentition of 32 natural teeth is a rarity, 8% of 16 to 24 year olds and 11% of 25 to 34 year olds had all 32 natural teeth. The reason for the slightly higher numbers in the older of these two age groups is likely to be because the third molars (wisdom teeth) would not yet have erupted in some of the younger participants. Looking across the table, as age increased, the number of teeth reduced. For 16 to 44 year olds the largest proportion had 27 to 31 teeth, for the groups aged 45 to 64 the highest proportion of the population had 21 to 26 teeth and by 75 years of age and over the peak had dropped below 21 teeth, with the highest proportion having 15 to 20 teeth and only 3% having 27 or more. The mean number of teeth also showed a steady reduction from a peak of 28.1 in 25 to 34 year olds down to 15.4 in the 75 years and over age group. The group with 1 to 8 natural teeth showed an age related trend very similar to those with teeth in one arch, and in many cases will represent the same individuals.

Figure 2.2.2, Table 2.2.8

| Fig 2.2.1 | **Teeth in only one jaw by age and usual reason for dental attendance** |

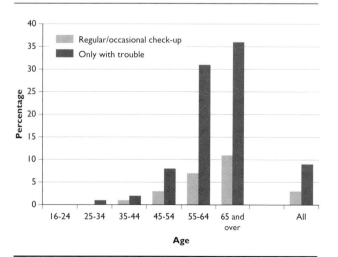

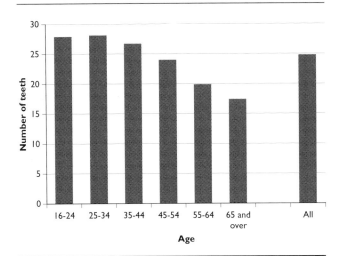

Fig 2.2.2   **Mean number of teeth by age**

To observe the relationship between the number of teeth retained and social and demographic factors the mean number of teeth are presented by age. In the two youngest groups there was very little variation according to gender, social class, geography or usual reason for attendance. The mean at this age was consistently around 28 teeth. Among those aged 35 to 54 years, there was still relatively little variation with around 27 teeth and 24 teeth for the two age groups respectively. It was among people aged 55 years and over that differences according to social or demographic factors became much more apparent; at the same time there was a sharp drop in the overall mean number of teeth for all groups. For example, the mean value for the group aged 65 years and over was 17.4 teeth. At this age, gender differences were still small, but there was a marked and significant variation according to social class of head of household, ranging from 19.0 for those from non-manual backgrounds to 15.7 for those from unskilled manual households. There was some geographical variation, with dentate adults in the South standing out as having a noticeably higher mean number of teeth (18.9) than those from the rest of the United Kingdom (14.8 to 16.3).

The biggest difference occurred with usual reason for attendance but only among older dentate adults. Figure 2.2.3 illustrates how the difference between the mean number of teeth among those who reported attending for check-ups and those who reported attending with trouble opened up with age, particularly in late middle age. Between the ages of 55 and 64 years the difference was up to 6 teeth (21.5 for those who reported check-ups and 15.5 for those who reported attending with trouble). Up to the age of 55 years the difference was smaller, indeed there was no difference at all in the youngest age groups, while among those aged 65 and over the difference was also less, although the smaller base numbers mean that there is a greater margin for error. This narrowing of the gap in the oldest groups may partly reflect

a tendency to greater loss through extraction of teeth and transition to being edentate of the more disease prone individuals from the group who only attend with trouble at an early age. This will leave the individuals for whom disease risk is lowest with natural teeth into old age. In those who said that they attended for dental check-ups (and by implication treatment) this process of extraction and tooth loss is delayed.

Figure 2.2.3, Table 2.2.9

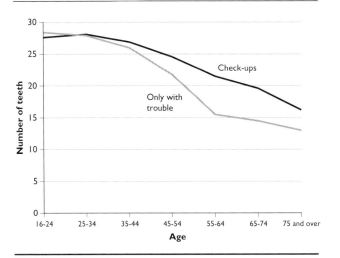

Fig 2.2.3   **Mean number of teeth at different ages for people who attend for check-ups and those who attend only when they have trouble**

2.2.4 The functional dentition

The retention of 21 or more natural teeth has been widely used as a method of defining a minimum functional dentition. The WHO adopted this threshold in its definition of oral health in 1982, on the basis of limited evidence available at that time. Subsequent research also suggests that where there are 21 or more natural teeth, the functional, dietary and aesthetic needs of most people are met with natural teeth alone, without the need for removable partial dentures[2]. The importance of partial dentures, in this context, is that they are associated with higher levels of dental disease, particularly decay of the roots, so may pose an additional risk to the longer term survival of a functional dentition[3].

Figure 2.2.4 shows the probability of wearing partial dentures where different numbers of natural teeth are present. Where there were less than 21 teeth, 56% or more of the sample wore partial dentures. This reduced to 17% where there were 21 to 24 teeth and 3% or less where there were more than 24 teeth. If a functional natural dentition is not only to provide a benefit in the short term but to be retained for a lifetime, it would be desirable to maintain it without the use of partial dentures where possible. Maintaining 21 or more natural teeth would appear to improve the chances of relying on a natural dentition alone.

Figure 2.2.4

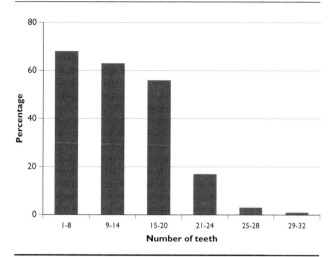

Fig 2.2.4   **Proportion of dentate adults who have partial dentures by the number of natural teeth**

Subsequent sections of this chapter will examine in some detail two other important determinants of oral function in people with 21 or more teeth: contacting back teeth, which contribute to chewing, and spaces at the front of the mouth which make an important impact on appearance.

2.2.5 Contacting back teeth where there were 21 or more natural teeth

It is not just the number of teeth which is important, their distribution can also make a significant contribution to the ability to function. Generally speaking, the ability to chew food is improved where more pairs of contacting back teeth (molars and premolars) are present. The National Diet and Nutrition Survey of adults aged 65 years and over[2] demonstrated a relationship between the number of posterior contacts and freedom of dietary choice, resulting in nutritional benefits. There was almost complete dietary freedom (that is, no major limitation in choice) where there were more than 4 pairs of posterior contacting teeth.

In this survey, the number of posterior contacting pairs of teeth was counted during the clinical examination by the dentist. This was recorded by looking to see which natural lower teeth made contact with a natural upper tooth when the teeth were fully closed together. Only natural tooth contacts were counted, contacts between natural teeth and a denture in the opposing jaw were not recorded. The details of how this was undertaken are given in Appendix C.

Table 2.2.10 shows the number of posterior contacts by age for dentate adults. This shows the same age related pattern as was apparent for number of teeth, and these two indicators of oral health will generally be closely correlated. Table 2.2.11 looks at the relationship between the number of posterior contacts and total number of teeth. People with 21 to 26 teeth were considerably more likely to have 5 or more posterior contacting pairs than those who had 15 to 20 teeth (88% compared with 14%). Having 21 or more teeth is almost invariably associated with having, at the very least, some posterior pairs of functioning natural teeth with which to chew.

Figure 2.2.5, Tables 2.2.10-11

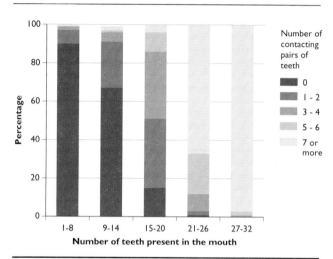

Fig 2.2.5   **Number of contacting pairs of posterior teeth by total number of teeth present in the mouth**

The previous tables have shown that the number of posterior contacts and the number of teeth both decreased with age but Table 2.2.12 shows that the relationship between the two also varied with age. Among those with 21 or more natural teeth, older people were still more likely than younger people to have 4 or fewer pairs of contacting back teeth; 18% of those aged 55 and over compared with 1% of those aged 16 to 34.

Table 2.2.12

2.2.6 Spaces at the front of the mouth where there were 21 or more natural teeth

The next chapter (2.3) discusses in some detail the wearing and distribution of partial dentures and bridges. In this section, however, the relationship between anterior spaces and partial dentures or bridges is investigated in more detail, with particular reference to the possibilities of retaining a natural dentition without the need for removable partial dentures.

Despite the strong associations with the lack of dentures, and with the presence of posterior contacting pairs of teeth, the retention of 21 or more teeth did not always guarantee a functional natural dentition without the need for a partial denture. Even where there were 25 to 28 teeth, some people said that they wore a partial denture. In many such cases the partial dentures were presumably worn to address an aesthetic need for a satisfactory appearance. This will usually result from spaces at the front of the mouth which are obvious during smiling, speaking and normal function. Where there are spaces present these may be tolerated by the individual, particularly if they are in the premolar area (at the sides of the mouth) which is less visible, or they may be filled with an artificial tooth. Removable partial dentures may fulfill this role but, a bridge, permanently fixed to the adjacent teeth, provides an alternative option.

The presence of spaces at the front of the mouth were recorded by the examining dentist during the dental examination. Spaces were recorded if they occurred in the anterior or premolar regions of the mouth (i.e. in the incisor, canine first or second premolar positions) and if they were of at least half a tooth in width. Different categories of space (unfilled, filled with a bridge and filled with a denture) were counted separately. Further details can be found in Appendix C.

Table 2.2.13 shows the proportions of the United Kingdom population who had spaces, unfilled or restored, in the front of the mouth. There were marked differences between people with 21 or more teeth and those with fewer teeth. Among people with fewer than 21 teeth a large majority had spaces at the front of the mouth; 96% had a space at the front of the upper arch and 84% a space at the front of the lower. People with 21 or more teeth were much less likely to have a space at the front of the mouth than those with fewer teeth; 77% had no upper spaces at the front of the mouth, while 23% had an upper space which was either filled or unfilled. In the lower jaw, 17% had a space at the front of the mouth.

Bridges were rarely used to fill anterior or premolar spaces and only 3% of the population had upper spaces at the front of the mouth restored with a bridge. Where there were 21 or more teeth though, bridges (3%) were just as likely to be used to fill spaces as partial dentures (3%). Where there were less than 21 teeth, partial dentures were a much more likely solution than bridges (46% of people with less than 21 teeth had dentures, compared with 3% who had bridges in the upper arch). Generally speaking, spaces in the lower jaw were less likely to be filled overall than they were in the upper, but in both arches the number of unfilled spaces was surprisingly high. Further analysis of the data showed that only 4% of people with 21 or more natural teeth had any unfilled spaces in the upper incisor or canine regions, while most of the spaces filled with dentures and bridges were at these anterior sites (data not in table). The majority of these unfilled spaces were in the premolar regions which are at the sides of the mouth and, although usually visible when smiling or talking are much less obvious than a space further forward. This indicates the importance to an individual of the front of the mouth and the presence of teeth. It also suggests that provided that function is adequate (with 21 or more teeth) some compromise on appearance because of a space at the side of the mouth may be acceptable, and often preferable to having the space restored.

Figure 2.2.6, Table 2.2.13

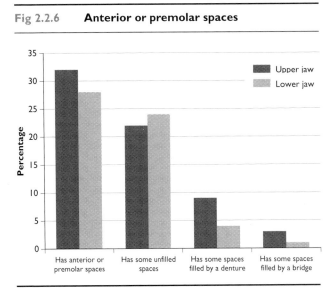

Fig 2.2.6   **Anterior or premolar spaces**

Among those who had 21 or more teeth, the proportion of people with unfilled spaces in the anterior or premolar regions increased markedly with age. For upper spaces the proportion with unfilled spaces increased from 7% of 16 to 34 year olds up to 43% of those aged 55 and over. The proportion of people who had filled spaces increased in a similar fashion; 1% with dentures and bridges respectively in the youngest age group, up to 10% with dentures and 7% with bridges in those aged 55 and over.

Women with 21 or more teeth were a little less likely than men to have unfilled spaces in the upper jaw, 15% compared with 19%. Otherwise there was little difference between men and women in the likelihood of having spaces or in having filled spaces.

People from manual backgrounds were more likely to have spaces at the front of the mouth than those from non-manual backgrounds (26% to 27% compared with 21% for the upper arch, 19% to 20% against 16% for the lower arch), and were slightly (but not significantly) less likely to have them filled in some way. Although there was no difference between the proportions of people from non-manual and manual backgrounds in the provision of dentures or bridges, the fact that there were more of the manual groups with spaces suggests a lower treatment experience among this group. There were no apparent differences between these groups in the provision of dentures as opposed to bridges.

People in Scotland (20%) and Wales (23%) were more likely to have unfilled spaces in the upper arch than those in the rest of the United Kingdom (16% to 18%), but there were no significant geographical trends in the provision of dentures or bridges in such cases.

When comparing people who said that they attend for check-ups with those who attend the dentist only when they have trouble, the differences were not significant. The differences are smaller than might be expected and probably illustrate the strength of motivation to seek treatment which results from an unsightly anterior space, even among people who would not normally attend the dentist.

Tables 2.2.14-18

This section and the previous one have looked in some detail at the relationship between having 21 or more teeth and the ability to rely on the natural teeth alone for function and aesthetics. In the large majority of cases, even where there are only just over 21 natural teeth, dentures are not required. Having 21 or more teeth is also generally associated with having several pairs of natural teeth at the back of the mouth which can be used for chewing, and having enough teeth at the front to ensure that dentures are rarely necessary to fill unsightly spaces. Using 21 or more teeth as an indicator of what constitutes a "functional dentition" for a population would seem to be well supported by the evidence.

2.2.7 The variation in the retention of a functional dentition

In this section the social and geographic variation in the retention of natural teeth is analysed to identify who are the people most likely to be above or below the 21 tooth threshold. Data are presented both as a percentage of the dentate sample and as a percentage of the whole sample (that is, including the edentate), in order to indicate how the whole population varies in terms of the retention of a functional dentition.

In 1998, 72% of the United Kingdom population had 21 or more natural teeth. Almost everybody in the 16 to 24 age group had 21 or more teeth, but both the proportion who were dentate and the proportion of dentate people with 21 or more teeth fell steadily through the age groups so that less

than half of adults aged 55 and over, and only 10% of those aged 65 and over had 21 or more natural teeth.

Figure 2.2.7, Table 2.2.19

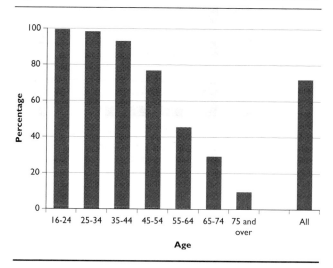

Fig 2.2.7 **Proportion of adults with 21 or more teeth**

There were variations in the retention of 21 or more teeth according to social and demographic factors. Differences between men and women were not large, with women having a slightly but not significantly higher overall percentage with 21 or more teeth from 25 years up to the age of 54 years, but with men having the slightly higher proportion with 21 or more teeth in the two oldest age groups. These trends reflect, and are partly influenced by, the trends for total tooth loss.

Table 2.2.20

When analysed according to social class, both the percentage who had natural teeth (see Table 2.1.7) and the percentage of dentate adults who had 21 or more natural teeth were higher in the non-manual than in the manual groups. The compound effect of both differences resulted in large variations in the proportions of all adults who had a dentition of 21 or more natural teeth, and this was particularly evident in the older age groups. Among those aged 65 and over, 31% of people from a non-manual background had 21 or more teeth, compared with 17% from a skilled manual and 10% from an unskilled manual background.

Figure 2.2.8, Table 2.2.21

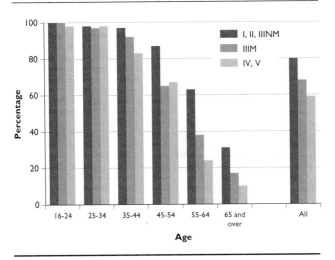

Fig 2.2.8   **Adults with 21 or more teeth by age and social class of head of household**

When comparing different countries, Scotland had the lowest percentage with 21 or more natural teeth among the dentate and the population as a whole, particularly in the 35 to 54 year old age bands. In Scotland 83% of 35 to 44 year olds and 58% of 45 to 54 year olds had 21 or more teeth compared with 91 to 95% of 35 to 44 year olds and 61 to 79% of 45 to 54 year olds in the rest of the United Kingdom. England had the highest probability of older adults retaining a functional dentition, although the differences between England and the other countries were not large (2 to 9 percentage points for the whole population). This difference is largely attributable to the higher levels of tooth retention found in the South, with the proportion in the North and the Midlands closer to the other countries of the United Kingdom. Among those aged 65 and over in the South, 28% of all adults had 21 or more natural teeth (40% in the 65 to 74 age band, data not in table). If the trends in the natural teeth continue, the prospects for many people retaining a functional dentition into old age in the South must be quite good.

Tables 2.2.22–23

Some of the most marked variation in the retention of 21 or more natural teeth occurs with reported usual reason for dental attendance. The usual reason for dental attendance is fundamentally different between those with and without teeth. The edentate show very low levels of attendance for check-ups because, once dentures have been provided and are satisfactory, there is rarely perceived to be any need for ongoing dental treatment[1] . In this survey, the current usual reason for attendance was only asked of those with some natural teeth. Up until the age of 44 years little difference is evident between those who attend the dentist for check-ups and those who attend only when they have trouble, with over 90% of both groups having 21 or more natural teeth. Over that age a marked difference opens up, with a difference of

38 percentage points between the groups at age 55 to 64 years. Among those who attended for a check-up, two thirds in that age band and 44% of people aged 65 and over had retained 21 or more natural teeth in 1998. The equivalent values for those who said they attended only when they had trouble were 29% and 23% respectively. The pattern is very similar to that for the mean number of teeth (see Figure 2.2.3), and it is evident that dental attendance, or at least reported dental attendance, is strongly associated with the ability to maintain a functional dentition throughout life.

Table 2.2.24

Notes and references

1.   Todd JE and Lader D. *Adult dental health 1988.United Kingdom*. London: HMSO 1991.

2.   Steele JG et al. *National diet and nutrition survey: people aged 65 years and over. Volume 2: Report of the oral health survey*. London: The Stationery Office 1998.

3   Drake CW, Beck JD. *The oral status of elderly removable partial denture wearers*. J Oral Rehabil 1993; 20:53-60.
      Locker D. *Incidence of root caries in an older Canadian Population*. Comm Dent Oral Epidemiol 1996; 24: 403-407.
      Steele JG, Walls AWG, Murray JJ. *Partial dentures as an independent indicator of root caries risk in a group of older adults*. Gerodontology 1998b; 14: 67-74
      Wright PS, Hellyer PM, Beighton D, Heath MR, Lynch E. *Relationship of removable partial denture use to root caries in an older population*. Int J Prosthodont 1992; 5: 39-46.

Table 2.2.1 **Whether had natural teeth in both jaws or only one jaw and which jaw was dentate by age**

Dentate adults *United Kingdom*

Age		Teeth in:		Teeth in:		Base
		both jaws	one jaw	upper jaw only	lower jaw only	
16-24	%	100	0	0	0	491
25-34	%	100	0	0	0	854
35-44	%	99	1	0	1	781
45-54	%	96	4	0	4	746
55-64	%	86	14	0	13	461
65-74	%	83	17	2	16	344
75 and over	%	81	19	2	17	140
All	%	95	5	0	4	3817

Table 2.2.2 **Proportion of all adults with natural teeth in only one jaw by age**

All adults *United Kingdom*

Age			Base
16-24	%	0	702
25-34	%	0	1158
35-44	%	1	1114
45-54	%	4	1106
55-64	%	11	833
65-74	%	11	758
75 and over	%	8	533
All	%	3	6204

Table 2.2.3 **Natural teeth in only one jaw by age and gender**

Dentate adults *United Kingdom*

Age	Gender	
	Men	Women
	percentage with natural teeth in only one jaw	
16-24	0	0
25-34	0	0
35-44	2	0
45-54	2	6
55-64	14	12
65 and over	17	19
All	4	5
	Base	
16-24	227	264
25-34	339	515
35-44	364	417
45-54	356	390
55-64	216	245
65 and over	243	241
All	1745	2072

Table 2.2.4 **Natural teeth in only one jaw by age and social class of head of household**

Dentate adults *United Kingdom*

Age	Social class of head of household		
	I, II, IIINM	IIIM	IV,V
	percentage with natural teeth in only one jaw		
16-24	0	0	0
25-34	0	0	0
35-44	0	1	3
45-54	2	4	5
55-64	10	15	22
65 and over	11	25	20
All	3	6	7
	Base		
16-24	215	116	90
25-34	402	254	148
35-44	395	214	121
45-54	394	204	107
55-64	247	115	85
65 and over	273	122	74
All	1926	1025	625

Table 2.2.5 **Natural teeth in only one jaw by age and by country**

Dentate adults

Age	Country			
	England	**Wales**	**Scotland**	**Northern Ireland**
	percentage with natural teeth in only one jaw			
16-24	0	0	0	0
25-34	0	0	1	0
35-44	1	0	2	0
45-54	3	6	9	7
55-64	14	10	17	12
65 and over	16	27	30	28
All	4	6	7	4
	Base			
16-24	246	69	80	96
25-34	517	90	157	90
35-44	447	96	142	96
45-54	423	109	132	82
55-64	259	73	75	54
65 and over	294	65	82	43
All	2186	502	668	461

Table 2.2.6 **Natural teeth in only one jaw by age and English region**

Dentate adults   *England*

Age	English region		
	North	**Midlands**	**South**
	percentage with natural teeth in only one jaw		
16-24	0	0	0
25-34	1	0	0
35-44	2	1	1
45-54	4	5	2
55-64	20	10	12
65 and over	20	26	10
All	5	6	4
	Base		
16-24	82	42	122
25-34	146	125	246
35-44	126	114	207
45-54	121	101	201
55-64	67	47	145
65 and over	75	66	153
All	617	495	1074

Table 2.2.7 **Natural teeth in only one jaw by age and usual reason for dental attendance**

Dentate adults   *United Kingdom*

Age	Usual reason for dental attendance	
	Regular/occasional check-up	**Only with trouble**
	percentage with natural teeth in only one jaw	
16-24	0	0
25-34	0	1
35-44	1	2
45-54	3	8
55-64	7	31
65 and over	11	36
All	3	9
	Base	
16-24	336	155
25-34	581	271
35-44	596	184
45-54	597	148
55-64	349	112
65 and over	349	133
All	2808	1003

Table 2.2.8   **Distribution of number of teeth by age**

Dentate adults   *United Kingdom*

Number of teeth	Number missing	Age							All
		16-24	25-34	35-44	45-54	55-64	65-74	75 and over	
		%	%	%	%	%	%	%	%
1-8	(24+)	0	0	1	2	10	15	20	4
9-14	(18-23)	0	0	1	6	10	11	26	4
15-20	(12-17)	0	1	4	11	22	28	32	9
21-26	(6-11)	21	18	32	43	42	33	20	30
27-31	(1-5)	71	70	55	38	14	13	3	48
32	(none)	8	11	7	1	1	0	0	5
Mean		27.9	28.1	26.7	24.0	19.9	18.3	15.4	24.8
Base		491	854	781	746	461	344	140	3817

1988 Table 5.2 (part)

Table 2.2.9   **Mean number of teeth by gender, social class of head of household, country, English region, usual reason for dental attendance and age**

Dentate adults   *United Kingdom*

	Age						All
	16-24	25-34	35-44	45-54	55-64	65 and over	
	mean number of teeth						
All	27.9	28.1	26.7	24.0	19.9	17.4	24.8
Gender							
Men	28.3	28.4	26.8	23.8	20.2	17.5	25.0
Women	27.5	27.8	26.6	24.2	19.5	17.2	24.6
Social class of head of household							
I, II, IIINM	28.0	28.3	27.4	24.9	21.5	19.0	25.4
IIIM	27.6	27.9	26.2	23.2	19.2	15.8	24.2
IV,V	27.7	27.6	25.0	22.4	16.9	15.7	23.6
Country							
England	27.9	28.3	26.9	24.3	20.1	17.6	24.9
Wales	28.0	27.9	26.3	24.1	18.7	14.8	24.2
Scotland	27.9	26.9	25.1	21.4	18.6	16.2	23.8
Northern Ireland	28.2	27.0	26.4	21.5	18.5	14.9	24.5
English region							
North	27.7	27.6	26.0	23.5	18.8	16.3	24.6
Midlands	28.9	28.6	26.9	23.1	20.2	15.8	24.6
South	27.7	28.5	27.4	25.3	20.6	18.9	25.3
Usual reason for attendance							
Regular/occasional check-up	27.6	28.1	26.9	24.6	21.5	18.4	25.1
Only with trouble	28.4	28.0	26.0	21.9	15.5	14.0	24.0

For bases see Table Appendix

Table 2.2.10 **The number of pairs of posterior contacting natural teeth by age**

Dentate adults   *United Kingdom*

Number of contacts	Age							All
	16-24	25-34	35-44	45-54	55-64	65-74	75 and over	
	%	%	%	%	%	%	%	%
0	0	1	2	6	24	27	36	8
1-2	0	0	2	7	10	17	27	5
3-4	1	1	5	7	18	15	15	7
5-6	4	2	7	14	14	14	12	8
7 or more	95	95	84	66	35	27	10	72
Mean number of contacts	11.4	11.5	10.2	8.0	5.0	4.4	2.5	8.9
Base	491	854	781	746	461	344	140	3817

Table 2.2.11 **The number of pairs of posterior contacting natural teeth by the number of natural teeth**

Dentate adults   *United Kingdom*

Number of contacts	Number of teeth				
	1-8	9-14	15-20	21-26	27-32
	%	%	%	%	%
0	91	67	15	1	0
1-2	7	24	36	2	0
3-4	2	5	35	9	1
5-6	0	1	10	21	2
7 or more	1	2	4	67	98
Base	165	184	385	1218	1865

Table 2.2.12 **The number of pairs of posterior contacting natural teeth by age and number of natural teeth**

Dentate adults   *United Kingdom*

Number of contacts	Age					
	16-34		35-54		55 and over	
	Less than 21 teeth	21 or more teeth	Less than 21 teeth	21 or more teeth	Less than 21 teeth	21 or more teeth
	%	%	%	%	%	%
0	*	0	34	0	50	1
1-2	*	0	28	1	26	3
3-4	*	1	23	4	20	14
5-6	*	3	10	10	4	24
7 or more	*	96	5	85	1	58
4 or less	*	1	85	5	95	18
Base	20	1325	205	1322	509	436

* Base number too small to calculate percentage

Table 2.2.13 **Anterior or premolar spaces and whether spaces are filled by number of natural teeth**

Dentate adults†   *United Kingdom*

Whether has anterior or premolar spaces and how filled	Number of natural teeth		All
	Less than 21 teeth	21 or more teeth	
	%	%	%
Upper jaw			
Has upper anterior or premolar spaces	96	23	32
Unfilled	54	17	22
Restored with denture	46	3	9
Restored with bridge	3	3	3
No upper anterior or premolar spaces	4	77	68
Lower jaw	%	%	%
Has lower anterior or premolar spaces	84	17	28
Unfilled	64	16	24
Restored with denture	21	0	4
Restored with bridge	1	1	1
No upper anterior or premolar spaces	16	83	72
		Base	
Dentate adults with some teeth in upper jaw	719	3083	3802
Dentate adults with some teeth in lower jaw	552	3083	3635

Percentages may add to more than 100% because respondents may have both unfilled and restored spaces
† Percentages are based on those respondents who had teeth present in the relevant arch, as some dentate adults had teeth present in only one jaw

Table 2.2.14 **Anterior or premolar spaces and whether spaces are filled by age**

Dentate adults with 21 or more natural teeth   *United Kingdom*

Whether has anterior or premolar spaces and how filled	Age			All
	16-34	35-44	55 and over	
	%	%	%	%
Upper jaw				
Has upper anterior or premolar spaces	9	26	58	23
Unfilled	7	19	43	17
Restored with denture	1	4	10	3
Restored with bridge	1	5	7	3
No upper anterior or premolar spaces	91	74	42	77
Lower jaw	%	%	%	%
Has lower anterior or premolar spaces	7	21	38	17
Unfilled	7	20	35	16
Restored with denture	0	0	1	0
Restored with bridge	0	1	2	1
No upper anterior or premolar spaces	93	79	62	83
Base	1325	1322	436	3083

Percentages may add to more than 100% because respondents may have both unfilled and restored spaces

Table 2.2.15 **Anterior or premolar spaces and whether spaces are filled by gender**

Dentate adults with 21 or more natural teeth   *United Kingdom*

Whether has anterior or premolar spaces and how filled	Gender		All
	Men	Women	
	%	%	%
Upper jaw			
Has upper anterior or premolar spaces	24	21	23
Unfilled	19	15	17
Restored with denture	3	3	3
Restored with bridge	3	3	3
No upper anterior or premolar spaces	76	79	77
Lower jaw	%	%	%
Has lower anterior or premolar spaces	17	16	17
Unfilled	16	15	16
Restored with denture	0	0	0
Restored with bridge	1	1	1
No upper anterior or premolar spaces	83	84	83
Base	1381	1702	3083

Percentages may add to more than 100% because respondents may have both unfilled and restored spaces

Table 2.2.16 **Anterior or premolar spaces and whether spaces are filled by social class of head of household**

Dentate adults with 21 or more natural teeth   *United Kingdom*

Whether has anterior or premolar spaces and how filled	Social class of head of household			All
	I, II, IIINM	IIIM	IV,V	
	%	%	%	%
Upper jaw				
Has upper anterior or premolar spaces	21	26	27	23
Unfilled	15	21	22	17
Restored with denture	4	3	3	3
Restored with bridge	4	2	3	3
No upper anterior or premolar spaces	79	74	73	77
Lower jaw	%	%	%	%
Has lower anterior or premolar spaces	16	20	19	17
Unfilled	15	19	18	16
Restored with denture	0	0	0	0
Restored with bridge	1	0	1	1
No upper anterior or premolar spaces	84	80	81	83
Base	1623	793	456	3083

Percentages may add to more than 100% because respondents may have both unfilled and restored spaces

Table 2.2.17 **Anterior or premolar spaces and whether spaces are filled by English region and country**

Dentate adults with 21 or more natural teeth

Whether has anterior or premolar spaces and how filled	English region			Country				
	North	Midlands	South	England	Wales	Scotland	Northern Ireland	United Kingdom
	%	%	%	%	%	%	%	%
Upper jaw								
Has upper anterior or premolar spaces	22	21	22	22	30	25	22	23
Unfilled	17	15	16	16	23	20	18	17
Restored with denture	3	4	3	3	2	3	4	3
Restored with bridge	3	2	4	3	7	3	2	3
No upper anterior or premolar spaces	78	79	78	78	70	75	78	77
Lower jaw	%	%	%	%	%	%	%	%
Has lower anterior or premolar spaces	18	15	16	16	18	19	24	17
Unfilled	17	14	15	15	18	18	23	16
Restored with denture	0	0	0	0	0	0	0	0
Restored with bridge	1	1	1	1	1	1	1	1
No upper anterior or premolar spaces	82	85	84	84	82	81	76	83
Base	*494*	*399*	*920*	*1813*	*401*	*506*	*363*	*3083*

Percentages may add to more than 100% because respondents may have both unfilled and restored spaces

Table 2.2.18 **Anterior or premolar spaces and whether spaces are filled by usual reason for dental attendance**

Dentate adults with 21 or more natural teeth   *United Kingdom*

Whether has anterior or premolar spaces and how filled	Usual reason for dental attendance		All
	Regular/occasional check-up	Only with trouble	
	%	%	%
Upper jaw			
Has upper anterior or premolar spaces	22	24	23
Unfilled	16	20	17
Restored with denture	3	2	3
Restored with bridge	4	3	3
No upper anterior or premolar spaces	78	76	77
Lower jaw	%	%	%
Has lower anterior or premolar spaces	17	17	17
Unfilled	16	16	16
Restored with denture	0	0	0
Restored with bridge	1	0	1
No upper anterior or premolar spaces	83	83	83
Base	*2343*	*734*	*3083*

Percentages may add to more than 100% because respondents may have both unfilled and restored spaces

Table 2.2.19 **Adults with 21 or more natural teeth by age**

United Kingdom

Age	Percentage of adults who are dentate	Percentage of dentate adults with 21 or more natural teeth	Percentage of all adults with 21 or more natural teeth
	a	b	a x b
16-24	100	100	100
25-34	100	98	98
35-44	99	94	93
45-54	94	82	77
55-64	80	57	45
65-74	64	46	29
75 and over	42	23	10
All	87	83	72
	Base†		
16-24	702	491	
25-34	1158	854	
35-44	1114	781	
45-54	1106	746	
55-64	833	461	
65-74	758	344	
75 and over	533	140	
All	6204	3817	

† *The percentage of all adults with 21 or more teeth was calculated by multiplying the percentage of adults who are dentate (taken from the interview sample) by the percentage of dentate adults with 21 or more teeth (taken from the examination sample). As these figures come from different samples the base for the resulting percentage is not shown*

1988 Table 4.2

Table 2.2.20 **Adults with 21 or more natural teeth among dentate adults and all adults by age and gender**

United Kingdom

Age	Men		Women	
	Dentate	All	Dentate	All
	percentage with 21 or more natural teeth			
16-24	100	100	99	99
25-34	98	97	99	99
35-44	93	92	95	94
45-54	81	75	82	78
55-64	59	48	55	43
65 and over	39	24	38	19
All	82	73	83	70
	Base†			
16-24	227		264	
25-34	339		515	
35-44	364		417	
45-54	356		390	
55-64	216		245	
65 and over	243		241	
All	1745		2072	

† *The percentage of all adults with 21 or more teeth was calculated by multiplying the percentage of adults who are dentate (taken from the interview sample) by the percentage of dentate adults with 21 or more teeth (taken from the examination sample). As these figures come from different samples the base for the resulting percentage is not shown.*

1988 Table 4.5 (part)

Table 2.2.21 **Adults with 21 or more natural teeth by age and social class of head of household**

United Kingdom

Age	Social class of head of household					
	I, II, IIINM		IIIM		IV,V	
	Dentate	**All**	**Dentate**	**All**	**Dentate**	**All**
	percentage with 21 or more natural teeth					
16-24	100	100	100	100	98	98
25-34	99	98	98	97	99	98
35-44	98	97	94	92	85	83
45-54	89	87	72	65	75	67
55-64	68	63	52	38	37	24
65 and over	47	31	33	17	25	10
All	86	80	79	68	76	59
	Base†					
16-24	215		116		90	
25-34	402		254		148	
35-44	395		214		121	
45-54	394		204		107	
55-64	247		115		85	
65 and over	273		122		74	
All	1926		1025		625	

† The percentage of all adults with 21 or more teeth was calculated by multiplying the percentage of adults who are dentate (taken from the interview sample) by the percentage of dentate adults with 21 or more teeth (taken from the examination sample). As these figures come from different samples the base for the resulting percentage is not shown.

1988 Table 4.5 (part)

Table 2.2.22 **Adults with 21 or more natural teeth by age and country**

Age	Country							
	England		**Wales**		**Scotland**		**Northern Ireland**	
	Dentate	**All**	**Dentate**	**All**	**Dentate**	**All**	**Dentate**	**All**
	percentage with 21 or more natural teeth							
16-24	100	100	100	100	98	98	100	100
25-34	99	98	100	99	95	95	94	93
35-44	95	94	96	95	86	83	93	91
45-54	84	79	86	74	66	58	65	61
55-64	58	48	48	36	48	32	46	36
65 and over	40	22	28	14	31	14	30	15
All	83	73	81	68	78	64	81	71
	Base†							
16-24	246		69		80		96	
25-34	517		90		157		90	
35-44	447		96		142		96	
45-54	423		109		132		82	
55-64	259		73		75		54	
65 and over	294		65		82		43	
All	2186		502		668		461	

† The percentage of all adults with 21 or more teeth was calculated by multiplying the percentage of adults who are dentate (taken from the interview sample) by the percentage of dentate adults with 21 or more teeth (taken from the examination sample). As these figures come from different samples the base for the resulting percentage is not shown

1988 Table 4.5 (part)

Table 2.2.23 **Adults with 21 or more natural teeth by age and English region**

England

Age	English region					
	North		Midlands		South	
	Dentate	All	Dentate	All	Dentate	All
	percentage with 21 or more natural teeth					
16-24	99	99	100	100	100	100
25-34	96	95	99	98	100	100
35-44	91	91	94	93	97	96
45-54	82	77	75	66	88	86
55-64	46	34	60	45	63	56
65 and over	33	17	28	16	47	28
All	82	71	80	68	85	77
	Base[†]					
16-24	82		42		122	
25-34	146		125		246	
35-44	126		114		207	
45-54	121		101		201	
55-64	67		47		145	
65 and over	75		66		153	
All	617		495		1074	

† The percentage of all adults with 21 or more teeth was calculated by multiplying the percentage of adults who are dentate (taken from the interview sample) by the percentage of dentate adults with 21 or more teeth (taken from the examination sample). As these figures come from different samples the base for the resulting percentage is not shown

1988 Table 4.5 (part)

Table 2.2.24 **Adults with 21 or more natural teeth by age and usual reason for dental attendance**

Dentate adults   *United Kingdom*

Age	Usual reason for dental attendance	
	Regular/occasional check-up	Only with trouble
	percentage with 21 or more natural teeth	
16-24	100	100
25-34	98	98
35-44	95	91
45-54	86	67
55-64	67	29
65 or over	44	22
All	85	77
	Base	
16-24	336	155
25-34	581	221
35-44	596	184
45-54	597	148
55-64	349	112
65 or over	349	133
All	2808	1003

1988 Table 4.8

2.3 Replacement of missing teeth among dentate adults: dentures and bridges

Summary

- Among dentate adults in the United Kingdom in 1998, 16% had dentures in combination with natural teeth. This was age related with a peak in late middle age to early old age (55 to 74 years of age).

- There were differences according to geography, social class and attendance pattern, many of which reflected the variation in the number of teeth.

- People with a partial upper denture and no lower denture accounted for nearly half (48%) of all those who had dentures in combination with natural teeth. The type of provision varied markedly with age, with a greater proportion of people having complete upper dentures in conjunction with natural lower teeth with increasing age.

- There were indications that some of the social and geographical differences in having dentures may have been related to differences in attitude.

- Around three quarters of all partial dentures (uppers and lowers) examined were acrylic based, the rest being metal based. Just under three quarters of these dentures were supported entirely by the soft tissues, or by a combination of teeth and soft tissues.

- In 1998 4% of dentate adults wore a bridge to replace missing natural teeth. This proportion was highest among adults aged 55 to 64 (8%). Most bridges replaced just a single tooth and most were of a conventional, rather than an adhesive, type.

2.3.1 Introduction

In previous UK Adult Dental Health Surveys considerable emphasis has been placed on the reporting of denture provision and wearing. In this survey, increased retention of natural teeth has reduced the number of people who have complete dentures, particularly among younger groups. As the previous two chapters have shown, a large majority of the adult population now have teeth and most also have a dentition which will allow them to function without the need for dentures. Furthermore, the prospects for both of these proportions to increase further appear to be good (see Part 5 of this report). The emphasis in this chapter reflects this different environment.

Data on dentures and bridges are still important, for a number of reasons. Firstly, over one hundred thousand partial dentures and many thousands of bridges are placed in the United Kingdom each year within the National Health Service alone, with substantial implications in terms of cost and provision for the dental services. Also, partial dentures (and bridges) are only a partial solution to the problems of function and appearance which result from tooth loss. Partial dentures in particular have been shown to increase the risk of disease and plaque accumulation.

The data in this section are drawn mostly from the survey interview. The data on bridges and the more detailed analysis of partial denture design and materials were obtained from the dental examination. There are minor differences in the base numbers reported because not all people who took part in the interview had a dental examination, and not all those who stated that they had partial dentures were wearing them at the time of the dental examination.

2.3.2 The combination of natural teeth and dentures

Among adults aged 16 and over, 72% reported that they had natural teeth only, 13% had no natural teeth and 16% reported having a combination of natural teeth and dentures. When looked at as a proportion of dentate adults, 18% had dentures of some sort to augment their natural dentition. The greatest variation was as a result of age, with older people more likely than younger people to have reported having a denture in combination with natural teeth up to the age of 74, but with a small reduction in the oldest age group. When the same age comparisons were made, but based only on dentate adults, the proportion who had a combination of dentures and natural teeth increased steadily with age, with 63% of dentate people in the oldest age group having dentures in conjunction with their natural teeth.

Table 2.3.1

Subsequent analysis is based on the people in the sample who were dentate. Having fewer than 21 natural teeth is strongly associated with having dentures in combination with natural teeth (see Chapter 2.2). The results reported in this section shows that, in general, the proportions of participants in different groups who had dentures in combination with natural teeth mirrored the proportion who had 21 or more natural teeth.

There was no significant difference between the overall proportions of dentate men and women who had a combination of natural teeth and dentures but there was some variation by country and English region. People in Scotland were the most likely to have dentures in combination with natural teeth, those in the South of England the least likely. The provision of dentures also varied with the social class of the head of household. Overall, the proportion of dentate people who also had dentures ranged from 16% of those in households headed by a non-manual worker, through 20% of those in skilled manual households to 23% of those in households headed by an unskilled worker. The differences according to social class were most apparent among people aged 35 or more (Figure 2.3.1).

Figure 2.3.1, Tables 2.3.2–4

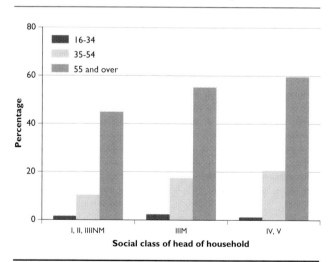

Fig 2.3.1 **The proportion of dentate adults with dentures and natural teeth by social class of head of household and age**

Overall, there was no significant difference in the proportion who had a combination of natural teeth and dentures between dentate adults who reported attending for a regular check-up and those who reported attending only with trouble; 18% and 20% respectively. However, among the two older age groups, regular attenders were less likely than those who attended with trouble to have a combination of natural teeth and dentures; 13% and 47% compared with 19% and 56% respectively. Those who reported attending for an occasional check-up were significantly less likely to have natural teeth and dentures (12%). Chapter 2.2 (Table 2.2.24) showed that among dentate adults who attended only with trouble, around three quarters of those aged 55 and over had fewer than 21 teeth, while Table 2.3.5 shows that just over half (56%) had dentures. Despite the strong association already seen between having fewer than 21 teeth and having a denture, among people who attend the dentist only with trouble, there are many (just under a fifth) who have fewer than 21 teeth and who do not have a denture.

Table 2.3.5

2.3.3 The type of dentures used in conjunction with natural teeth

Having a combination of natural teeth and dentures does not necessarily mean that the denture which the person has is a partial denture (i.e. one which fills the gaps where only some of the teeth in a jaw have been lost). There may be a complete denture in one jaw where all of the teeth have been lost, and natural teeth in the other, or any one of a number of different combinations of partial or complete dentures with natural teeth. This section illustrates the frequencies of different combinations. The data presented in most of the analyses which follow relate only to people who reported having dentures with natural teeth.

Overall, 48% of people with a combination of natural teeth and dentures had a partial denture in the upper jaw and no denture in the lower jaw. This was the most common finding, followed by both an upper and lower partial denture together (18%) and then a complete upper denture opposed by natural teeth (14%). An upper complete denture with a lower partial denture was reported by 11% of people using dentures and all other combinations of denture accounted for less that 10%. These data are very similar to the findings in the 1988 survey. In 1988, 49% of people with a combination of natural teeth and dentures had a partial upper denture alone, compared with 48% in this survey, a slightly higher proportion had a complete upper alone in 1988 (17% compared with 14%) and a slightly lower proportion had partial upper and lower dentures (15% in 1988 compared with 18%).

When calculated as a proportion of all dentate adults, 18% had a partial or complete denture in combination with natural teeth, 11% had a partial denture, either alone or in combination with complete denture, and 5% of dentate adults had a complete denture alone or with a partial denture. Only 3% of dentate adults had partial dentures in both jaws.

Table 2.3.6

Type of dentures was strongly related to age. Although the proportion of dentate adults who were aged less than 35 and

had a denture was very small, the vast majority of dentures among this group were partial upper dentures, either alone (78%) or with a partial lower denture (6%). This reflects a number of factors, but particularly the importance of the upper arch in terms of appearance, combined with a pattern of tooth loss in which upper anterior teeth are fairly frequently lost early in life as a result of trauma and disease. There was a reduction with increasing age in the proportion with an upper partial denture alone, to 60% for the 35 to 54 year old age group and 40% for those aged 55 and over. The proportions of all other combinations of dentures and natural teeth increased with age, particularly those having partial dentures in both jaws which increased from 6% of those aged 16 to 24 up to 21% of those aged 55 and over, reflecting greater tooth loss. There was little difference between men and women in the type of denture provision.

Figure 2.3.2, Tables 2.3.7–8

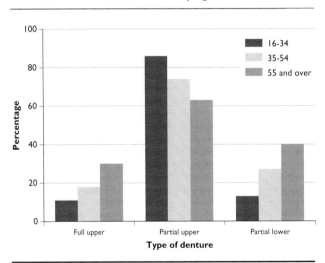

Fig 2.3.2 **The type of dentures used in conjunction with natural teeth by age**

When looked at by country and English region, some marked differences in the type of reported denture provision emerged. Of all of the people with a combination of natural teeth and dentures, 25% nationally had a complete upper denture with some natural lower teeth. However, in the North of England and in Scotland 32% had this combination compared with 20% in the South of England and 22% in the Midlands. Small base sizes meant these differences were not significant, but much of this variation is likely to be related to the proportion of the population with teeth in one jaw (see Chapter 2.2). In each country and English region a partial upper denture alone was the most common type of denture in combination with natural teeth. The range was from 46% in Wales to 52% in Northern Ireland and from 44% in the North of England to 50% in the South. People in Scotland with a combination of natural teeth and dentures were much less likely to have a lower partial denture (25%) than those in England (37%) and Wales (36%). Lower partial dentures are often the least essential denture in terms of both aesthetics and function and this difference may indicate a different attitude on the part of patients, or a different management philosophy on the part of dentists in different regions.

Figure 2.3.3 Table 2.3.9

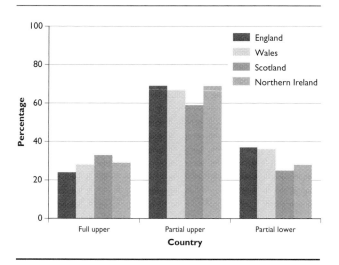

| Fig 2.3.3 | **The type of dentures used in conjunction with natural teeth by country** |

There were also variations in the type of denture provision in conjunction with natural teeth between people from different social class groups. Among people from non-manual backgrounds who had dentures in conjunction with natural teeth 52% had an upper partial denture and no lower denture compared with 39% of people from unskilled backgrounds. The trend was reversed for those with partial dentures in both arches; 23% of those from unskilled manual households had partial dentures in both arches compared with 16% of people from skilled manual, and 17% of people from non-manual backgrounds. This will probably reflect in large part, the total number of natural teeth lost.

Table 2.3.10

The reported pattern of dental attendance had a major influence on the type of denture provision in conjunction with natural teeth. Among people who attended only with trouble, there was a much higher proportion of people with a complete upper denture (42%) and far fewer with a partial upper denture alone or in combination with a lower (53%) than among occasional attenders (22% and 69%) and regular attenders (17% and 75%). This will partly reflect the different number of teeth in the three groups, but may also reflect the pattern of tooth loss, with people who attend only with trouble perhaps more likely to miss the stage of having a partial denture before losing all of their teeth in one jaw.

Table 2.3.11

2.3.4 Length of time since the most recent visit to the dentist among people with a combination of natural teeth and dentures

Among adults with natural teeth and dentures just over half (55%) had visited a dentist within the last 6 months, 15% had last visited between 6 months and a year prior to interview, while 8% had not attended within the past 10 years. There was little variation with age. Comparison with the distribution for all dentate adults (Table 6.1.12) shows very similar figures with 56% who had visited within the past 6 months and a further 15% who had visited within the past year.

Table 2.3.12

2.3.5 The type and condition of dentures assessed during the dental examination

Among people who had natural teeth and dentures and who were wearing upper dentures at the time of the examination, 30% had complete upper dentures and 69% had partial upper dentures. In the lower arch the proportions were very different and the base number of dentures was much smaller (less than one third of the number of upper dentures). Only 7% of lower dentures worn by those who took part in the dental examination were complete dentures, 91% were partial dentures, and 2% were complete overdentures (see Appendix A for definition). Examiners were asked to judge whether the dentures they found were intact or in need of repair. This can be a highly subjective judgment, but 8% of both upper and lower dentures were said to be in need of replacement.

A variety of materials and designs can be used to construct partial dentures. Of upper partial dentures, 26% had metal bases, these are usually cast cobalt chromium alloy with acrylic teeth and gums. The rest of the upper partial dentures (74%) were acrylic based. Partial lower dentures were slightly less likely to have metal bases (22%) and more likely to be acrylic (78%). Partial dentures can be constructed to be supported by the teeth, the soft tissues of the mouth (the gums and palate) or to be supported partly by the teeth and partly by the soft tissues. Where tooth support is possible this is generally seen as the preferable option, but often there are too few teeth, or teeth in the wrong places, to allow this. One in ten (10%) upper partial dentures were tooth borne, 64% were completely tissue borne and in 26% of cases support was judged to be shared between teeth and tissues. For lower dentures the proportions were very similar (8%, 66% and 26%).

Table 2.3.13

2.3.6 Bridges

The major alternative to dentures for the replacement of natural teeth is to use bridges. These prostheses are firmly cemented to a natural tooth or teeth adjacent to the space requiring replacement. In general, they are only appropriate for smaller gaps. Other alternatives include implant retained prostheses which rely on a surgically placed device which is fixed to the bone of the jaw, but this is a relatively new technique and is still uncommon. None of the dentate adults who were examined in the 1998 Adult Dental Health Survey had an implant retained prosthesis.

Dental bridges rely either on adjacent teeth being cut down for crowns (referred to here as conventional bridges) or the use of adhesive techniques (referred to here as adhesive bridges) where the bridge is attached to the teeth with an adhesive, reducing the need to prepare the tooth. The latter is a more recently developed technique. In 1998, 4% of the dentate population had a bridge of some sort and less than 1% had an adhesive bridge. In the large majority of cases (3% of the whole dentate population) only one tooth was replaced by a bridge. The proportion of the population with more teeth than this replaced by bridges decreased steeply up to a maximum of 10 teeth, but this was exceptional.

Table not shown

The proportion with and without bridges varied markedly with age. The lowest proportion was in the 16 to 24 year olds among whom 1% had a bridge, in this instance reflecting the lack of spaces at the front of the mouth. The highest proportion was among dentate 55 to 64 year olds, of whom 8% had a bridge. This dropped sharply to only 2% of dentate people aged 65 and over.

As there were so few dentate people with bridges, none of the differences by socio-demographic factors were significant but there were indications that women were slightly more likely than men to have a bridge (5% compared with 3% for men), and people from non-manual backgrounds were more likely to have a bridge (5%) than those from skilled (4%) or unskilled (3%) manual backgrounds. There was a significant difference in the proportion of people who had a bridge between those who said they attended the dentist for a check-up (5% of regular and 4% of occasional attenders) and those who attend only when they are having trouble (2%).

Table 2.3.14

Table 2.3.1 **The reliance on natural teeth and dentures by age**

United Kingdom

Reliance on both natural teeth and dentures	Age							All
	16-24	25-34	35-44	45-54	55-64	65-74	75 and over	
	%	%	%	%	%	%	%	%
All adults								
Natural teeth only	99	97	89	75	46	29	16	72
Dentures and natural teeth	1	3	10	18	34	35	27	16
No remaining natural teeth (edentate)	0	0	1	6	20	36	58	13
Dentate adults								
Proportion with natural teeth and dentures	1	3	10	20	43	54	63	18
				Base				
All adults	*702*	*1158*	*1114*	*1106*	*833*	*758*	*533*	*6204*
Dentate adults	*701*	*1151*	*1099*	*1016*	*645*	*456*	*213*	*5281*

1988 Table 16.1

Table 2.3.2 **The reliance on natural teeth and dentures by gender and age**

Dentate adults   *United Kingdom*

Gender	Age			All
	16-34	35-54	55 and over	
	%	%	%	%
Men				
Natural teeth only	98	86	48	82
Dentures and natural teeth	2	14	52	18
Women				
Natural teeth only	98	85	52	82
Dentures and natural teeth	2	15	48	18
		Base		
Men	*803*	*994*	*609*	*2406*
Women	*1049*	*1121*	*705*	*2875*

1988 Table 16.3

Table 2.3.3 **The reliance on natural teeth and dentures by country, English region and age**

Dentate adults

Country and English region	Age			All
	16-34	35-54	55 and over	
	%	%	%	%
United Kingdom				
Natural teeth only	98	86	50	82
Dentures and natural teeth	2	14	50	18
Country				
England				
Natural teeth only	99	87	51	83
Dentures and natural teeth	1	13	49	17
Wales				
Natural teeth only	98	83	44	79
Dentures and natural teeth	2	17	56	21
Scotland				
Natural teeth only	94	73	47	76
Dentures and natural teeth	6	27	53	24
Northern Ireland				
Natural teeth only	98	81	40	79
Dentures and natural teeth	2	19	60	21
English region				
North				
Natural teeth only	98	83	44	80
Dentures and natural teeth	2	17	56	20
Midlands				
Natural teeth only	99	82	48	80
Dentures and natural teeth	1	18	52	20
South				
Natural teeth only	99	92	55	85
Dentures and natural teeth	1	8	45	15
	Base			
United Kingdom	1852	2115	1314	5281
Country				
England	1053	1205	752	3010
Wales	221	279	182	682
Scotland	334	384	235	953
Northern Ireland	244	247	145	636
English region				
North	293	319	211	823
Midlands	244	320	167	731
South	516	566	374	1456

1988 Table 16.2

Table 2.3.4 **The reliance on natural teeth and dentures by social class of head of household and age**

Dentate adults   *United Kingdom*

Social class of head of household	Age			All
	16-34	35-54	55 and over	
	%	%	%	%
I, II, IIINM				
Natural teeth only	98	90	55	84
Dentures and natural teeth	2	10	45	16
IIIM				
Natural teeth only	98	83	45	80
Dentures and natural teeth	2	17	55	20
IV, V				
Natural teeth only	99	79	40	77
Dentures and natural teeth	1	21	60	23
	Base			
I, II, IIINM	799	1006	678	2483
IIIM	512	587	332	1431
IV, V	335	342	238	915

1988 Table 16.4

Table 2.3.5 **The reliance on natural teeth and dentures by usual reason for dental attendance and age**

Dentate adults   *United Kingdom*

Usual reason for dental attendance	Age			All
	16-34	35-54	55 and over	
	%	%	%	%
Regular check-up				
Natural teeth only	98	87	53	82
Dentures and natural teeth	2	13	47	18
Occasional check-up				
Natural teeth only	100	90	37	88
Dentures and natural teeth	0	10	63	12
Only with trouble				
Natural teeth only	97	81	44	80
Dentures and natural teeth	3	19	56	20
	Base			
Regular check-up	955	1328	838	3121
Occasional check-up	277	223	75	575
Only with trouble	617	560	395	1572

1988 Table 16.5

Table 2.3.6   The type of dentures used in conjunction with natural teeth

United Kingdom

The type of denture provision		Adults with natural teeth and dentures	All dentate adults
Upper jaw	**Lower jaw**		
		%	%
Complete denture	None	14 ⎫ 25	3 ⎫ 5
Complete denture	Partial denture	11 ⎭	2 ⎭
Partial denture	Partial denture	18	3
Partial denture	None	48	8
None	Partial denture	6 ⎫	1
Partial denture	Complete denture	2 ⎬ 9	0
None	Complete denture	1 ⎭	0
None	None	0	82
Base		1093	5281

Table 2.3.7   The type of dentures used in conjunction with natural teeth by age

Adults with natural teeth and dentures   *United Kingdom*

Type of denture provision	Age			All
Upper jaw - lower jaw	**16-34**	**34-54**	**55 and over**	
	%	%	%	%
Complete – none	6	12	16	14
Complete – partial	5	6	14	11
Partial – partial	6	14	21	18
Partial – none	78	60	40	48
Other	6	8	9	9
Summary				
Complete upper denture	11	18	30	25
Partial upper denture	86	75	63	68
Partial lower denture	13	27	40	35
Base	46	361	686	1093

1988 Tables 16.6, 16.7

Table 2.3.8   The type of dentures used in conjunction with natural teeth by gender

Adults with natural teeth and dentures   *United Kingdom*

Type of denture provision	Gender		All
Upper jaw - lower jaw	**Men**	**Women**	
	%	%	%
Complete – none	14	15	14
Complete – partial	11	11	11
Partial – partial	18	18	18
Partial – none	49	47	48
Other	8	10	9
Summary			
Complete upper denture	26	25	25
Partial upper denture	68	67	68
Partial lower denture	35	36	35
Base	513	580	1093

Table 2.3.9   **The type of dentures used in conjunction with natural teeth by English region and country**

Adults with natural teeth and dentures

Type of denture provision	English region			Country				United Kingdom
Upper jaw – lower jaw	North	Midlands	South	England	Wales	Scotland	Northern Ireland	
	%	%	%	%	%	%	%	%
Complete – none	19	11	10	13	14	23	18	14
Complete – partial	13	11	9	11	13	9	11	11
Partial – partial	17	19	21	20	19	9	16	18
Partial – none	44	48	50	47	46	50	52	48
Other	8	10	9	9	8	9	4	9
Summary								
Complete upper denture	32	23	20	24	28	33	29	25
Partial upper denture	61	67	74	69	67	59	69	68
Partial lower denture	36	40	36	37	36	25	28	35
Base	182	147	218	547	152	253	141	1093

1988 Table 16.8

Table 2.3.10   **The type of dentures used in conjunction with natural teeth by social class of head of household**

Adults with natural teeth and dentures   *United Kingdom*

Type of denture provision	Social class of head of household			All
Upper jaw – lower jaw	I, II, IIINM	IIIM	IV, V	
	%	%	%	%
Complete – none	12	15	18	14
Complete – partial	9	12	10	11
Partial – partial	17	16	23	18
Partial – none	52	50	39	48
Other	10	7	11	9
Summary				
Complete upper denture	22	27	28	25
Partial upper denture	70	66	64	68
Partial lower denture	33	34	40	35
Base	461	332	231	1093

1988 Table 16.9

Table 2.3.11 The type of dentures used in conjunction with natural teeth by usual reason for dental attendance

Adults with natural teeth and dentures   *United Kingdom*

Type of denture provision	Usual reason for dental attendance			All
Upper jaw – lower jaw	Regular check-up	Occasional check-up	Only with trouble	
	%	%	%	%
Complete – none	9	17	24	14
Complete – partial	7	5	18	11
Partial – partial	20	23	15	18
Partial – none	55	43	36	48
Other	10	12	7	9
Summary				
Complete upper denture	17	22	42	25
Partial upper denture	75	69	53	68
Partial lower denture	34	36	37	35
Base	*609*	*90*	*392*	*1093*

1988 Table 16.10

Table 2.3.12 Time since most recent visit to the dentist by age

Adults with natural teeth and dentures   *United Kingdom*

Time since most recent visit	Age			All
	16-34	35-54	55 and over	
	%	%	%	%
Less than 6 months	54	54	56	55
6 months, less than 1 year	23	16	15	15
1 year, less than 2 years	4	6	7	6
2 years, less than 5 years	8	12	10	11
5 years, less than 10 years	4	4	4	4
10 years or more	8	10	8	8
Base	*46*	*361*	*686*	*1093*

1988 Table 16.23

Table 2.3.13 The type and condition of upper and lower dentures

Adults with natural teeth who were wearing dentures during the dental examination   *United Kingdom*

Type and condition of dentures	Upper denture	Lower denture
	%	%
Type of denture		
Complete	30	7
Partial	69	91
Complete overdenture	0	2
Condition of denture		
Intact	92	92
In need of replacement	8	8
	%	%
Composition and design of partial dentures		
Material		
Metal	26	22
Plastic (acrylic)	74	78
Support		
Tooth borne	10	8
Tissue borne	64	66
Both	26	26
	Base	
Dentate adults with any type of denture	*612*	*203*
Dental adults with a partial denture	*427*	*179*

1998 Table 16.31

Table 2.3.14 **Bridges by age, gender, country, English region, social class of head of household and usual reason for dental attendance**

Dentate adults   *United Kingdom*

	Percentage with bridges	Base
All	4	*3817*
Age		
16-24	1	*491*
25-34	2	*854*
35-44	5	*781*
45-54	7	*746*
55-64	8	*461*
65 and over	2	*484*
Gender		
Men	3	*1745*
Women	5	*2072*
Country		
England	4	*2186*
Wales	6	*502*
Scotland	4	*668*
Northern Ireland	3	*461*
English region		
North	5	*617*
Midlands	2	*495*
South	4	*1074*
Social class of head of household		
I, II, IIINM	5	*1926*
IIIM	4	*1025*
IV,V	3	*625*
Usual reason for dental attendance		
Regular check-up	5	*2400*
Occasional check-up	4	*408*
Only with trouble	2	*1003*

The condition of the natural teeth, restorative treatment and supporting structures

3.0 Overview of the condition of the natural teeth, restorative treatment and supporting structures

Part 2 of this report described in detail the presence, number and distribution of natural teeth. Part 3 deals in detail with the condition of these teeth.

The presence or absence of natural teeth is in large part determined by the disease and treatment they have experienced. Two major conditions, dental caries (tooth decay) and periodontal (gum) diseases, are responsible, either directly or indirectly, for most dental treatment and for most tooth loss. Dental caries is a common condition which, if it progresses unchecked, may result in the need for a tooth to be restored with a filling or other restoration. The prevalence of dental decay is a fundamental measure of oral health and an indicator of the long term prospects for a natural functional dentition (see Glossary), while the extent to which teeth have been restored gives an indication of the past history of treatment for dental caries. Chapter 3.1 covers the prevalence and distribution of dental caries and, conversely, the prevalence of sound and untreated teeth, while Chapter 3.2 describes the pattern of fillings and other restorations among adults in the United Kingdom. The other major threat to the natural dentition comes from the periodontal (gum) diseases. Chapter 3.3 describes in detail the extent and severity of periodontal diseases among adults in the United Kingdom. Data on oral cleanliness, which is an important factor for the development and management of the periodontal diseases, are also reported.

Before presenting detailed data on these dental conditions, some summary data drawn from Parts 2 and 3 of this report are presented. These give an overview of oral health and disease in the adult population of the United Kingdom in 1998 in the form of three tables and a brief synopsis of the results.

Across the United Kingdom, 13% of all adults were edentate, with a slightly higher proportion in Scotland and Wales, and fewer in the South of England compared with the North of England. Under 1% of those aged less than 45 had lost all their natural teeth; this increased to 6% of those aged 45 to 54 and above this age increased steeply up to 58% of those aged 75 and over. Of those who retained teeth, possessing 21 or more is taken as an indicator of a functional dentition. Over four fifths (83%) of dentate adults had 21 or more natural teeth, ranging from 100% of 16 to 24 year olds to 23% of those aged 75 years and over.

In each of the constituent countries of the United Kingdom, dentate adults in 1998 had about three quarters of the 32 teeth that are potentially present in adults. This figure varied with age as one might expect but even among the oldest age group shown, those aged 65 and over, an average of just over half of the potential 32 teeth were present (17.3).

Considering the condition of the surfaces of the natural crowns, it is encouraging to see that the majority were sound and untreated and that, of the remainder, many more were filled than were actively decayed, irrespective of age, gender, country, social class or usual reason for dental attendance. The mean numbers of sound and untreated teeth decreased with age while, with the exception of the two oldest age groups, the mean number of filled (otherwise sound) teeth increased with age.

With age come other changes to the natural dentition and to the supporting tissues of the teeth. As well as the surfaces of the crowns of the teeth, the roots can also become vulnerable depending on their exposure in the mouth. Across nearly all the socio-demographic factors reported, less than 10% of dentate adults were recorded as showing loss of attachment of gingival tissues of 6mm or more. Only in the oldest two age groups, those aged 55 to 64 and those aged 65 and over, were the proportions with this degree of loss of attachment higher (17% and 31% respectively). Slightly lower proportions of adults had pocketing of periodontal tissues of 6mm or more, the highest being among those aged 65 and over with a prevalence of 15%.

Such disease of the soft tissues exposes the root surface; approximately two thirds of all dentate adults in the United Kingdom had either exposed, worn, filled or decayed root surfaces. This ranged from 20% of dentate 16 to 24 year olds to 97% of those aged 65 and over. In comparison there was little variation with respect to other socio-demographic factors.

An average of 6.4 teeth had root surfaces which were either exposed, worn, filled or decayed, slightly less than the number of teeth with coronal surfaces affected by dental disease as estimated from the number of teeth which were filled (otherwise sound) and the number which were decayed or unsound (8.5). This relationship was relatively stable across all socio-demographic groups (although the differences were not always significant) with three exceptions. Dentate adults in the Midlands and those who reported only attending the dentist with trouble both had a higher mean number of teeth with root surfaces which were either exposed, worn, filled or decayed than the mean number of teeth with coronal surfaces affected by dental disease. However the largest variation was seen with age; teeth with exposed, worn, filled or decayed root surfaces were far less prevalent than teeth with coronal surfaces affected by dental disease among those aged 25 to 44 while among those age 55 and over the mean number of teeth with vulnerable and affected root surfaces was higher than the number of teeth with affected coronal surfaces.

Tables 3.0.1-3

Table 3.0.1   **The overall condition of the teeth in 1998**

United Kingdom

	All adults		Dentate adults				
	Edentate	Dentate	Natural teeth and dentures	21 or more natural teeth	At least one artificial crown	At least moderate tooth wear	Exposed, worn, filled or decayed roots
			percentage with each condition				
All	13	87	18	83	34	11	66
Age							
16-24	0	100	1	100	7	2	20
25-34	0	100	3	98	24	5	50
35-44	1	99	10	94	41	8	71
45-54	6	94	20	82	49	12	86
55-64	20	80	43	57	48	18	90
65-74	36	64	54	46	39	} 34	97
75 and over	58	42	63	23	36		97
Gender							
Men	10	90	18	82	31	15	70
Women	15	85	18	83	37	8	63
Country							
England	12	88	17	83	34	12	66
Wales	17	83	21	81	36	15	62
Scotland	18	82	24	78	35	9	70
Northern Ireland	12	88	21	81	33	9	67
English region							
North	14	86	20	82	31	12	63
Midlands	15	85	20	80	32	8	74
South	10	90	15	85	35	13	65
Social class of head of household							
I, II, IIINM	8	92	16	86	38	10	67
IIIM	15	85	20	79	30	14	68
IV,V	22	78	23	76	28	12	66
Usual reason for dental attendance†							
Regular check up	-	-	18	84	40	11	68
Occasional check up	-	-	12	88	28	7	59
Only with trouble	-	-	20	77	23	13	66

† This question was only asked of dentate adults. Those who were edentate were asked about their attendance when they still had teeth, but not their current reasons for attendance

Table 3.0.2   **The mean number of teeth in each condition in 1998**

Dentate adults   *United Kingdom*

	Missing	Present	Decayed and unsound	Filled (otherwise sound)	With artificial crowns	Sound and untreated	With exposed, worn, filled or decayed roots
				mean number of teeth			
All	7.2	24.8	1.5	7.0	1.0	15.3	6.4
Age							
16-24	4.1	27.9	1.6	2.6	0.1	23.4	1.2
25-34	3.9	28.1	1.8	6.6	0.6	19.1	3.6
35-44	5.3	26.7	1.4	8.8	1.1	15.4	6.4
45-54	8.0	24.0	1.4	9.3	1.6	11.7	9.3
55-64	12.1	19.9	1.3	7.4	1.6	9.6	9.9
65 and over	14.7	17.3	1.2	6.3	1.3	8.5	10.6
Gender							
Men	7.0	25.0	1.7	6.6	0.8	15.7	7.0
Women	7.4	24.6	1.3	7.3	1.2	14.9	5.7
Country							
England	7.1	24.9	1.5	6.9	1.0	15.6	6.5
Wales	7.8	24.2	1.2	6.9	1.0	15.1	5.1
Scotland	8.2	23.8	1.8	7.4	1.0	13.6	5.9
Northern Ireland	7.5	24.5	1.5	8.2	0.8	13.9	5.6
English region							
North	7.4	24.6	1.9	6.5	0.9	15.4	5.5
Midlands	7.4	24.6	1.4	6.7	0.9	15.6	8.5
South	6.7	25.3	1.3	7.2	1.1	15.7	6.3
Social class of head of household							
I, II, IIINM	6.6	25.4	1.2	7.6	1.2	15.4	6.8
IIIM	7.8	24.2	1.7	6.9	0.9	14.8	6.2
IV,V	8.4	23.6	1.9	5.7	0.7	15.3	6.3
Usual reason for dental attendance							
Regular check up	7.1	24.9	1.1	8.0	1.2	14.6	6.8
Occasional check up	5.7	26.3	1.4	6.4	0.9	17.5	4.7
Only with trouble	8.0	24.0	2.3	5.1	0.6	16.0	6.1

Table 3.0.3   **The presence of visible plaque and periodontal conditions in 1998**

Dentate adults   *United Kingdom*

	Visible plaque	Calculus	Pocketing 4mm or greater	Pocketing 6mm or greater	LOA 4mm or greater	LOA 6mm or greater
			percentage with each condition			
All	72	73	54	5	43	8
Age						
16-24	72	61	34	1	14	0
25-34	70	71	47	2	26	2
35-44	72	74	59	5	42	3
45-54	69	77	61	6	52	10
55-64	75	77	62	9	70	17
65 and over	78	83	67	15	85	31
Gender						
Men	76	76	57	6	46	9
Women	68	70	51	5	40	7
Country						
England	75	75	55	6	44	8
Wales	52	61	47	4	34	7
Scotland	60	62	47	5	44	8
Northern Ireland	66	67	48	4	39	6
English region						
North	80	80	50	4	45	7
Midlands	70	76	58	3	45	6
South	74	72	57	8	42	9
Social class of head of household						
I, II, IIINM	70	71	52	6	42	7
IIIM	75	75	56	5	44	9
IV,V	78	76	57	7	47	11
Usual reason for dental attendance						
Regular check up	68	68	52	5	43	7
Occasional check up	72	75	56	4	39	5
Only with trouble	80	82	57	6	44	10

3.1   The condition of natural teeth

Summary

Sound and untreated teeth

- Overall, in 1998, dentate adults had an average of 15.3 sound and untreated teeth. This varied from 23.4 teeth among those aged 16 to 24 years to 9.1 teeth among those aged 55 and over. Four out of ten (40%) had 18 or more sound and untreated teeth. This proportion declined steeply with age from 89% of 16 to 24 year olds to 37% of those aged 35 to 44 and 5% of those aged 55 and over.

- Men had on average 15.7 sound and untreated teeth which was significantly higher than the average of 14.9 for women.

- The proportion with 18 or more sound and untreated teeth varied from 41% of dentate adults in England to 29% of dentate adults in Scotland; Wales and Northern Ireland were intermediate in the range, with 37% and 32% respectively.

Fissure sealants

- Five per cent of dentate adults had teeth which had been fissure sealed; the majority of these sealants were found in the youngest age group.

Decayed and unsound teeth

- Dentate adults had an average of 1.5 decayed or unsound teeth and over half (55%) had at least one such tooth. The numbers of decayed and unsound teeth varied little with age.

- Dentate adults in Wales were less likely to have at least one decayed or unsound tooth than people in the other countries; 48% compared with England (55%), Scotland (58%) and Northern Ireland (59%).

- Within the individual regions of England, decayed and unsound teeth were more prevalent in the North than in the Midlands and the South. This was particularly true among 16 to 24 year olds; 70% in the North had at least one decayed or unsound tooth compared with 46% in the Midlands and 38% in the South.

- Adults who reported going to the dentist only with trouble had twice as many decayed or unsound teeth as those who reported attending for regular check-ups; 2.3 compared with 1.1 teeth on average were decayed or unsound.

Primary and recurrent decay

- The classification of 'decayed and unsound' includes teeth with restorations which were unsound but not affected by caries. Further analysis separately identified teeth which were decayed and investigated both the type and the extent of decay.

- Overall the mean number of teeth with decay was 1.2 of which the majority was primary decay, 0.9 teeth on average, compared with an average of 0.2 teeth with recurrent decay.

- Primary decay was more prevalent among the younger adults. Of those aged 25 to 34, 51% had some primary decay compared with 34% of those aged 35 to 44.

- The largest variation in the prevalence of recurrent decay was found among the regions of England. Twenty-two per cent of dentate adults in the North had recurrent decay compared with 12% of those in the Midlands and the South.

- Teeth which had unrestorable decay were more likely to be found among men than women (11% compared with 5%) and those from skilled or unskilled manual backgrounds (10% and 13% compared with 5% of those from non-manual backgrounds). As many as 19% of those who attended the dentist only with trouble had unrestorable teeth compared with 3% and 4% respectively of those who reported attending for regular or occasional check-ups. There was little difference in the proportions of people with unrestorable teeth with respect to country or English region.

- Overall, 24% of dentate adults had primary visual decay, 22% had primary cavitated decay, 8% had recurrent visual and 8% had recurrent cavitated decay.

- There was little difference between the mean number of teeth with primary visual decay, 0.5, and the number with primary cavitated decay, 0.4 teeth.

- Dentate adults who reported attending the dentist only with trouble were more likely than those who attended for regular check-ups to have primary visual or primary cavitated decay; 30% compared with 20% and 37% compared with 15% respectively.

Condition of individual teeth

- For dentate adults of all ages, the lower canines and incisors were the least decay prone teeth, with 10% or fewer missing, decayed, unsound or restored.

- First molars showed high levels of decay experience with more than 85% of them missing, decayed, unsound or restored.

Tooth wear

- Two thirds (66%) of dentate adults had some wear (loss of tooth substance due to a non-bacterial cause) in their anterior teeth that involved at least some dentine, with 11% having wear that was moderate (extensive involvement of dentine) and 1% with severe wear.

- The proportion of dentate adults with tooth wear increased with age, from 36% of 16 to 24 year olds to 89% of those aged 65 and over.

Status of root surfaces

- Overall, two thirds of dentate adults (66%) had at least one tooth with a root surface that was either exposed, worn, filled or decayed, with an average of 6.4 teeth involved. Nearly a quarter (24%) had 12 or more teeth in this condition.

- Root surface problems increased with age. The mean number of teeth with root surfaces which were exposed, worn, filled or decayed increased from 1.2 among 16 to 24 year olds to 10.6 among those aged 65 and over, while the mean number of teeth overall decreased from 27.9 to 17.5.

3.1.1 Introduction

The lifetime retention of sufficient natural teeth to maintain a functional dentition, with little history of disease to either the teeth or their supporting structures, could be considered as a desirable objective for the population. However, the overall rate of progress towards such an objective will be tempered by the existing burden of disease experience, both treated and untreated, among older adults. Decrease in the experience of decay will lead to overall figures which mask a large variation with age.

The changing pattern of dental disease over the decades leads to changes in the way criteria for the assessment of dental disease are derived and applied; changes in knowledge as well as the appearance of new threats to oral health mean thresholds become refined and new criteria are introduced for assessment.

One of the more significant changes in the assessment of oral health in this survey has been the introduction of a measure intended to improve the sensitivity of recording dental caries.

In the 1998 survey the criteria for measuring decay were changed from those used in earlier surveys in this series. In the previous surveys, a tooth was recorded as decayed only if cavitated caries was present. At the same time an assessment was made of the extent of the decay. Teeth were classified either as having cavitated decay or decay that was so extensive that the tooth was broken down, perhaps with involvement of the dental pulp, and unrestorable as a consequence. While both these categories were used in the 1998 survey, for the first time an assessment of visual caries (that is early caries that has caused demineralisation of the tooth but without cavitation) was also included in the assessment of decay. It should be noted that neither set of criteria included lesions confined to the enamel of the tooth.

As has been the case in all the surveys in this series, the results are not intended to exactly duplicate the findings of a conventional dental examination. The survey examinations took place in the respondent's home, in somewhat different circumstances from those which pertain in a dental surgery (in terms of lighting, the position of the respondent, use of compressed air for drying teeth and availability of X-rays and other diagnostic aids). However, with the inclusion of criteria for recording tooth decay that can be seen to extend into dentine the survey results are a step closer to those obtained in the dental practice. Nevertheless, the examination

and recording for dental caries remains significantly less sensitive than that which can be carried out in the dental surgery.

The change in criteria for assessing dental caries has implications for each of the definitions of tooth condition. In 1998, teeth with untreated visual caries were classified as decayed but in the 1988 survey and earlier, such teeth were recorded as sound and untreated. Filled teeth with recurrent visual caries were assessed as decayed in the 1998 classification whereas they were defined as teeth with sound restorations according to the pre-1998 criteria. Full definitions and the survey examination criteria can be found in Appendices A and C.

This chapter presents data based on the 1998 definition of decay. Some analyses identify the three levels of decay. Section 3.1.5 discusses the effect on the findings of these changes to the criteria for decay. In Chapter 5.2 where the 1998 data are compared with previous survey data, the analyses are based on the criteria used in the 1988 survey.

CORONAL STATUS

3.1.2 Sound and untreated teeth

Sound and untreated teeth are those which were assessed as having no visual dental caries, no cavitated dental caries nor any restoration on any surface. One of the indicators of dental health used in the previous dental surveys was the proportion of dentate adults who had 18 or more teeth that were sound and untreated. The same indicator is used in the present survey. As dental health continues to improve, consideration of a higher threshold of the number of sound and untreated teeth may prove informative so tables giving the full distribution of the numbers of teeth also show the proportion with 24 or more teeth in this condition.

Overall, in 1998, dentate adults had an average of 15.3 sound and untreated teeth. This varied from 23.4 teeth among those aged 16 to 24 years to 8.5 teeth among those aged 65 and over. Four out of ten (40%) had 18 or more sound and untreated teeth. This proportion declined steeply with age from 89% of 16 to 24 year olds to 37% of those aged 35 to 44 and 5% of those aged 65 and over.

Figure 3.1.1, Table 3.1.1

Fig 3.1.1 **Proportion of dentate adults with 18 or more sound and untreated teeth by age**

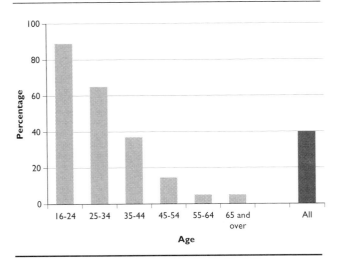

Fig 3.1.2 **Proportion of dentate adults with 18 or more sound and untreated teeth by country**

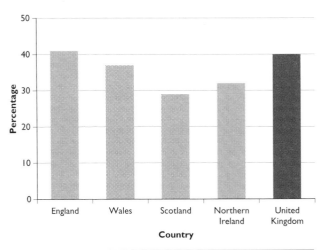

Men had on average 15.7 sound and untreated teeth which was significantly higher than the average of 14.9 for women. This was reflected in the proportion who had 18 or more sound and untreated teeth; 42% of men and 38% of women, although this difference was not significant. Within age groups there was no difference between the proportions of men and women with 18 or more sound and untreated teeth below the age of 35. For those aged 35 to 44, men were significantly more likely than women to have 18 or more sound and untreated teeth (42% compared with 31%), a difference which was repeated in the other older age groups although not at a level to be statistically significant.

Tables 3.1.2–3

There were marked differences between the data from the constituent countries of the United Kingdom in terms of both the average number and the proportion with 18 or more sound and untreated teeth. The average number of teeth in this condition was higher in England (15.6) and in Wales (15.1) than in Scotland (13.6) and in Northern Ireland (13.9). The proportion with 18 or more such teeth varied from 41% of dentate adults in England to 29% of dentate adults in Scotland with Wales and Northern Ireland intermediate in the range, 37% and 32% respectively. These differences were most apparent among those aged 25 to 44. For example 38% of dentate adults in Scotland aged 25 to 34 had 18 or more sound and untreated teeth compared with 68% and 69% of those in England and Wales respectively. However among the youngest age group the proportion for Scotland was considerably higher at 79%. In Northern Ireland, people aged 45 to 64 years had a similar dental status to people ten years older in the other countries of the UK, that is, 5% of 45 to 54 year olds in Northern Ireland had 18 or more sound, untreated teeth compared with 4%, 8% and 5% of 55 to 64 year olds in England, Wales and Scotland, respectively.

Overall, there was very little difference between the English regions in numbers of sound and untreated teeth. However, among those aged 25 to 44, dentate adults in the North were less likely than those elsewhere to have 18 or more sound and untreated teeth.

Tables 3.1.4–5

The numbers of sound and untreated teeth showed little variation between those from different social class backgrounds. The only significant differences within age groups were among dentate adults aged 16 to 24 and 35 to 44 among whom those from a non-manual background were more likely to have 18 or more sound and untreated teeth. In the case of the youngest adults this was in comparison with those from a skilled manual background (93% and 79%), while for 35 to 44 year olds 38% of those from a non-manual background had 18 or more such teeth compared with 29% of those from an unskilled manual background.

Tables 3.1.6–7

Dentate adults who attended the dentist for occasional check-ups and those who only attended with trouble were more likely to have sound and untreated teeth than those who attended for a regular check-up; these groups had an average of 17.5, 16.0 and 14.6 such teeth respectively. The proportions with 18 or more sound and untreated teeth were 52%, 45% and 35% respectively. Such differences are a reflection of the amount of restorative treatment received by these three groups of people (see Chapter 3.2). Regular dental attenders are likely to have had some restorative treatment for disease at an earlier stage than that identified by the survey examination (see 3.1.1). This leads to them having fewer teeth which can be classified as sound and untreated. Thus these differences are likely to be more obvious among those with more extensive experience of restorative treatment -

the older adults - which can be seen in Table 3.1.9. The differences between people with different attendance patterns are more evident among those aged 35 to 54 than among the younger age groups for whom the experience of restorative treatment is lower.

Tables 3.1.8–9

3.1.3 Fissure sealants

The use of sealants as a preventive treatment was introduced on a widespread basis in the 1970's. The pits and fissures of teeth are the most vulnerable sites for the onset of caries. The application of resin to these surfaces soon after eruption of the teeth protects against the development of caries. Such sealants are also used in conjunction with composite restorations in small carious cavities in posterior teeth. Teeth with these small composite restorations/sealants have obviously experienced caries while those with sealants only, have, as far as can be discerned, not been affected by caries and are included in the count of sound and untreated teeth. This is the first time an assessment of the presence of sealants has been included in the surveys of Adult Dental Health.

The prevalence of sealants among dentate adults was 5%. Sealants were found predominantly in the youngest age group, the 16 to 24 year olds, of whom 23% had sealants. The only other significant difference with respect to socio-demographic factors was between the countries. In Northern Ireland 13% of dentate adults were found to have sealants compared with 5% in England and Wales and 7% in Scotland.

Table 3.1.10

3.1.4 Decayed and unsound teeth

The changes to the criteria for the assessment of caries were described in the introduction to this chapter. All data in this section refer to teeth with visual caries or cavitated caries (including unrestorable teeth). 'Unsound' refers to teeth with restorations which were defective but not carious. Teeth with an unsound restoration and visual or cavitated recurrent caries were classified as decayed.

Over half (55%) of dentate adults had one or more decayed or unsound teeth, with an average of 1.5 teeth affected. The numbers of decayed and unsound teeth varied little with age. Younger adults had slightly higher average numbers of decayed or unsound teeth (ranging from 1.8 among 25 to 34 year olds to 1.2 among those aged 65 and over) although the differences between individual age groups were not significant. The proportion with one or more decayed and unsound teeth showed some variation but no discernible trend with age.

The average number of decayed and unsound teeth is an important statistic but in assessing the variation with age, the number of teeth retained in the mouth also needs to be considered. Dividing the mean number of decayed and unsound teeth by the number of teeth present shows that the proportion of teeth affected was highest among the older age groups.

Table 3.1.11

Men had more decayed or unsound teeth on average than women (1.7 compared with 1.3) and they were more likely than women to have at least one decayed or unsound tooth, 58% and 52% respectively. This relationship was more evident among the younger age groups.

Tables 3.1.12–13

Dentate adults in Wales had significantly fewer decayed and unsound teeth on average than those in other countries; 1.2 compared with 1.5 in England and Northern Ireland and 1.8 in Scotland. They were also less likely to have one or more such teeth; 48% compared with England (55%), Scotland (58%) and Northern Ireland (60%). This difference was repeated within all the age groups except for 35 to 44 year olds, although none reached the level of statistical significance. Among 16 to 24 year olds those in Northern Ireland had the highest proportion of adults with at least one decayed or unsound tooth (72%).

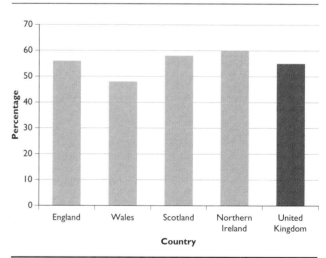

Fig 3.1.3 **Proportion of dentate adults with some decayed or unsound teeth by country**

Within the individual regions of England, decayed and unsound teeth were more prevalent in the North than in the Midlands and the South. This was particularly true among 16 to 24 year olds, among whom 70% of those in the North had at least one decayed or unsound tooth compared with 46% in the Midlands and 38% in the South. These differences were maintained with increasing age although the contrast was less marked.

Figure 3.1.3, Tables 3.1.14–15

Dentate adults from non-manual backgrounds had fewer decayed or unsound teeth on average than those from skilled and unskilled manual backgrounds (1.2 compared with 1.7 and 1.9 respectively) and were less likely to have any decayed or unsound teeth (50% compared with 57% and 62% respectively). These differences were found among those aged less than 45 but above this age there was little significant difference with respect to social class.

Figure 3.1.4, Tables 3.1.16–17

Fig 3.1.4   **Proportion of dentate adults with some decayed or unsound teeth by age and social class**

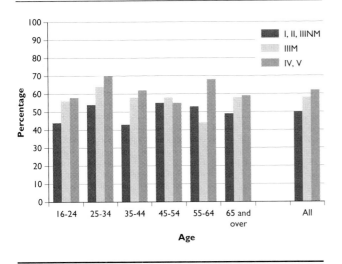

Adults who reported going to the dentist only with trouble had around twice as many decayed or unsound teeth as those who reported attending for check-ups; 2.3 compared with 1.1 among those who attended for regular check-ups and 1.4 among those who attended for occasional check-ups. They were also more likely to have at least one decayed or unsound tooth compared with those who reported attending for regular or occasional check-ups (67%, 48% and 56% respectively). These differences were repeated within each age group. Although these are differences that might be expected it should be noted that almost one half of those who reported attending for regular check-ups had at least one decayed or unsound tooth. The final table in this section shows the variation in the proportion with any decayed or unsound teeth by the length of time since the most recent visit to the dentist and this supports the comparatively high level of decay found among attenders for check-ups. One half of those who reported attending the dentist in the previous six months had at least one decayed or unsound tooth.

Tables 3.1.18–20

3.1.5 The effect of including visual caries in the criteria for decayed and unsound teeth

For the reasons outlined in Section 3.1.1, an assessment of visual caries was included in the criteria for the first time in the 1998 Adult Dental Health Survey. The impact of including this criterion is shown in Table 3.1.21. The inclusion of visual caries increased the prevalence of caries detected overall from 42% to 55% and the mean number of teeth affected from 1.0 to 1.5. These increases were seen for all socio-demographic groups but the size of the effect varied. The inclusion of visual caries had the greatest effect on assessment of prevalence levels among those aged less than 35 and those living in Scotland, Northern Ireland and the North of England. For example, the proportion of dentate adults aged 16 to 24 with at least one decayed or unsound tooth was 31% when visual caries was not included and 51% when it was included. Among those in the North of England the inclusion of visual caries added 19 percentage points to the prevalence levels from 46% to 65%. The mean number of teeth affected also showed similar differences.

Table 3.1.21

3.1.6 Decayed teeth - the type and extent of decay

The classification 'decayed and unsound teeth' includes teeth with restorations which were unsound but not affected by caries. An assessment of restorations is covered in Chapter 3.2. Further analysis separately identified teeth with decay and investigated both the type and the extent of decay.

In the 1998 survey, decay was assessed as primary (new) decay or as recurrent decay, that is, decay which occurred around restorations. This is referred to as the 'type' of decay found. As explained previously, teeth were also assessed for three levels of decay:
- teeth with visual caries only,
- teeth with cavitated caries
- teeth that were so badly broken down as to be unrestorable.

In order to distinguish the term 'level' from the levels of decay found in the population these three categories are referred to as the 'extent' of caries involvement. Decay in teeth coded as unrestorable was not assessed for its primary or recurrent status. It should be noted that mean values for the separate categories of decay do not always exactly sum to the overall mean number of decayed teeth because of rounding.

3.1.7 Primary and recurrent decay

Overall the mean number of teeth with decay was 1.2 of which the majority was primary decay, 0.9 teeth on average, compared with an average of 0.2 teeth with recurrent decay. This distribution varied with age. Among the younger adults most of the decay was primary, while among older adults a greater contribution was made by recurrent decay. For example, 25 to 34 year olds had an average of 1.2 teeth with primary decay and 0.2 teeth with recurrent decay compared with 0.6 and 0.2 teeth respectively for those aged 55 to 64. As noted earlier, the overall decline with age in the number of teeth with either type of decay is related to the decrease in the number of teeth present in the mouth (see Section 3.1.4).

Table 3.1.22

Comparison of these figures with those for decayed and unsound teeth (Table 3.1.11) also shows the greater contribution made by the category 'unsound teeth' among older adults. For example, there was no apparent difference between the mean number of decayed and unsound teeth and the mean number of decayed teeth for 16 to 24 year olds (1.6 teeth in both cases) while among those aged 45 to 54 the equivalent figures were 1.4 and 1.0 teeth on average . Since the categories of recurrent decay and unsound teeth both relate to restorations, these variations with age are to be expected given the greater experience of restorative treatment among the older adults.

Tables 3.1.11 and 22

The proportion of dentate adults who had primary decay was higher among those aged less than 35. For example, 51% of those aged 25 to 34 had some primary decay compared with 34% of those aged 35 to 44. However, there was less variation with age in the prevalence of recurrent decay. Thus, the highest proportions of those with any decay were found among the younger age groups, 48% of those aged 16 to 24 and 55% of those aged 25 to 34 compared with between 41% and 44% of the older age groups. The proportion of dentate adults with some unrestorable teeth increased gradually with age from 6% of 16 to 24 year olds to 12% of those aged 65 and over.

Figure 3.1.5

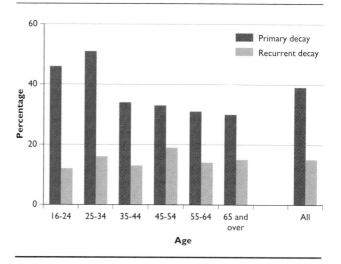

Fig 3.1.5   **Proportion of dentate adults with primary decay and with recurrent decay by age**

Primary decay was more prevalent than recurrent decay for each of the socio-demographic factors and the usual reason for dental attendance but the size of the difference varied. This was mainly due to the variation in the prevalence of primary decay with respect to these factors.

Dentate adults in Scotland and Northern Ireland were more likely to have some decay than their English or Welsh counterparts, 52%, 52%, 46% and 41% respectively, and also more likely to have primary decay; 46% and 45% compared with 38% and 34% respectively. There was little or no variation with respect to gender or country in the proportions of dentate adults with recurrent decay. With these exceptions, the variation seen in the proportions of dentate adults with primary decay, recurrent decay and the overall decay levels were generally similar to those described in Section 3.1.4 with regard to decayed and unsound teeth. Thus, those in the North of England, those from manual backgrounds and those who attended the dentist only with trouble were all more likely than their counterparts to have each type of decay. For example, the largest variation in the prevalence of recurrent decay was found among the English regions. Among dentate adults in the North 22% had recurrent decay compared with 12% of those in the Midlands and the South.

Teeth with unrestorable decay were more likely to be found among men than women (11% compared with 5%) and those from skilled and unskilled manual backgrounds (10% and 13% respectively compared with 5% of those from non-manual households). As many as 19% of those who attended the dentist only with trouble had unrestorable teeth compared with 3% and 4% of those who went for regular or occasional check-ups. There was little variation in the proportions of people with unrestorable teeth by country or English region.

Table 3.1.23

3.1.8 The variation in the extent of primary and recurrent decay: visual and cavitated decay

Teeth with primary and recurrent decay were further categorised according to the extent of the decay found. Two categories were used: visual caries and cavitated caries. Teeth which were identified as having decay with involvement of the dental pulp were separately classified as unrestorable and have already been discussed.

The previous section showed that the bulk of caries affecting adults, irrespective of age, was primary caries. There was little difference between the mean number of teeth with primary visual decay, 0.5, and those with primary cavitated decay, 0.4 teeth. Recurrent caries, either visual or cavitated, was much less prevalent, with a mean of 0.1 for both visual and cavitated decay. Younger dentate adults had slightly more primary visual decay than cavitated decay but the differences were not significant. There was no variation with age in the mean number of teeth with recurrent visual or cavitated decay.

Table 3.1.24

Overall, 24% of dentate adults had primary visual decay, 22% had primary cavitated decay, 8% had recurrent visual and 8% had recurrent cavitated decay. For the younger age groups, primary visual decay was more prevalent than primary cavitated decay; 33% of 16 to 24 year olds had visual decay compared with 21% who had cavitated caries. By contrast, the figures for those aged 65 years and over were 12% and 23% respectively. Among those aged 35 to 44 there was no significant difference between the proportions with primary visual or primary cavitated decay. There was little difference between the prevalence of recurrent visual and recurrent cavitated decay for any age group.

Figure 3.1.6

Women were more likely to have primary visual decay than primary cavitated decay, 22% compared with 17% but there was no difference for men. There was no significant difference between the proportions of men or women with recurrent visual or cavitated decay.

The prevalence of primary visual decay was higher than that of primary cavitated decay in both Scotland and Northern Ireland while for England and Wales there was no significant difference between the two proportions; 35% of dentate adults in both Scotland and Northern Ireland had primary visual

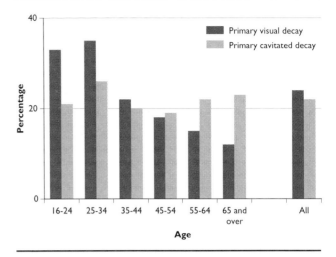

decay compared with 23% and 21% with primary cavitated decay. There was little variation with recurrent visual or cavitated decay.

With the exception of primary cavitated decay, dentate adults in the North of England were more likely than those elsewhere to have each type of decay. Dentate adults in the North were more likely to have primary visual decay than cavitated while the opposite was true for those in the South (34% compared with 22% respectively for the North and 16% compared with 21% respectively for the South).

Dentate adults from non-manual backgrounds were slightly more likely to have primary visual decay than primary cavitated decay (21% compared with 17%) while there was no difference in the prevalence levels for these two types of decay among people from either of the other two social class groups. Nor was there any significant variation in the proportions of dentate adults in different social classes with recurrent visual or recurrent cavitated decay.

Among those who reported attending for check-ups, primary visual decay was more prevalent than primary cavitated decay while the opposite was true for those who attended only with trouble. For example 28% of those who reported occasional check-ups had primary visual decay compared with 18% who had primary cavitated decay while the equivalent figures for those who only attended with trouble were 30% and 37%.

Table 3.1.25

3.1.9 Distribution of tooth conditions around the mouth

The assessment of the different conditions that may affect the coronal surfaces of teeth were recorded at the time of the examination, by surface, for each permanent tooth. The tables presented so far in this chapter have used these data aggregated for all teeth in the mouth. The position of the teeth in the mouth, their function and shape result in different predisposition to disease. This section indicates where in the mouth decay tends to occur and the different outcomes for people of different ages. Analysis of these data provides an overall picture of the state of the dentition in both jaws for each age group. Chapter 5.2 presents data on individual tooth conditions for people with different dental attendance patterns for comparison with data from the 1988 survey.

Figure 3.1.7 shows for dentate adults of all ages and in ten year age groups the proportion of each tooth type that was missing, decayed or unsound, restored, or sound and untreated. Decayed teeth includes those which were recorded as having visual decay, cavitated decay and those which were considered to be unrestorable.

The diagrams number the tooth positions in the mouth and refer to the different types of teeth as follows:

Tooth type	Tooth Number
Incisor	1 and 2
Canine	3
Premolar	4 and 5
Molar	6,7 and 8

For dentate adults of all ages, the lower canines and incisors were the least decay prone teeth, with 10% or fewer missing, decayed or restored. The upper canines were next with around 30% missing, decayed or restored. The lower first premolars had 36% missing, decayed or restored and the upper incisors ranged between 36% and 41% which were missing, decayed or restored. For all other tooth types more than half were missing or diseased. It should be noted that no distinction was made for third molars (the wisdom teeth) as to whether they had been lost or had not (yet) erupted. First molars showed high levels of decay experience with less than 15% sound and untreated.

Figure 3.1.7, Table 3.1.26

The overall picture for dentate adults of all ages conceals large differences in the disease and treatment experience for the different age cohorts. Among young people the level of disease experience is low, while in the middle age groups greater reliance can be seen on restorative treatment. In the oldest age group (those aged 55 or more) missing teeth form a large part of the overall tooth condition.

For the youngest age group, the 16 to 24 year olds, the majority of the erupted teeth, particularly in the lower jaw, were sound. Where there was untreated decay or unsound restorations, this was most commonly seen on first and second

molar teeth; ranging from 9% to 14% in the upper jaw and 11% to 16% in the lower. Likewise, the teeth most likely to be restored were the first and second permanent molar teeth (around a third of first molars were restored), and to a lesser extent, premolars and incisors. In this age group, the majority of third molar teeth (66% to 69%) were missing, presumably unerupted but possibly extracted due to impaction, lack of space or disease. The other teeth that were most likely to be missing were the first premolar teeth (around 20% in the upper jaw and 12% in the lower), commonly removed as part of an orthodontic treatment plan. A smaller number of second premolar teeth were missing, most probably for similar reasons.

With increasing age the pattern identified for the 16 to 24 year olds was confirmed in the 25 to 34 year olds, but this age group had more of their premolars (around 20% in the upper jaw), upper incisors, as well as molar teeth, restored. They were also more likely to have missing first molars (between 12% and 16%).

Among 35 to 44 year olds there was generally more evidence of restorative treatment than seen for the two younger age groups. For example, around a quarter of upper incisors were restored (24% to 27%) as were around 40% of the upper and lower second premolars. Just over a quarter of the of the first molars were missing.

Restorative treatment was also much in evidence among 45 to 54 year olds and this was also the age group among whom the proportion of missing teeth first becomes more marked. Between 42% and 49% of lower first molars were missing as were just over a third of upper first molars.

The oldest age group, those adults aged 55 years and over, had a significant proportion of their molar teeth missing, in both jaws. While there were some carious or untreated teeth, the proportion of the teeth affected in this way was smaller compared with that in previous age groups, presumably because the carious or unsound teeth had been removed. Around a third (29% to 36%) of upper incisor teeth were missing in this age group.

Figure 3.1.7, Tables 3.1.27–31

Fig 3.1.7   Distribution of tooth conditions around the mouth

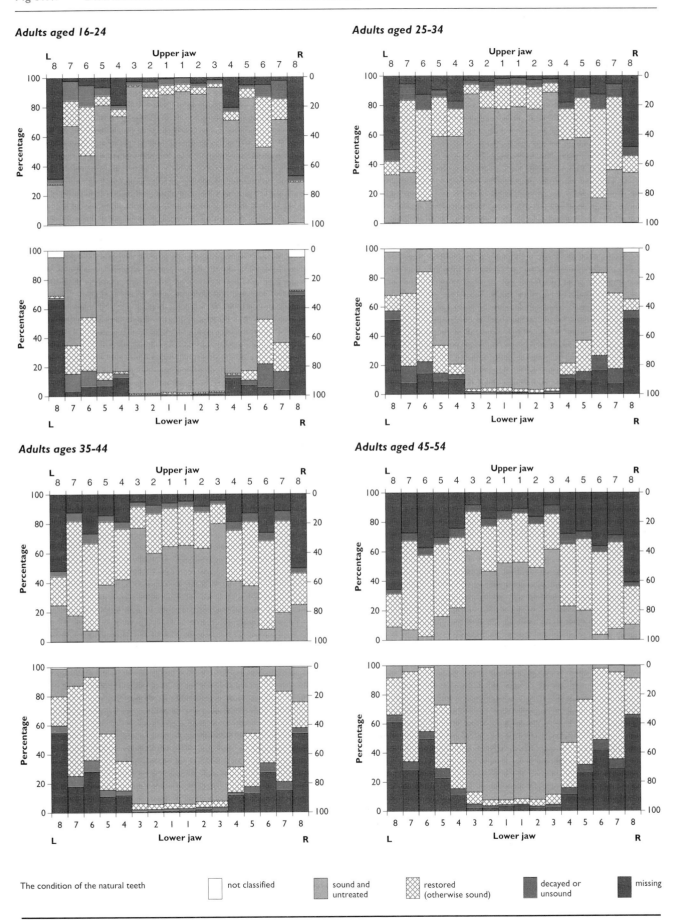

The condition of the natural teeth

not classified   sound and untreated   restored (otherwise sound)   decayed or unsound   missing

Fig 3.1.7   **Distribution of tooth conditions around the mouth** (continued)

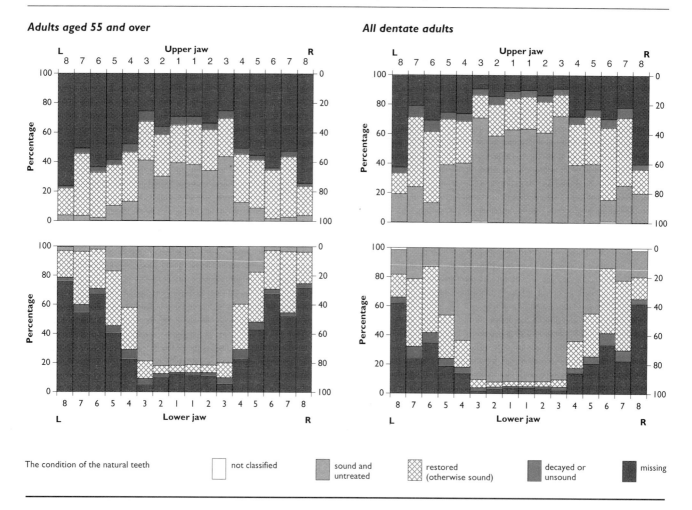

Adults aged 55 and over

All dentate adults

The condition of the natural teeth   □ not classified   ▒ sound and untreated   ▨ restored (otherwise sound)   ▓ decayed or unsound   ■ missing

3.1.10 Tooth wear

Since the Adult Dental Health Survey in 1988 there has been increasing interest in the wear of the dentition. As people keep more of their natural teeth for longer, some tooth wear is to be expected but if wear is substantial it can involve the need for treatment. Tooth wear was not assessed in the 1988 survey but was introduced in the 1993 UK survey of children's dental health (as a measure of erosion)[1]. Tooth wear was also assessed as part of the 1995 National Diet and Nutrition Survey among adults aged 65 years and over[2] and the criteria employed in that survey was used for the 1998 survey.

The assessment of tooth wear in adults in 1998 included wear through attrition, abrasion and erosion (non-carious loss of tooth substance caused by chemical action). Wear was recorded for the three surfaces of the upper six anterior teeth: buccal (outward surface), palatal (inward surface) and incisal (cutting edge). The worst affected surface of each of the six lower anterior teeth was also recorded. Wear on root surfaces is described in Section 3.1.12.

Wear was assessed on the surfaces indicated above as:
● restricted to the enamel
● loss of enamel just exposing dentine
● more extensive exposure of dentine (more than one third of the buccal or palatal surface) or loss of dentine (incisal surface)
● complete enamel loss with exposure of dental pulp or secondary dentine.

An assessment of fractured teeth as a consequence of trauma, as distinct from loss of tooth tissue because of wear, was also included in this section.

Three categories of wear are presented in the tables:
● any tooth wear - any wear except that which involved only the enamel
● moderate wear - wear involving more extensive exposure of dentine
● severe wear - complete enamel loss with exposure of the dental pulp or secondary dentine.

Figures are presented for the proportion of dentate adults who had evidence of each degree of wear and the proportion of teeth assessed with each degree of wear averaged over the population.

Two thirds (66%) of adults examined had wear in their anterior teeth that involved at least some dentine, with 11% having wear that was moderate (more extensive involvement of dentine) and 1% with severe wear. Four per cent of dentate adults had one or more anterior teeth with fractures as a result of trauma.

The proportion of dentate adults with tooth wear increased with age, from 36% of 16 to 24 year olds to 89% of those aged 65 and over. Accordingly, the proportion of adults with both moderate and severe wear increased with age so that by 65 years of age and older, one third of dentate adults (33%) had some teeth with moderate wear and 6% had some severe wear of anterior teeth. There was no significant variation with respect to age in the proportion with fractured anterior teeth.

Tooth wear was more prevalent among men than women, 70% and 61% respectively. This difference was even more marked when the data are presented by severity of wear. Moderate wear was seen in 14% of men and 8% of women; severe wear was recorded for 2% of men and less than 1% of women. Six per cent of men had fractured anterior teeth compared with 4% of women.

Dentate adults in Scotland were the least likely to have tooth wear (61%) although there was no difference between the Scotland and Northern Ireland in the proportions of people with moderate wear (9%). There was little variation between the English regions in overall experience of tooth wear but those in the Midlands were less likely to have moderate wear (7% compared with 12% in the North and the South).

Tooth wear was more prevalent among those from skilled manual households than those from unskilled manual households (70% compared with 62%) but there was no significant difference between the proportions with some moderate wear or severe wear. Nor was there any significant variation in levels of overall wear between those with different reasons for dental attendance.

Table 3.1.32

Table 3.1.33 shows the average proportion of teeth with wear. This measure was calculated at an individual level by dividing the number of teeth with each degree of tooth wear by the number of teeth present to be assessed. As explained in the introduction to this section, wear was assessed only on the upper and lower incisors and canines. The result was then averaged for the relevant population. Presentation of this measure allows account to be taken of the variation in the number of teeth which have been lost by different groups of people. However, the figures in Table 3.1.33 show very similar variations to those seen in the previous table.

On average, dentate adults had some wear on 34% of their anterior teeth, which was classed as moderate on 3% of their anterior teeth. Some severe wear was found on less than 1% of anterior teeth. Tooth wear was comparatively more extensive among those aged 65 and over (58% of anterior teeth on average) than among those aged 55 to 64 (49%) whereas the prevalence among these two age groups was shown to be very similar in Table 3.1.32. The only other deviation from the prevalence levels was with respect to English region where those in the Midlands were recorded as having the same extent of moderate wear (2% on average compared with 4% in the North and 3% in the South) while prevalence levels were shown to be lower in the Midlands than elsewhere in Table 3.1.32.

Table 3.1.33

Fig 3.1.8   **Degree of tooth wear on individual upper anterior teeth by surface of tooth (buccal, incisal, palatal)**

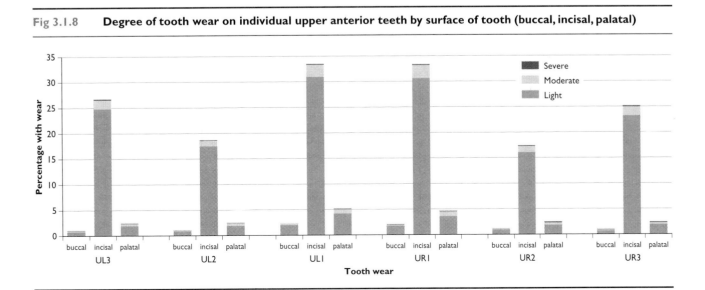

3.1.11 Tooth wear on individual teeth

Of the three surfaces on which tooth wear was assessed, it was not surprising to find that the incisal surfaces of all six upper anterior teeth showed the highest prevalence of wear. A third of central incisors showed some evidence of wear of the incisal edge but in only a small minority was this moderate (3%) or severe (less than 1 %). The canines had the next highest level of wear but again only a minority had evidence of moderate or severe wear.

Figure 3.1.8, Table 3.1.34

STATUS OF THE ROOT SURFACES

3.1.12 Condition of the root surfaces

With increasing maturity, the root surface of a tooth may become exposed. This is often attributed by patients to part of the natural ageing process or by dentists and dental hygienists to incorrect brushing techniques. With exposure of the root surface comes the potential for carious attack as well as wear of the root surface. As adults retain their teeth for longer, problems with the root surfaces are likely to become more common. This section examines the prevalence of these conditions in 1998.

The conditions assessed for root surfaces were whether the roots were exposed at all and whether the surfaces were worn or decayed or filled.

Overall, two thirds of dentate adults (66%) had at least one tooth with a root surface that was either exposed, worn, filled or decayed, with an average of 6.4 teeth involved. Nearly a quarter (24%) had 12 or more teeth in this condition.

The proportion of adults with decayed root surfaces increased with increasing age, from less than 2% in the youngest group, up to 29% among those aged 65 and over. This reflected, in large part, the proportion of people who had root surfaces which were vulnerable (this includes exposed, worn, filled or decayed roots) which accounted for only 20% of people in the youngest group but 97% of people aged 65 and over. The mean number of teeth with vulnerable (exposed, worn, filled or decayed) root surfaces also increased with increasing age, up to an average of 10.6 teeth among adults aged 65 years and over. The number of teeth with decayed root surfaces showed a similar increase, from almost none in the youngest group up to 0.7 for those aged 65 and over. When the mean was recalculated to find the average number of affected roots amongst the minority of the population who are affected by root decay, overall, 2.3 teeth were affected.

Over a half of the root surface lesions recorded were in adults aged 55 and over (data not in table).

Table 3.1.35

There was a marked and significant difference according to gender, both in the proportions of each group with vulnerable roots and in the mean number of teeth affected. Men had on average 7.0 teeth which had exposed, worn, filled or decayed root surfaces, and women had 5.7 teeth on average which were vulnerable. Similar differences were seen between the average numbers of teeth with root surfaces which were decayed but these did not reach the level of statistical significance (0.4 compared with 0.2 teeth with decayed root surfaces). Among those affected the mean number of teeth with decayed root surfaces was 2.4 for men and 2.0 for women although these were not significantly different.

The differences in the prevalence of root surface decay according to country and English region were small. At the upper end of the range, dentate adults in England had on average 6.5 teeth with vulnerable root surfaces while those in Wales had 5.1, in Northern Ireland 5.6 and in Scotland 5.9. Wales had a slightly smaller proportion of the people affected by root surface decay (7% compared with 12% in England and Northern Ireland and 13% in Scotland). Within England, the Midlands had a higher proportion of dentate adults with vulnerable root surfaces (74% compared with 63% in the North and 65% in the South).

The proportion of people with some teeth vulnerable to root surface decay (those with exposed, worn, filled or decayed root surfaces) and the mean number of such teeth varied little according to the social class of the head of the household. However, the proportion with decayed root surfaces varied significantly. Among people from a non-manual background 10% had some teeth with decayed root surfaces with an overall average of 0.2 affected teeth. Among those from an unskilled background, 16% were affected and the overall average was 0.4 teeth per person. Among those affected, the mean number of teeth with decayed root surfaces were all very similar at 2.1, 2.3 and 2.6 for the three social class groups respectively.

People who said they attended occasionally for check-ups were significantly less likely to have teeth with root surfaces which were potentially vulnerable to decay (exposed, worn, filled or decayed) than those who reported attendance for regular check-ups; 59% had one or more such teeth compared with 68% of the latter group. However, when looking at root surface decay only, people who attended only with trouble were significantly more likely to have root surface decay (19% compared with 7% for occasional attenders and 10% for regular attenders) and had more teeth affected on average than people who attended for check-ups (0.5 compared with 0.1 and 0.2 teeth) although these differences were not significant. There were similar variations in the mean number of teeth affected when only those with root surface decay were considered (2.7 and 2.0 teeth respectively). Part of the reason for these differences may be related to decay remaining untreated among those who do not attend the dentist, but there may be other behavioural factors at work too, including oral hygiene and dietary habits.

Tables 3.1.36–39

In addition to the assessment of the presence of decay in the root surfaces, the type of decay was also recorded. Overall, 12% of dentate adults had some decay affecting the root surfaces. Active decay was more prevalent than decay that was no longer active (arrested decay); 9% compared with 2%; 3% of people had unrestorable roots and 1% had recurrent decay associated with a root restoration.

Table 3.1.40

3.1.13 Condition of the roots of individual teeth

Figure 3.1.9 shows the condition of the root surfaces for individual teeth for all dentate adults and by ten year age groups. For each tooth, proportions are shown for sound root surfaces, teeth with root surfaces that were exposed only and those which were worn, filled or decayed. In cases where it was not possible to assess the root surface, teeth were recorded as unscored.

Overall, teeth with vulnerable root surfaces were fairly evenly distributed around the mouth. Teeth in the lower jaw were more likely to be affected than those in the upper. In the lower jaw those most likely to have vulnerable root surfaces were the premolars, canines and incisors, but this difference was due mainly to the higher proportions of missing molars. For example, around 30% of lower first premolars had vulnerable root surfaces compared with around 18% of lower first molars but more than twice as many of the latter tooth type were missing.

Figure 3.1.9, Table 3.1.41

Among dentate adults aged 16 to 24, teeth with vulnerable root surfaces were fairly rare both in absolute terms and relative to the overall number of teeth present. Those teeth found to have vulnerable root surfaces were fairly evenly distributed between the two jaws and around the mouth.

There was a greater prevalence of teeth with vulnerable root surfaces among dentate adults aged 25 to 34 than among the youngest age group, although this was mostly because more of the teeth had exposed roots. In the upper jaw the first premolars and the first molars were most likely to be affected particularly in relation to the number of teeth present. In the lower jaw, the first molars were most likely to be affected. The 1988 report noted a tendency for a greater proportion of teeth to have exposed roots on the left side of the mouth but the 1998 results indicate this only for the premolars and first molars in the lower jaw.

Among dentate adults aged 35 to 44 the proportion of teeth with vulnerable roots was greater than for the younger adults and given the higher proportion of missing teeth in this age group this difference was even larger than shown by these figures. First premolars were the most likely to have vulnerable root surfaces particularly in the lower jaw with around 29% with exposed, worn, filled or decayed root surfaces .

Among dentate adults aged 45 to 54, around two fifths of the lower first premolars had exposed root surfaces and 7% were worn, filled or decayed. Vulnerable root surfaces were more apparent in the lower incisors and canines than for the younger age groups. Around a third of these teeth had exposed root surfaces.

Among those aged 55 or more the decrease in the numbers of teeth present made the prevalence of teeth with vulnerable root surfaces even more marked. The proportion of teeth present with vulnerable root surfaces was higher than the proportion without them for all tooth types except for the upper incisors and the wisdom teeth.

Tables 3.1.42–46

Notes and references

1. O'Brien M. *Children's dental health in the United Kingdom 1993* HMSO London 1994
2. See Chapter 1.1

Fig 3.1.9   Condition of the roots around the mouth

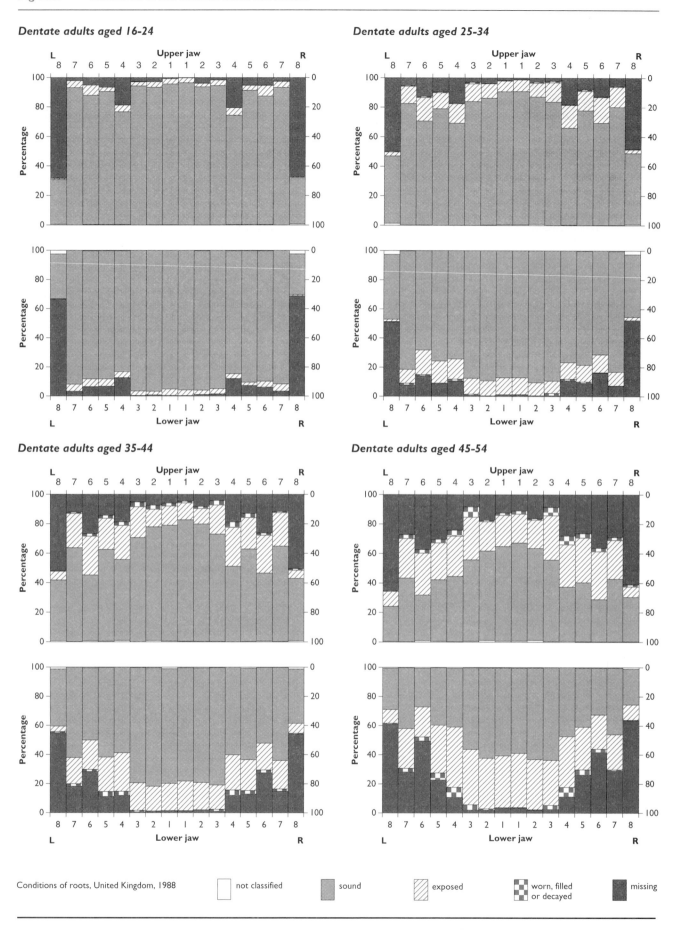

Conditions of roots, United Kingdom, 1988   ☐ not classified   ▨ sound   ▨ exposed   ▨ worn, filled or decayed   ■ missing

Fig 3.1.9   **Condition of the roots around the mouth**

Dentate adults aged 55 and over

All dentate adults

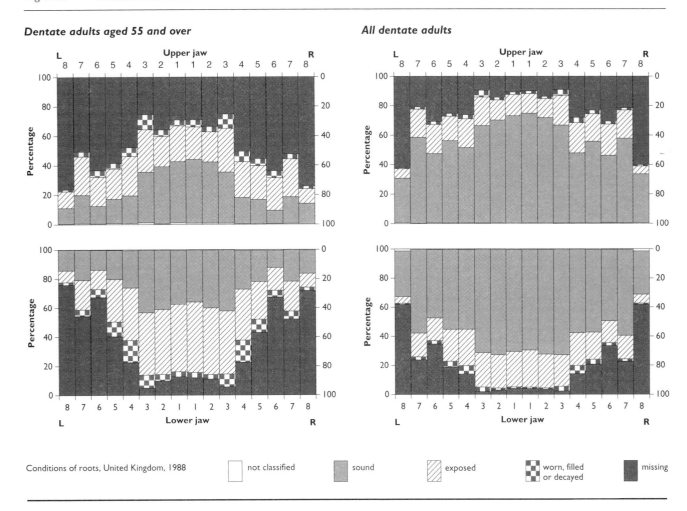

Conditions of roots, United Kingdom, 1988   ☐ not classified   ▨ sound   ▧ exposed   ▨ worn, filled or decayed   ■ missing

Table 3.1.1   Distribution of the numbers of sound and untreated teeth by age

Dentate adults   *United Kingdom*

Number of sound and untreated teeth	Age						All
	16-24	25-34	35-44	45-54	55-64	65 and over	
	%	%	%	%	%	%	%
0	0	0	1	0	2	4	1
1-5	1	1	5	9	13	23	7
6-11	2	9	20	40	51	45	25
12-17	8	25	38	35	29	22	27
18-23	33	42	28	14	4	4	24
24 or more	56	23	8	1	1	0	16
Proportion with 18 or more	**89**	**65**	**37**	**15**	**5**	**5**	**40**
Mean number of sound and untreated teeth	23.4	19.1	15.4	11.7	9.6	8.5	15.3
Mean number of teeth	27.9	28.1	26.7	24.0	19.9	17.3	24.8
Base	491	854	781	746	461	484	3817

1988 Table 6.16

Table 3.1.2   Distribution of the numbers of sound and untreated teeth by gender

Dentate adults   *United Kingdom*

Number of sound and untreated teeth	Males	Females	All
	%	%	%
0	1	1	1
1-5	7	8	7
6-11	23	26	25
12-17	27	27	27
18-23	24	24	24
24 or more	18	14	16
Proportion with 18 or more	**42**	**38**	**40**
Mean	15.7	14.9	15.3
Base	1745	2072	3817

1988 Table 6.17

Table 3.1.3   Eighteen or more sound and untreated teeth by gender and age

Dentate adults   *United Kingdom*

Age	Males	Females	All
	percentage with 18 or more sound and untreated teeth		
16-24	90	88	89
25-34	65	65	65
35-44	42	31	37
45-54	18	13	15
55-64	6	3	5
65 and over	7	3	5
All	42	38	40
	Base		
16-24	227	264	491
25-34	339	515	854
35-44	364	417	781
45-54	356	390	746
55-64	216	245	461
65 and over	243	241	484
All	1745	2072	3817

1988 Table 6.5

Table 3.1.4 **Distribution of the numbers of sound and untreated teeth among dentate adults by English region and country**

Dentate adults

Number of sound and untreated	English region			Country				
	North	Midlands	South	England	Wales	Scotland	Northern Ireland	United Kingdom
	%	%	%	%	%	%	%	%
0	I	I	I	I	0	I	2	I
1-5	8	8	6	7	8	II	8	7
6-11	26	20	25	24	25	29	32	25
12-17	25	28	27	26	30	30	27	27
18-23	26	24	24	24	22	21	23	24
24 or more	15	18	17	17	15	8	9	16
Proportion with 18 or more	**41**	**42**	**41**	**41**	**37**	**29**	**32**	**40**
Mean	15.4	15.6	15.7	15.6	15.1	13.6	13.9	15.3
Base	*617*	*495*	*1074*	*2186*	*502*	*668*	*461*	*3817*

Table 3.1.5 **Eighteen or more sound and untreated teeth by English region, country and age**

Dentate adults

Age	English region			Country				
	North	Midlands	South	England	Wales	Scotland	Northern Ireland	United Kingdom
	percentage with 18 or more sound and untreated teeth							
16-24	87	95	93	92	85	79	71	89
25-34	58	72	72	68	69	38	45	65
35-44	33	47	40	40	19	22	21	37
45-54	16	II	18	16	14	10	5	15
55-64	8	2	4	4	8	5	2	5
65 and over	4	5	6	5	7	4	2	5
All	41	42	41	41	37	29	32	40
	Base							
16-24	*82*	*42*	*122*	*246*	*69*	*80*	*96*	*491*
25-34	*146*	*125*	*246*	*517*	*90*	*157*	*90*	*854*
35-44	*126*	*114*	*207*	*447*	*96*	*142*	*96*	*781*
45-54	*121*	*101*	*201*	*423*	*109*	*132*	*82*	*746*
55-64	*67*	*47*	*145*	*259*	*73*	*75*	*54*	*461*
65 and over	*75*	*66*	*153*	*294*	*65*	*82*	*43*	*484*
All	*617*	*495*	*1074*	*2186*	*502*	*668*	*461*	*3817*

1988 Table 6.6

Table 3.1.6 **Distribution of the numbers of sound and untreated teeth by social class of head of household**

Dentate adults   *United Kingdom*

Number of sound and untreated teeth	Social class of head of household			All*
	I, II, IIINM	**IIIM**	**IV,V**	
	%	%	%	%
0	1	1	1	1
1-5	7	8	7	7
6-11	25	25	25	25
12-17	28	28	26	27
18-23	23	26	24	24
24 or more	17	12	16	16
Proportion with 18 or more	**40**	**37**	**40**	**40**
Mean	15.4	15.8	15.3	15.2
Base	1926	1025	625	3817

* Includes those for whom social class was not known and armed forces.

1988 Table 6.18

Table 3.1.7 **Eighteen or more sound and untreated teeth by social class of the head of household and age**

Dentate adults   *United Kingdom*

Age	Social class of head of household			All*
	I, II, IIINM	**IIIM**	**IV,V**	
	percentage with 18 or more sound and untreated teeth			
16-24	93	79	89	89
25-34	67	62	62	65
35-44	38	40	29	37
45-54	14	15	22	15
55-64	5	5	4	5
65 and over	6	4	4	5
All	40	37	40	40
	Base			
16-24	215	116	90	491
25-34	402	254	148	854
35-44	395	214	121	781
45-54	394	204	107	746
55-64	247	115	85	461
65 and over	273	122	74	484
All	1926	1025	625	3817

* Includes those for whom social class was not known and armed forces.

1988 Table 6.7

Table 3.1.8 **Distribution of the numbers of sound and untreated teeth by usual reason for dental attendance**

Dentate adults   *United Kingdom*

Number of sound and untreated teeth	Usual reason for attendance			All
	Regular check-up	**Occasional check-up**	**Only with trouble**	
	%	%	%	%
0	1	0	1	1
1-5	8	3	9	7
6-11	28	22	20	25
12-17	29	23	25	27
18-23	22	29	26	24
24 or more	13	24	19	16
Proportion with 18 or more	**35**	**52**	**45**	**40**
Mean	14.6	17.5	16.0	15.3
Base	2400	408	1003	3817

1988 Table 6.19

Table 3.1.9 **Eighteen or more sound and untreated teeth by usual reason for dental attendance and age**

Dentate adults　　　　　　　　　　　　*United Kingdom*

Age	Usual reason for attendance			All
	Regular check-up	Occasional check-up	Only with trouble	
	percentage with 18 or more sound and untreated teeth			
16-24	92	85	88	89
25-34	66	65	63	65
35-44	33	42	43	37
45-54	12	31	16	15
55-64	4	5	6	5
65 and over	4	11	6	5
All	35	52	45	40
	Base			
16-24	*241*	*95*	*155*	*491*
25-34	*479*	*102*	*271*	*854*
35-44	*515*	*81*	*184*	*781*
45-54	*516*	*81*	*148*	*746*
55-64	*325*	*24*	*112*	*461*
65 and over	*324*	*25*	*133*	*484*
All	*2400*	*408*	*1003*	*3817*

1988 Table 6.8

Table 3.1.10 **Sealants by age, gender, English region, country, social class of head of household and usual reason for dental attendance**

Dentate adults　　　　　　　　　　　　*United Kingdom*

	Percentage of adults with sealants	Base
All	5	*3817*
Age		
16-24	23	*491*
25-34	5	*854*
35-44	1	*781*
45-54	0	*746*
55-64	0	*461*
65 and over	0	*484*
Gender		
Men	4	*1745*
Women	6	*2072*
English region		
North	6	*617*
Midlands	2	*495*
South	5	*1074*
Country		
England	5	*2186*
Wales	5	*502*
Scotland	7	*668*
Northern Ireland	13	*461*
Social class of head of household		
I, II, IIINM	6	*1926*
IIIM	4	*1025*
IV,V	4	*625*
Usual reason for dental attendance		
Regular check-up	5	*2400*
Occasional check-up	7	*408*
Only with trouble	4	*1003*

Table 3.1.11 **Distribution of the numbers of decayed and unsound teeth by age**

Dentate adults   *United Kingdom*

Number of decayed and unsound teeth	Age						All
	16-24	25-34	35-44	45-54	55-64	65 and over	
	%	%	%	%	%	%	%
0	49	40	49	43	46	46	45
1-5	44	51	45	52	50	50	49
6 or more	7	8	6	5	4	4	6
Proportion with any decayed or unsound teeth	**51**	**60**	**51**	**57**	**54**	**54**	**55**
Mean number of decayed and unsound teeth	1.6	1.8	1.4	1.4	1.3	1.2	1.5
Mean number of teeth	27.9	28.1	26.7	24.0	19.9	17.3	24.8
Base	*491*	*854*	*781*	*746*	*461*	*484*	*3817*

1988 Table 8.22

Table 3.1.12 **Distribution of the numbers of decayed and unsound teeth by gender**

Dentate adults   *United Kingdom*

Number of decayed and unsound teeth	Males	Females	All
	%	%	%
0	42	48	45
1-5	50	48	49
6 or more	8	4	6
Mean	1.7	1.3	1.5
Base	*1745*	*2072*	*3817*

1988 Table 8.23

Table 3.1.13 **Decayed or unsound teeth by gender and age**

Dentate adults   *United Kingdom*

Age	Males	Females	All
	percentage with any decayed or unsound teeth		
16-24	54	48	51
25-34	64	56	60
35-44	55	47	51
45-54	59	54	57
55-64	55	54	54
65 and over	55	52	54
All	58	52	55
		Base	
16-24	*227*	*264*	*491*
25-34	*339*	*515*	*854*
35-44	*364*	*417*	*781*
45-54	*356*	*390*	*746*
55-64	*216*	*245*	*461*
65 and over	*243*	*241*	*484*
All	*1745*	*2072*	*3817*

1988 Table 8.5

Table 3.1.14 **Distribution of the numbers of decayed or unsound teeth by English region and country**

Dentate adults

Number of decayed or unsound teeth	English region			Country				
	North	Midlands	South	England	Wales	Scotland	Northern Ireland	United Kingdom
	%	%	%	%	%	%	%	%
0	35	48	49	45	52	42	40	45
1-5	57	47	46	49	43	49	54	49
6 or more	8	5	5	6	5	9	5	6
Mean	1.9	1.4	1.3	1.5	1.2	1.8	1.5	1.5
Base	617	495	1074	2186	502	668	461	3817

Table 3.1.15 **Decayed or unsound teeth by English region and country, and age**

Dentate adults

Age	English region			Country				
	North	Midlands	South	England	Wales	Scotland	Northern Ireland	United Kingdom
	percentage with any decayed or unsound teeth							
16-24	70	46	38	50	46	56	72	51
25-34	67	59	58	61	47	58	57	60
35-44	62	44	46	50	51	61	57	51
45-54	64	54	54	56	50	61	58	57
55-64	58	51	55	55	52	53	57	54
65 and over	63	55	52	55	39	56	42	54
All	65	52	51	55	48	58	60	55
	Base							
16-24	82	42	122	246	69	80	96	491
25-34	146	125	246	517	90	157	90	854
35-44	126	114	207	447	96	142	96	781
45-54	121	101	201	423	109	132	82	746
55-64	67	47	145	259	73	75	54	461
65 and over	75	66	153	294	65	82	43	484
All	617	495	1074	2186	502	668	461	3817

1988 Table 8.4

Table 3.1.16 **Distribution of the numbers of decayed or unsound teeth by social class of head of household**

Dentate adults   *United Kingdom*

Number of decayed or unsound teeth	Social class of head of household			All*
	I, II, IIINM	IIIM	IV,V	
	%	%	%	%
0	50	43	38	45
1-5	46	50	52	48
6 or more	4	7	10	6
Mean	1.2	1.7	1.9	1.5
Base	1926	1025	625	3817

* Includes those for whom social class was not known and armed forces.

1988 Table 8.24

Table 3.1.17 **Decayed or unsound teeth by social class of head of household as defined by occupation, and age**

Dentate adults   *United Kingdom*

	Social class of head of household			All*
	I, II, IIINM	IIIM	IV,V	
	percentage with any decayed or unsound teeth			
16-24	44	56	58	51
25-34	54	64	70	60
35-44	43	58	62	51
45-54	55	58	55	57
55-64	53	44	68	54
65 and over	49	58	59	54
All	50	57	62	55
	Base			
16-24	215	116	90	491
25-34	402	254	148	854
35-44	395	214	121	781
45-54	394	204	107	746
55-64	247	115	85	461
65 and over	273	122	74	484
All	1926	1025	625	3817

* Includes those for whom social class was not known and armed forces.

1988 Table 8.6

Table 3.1.18 **Distribution of the numbers of decayed or unsound teeth by usual reason for dental attendance**

Dentate adults   *United Kingdom*

Number of decayed or unsound teeth	Usual reason for attendance			All
	Regular check-up	Occasional check-up	Only with trouble	
	%	%	%	%
0	52	44	33	45
1-5	45	51	55	48
6 or more	3	5	12	6
Mean	1.1	1.4	2.3	1.5
Base	2400	408	1003	3817

1988 Table 8.25

Table 3.1.19 Decayed or unsound teeth by usual reason for dental attendance and age

Dentate adults   *United Kingdom*

Age	Usual reason for attendance			All
	Regular check-up	Occasional check-up	Only with trouble	
	percentage with any decayed or unsound teeth			
16-24	43	50	63	51
25-34	51	60	74	60
35-44	47	49	62	51
45-54	51	64	69	57
55-64	48	63	68	54
65 and over	49	55	66	54
All	48	56	67	55
	Base			
16-24	*241*	*95*	*155*	*491*
25-34	*479*	*102*	*271*	*854*
35-44	*515*	*81*	*184*	*781*
45-54	*516*	*81*	*148*	*746*
55-64	*325*	*24*	*112*	*461*
65 and over	*324*	*25*	*133*	*484*
All	*2400*	*408*	*1003*	*3817*

1988 Table 8.7

Table 3.1.20 Decayed and unsound teeth by age and most recent visit to the dentist

Dentate adults

Age	Time since most recent visit to the dentist		
	Less than 6 months	6 months, less than 2 years	2 years or more
	percentage with any decayed or unsound teeth		
16-24	46	55	55
25-34	56	61	68
35-44	47	54	61
45-54	54	55	70
55-64	48	52	75
65 and over	49	56	66
All	50	56	65
	Base		
16-24	*243*	*153*	*94*
25-34	*476*	*197*	*179*
35-44	*489*	*172*	*119*
45-54	*488*	*152*	*105*
55-64	*299*	*88*	*74*
65 and over	*303*	*91*	*89*
All	*2298*	*853*	*660*

Table 3.1.21 **The effect of including visual caries in the criteria for decayed and unsound teeth**

Dentate adults   *United Kingdom*

	Proportion with one or more decayed or unsound teeth			Mean number of decayed or unsound teeth			Base
	Includes visual caries*	Does not include visual caries**	Difference	Includes visual caries*	Does not include visual caries**	Difference	
All	55	42	13	1.5	1.0	0.5	3817
Age							
16-24	51	31	20	1.6	0.8	0.8	491
25-34	60	42	18	1.8	1.0	0.8	854
35-44	51	40	11	1.4	0.9	0.5	781
45-54	57	46	11	1.4	1.0	0.4	746
55-64	54	48	6	1.3	1.0	0.3	461
65 and over	54	48	6	1.2	1.0	0.2	484
Gender							
Males	58	45	13	1.7	1.1	0.6	1745
Females	52	38	14	1.3	0.8	0.5	2072
Country							
England	55	42	13	1.5	1.0	0.5	2186
Wales	48	34	14	1.2	0.7	0.5	502
Scotland	58	41	17	1.8	0.9	0.9	668
Northern Ireland	60	41	19	1.5	0.8	0.7	461
English region							
North	65	46	19	1.9	1.0	0.9	617
Midlands	52	38	14	1.4	0.9	0.5	495
South	51	42	9	1.3	1.0	0.3	1074

* Criteria used for 1998 survey.
** Criteria used for 1988 survey.

Table 3.1.22 **The mean number of teeth with different types of decay by age**

Dentate adults   *United Kingdom*

	Any decay	Primary decay*	Recurrent decay**	Some unrestorable teeth	Base
			mean number of teeth		
All	1.2	0.9	0.2	0.2	3817
Age					
16-24	1.6	1.3	0.2	0.1	491
25-34	1.5	1.2	0.2	0.1	854
35-44	1.1	0.8	0.2	0.2	781
45-54	1.0	0.6	0.3	0.1	746
55-64	0.9	0.6	0.2	0.2	461
65 and over	0.9	0.5	0.2	0.2	484

* Decayed not previously treated.
** Filled and decayed.

Table 3.1.23 **Different types of decay by age, gender, English region, country, social class of head of household and usual reason for dental attendance**

Dentate adults   *United Kingdom*

	Any decay	Primary decay*	Recurrent decay**	Some unrestorable teeth	Base
			percentage with each type of decay		
All	46	39	15	8	3817
Age					
16-24	48	46	12	6	491
25-34	55	51	16	7	854
35-44	41	34	13	7	781
45-54	44	33	19	8	746
55-64	41	31	14	10	461
65 and over	41	30	15	12	484
Gender					
Men	50	43	15	11	1745
Women	42	34	15	5	2072
Country					
England	46	38	15	8	2186
Wales	41	34	15	5	502
Scotland	52	46	18	6	668
Northern Ireland	52	45	17	9	461
English region					
North	57	48	22	9	617
Midlands	44	39	12	6	495
South	41	32	12	9	1074
Social class of head of household					
I, II, IIINM	40	33	13	5	1926
IIIM	49	42	17	10	1025
IV, V	55	48	16	13	625
Usual reason for attendance					
Regular check-up	38	31	14	3	2400
Occasional check-up	47	40	13	4	408
Only with trouble	62	54	18	19	1003

* Decayed not previously treated.
** Filled and decayed.

Table 3.1.24 **The mean number of teeth with different types of decay by degree of decay and by age**

Dentate adults   *United Kingdom*

	Primary decay*		Recurrent decay**		Base
	Visual	**Cavitated**	**Visual**	**Cavitated**	
	mean number of teeth				
All	0.5	0.4	0.1	0.1	3817
Age					
16-24	0.7	0.5	0.1	0.1	491
25-34	0.7	0.5	0.1	0.1	854
35-44	0.4	0.3	0.1	0.1	781
45-54	0.3	0.3	0.1	0.1	746
55-64	0.3	0.3	0.1	0.1	461
65 and over	0.2	0.3	0.1	0.1	484

* Decayed not previously treated.
** Filled and decayed.

Table 3.1.25 **Different types of decay by degree of decay and by age, gender, English region, country, social class of head of household and usual reason for dental attendance**

Dentate adults   *United Kingdom*

	Primary decay*		Recurrent decay**		Base
	Visual	**Cavitated**	**Visual**	**Cavitated**	
	percentage with each type of decay				
All	24	22	8	8	*3817*
Age					
16-24	33	21	7	6	*491*
25-34	35	26	9	9	*854*
35-44	22	20	9	6	*781*
45-54	18	19	10	10	*746*
55-64	15	22	5	9	*461*
65 and over	12	23	7	9	*484*
Gender					
Men	26	26	9	8	*1745*
Women	22	17	8	8	*2072*
Country					
England	22	22	8	8	*2186*
Wales	23	18	9	8	*502*
Scotland	35	23	12	9	*668*
Northern Ireland	35	21	11	6	*461*
English region					
North	34	22	14	11	*617*
Midlands	25	22	6	6	*495*
South	16	21	5	8	*1074*
Social class of head of household					
I, II, IIINM	21	17	8	7	*1926*
IIIM	25	25	9	9	*1025*
IV,V	30	30	9	9	*625*
Usual reason for dental attendance					
Regular check-up	20	15	8	8	*2400*
Occasional check-up	28	18	9	6	*408*
Only with trouble	30	37	10	10	*1003*

* *Decayed not previously treated.*
** *Filled and decayed.*

Table 3.1.26 **Condition of the individual teeth for dentate adults of all ages**

Dentate adults   *United Kingdom*

Condition	Upper jaw															
	Left								Right							
	8	7	6	5	4	3	2	1	1	2	3	4	5	6	7	8
	%	%	%	%	%	%	%	%	%	%	%	%	%	%	%	%
Sound and untreated	19	24	13	39	40	70	59	63	64	61	72	39	40	15	25	20
Restored	14	47	48	31	28	16	21	21	22	21	15	28	32	49	46	16
Decayed or unsound	4	8	8	5	5	4	6	5	4	4	4	5	5	6	7	4
broken filling	0	2	2	1	1	1	1	1	1	1	1	1	1	2	1	0
decayed and filled	1	2	3	1	1	1	2	1	1	1	1	1	2	2	2	1
decayed	2	4	2	2	2	2	2	2	2	2	2	2	1	2	4	2
unrestorable	0	0	1	1	1	1	1	0	0	1	1	1	1	0	0	0
Missing	62	21	31	25	26	10	14	11	10	14	9	28	23	30	22	60

Condition	Lower jaw															
	Left								Right							
	8	7	6	5	4	3	2	1	1	2	3	4	5	6	7	8
	%	%	%	%	%	%	%	%	%	%	%	%	%	%	%	%
Sound and untreated	17	21	13	46	64	90	92	92	92	92	90	64	45	14	22	18
Restored	16	46	45	30	18	6	3	3	3	3	5	18	29	45	48	15
Decayed or unsound	4	8	7	5	4	2	2	1	1	1	2	4	5	8	7	3
broken filling	0	2	2	1	1	0	0	0	0	0	0	1	1	2	1	0
decayed and filled	1	2	2	1	1	1	0	0	0	0	1	1	2	3	2	1
decayed	2	4	2	2	2	1	1	0	1	1	1	2	2	2	3	2
unrestorable	0	0	0	1	1	0	0	0	0	0	0	0	1	1	0	0
Missing	62	24	35	19	14	2	3	4	4	4	2	14	21	34	23	62

1988 Table 10.6

Table 3.1.27 **Condition of the individual teeth for dentate adults aged 16-24**

Dentate adults *United Kingdom*

Condition	Upper jaw															
	Left								Right							
	8	7	6	5	4	3	2	1	1	2	3	4	5	6	7	8
	%	%	%	%	%	%	%	%	%	%	%	%	%	%	%	%
Sound and untreated	27	67	47	81	74	94	87	88	90	89	93	71	86	52	71	28
Restored	1	17	33	6	5	1	6	6	5	5	3	6	6	34	14	1
Decayed or unsound	3	14	14	6	3	2	5	4	4	3	3	3	2	9	13	3
broken filling	0	0	0	0	0	0	0	0	0	0	0	0	0	1	0	0
decayed and filled	0	2	4	1	0	1	0	1	1	0	0	0	1	2	2	0
decayed	3	10	10	2	2	2	4	3	3	2	2	3	1	5	10	3
unrestorable	0	1	0	2	1	0	0	0	0	0	0	0	0	1	1	0
Missing	68	2	5	6	19	3	3	1	0	4	2	20	5	5	2	67

Condition	Lower jaw															
	Left								Right							
	8	7	6	5	4	3	2	1	1	2	3	4	5	6	7	8
	%	%	%	%	%	%	%	%	%	%	%	%	%	%	%	%
Sound and untreated	26	65	46	84	83	98	98	98	98	98	97	85	83	48	64	23
Restored	1	20	36	5	2	1	1	1	1	1	1	1	7	30	20	1
Decayed or unsound	1	12	11	4	3	1	1	1	1	0	1	2	3	16	13	3
broken filling	0	1	1	1	0	0	0	0	0	0	0	0	0	2	1	0
decayed and filled	0	2	4	0	0	0	0	0	0	0	0	0	1	4	2	0
decayed	1	8	5	3	3	0	0	0	1	0	1	2	2	8	10	2
unrestorable	0	0	1	0	0	0	0	0	0	0	0	0	0	2	1	0
Missing	66	3	6	7	12	0	0	0	0	1	1	12	7	6	3	69

1988 Table 10.1

Table 3.1.28 Condition of the individual teeth for dentate adults aged 25-34

Dentate adults   *United Kingdom*

Condition	Upper jaw															
	Left								Right							
	8	7	6	5	4	3	2	1	1	2	3	4	5	6	7	8
	%	%	%	%	%	%	%	%	%	%	%	%	%	%	%	%
Sound and untreated	32	34	15	59	59	88	78	78	79	77	89	56	58	17	36	34
Restored	9	49	62	26	19	6	12	16	15	15	6	21	27	60	49	12
Decayed or unsound	8	11	10	5	5	3	6	5	5	5	3	4	7	10	9	6
broken filling	0	2	3	1	1	0	1	1	1	1	0	1	1	2	1	0
decayed and filled	2	3	4	2	2	1	2	1	1	1	0	2	3	4	2	1
decayed	6	7	2	2	2	1	4	3	3	3	2	2	3	3	7	5
unrestorable	1	1	0	0	0	0	0	0	0	1	0	0	1	0	0	1
Missing	50	5	12	10	17	3	4	2	1	3	2	18	8	13	6	48

Condition	Lower jaw															
	Left								Right							
	8	7	6	5	4	3	2	1	1	2	3	4	5	6	7	8
	%	%	%	%	%	%	%	%	%	%	%	%	%	%	%	%
Sound and untreated	30	31	16	66	79	96	96	96	96	97	96	79	63	17	31	32
Restored	11	50	62	19	7	2	3	3	2	2	2	8	22	57	52	8
Decayed or unsound	6	12	8	6	3	1	1	0	0	1	2	2	6	10	10	5
broken filling	0	2	2	0	0	0	0	0	0	0	0	0	1	2	1	0
decayed and filled	1	3	4	1	1	0	0	0	0	0	0	0	2	4	4	1
decayed	5	6	3	4	1	0	1	0	0	0	1	2	3	2	5	4
unrestorable	0	1	0	0	0	0	0	0	0	0	0	0	1	1	1	0
Missing	51	8	14	9	10	1	0	1	1	1	1	11	9	16	7	52

1988 Table 10.2

Table 3.1.29 **Condition of the individual teeth for dentate adults aged 35-44**

Dentate adults   *United Kingdom*

Condition	Upper jaw															
	Left								Right							
	8	7	6	5	4	3	2	1	1	2	3	4	5	6	7	8
	%	%	%	%	%	%	%	%	%	%	%	%	%	%	%	%
Sound and untreated	25	18	8	38	42	77	60	65	65	63	80	41	38	8	19	24
Restored	20	64	59	42	34	15	27	26	26	24	13	34	43	60	62	22
Decayed or unsound	4	6	7	5	5	3	6	4	4	4	3	6	6	6	7	4
broken filling	1	1	2	0	1	0	1	1	0	0	0	2	1	2	1	0
decayed and filled	1	3	2	0	2	1	2	1	1	1	0	2	2	1	4	1
decayed	2	1	1	2	2	1	2	2	2	2	2	2	2	1	2	1
unrestorable	1	1	1	1	1	1	0	0	0	0	0	1	2	1	0	0
Missing	52	12	27	14	19	5	7	6	5	8	4	19	13	26	12	50

Condition	Lower jaw															
	Left								Right							
	8	7	6	5	4	3	2	1	1	2	3	4	5	6	7	8
	%	%	%	%	%	%	%	%	%	%	%	%	%	%	%	%
Sound and untreated	19	13	7	45	65	94	94	94	94	93	92	69	46	6	17	24
Restored	20	62	57	38	20	4	4	4	3	4	4	18	36	60	62	18
Decayed or unsound	5	7	8	5	3	2	1	2	1	2	2	2	5	6	6	4
broken filling	0	2	4	1	0	0	0	0	0	0	0	0	2	2	2	0
decayed and filled	0	2	3	1	1	0	0	0	0	1	2	0	1	3	2	1
decayed	4	2	0	2	2	1	1	1	1	0	1	1	2	0	1	3
unrestorable	1	0	0	0	0	0	0	0	0	0	0	0	0	0	0	0
Missing	55	18	28	11	12	1	1	1	1	2	1	12	13	28	15	54

1988 Table 10.3

Table 3.1.30 **Condition of the individual teeth for dentate adults aged 45-54**

Dentate adults   *United Kingdom*

Condition	Upper jaw															
	Left								Right							
	8	7	6	5	4	3	2	1	1	2	3	4	5	6	7	8
	%	%	%	%	%	%	%	%	%	%	%	%	%	%	%	%
Sound and untreated	9	7	2	16	22	60	46	52	53	49	61	23	20	4	8	10
Restored	22	60	55	49	48	26	31	30	33	30	24	42	48	56	59	26
Decayed or unsound	3	6	5	5	6	5	5	6	3	5	6	7	5	4	5	3
broken filling	1	2	2	2	2	1	1	2	1	2	2	3	2	2	1	0
decayed and filled	1	2	2	1	2	2	2	2	1	2	2	2	2	2	3	1
decayed	1	1	0	1	1	2	1	1	1	2	2	1	1	0	1	0
unrestorable	0	0	1	1	1	0	1	0	0	0	1	2	0	0	0	1
Missing	65	27	37	30	24	8	18	12	11	16	9	28	26	36	29	61

Condition	Lower jaw															
	Left								Right							
	8	7	6	5	4	3	2	1	1	2	3	4	5	6	7	8
	%	%	%	%	%	%	%	%	%	%	%	%	%	%	%	%
Sound and untreated	8	4	1	27	54	87	92	92	92	92	89	53	24	2	5	8
Restored	25	62	44	44	31	8	4	3	4	4	7	31	45	49	60	25
Decayed or unsound	5	6	5	7	5	3	2	1	1	1	2	4	6	7	7	2
broken filling	2	2	2	2	1	0	0	1	0	0	0	1	2	3	2	0
decayed and filled	2	2	2	2	2	1	0	0	0	0	1	2	2	2	4	1
decayed	1	2	1	1	1	2	1	0	0	0	1	2	1	1	0	1
unrestorable	0	1	1	1	0	0	0	0	0	0	0	0	0	1	0	0
Missing	61	28	49	23	11	2	2	3	4	2	2	11	26	42	29	64

1988 Table 10.4

Table 3.1.31 Condition of the individual teeth for dentate adults aged 55 or more

Dentate adults   *United Kingdom*

Condition	Upper jaw															
	Left								Right							
	8	7	6	5	4	3	2	1	1	2	3	4	5	6	7	8
	%	%	%	%	%	%	%	%	%	%	%	%	%	%	%	%
Sound and untreated	4	4	2	10	13	41	30	39	38	34	43	13	9	2	3	4
Restored	18	42	31	27	33	26	28	26	27	28	26	33	33	33	41	20
Decayed or unsound	2	4	4	4	6	7	6	6	6	4	5	4	3	2	4	2
broken filling	0	2	1	2	2	2	1	2	2	1	1	2	1	1	1	0
decayed and filled	0	1	2	1	1	2	1	1	1	1	2	1	1	0	2	1
decayed	1	1	1	1	2	2	2	3	2	1	1	1	1	0	0	0
unrestorable	0	0	0	0	1	1	1	0	1	1	1	1	0	0	1	0
Missing	76	51	64	58	48	26	36	29	29	34	25	50	56	64	52	74

Condition	Lower jaw															
	Left								Right							
	8	7	6	5	4	3	2	1	1	2	3	4	5	6	7	8
	%	%	%	%	%	%	%	%	%	%	%	%	%	%	%	%
Sound and untreated	3	4	2	17	42	79	82	81	81	81	80	40	18	3	3	4
Restored	18	36	27	38	29	12	6	5	6	5	10	31	34	26	42	22
Decayed or unsound	3	6	4	5	7	4	3	1	1	2	4	7	5	4	2	3
broken filling	0	3	2	2	1	0	0	0	0	0	1	2	1	1	1	0
decayed and filled	1	2	1	1	2	2	0	0	0	0	1	2	1	1	1	1
decayed	0	0	0	1	2	2	2	1	1	2	2	2	2	0	0	1
unrestorable	1	0	0	1	2	0	0	0	0	0	0	1	1	1	0	0
Missing	76	54	67	40	22	5	10	12	12	11	6	23	43	67	52	72

1988 Table 10.5

Table 3.1.32 **Degree of tooth wear and presence of fractures in anterior teeth by age, gender, country, English region, social class of head of household, and usual reason for dental attendance**

Dentate adults   *United Kingdom*

	Proportion of dentate adults with:				Base
	Any tooth wear	Some moderate wear	Some severe wear	Fractured teeth	
All	66	11	1	4	3817
Age					
16-24	36	1	0	4	491
25-34	58	5	0	4	854
35-44	63	8	0	6	781
45-54	77	12	1	4	746
55-64	83	18	1	6	461
65 and over	89	33	6	4	484
Gender					
Men	70	14	2	6	1745
Women	61	8	0	4	2072
Country					
England	66	11	1	5	2186
Wales	67	15	0	5	502
Scotland	61	9	1	3	668
Northern Ireland	73	9	1	1	461
English region					
North	68	12	2	3	617
Midlands	65	7	1	3	495
South	65	12	1	6	1074
Social class of head of household					
I, II, IIINM	66	10	1	4	1926
IIIM	70	14	2	5	1025
IV,V	62	11	2	5	625
Usual reason for dental attendance					
Regular check-up	66	11	1	4	2400
Occasional check-up	62	7	0	3	408
Only with trouble	66	12	2	7	1003

Table 3.1.33 **Amount and degree of tooth wear and fractures by age, gender, country, English region, social class of head of household, and usual reason for dental attendance**

Dentate adults   *United Kingdom*

	Mean proportion of teeth with:			Base
	Any tooth wear	Some moderate wear	Some severe wear	
All	34	3	0	*3817*
Age				
16-24	12	0	0	*491*
25-34	24	1	0	*854*
35-44	29	2	0	*781*
45-54	41	4	0	*746*
55-64	49	5	0	*461*
65 and over	58	9	2	*484*
Gender				
Men	40	4	1	*1745*
Women	28	2	0	*2072*
Country				
England	34	3	0	*2186*
Wales	37	4	0	*502*
Scotland	31	2	0	*668*
Northern Ireland	37	2	0	*461*
English region				
North	37	4	1	*617*
Midlands	33	2	0	*495*
South	32	3	0	*1074*
Social class of head of household				
I, II, IIINM	34	2	0	*1926*
IIIM	37	4	0	*1025*
IV,V	34	4	0	*625*
Usual reason for dental attendance				
Regular check-up	34	3	0	*2400*
Occasional check-up	30	2	0	*408*
Only with trouble	37	4	1	*1003*

Table 3.1.34 **Degree of tooth wear on individual upper anterior teeth by surface of tooth (buccal, incisal, palatal)**

Dentate adults   *United Kingdom*

Degree of tooth wear	UL3			UL2			UL1			UR1			UR2			UR3		
	b†	i	p	b	i	p	b	i	p	b	i	p	b	i	p	b	i	p
	%	%	%	%	%	%	%	%	%	%	%	%	%	%	%	%	%	%
Light	1	25	2	1	18	2	2	31	4	2	31	4	1	16	2	1	23	2
Moderate	0	2	0	0	1	1	0	2	1	0	3	1	0	1	0	0	2	0
Severe	0	0	0	0	0	0	0	0	0	0	0	0	0	0	0	0	0	0

† b = buccal,   i = incisal,   p = palatal

Table 3.1.35 **Condition of roots by age**

Dentate adults   *United Kingdom*

	Age						All
	16-24	25-34	35-44	45-54	55-64	65 and over	
Exposed, worn, filled or decayed roots	%	%	%	%	%	%	%
None	80	50	29	14	10	3	34
1-5	14	26	29	27	23	20	24
6-11	3	13	20	22	28	32	18
12 or more	3	11	23	37	39	45	24
Mean	1.2	3.6	6.4	9.3	9.9	10.6	6.4
Decayed roots	%	%	%	%	%	%	%
None	98	95	90	86	76	71	88
1	1	3	4	8	11	13	6
2	0	1	2	3	7	6	3
3 or more	0	2	4	4	6	10	4
Mean	0.0	0.1	0.3	0.3	0.5	0.7	0.3
Mean number among those with decayed roots	-	2.1	2.9	2.1	2.2	2.3	2.3
				Base			
All dentate adults	*491*	*854*	*781*	*746*	*461*	*484*	*3817*
Dentate adults with decayed roots	*8*	*37*	*70*	*105*	*100*	*142*	*462*

1988 Table 11.1

Table 3.1.36 **Condition of roots by gender**

Dentate adults   *United Kingdom*

	Men	**Women**	**All**
Exposed, worn, filled or decayed roots	%	%	%
None	30	37	34
1-5	23	25	24
6-11	18	18	18
12 or more	28	20	24
Mean	7.0	5.7	6.4
Decayed roots	%	%	%
None	84	92	88
1	7	4	6
2	3	2	3
3 or more	5	2	4
Mean	0.4	0.2	0.3
Mean number among those with decayed roots	2.4	2.0	2.3
		Base	
All dentate adults	*1745*	*2072*	*3817*
Dentate adults with decayed roots	*289*	*173*	*462*

1988 Table 11.2

Table 3.1.37 **Condition of roots by English region and country**

Dentate adults

	English region			Country				United Kingdom
	North	Midlands	South	England	Wales	Scotland	Northern Ireland	United Kingdom
Exposed, worn, filled or decayed roots	%	%	%	%	%	%	%	%
None	37	26	35	34	38	30	33	34
1-5	25	22	23	23	25	28	27	24
6-11	20	20	16	18	20	20	23	18
12 or more	18	33	26	25	17	22	17	24
Mean	5.5	8.5	6.3	6.5	5.1	5.9	5.6	6.4
Decayed roots	%	%	%	%	%	%	%	%
None	87	89	87	88	93	87	88	88
1	6	4	7	6	3	5	7	6
2	3	1	3	3	1	3	2	3
3 or more	4	6	3	4	2	5	3	4
Mean	0.3	0.4	0.3	0.3	0.2	0.3	0.3	0.3
Mean number among those with decayed roots	2.2	3.2	2.0	2.3	2.6	2.6	2.1	2.3
				Base				
All dentate adults	*617*	*495*	*1074*	*2186*	*502*	*668*	*461*	*3817*
Dentate adults with decayed roots	*88*	*52*	*137*	*277*	*36*	*90*	*59*	*462*

Table 3.1.38 **Condition of roots by social class of head of household**

Dentate adults   *United Kingdom*

	Social class of head of household			All†
	I, II, IIINM	**IIIM**	**IV,V**	
Exposed, worn, filled or decayed roots	%	%	%	%
None	33	32	34	34
1-5	22	26	21	24
6-11	18	19	21	18
12 or more	27	23	24	24
Mean	6.8	6.2	6.3	6.4
Decayed roots	%	%	%	%
None	90	86	84	88
1	6	7	6	6
2	2	3	4	3
3 or more	3	4	6	4
Mean	0.2	0.3	0.4	0.3
Mean number among those with decayed roots	2.1	2.3	2.6	2.3
			Base	
All dentate adults	*1926*	*1025*	*625*	*3817*
Dentate adults with decayed roots	*205*	*138*	*99*	*462*

† Includes those for whom social class of head of household was not known and armed forces.
1988 Table 11.3

Table 3.1.39 **Condition of roots by usual reason for dental attendance**

Dentate adults   *United Kingdom*

	Usual reason for attendance			All
	Regular check-up	**Occasional check-up**	**Only with trouble**	
Exposed, worn, filled or decayed roots	%	%	%	%
None	32	41	34	34
1-5	23	25	24	24
6-11	18	18	20	18
12 or more	27	15	22	24
Mean	6.8	4.7	6.1	6.4
Decayed roots	%	%	%	%
None	90	93	81	88
1	6	4	7	6
2	2	1	4	3
3 or more	2	1	8	4
Mean	0.2	0.1	0.5	0.3
Mean number among those with decayed roots	2.0	-	2.7	2.3
	Base			
All dentate adults	*2400*	*408*	*1003*	*3817*
Dentate adults with decayed roots	*242*	*24*	*196*	*462*

1988 Table 11.4

Table 3.1.40 **Proportion of dentate adults with roots with different kinds of decay**

Dentate adults   *United Kingdom*

Type of decay	*percentage with each type of decay*
Active decay	9
Arrested decay	2
Recurrent decay	1
Unrestorable roots	3
Base	*3817*

Table 3.1.41 **Condition of roots around the mouth for dentate adults of all ages**

Dentate adults   *United Kingdom*

Condition	Upper jaw – left								Upper jaw – right							
	8	7	6	5	4	3	2	1	1	2	3	4	5	6	7	8
	%	%	%	%	%	%	%	%	%	%	%	%	%	%	%	%
Sound	30	58	47	56	51	66	70	72	74	71	66	48	56	46	58	33
Exposed	6	19	20	16	20	19	14	14	13	13	20	20	19	21	19	5
Worn, filled or decayed	1	2	2	2	3	5	2	2	2	2	4	4	2	2	2	1
Missing	62	21	31	25	26	10	14	11	10	14	9	28	23	30	22	60
Unscored	1	0	1	0	0	0	1	1	1	1	0	0	0	0	0	1

Condition	Lower jaw – left								Lower jaw – right							
	8	7	6	5	4	3	2	1	1	2	3	4	5	6	7	8
	%	%	%	%	%	%	%	%	%	%	%	%	%	%	%	%
Sound	31	58	47	55	55	71	72	70	69	72	73	58	57	49	59	30
Exposed	5	16	16	22	25	24	23	24	26	23	22	22	19	15	16	6
Worn, filled or decayed	1	2	2	4	6	3	1	1	1	1	3	6	4	2	2	1
Missing	62	24	35	19	14	2	3	4	4	4	2	14	21	34	23	62
Unscored	1	0	0	0	0	0	0	0	0	0	0	0	0	0	0	2

1988 Table 11.11

Table 3.1.42 **Condition of roots around the mouth for dentate adults aged 16-24**

Dentate adults   *United Kingdom*

Condition	Upper jaw – left								Upper jaw – right							
	8	7	6	5	4	3	2	1	1	2	3	4	5	6	7	8
	%	%	%	%	%	%	%	%	%	%	%	%	%	%	%	%
Sound	30	93	88	90	76	94	93	96	96	94	94	74	91	87	93	32
Exposed	1	5	7	3	4	3	4	4	3	2	4	5	3	7	4	0
Worn, filled or decayed	0	0	0	0	0	0	0	0	0	0	0	0	0	0	0	0
Missing	68	2	5	6	19	3	3	1	0	4	2	20	5	5	2	67
Unscored	1	0	0	0	0	0	0	0	0	0	0	0	0	0	0	1

Condition	Lower jaw – left								Lower jaw – right							
	8	7	6	5	4	3	2	1	1	2	3	4	5	6	7	8
	%	%	%	%	%	%	%	%	%	%	%	%	%	%	%	%
Sound	31	92	88	88	83	96	97	95	96	96	95	84	90	90	92	28
Exposed	0	5	5	5	4	3	2	4	4	3	4	3	2	4	5	1
Worn, filled or decayed	0	0	0	0	0	0	0	0	0	0	0	0	0	0	0	0
Missing	66	3	6	7	12	0	0	1	0	1	1	12	7	6	3	69
Unscored	3	0	0	0	0	0	0	0	0	0	0	0	0	0	0	2

1988 Table 11.6

Table 3.1.43 Condition of roots around the mouth for dentate adults aged 25-34

Dentate adults   *United Kingdom*

Condition	Upper jaw – left								Upper jaw – right							
	8	7	6	5	4	3	2	1	1	2	3	4	5	6	7	8
	%	%	%	%	%	%	%	%	%	%	%	%	%	%	%	%
Sound	46	82	70	79	69	84	86	90	90	86	84	66	78	69	80	48
Exposed	3	12	16	11	13	12	10	7	8	9	13	16	13	18	14	2
Worn, filled or decayed	0	0	1	1	0	1	0	0	0	1	1	0	1	0	0	0
Missing	50	5	12	10	17	3	4	2	1	3	2	18	8	13	6	48
Unscored	1	0	0	0	0	0	0	1	1	1	0	0	0	0	0	1

Condition	Lower jaw – left								Lower jaw – right							
	8	7	6	5	4	3	2	1	1	2	3	4	5	6	7	8
	%	%	%	%	%	%	%	%	%	%	%	%	%	%	%	%
Sound	44	81	68	76	74	88	89	87	87	90	89	77	78	71	83	43
Exposed	2	10	17	15	14	11	10	12	12	9	8	11	12	12	10	2
Worn, filled or decayed	0	1	1	0	1	1	0	0	0	0	2	1	1	0	0	0
Missing	51	8	14	9	10	1	0	1	1	0	0	11	9	16	7	52
Unscored	3	0	0	0	0	0	0	0	0	0	0	0	0	0	0	3

1988 Table 11.7

Table 3.1.44 Condition of roots around the mouth for dentate adults aged 35-44

Dentate adults   *United Kingdom*

Condition	Upper jaw – left								Upper jaw – right							
	8	7	6	5	4	3	2	1	1	2	3	4	5	6	7	8
	%	%	%	%	%	%	%	%	%	%	%	%	%	%	%	%
Sound	41	64	45	62	55	71	77	78	82	80	73	51	63	46	65	42
Exposed	6	23	26	21	23	21	12	13	12	11	20	26	21	26	23	6
Worn, filled or decayed	0	1	2	2	2	3	3	2	1	2	3	4	3	2	1	1
Missing	52	12	27	14	19	5	7	6	5	8	4	19	13	26	12	50
Unscored	1	0	1	1	1	0	1	1	0	0	0	0	0	1	0	1

Condition	Lower jaw – left								Lower jaw – right							
	8	7	6	5	4	3	2	1	1	2	3	4	5	6	7	8
	%	%	%	%	%	%	%	%	%	%	%	%	%	%	%	%
Sound	39	62	50	62	59	79	81	79	78	79	80	60	63	52	64	37
Exposed	4	18	20	24	26	19	17	19	20	19	17	24	21	18	20	7
Worn, filled or decayed	1	2	1	4	3	1	0	0	0	0	1	4	2	2	2	0
Missing	55	18	28	11	12	1	1	1	1	2	1	12	13	28	15	54
Unscored	1	0	1	0	0	0	0	1	0	1	0	0	1	0	0	1

1988 Table 11.8

Table 3.1.45 Condition of roots around the mouth for dentate adults aged 45-54

Dentate adults   *United Kingdom*

Condition	Upper jaw – left								Upper jaw – right							
	8	7	6	5	4	3	2	1	1	2	3	4	5	6	7	8
	%	%	%	%	%	%	%	%	%	%	%	%	%	%	%	%
Sound	24	43	31	42	44	55	61	64	66	62	55	37	40	29	43	30
Exposed	10	27	28	25	27	29	20	21	20	19	30	29	30	32	26	7
Worn, filled or decayed	0	3	2	2	4	7	1	2	2	1	6	6	3	2	2	1
Missing	65	27	37	30	24	8	18	12	11	16	9	28	26	36	29	61
Unscored	0	0	1	1	0	1	1	1	1	1	1	1	0	0	0	0

Condition	Lower jaw – left								Lower jaw – right							
	8	7	6	5	4	3	2	1	1	2	3	4	5	6	7	8
	%	%	%	%	%	%	%	%	%	%	%	%	%	%	%	%
Sound	28	42	26	39	40	56	62	60	58	63	64	47	40	32	46	24
Exposed	10	27	21	33	41	38	35	36	37	34	31	35	29	24	24	11
Worn, filled or decayed	1	2	3	5	7	4	1	0	0	0	3	7	4	2	1	0
Missing	61	28	49	23	11	2	2	3	4	2	2	11	26	42	29	64
Unscored	1	0	1	0	1	0	0	0	0	0	0	0	0	0	0	1

1988 Table 11.9

Table 3.1.46 Condition of roots around the mouth for dentate adults aged 55 and over

Dentate adults   *United Kingdom*

Condition	Upper jaw – left								Upper jaw – right							
	8	7	6	5	4	3	2	1	1	2	3	4	5	6	7	8
	%	%	%	%	%	%	%	%	%	%	%	%	%	%	%	%
Sound	11	19	12	16	19	34	39	42	43	42	35	18	17	9	18	14
Exposed	11	26	20	21	27	29	21	24	22	20	30	24	23	22	26	10
Worn, filled or decayed	1	3	4	4	6	10	4	4	5	4	10	7	5	5	3	2
Missing	76	51	64	58	48	26	36	29	29	34	25	50	56	64	52	74
Unscored	0	1	0	1	1	1	0	1	1	1	1	0	0	0	1	0

Condition	Lower jaw – left								Lower jaw – right							
	8	7	6	5	4	3	2	1	1	2	3	4	5	6	7	8
	%	%	%	%	%	%	%	%	%	%	%	%	%	%	%	%
Sound	14	21	14	20	26	43	41	37	36	40	42	27	22	12	22	16
Exposed	8	20	13	29	36	43	45	46	48	46	43	35	26	16	21	10
Worn, filled or decayed	2	4	6	10	15	9	4	4	4	3	9	15	9	5	6	2
Missing	76	54	67	40	22	5	10	12	12	11	6	23	43	67	52	72
Unscored	0	0	0	0	1	0	0	1	0	0	0	1	0	1	0	0

1988 Table 11.10

3.2   Restorative treatment

Summary

● This chapter presents data on filled teeth, teeth with artificial crowns and teeth with root surface fillings separately. Previous surveys of adult dental health have used the term 'filled teeth' to include teeth with artificial crowns.

● Dentate adults in the United Kingdom had an average of 7.0 filled (otherwise sound) teeth.

● Nearly a third (31%) of dentate adults aged 16 to 24 had no filled (otherwise sound) teeth, in sharp contrast to the equivalent proportions for those aged 25 to 54 which ranged between 3% and 9%.

● Among people aged 35 and over, around a third of all teeth were filled (otherwise sound).

● Eighteen per cent of dentate adults had 12 or more filled (otherwise sound) teeth.

● People aged 35 and over, women, people from non-manual backgrounds and those who reported regular dental attendance were the groups most likely to have 12 or more filled (otherwise sound) teeth.

● England and Wales had the lowest proportions of dentate adults with at least 12 teeth filled (otherwise sound); 18% in each country, compared with 23% in Scotland and 27% in Northern Ireland.

● The presence of root surface fillings was strongly related to age; no root surface fillings were found among those aged 16 to 24, but 43% of people aged 65 and over had teeth with such fillings.

● One third of dentate adults in the United Kingdom had artificially crowned teeth.

● Overall, 92% of dentate adults had at least one coronal restoration - either a filling or an artificial crown irrespective of condition.

● A third of dentate adults (32%) had at least 12 teeth with coronal restorations.

● The inclusion of teeth with root surface fillings in the overall total made very little difference to the overall figures indicating that the majority of teeth with such fillings also had some coronal restoration.

● Twenty-nine per cent of 16 to 24 year olds had no coronal or root surface restorations at all.

● On average about half of all teeth in people aged 45 and over were restored in some way.

3.2.1 Introduction

This chapter presents information about the type of dental care provided by dentists to repair natural teeth. Dental decay is the most common reason for restoring teeth. Most decay occurs on the crown of the tooth, but as people get older and retain their natural teeth (as is increasingly the case), their gums may recede to expose the tooth root surfaces, which are then also prone to decay. Restorations are also placed to manage other problems, such as fractured teeth, wear and to improve appearance.

Restorations include both fillings and artificial crowns. Fillings can be small, limited to a single surface of a tooth, or much bigger, extending over much of the tooth, depending partly on the extent of decay before treatment was provided. Fillings are made from any of a range of materials, but the most common type of filling in the United Kingdom is dental amalgam, usually a mixture of a silver tin alloy with mercury. If much of the tooth structure has been lost, an artificial crown may be provided to cover the tooth and maintain its form, function and strength.

The number and distribution of fillings and artificial crowns is partly a reflection of the incidence of dental caries, but also reflects a complex mix of other factors including the prevailing attitudes towards restorative treatment, at what stage an individual attends the dentist for treatment and variations in the type of treatment provided. This chapter describes the pattern of restoration of natural teeth recorded in 1998. The extent and type of restorations found, however, will reflect the treatment provided over the respondent's lifetime. Although the number and extent of restorations reflect a lifetime's treatment provision, restorations do not always last a lifetime, and replacement or repair can be required.

3.2.2 Presentation of the 1998 data on restorations

Previous surveys of adult dental health have used the term 'filled teeth' in general to describe teeth with fillings and teeth with artificial crowns. Thus, for example, 'filled (otherwise sound)' included all teeth with sound fillings or artificial crowns and had no other decay. Both the 1978 and 1988 reports present some analysis of crowned and filled teeth separately but the overall figures combine these two forms of restoration. These surveys have shown that the provision of artificial crowns has become more common, so in 1998, the data for these two different forms of restorative treatment were investigated separately, although a summary table is presented at the end of this chapter. Elsewhere in the

report the term 'restorations' is used to describe teeth with either fillings or artificial crowns. Thus:
● 'restored teeth' are those with fillings or artificial crowns
● 'filled teeth' refers only to teeth which have fillings

The survey dental examination recorded the presence and condition of fillings and artificial crowns for each surface of the tooth. The data were combined to provide an overall classification for each tooth. For the purposes of the analysis which follows, filled teeth have been classified into three groups:
● 'filled (otherwise sound)', representing a tooth with an undamaged filling with no active caries anywhere on that tooth
● 'filled but unsound', representing a tooth with a broken filling but no active caries anywhere on that tooth
● 'filled and decayed', representing a tooth with a filling and some active caries either as recurrent caries in association with the filling or as primary caries on another surface.

Similar details were recorded for artificially crowned teeth. Recurrent caries is investigated in more detail in Chapter 3.1 with other data on caries. It should be noted that decay includes visual caries, as defined by the criteria for the 1998 survey. Chapter 5.3, which presents data on change over time, includes both artificial crowns and fillings as restorations (as for previous surveys) and excludes visual caries in the comparison of decayed fillings.

3.2.3 Dentate adults with filled teeth which were otherwise sound

In 1998, 90% of all dentate adults had at least one filled (otherwise sound) tooth and the overall mean for the dentate population was 7.0 filled (otherwise sound) teeth per person. Among younger people aged 16 to 24, 31% had no filled (otherwise sound) teeth at all. This contrasted markedly with the proportion of dentate adults aged 25 to 54 who had no filled (otherwise sound) teeth, a figure which ranged between 3% and 9%. Only 16% of 16 to 24 year olds had more than 5 filled (otherwise sound) teeth. The number of fillings in younger adults was very low indeed, reflecting both lower levels of decay, and perhaps also changes in treatment behaviour among dentists. Forty-two per cent of 25 to 34 year olds had 5 or fewer filled (otherwise sound) teeth and only 25% of those aged 35 to 44 were in this category. Adults aged 45 to 54 had most filled teeth with 80% having more than 5 filled (otherwise sound) teeth. The percentage with more than 5 filled (otherwise sound) teeth was lower among those aged 55 or more; 60% of 55 to 64 year olds and 52% of dentate adults aged 65 and over, probably reflecting the loss of teeth through extraction among these older age groups.

These findings were reflected in the average number of filled (otherwise sound) teeth for each age group. The average number of filled (otherwise sound) teeth for young dentate adults (16 to 24) was 2.6, rising to 6.6 at 25 to 34 and 8.8 at 35 to 44. Each of these values was significantly higher than that for the age group below. The highest mean value was 9.3 for those aged 45 to 54. With increasing age and a reduced number of teeth, the mean values reduced to 6.3 for those aged 65 and above.

Figure 3.2.1, Table 3.2.1

The relationship between the number of filled (otherwise sound) teeth and the number of teeth remaining with increasing age is shown in Table 3.2.2. The mean proportion of teeth present in the mouth which were filled (otherwise sound) was calculated from the mean number of filled teeth and the mean number of natural teeth. This proportion was 9% for the youngest age group. It increased to a maximum of 39% by age 45 to 54, and then remained at around this level among older dentate adults, despite the sharp reduction in the number of teeth. In adults aged 35 years and over in 1998, more than one third of all teeth were filled (otherwise sound).

Figure 3.2.2, Table 3.2.2

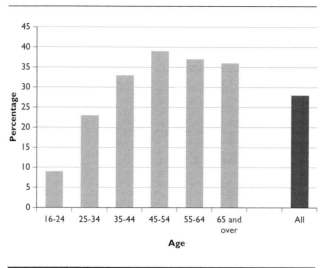

Fig 3.2.2 **Mean proportion of teeth which were filled (otherwise sound) by age**

Dentate women had more filled (otherwise sound) teeth on average than dentate men although the difference in the mean number was small (7.3 compared with 6.6). Only 8% of women of all ages had no filled teeth compared with 11% of men.

Table 3.2.3

Dentate adults had on average, 6.9 filled (otherwise sound) teeth in England and in Wales. People in Scotland had a slightly higher average at 7.4 and dentate adults living in Northern Ireland had significantly more filled (otherwise sound) teeth on average than their counterparts in England and Wales (8.2). Within England, dentate adults in the South had significantly more filled (otherwise sound) teeth on average than those in the North; 7.2 in the South compared with 6.5 in the North. In the Midlands the mean number was 6.7 teeth.

Table 3.2.4

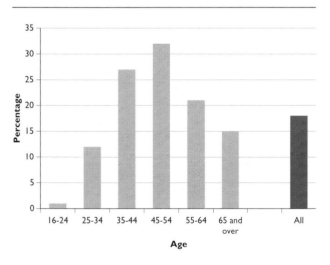

Fig 3.2.1 **Proportion of dentate adults with 12 or more filled (otherwise sound) teeth by age**

The mean number of filled (otherwise sound) teeth varied with social class of head of household. Those from unskilled manual backgrounds had the lowest number of filled teeth, 5.7 on average, increasing to 6.9 for those with skilled manual backgrounds and 7.6 for those in households headed by a non-manual worker. Over half of dentate adults (52%) from unskilled backgrounds had 5 or fewer filled (otherwise sound) teeth compared to just over a third of adults (36%) from non manual households.

Table 3.2.5

Dentate adults who reported attending for regular check-ups had more filled (otherwise sound) teeth on average than those who attended for occasional check-ups or those who attended only with trouble; 8.0, 6.4 and 5.1 respectively. Five per cent of attenders for regular check-ups had no filled (otherwise sound) teeth compared with 18% of those who attended only when they had trouble, while 68% of attenders for regular check-ups had more than 5 such teeth compared with 40% of those who attended with trouble. Dentate adults who only attend with trouble are likely to have more advanced stages of decay and thus have a greater likelihood of teeth being extracted rather than filled. The relationship between the usual reason for dental attendance and the social class and geographical differences discussed above are explored more fully in Chapter 4.1.

Table 3.2.6

3.2.4 Twelve or more filled (otherwise sound) teeth

The previous surveys of adult dental health used 12 or more filled (otherwise sound) teeth as a statistic to identify which groups have more experience of restorative dentistry. As noted at the beginning of this chapter, previous surveys included artificial crowns in this definition, while the analysis presented in this chapter does not.

In 1998, fewer than 1 in 5 people (18%) overall had 12 or more filled (otherwise sound) teeth. The proportions varied markedly with age. For those aged 16 to 24, only 1 in 100 had 12 or more such teeth, reflecting the low decay experience at this age. By 25 to 34 years, 12% were affected, and this more than doubled for the next ten year cohort to 27%. The effects of decay are cumulative and, as people age, their teeth have been exposed to the risk of decay for longer. The highest proportion was found among dentate adults aged 45 to 54 with just less than 1 in 3 people (32%) having at least 12 teeth filled (otherwise sound). Among older dentate adults the proportion was lower with only 15% of people aged 65 years and over having 12 or more teeth filled (otherwise sound). As noted earlier, this reduction with age partly reflects the loss of teeth through extraction.

There was no significant difference between the overall proportions of men and women with 12 or more filled (otherwise sound) teeth (17% and 20%). The only significant difference within age groups was for those aged 35 to 44 years with 33% of women having 12 or more such teeth compared with 20% of men.

Table 3.2.7

England and Wales had the lowest proportions of dentate adults with at least 12 teeth filled (otherwise sound); 18% in each country, compared with 23% in Scotland and 27% in Northern Ireland. These differences were most apparent among those aged 25 to 34. Within England, there was no

significant difference between the overall proportions of dentate adults with 12 or more filled (otherwise sound) teeth. However, within age the data indicated that those aged 55 or over were more likely to have 12 or more filled teeth if they lived in the South rather than the North but none of the differences were significant.

Figure 3.2.3, Table 3.2.8

Twice as many adults from non-manual backgrounds as from unskilled backgrounds had 12 or more filled (otherwise sound) teeth, 22% compared with 11% . Again, these differences were greater for older adults compared with those aged under 35 years. For those aged 55 to 64 years, 25% from non-manual backgrounds had 12 or more filled (otherwise sound) teeth compared with only 6% from unskilled manual backgrounds. In contrast, among 25 to 34 year olds, 13% from non-manual backgrounds had 12 or more such teeth compared with 10% from unskilled manual backgrounds. The social class related pattern of restoration appeared to be diminishing for younger adults in the United Kingdom.

Table 3.2.9

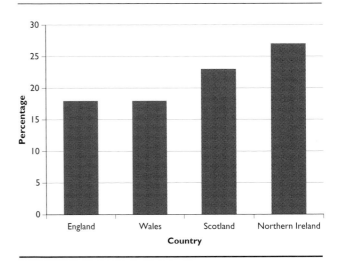

Fig 3.2.3 **Proportion of dentate adults with 12 or more filled (otherwise sound) teeth by country**

Among dentate adults who reported attending the dentist for regular check-ups 24% had at least 12 teeth filled (otherwise sound) and among attenders for an occasional check-up the figure was 15%. In contrast, 8% of people who reported attending a dentist only when they had trouble had 12 or more filled (otherwise sound) teeth. These differences persisted across all age groups except the youngest (16 to 24), among whom very few had 12 or more filled teeth irrespective of reported dental attendance.

Table 3.2.10

3.2.5 The condition and type of filling

As levels of dental decay reduce and cohorts of younger adults advance to middle age, fewer filled teeth may be expected in both younger and middle aged adults. Currently, however, a higher level of dental fillings among middle aged and older adults can be seen to reflect the successful retention of more teeth. These higher levels of fillings, however, may require maintenance to ensure continued retention of the teeth.

Fillings are subject to failure from time to time. This may be due to decay around the edges or it may be as a result of the filling breaking or falling out. The dental examinations recorded whether or not fillings were sound, unsound (damaged or portions missing) or decayed. This section looks at the condition of the fillings rather than the whole tooth as presented in the previous sections. For example, a tooth which was classified as 'filled and decayed' could have had a sound filling on one surface and decay on another surface. Thus, although the tooth would *not* have been classified as filled (otherwise sound), the restoration itself was sound. This explains the slight differences between the mean numbers of teeth which were classified as 'filled (otherwise sound)' shown in Table 3.2.2 and the mean number of sound fillings shown in Table 3.2.11.

On average, dentate adults had 7.6 teeth with fillings, of which 7.1 were sound fillings, 0.3 were unsound and 0.2 were decayed. Thus, of those teeth which had fillings with problems, a little over half were unsound without any decay (fractured or lost), and a little under half were unsound due to decay. There was no significant variation with age. The numbers of unsound fillings (without decay) were small compared with the total number of fillings that had been provided (4% averaged across dentate adults). More information on fillings with recurrent decay can be found in Chapter 3.1 with other data on decay.

Figure 3.2.4, Table 3.2.11

Among dentate adults, out of an average of 7.6 teeth which were filled, 6.4 had been filled with dental amalgam representing 84% of all filled teeth. Thus the overwhelming majority of filled teeth were restored with amalgam. The average proportion of amalgam fillings judged to be unsound was 3% which showed little variation with age.

Table 3.2.12

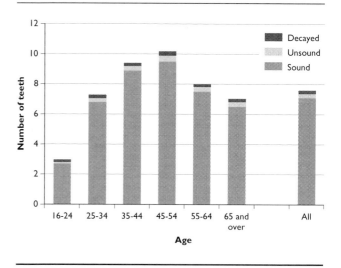

Fig 3.2.4 **Mean number of teeth with sound fillings, unsound fillings, and decayed fillings by age**

3.2.6 Artificial crowns

Artificial crowns are placed for a variety of reasons, these include restoring the form, function and appearance of the underlying natural tooth.

One third of dentate adults in the United Kingdom had at least one tooth with an artificial crown (34%). Most of these people had one or two artificial crowns (20% of the dentate population), but 5% had at least 6 artificial crowns. The proportion of dentate adults with crowned teeth increased significantly with age up to the age of 54, reflecting the cumulative nature of such restorative treatment. Among those aged 16 to 24, only 7% had any artificial crowns, this increased to 24% for those aged 25 to 34, and to 41% of those aged 35 to 44. Between the ages of 45 and 64 around half of the dentate adults examined had at least one crowned tooth.

Figure 3.2.5, Table 3.2.13

Fig 3.2.5 **Proportion of dentate adults wih one or more artificial crowns by age**

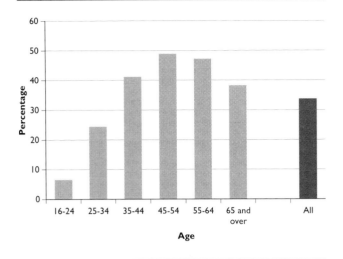

A greater proportion of dentate women than dentate men had artificial crowns; 37% compared with 31%. The range across the countries was small from 33% in Northern Ireland to 36% in Wales. There were only minor variations between the regions of England with 35% of people in the South having at least one artificial crown compared with 31% in the North.

Dentate adults living in households headed by people in non-manual occupations were more likely to have crowns (38%) than those from unskilled backgrounds (28%). Those who reported attending the dentist for regular check-ups were nearly twice as likely to have artificial crowns as those who only attended the dentist with trouble (40% compared with 23%).

Table 3.2.14

Thus, having artificial crowns varied according to age, social class and the usual reason for attendance. However, as with patterns of filling, these factors confound each other and are discussed in Chapter 4.1.

3.2.7 Root surface fillings

The presence of fillings in root surfaces does not reflect the pattern of decay alone as some fillings will have been placed to fill wear lesions. Tooth wear is not a disease process but occurs when the root is exposed to chemical or mechanical damage, such as dietary acids or abrasion from vigorous tooth brushing. To develop decay or wear lesions on root surfaces, the root of the tooth must first be exposed to the mouth. In health, root surfaces are normally not exposed and the only visible part of the tooth is the crown. If gums shrink back as age increases, root surfaces are exposed. It should also be

noted that some root decay does not progress and can be treated without a filling.

Overall, root surface fillings were found among 15% of dentate adults, resulting in an average of 0.4 teeth with root surface fillings. There was a steady increase in the average number with root surface fillings with increasing age. For example, no dentate adults aged 16 to 24 had root surface fillings, those aged 45 to 54 had an average of 0.5 teeth with root surface fillings and for those aged 65 years and older, the figure was 1.3. The proportion of dentate adults with such fillings also increased with age from 4% of those aged 25 to 34, to 43% of those aged 65 and over. Not only were more of the older adults so affected but they also had a higher proportion of teeth with root surface fillings. Among dentate adults aged 65 and over, 6% had 6 or more teeth with root surface fillings. There was no variation with respect to gender.

Tables 3.2.15–16

Dentate adults in England were almost twice as likely (16%) to have one or more teeth with root surface fillings compared with people from Northern Ireland (9%) although both average numbers of such teeth were small (0.4 and 0.2 respectively). Within England, regional variations were observed with 19% of those in the South having root surface fillings compared to 10% in the Midlands. It should be noted that detection and repair of affected root surfaces depends on attending for regular care by older adults and levels of reported attendance varied regionally (see Chapter 6.1) Furthermore, root surface fillings are much more likely to be required where teeth are retained into old age. The retention of teeth in older age groups also varied regionally (see Chapter 2.1).

Table 3.2.17

People who reported that they attended regularly for check-ups were more likely (19%) to have some root surface fillings than those who said they attended for occasional check-ups (8%) or only with trouble (11%).

Table 3.2.18

The prevalence of root surface restorations was relatively high overall in the older groups, and with the projected increases in the retention of natural teeth discussed in Chapter 5.1, this may be expected to form an increasing proportion of the filled component of natural teeth in the future.

3.2.8 The overall number of restored teeth

This chapter has so far considered coronal fillings, artificial crowns and root surface fillings[1] separately. An overall picture of restorative dental care is given in the Tables 3.2.19 and 20. Table 3.2.19 shows the mean number of teeth with any coronal restorations (fillings and crowns) irrespective of the condition of the restoration. Table 3.2.20 gives the combined total of all restorations including coronal fillings, root surface fillings and artificial crowns. This also includes those that were sound, those that were unsound and those with recurrent decay.

In 1998, 92% of all dentate adults had at least one coronal restoration (artificial crown or filling) irrespective of condition. A third of people (32%) had at least 12 teeth restored. On average, 8.6 teeth had a restoration of some kind, representing around a third of the natural dentition. Of those aged 16 to 24 years 29% had no evidence of any experience of restorative dentistry in the permanent dentition. In some cases restorations may have been provided and then the tooth extracted, but it is likely that most had no experience of restorative dental care. This will, in large part, be due to reduced levels of decay in childhood, but modern concepts of disease diagnosis and management by dentists may also have played a role. The proportion who were completely restoration free was low in all other groups and lowest in middle aged adults (2% of 35 to 54 year olds), but increased again in older groups to 8% of those aged 65 and over. This increase with age is likely to be as a result of two things. The first is that disease resistant people who have retained healthy natural teeth represent a greater proportion of the dentate population as others have been rendered edentate. The second is that restored teeth may have been more likely to be removed in some cases leaving only a few unrestored teeth in the mouth.

The highest numbers of restorations were found among middle aged adults. For those aged 45 to 54, over half (54%) had at least 12 teeth with a dental restoration. The mean number of teeth restored was only 3.1 among those aged 16 to 24 years but was nearly four times this level for the 45 to 54 year olds, at 11.8 teeth. The average proportion of all teeth which were restored was only 11% among 16 to 24 year olds, but increased to 49% of all teeth among 45 to 54 year olds, stabilising at this level in the older age groups. On average about half of all teeth in adults aged 45 or over were restored in some way.

The inclusion of teeth with root surface fillings made very little difference to the overall figures indicating that the majority of teeth with such fillings also had some coronal restoration.

Figure 3.2.6, Tables 3.2.19–20

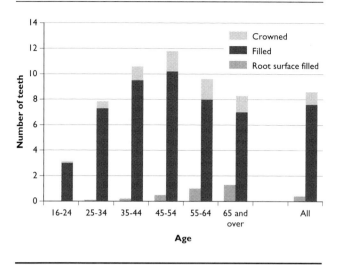

Fig 3.2.6   **Mean number of teeth which were filled, root surface filled or had artificial crowns**

Notes and references

1. Root restorations extending onto the natural crown of the tooth were also recorded as coronal restorations if they extended at least 3mm onto the crown, and the same was true for coronal restorations extending onto the root surface. Consequently, some root surface fillings will also have been recorded as coronal fillings, but only if the coronal involvement was substantial. Details of the criteria used can be found in Appendix C.

Table 3.2.1 **Distribution and mean number of filled (otherwise sound) teeth by age**

Dentate adults   *United Kingdom*

| Number of filled (otherwise sound) teeth | Age | | | | | | All |
	16-24	25-34	35-44	45-54	55-64	65 and over	
	%	%	%	%	%	%	%
None	31	4	3	3	9	14	10
1-5	53	38	22	17	31	34	32
6-11	15	45	48	48	39	37	40
12 or more	1	12	27	32	21	15	18
Mean number of filled (otherwise sound) teeth	2.6	6.6	8.8	9.3	7.4	6.3	7.0
Base	491	854	781	746	461	484	3817

1988 Table 7.17

Table 3.2.2 **The amount of filled (otherwise sound) teeth by age**

Dentate adults   *United Kingdom*

Age	Mean number of filled (otherwise sound) teeth	Mean number of teeth	Mean proportion of filled teeth	Base
16-24	2.6	27.9	9	491
25-34	6.6	28.1	23	854
35-44	8.8	26.7	33	781
45-54	9.3	24.0	39	746
55-64	7.4	19.9	37	461
65 and over	6.3	17.5	36	484
All	7.0	24.8	28	3817

Table 3.2.3 **Distribution and mean number of filled (otherwise sound) teeth by gender**

Dentate adults   *United Kingdom*

| Number of filled (otherwise sound) teeth | Gender | | |
	Men	Women	All
	%	%	%
None	11	8	10
1-5	34	30	32
6-11	38	41	40
12 or more	17	20	18
Mean	6.6	7.3	7.0
Base	1745	2072	3817

Table 3.2.4   **Distribution and mean number of filled (otherwise sound) teeth by English region and country**

Dentate adults

Number of filled (otherwise sound) teeth	English region			Country				
	North	Midlands	South	England	Wales	Scotland	Northern Ireland	United Kingdom
	%	%	%	%	%	%	%	%
None	11	11	9	10	8	10	6	10
1-5	36	32	31	32	34	29	28	32
6-11	38	40	41	40	40	38	39	40
12 or more	15	18	19	18	18	23	27	18
Mean	6.5	6.7	7.2	6.9	6.9	7.4	8.2	7.0
Base	617	495	1074	2186	502	668	461	3817

Table 3.2.5   **Distribution and mean number of filled (otherwise sound) teeth by social class of head of household**

Dentate adults   *United Kingdom*

Number of filled (otherwise sound) teeth	Social class of head of household			
	I, II, IIINM	IIIM	IV,V	All
	%	%	%	%
None	9	8	13	9
1-5	27	35	39	32
6-11	42	40	37	40
12 or more	22	17	11	18
Mean	7.6	6.9	5.7	7.0
Base	1926	1025	625	3817

Table 3.2.6   **Distribution and mean number of filled (otherwise sound) teeth by usual reason for dental attendance**

Dentate adults   *United Kingdom*

Number of filled (otherwise sound) teeth	Usual reason for dental attendance			
	Regular check-up	Occasional check-up	Only with trouble	All
	%	%	%	%
None	5	13	18	10
1-5	27	32	42	32
6-11	44	40	32	40
12 or more	24	15	8	18
Mean	8.0	6.4	5.1	7.0
Base	2400	408	1003	3817

1988 Table 7.20

Table 3.2.7 **12 or more filled (otherwise sound) teeth by age and gender**

Age	Men	Women	All
[percentage with 12 or more filled (otherwise sound) teeth]			
16-24	1	1	1
25-34	13	11	12
35-44	20	33	27
45-54	30	34	32
55-64	21	21	21
65 and over	16	14	15
All	17	20	18
		Base	
16-24	227	264	491
25-34	339	515	854
35-44	364	417	781
45-54	356	390	746
55-64	216	245	461
65 and over	243	241	484
All	1745	2072	3817

1988 Table 7.6

Table 3.2.8 **12 or more filled (otherwise sound) teeth by English region, country and age**

Dentate adults

Age	English region			Country				United Kingdom
	North	Midlands	South	England	Wales	Scotland	Northern Ireland	
percentage with 12 or more filled (otherwise sound) teeth								
16-24	1	0	0	0	3	4	8	1
25-34	12	13	8	10	10	26	26	12
35-44	20	26	27	25	32	34	49	27
45-54	30	30	33	32	31	29	37	32
55-64	13	24	23	21	18	21	20	21
65 and over	11	10	19	15	11	11	16	15
All	15	18	19	18	18	23	27	18
Base								
16-24	82	42	122	246	69	80	96	491
25-34	146	125	246	517	90	157	90	854
35-44	126	114	207	447	96	142	96	781
45-54	121	101	201	423	109	132	82	746
55-64	67	47	145	259	73	75	54	461
65 and over	75	66	153	294	65	82	43	484
All	617	495	1074	2186	502	668	461	3817

3.2: Restorative treatment

Table 3.2.9 12 or more filled (otherwise sound) teeth by social class of head of household and age

Dentate adults   *United Kingdom*

Age	Social class of head of household			
	I, II, IIINM	IIIM	IV,V	All[†]
percentage with 12 or more filled (otherwise sound) teeth				
16-24	1	2	0	1
25-34	13	13	10	12
35-44	33	21	17	27
45-54	36	27	24	32
55-64	25	26	6	21
65 and over	19	11	10	15
All ages	22	17	11	18
Base				
16-24	*215*	*116*	*90*	*491*
25-34	*402*	*254*	*148*	*854*
35-44	*395*	*214*	*121*	*781*
45-54	*394*	*204*	*107*	*746*
55-64	*247*	*115*	*85*	*461*
65 and over	*273*	*122*	*74*	*484*
All ages	*1926*	*1025*	*625*	*3817*

† Includes those for whom social class of head of household was unknown and armed forces

Table 3.2.10 12 or more filled (otherwise sound) teeth by usual reason for dental attendance and age

Dentate adults   *United Kingdom*

Age	Usual reason for attendance			
	Regular check-up	Occasional check-up	Only with trouble	All
percentage with 12 or more filled (otherwise sound) teeth				
16-24	0	2	2	1
25-34	15	12	8	12
35-44	31	31	14	27
45-54	38	23	18	32
55-64	29	-	4	21
65 and over	21	-	0	15
All	24	15	8	18
Base				
16-24	*241*	*95*	*155*	*491*
25-34	*479*	*102*	*271*	*854*
35-44	*515*	*81*	*184*	*781*
45-54	*516*	*81*	*148*	*746*
55-64	*325*	*24*	*112*	*461*
65 and over	*324*	*25*	*133*	*484*
All	*2400*	*408*	*1003*	*3817*

Table 3.2.11 The condition of filled teeth by age

Dentate adults   *United Kingdom*

Age	Mean number of teeth with:				Mean proportion of filled teeth that were unsound	Base
	Fillings	Sound fillings[†]	Unsound fillings	Decayed fillings		
16-24	3.0	2.7	0.1	0.2	3	491
25-34	7.3	6.8	0.2	0.2	3	854
35-44	9.5	8.9	0.3	0.2	3	781
45-54	10.2	9.5	0.4	0.3	4	746
55-64	8.0	7.5	0.3	0.2	4	461
65 and over	7.0	6.5	0.3	0.2	4	484
All	7.6	7.1	0.3	0.2	4	3817

† These values represent the nubmer of teeth with sound fillings irrespective of the condition of the other surfaces of the tooth

Table 3.2.12 **The condition of teeth with amalgam fillings by age**

Dentate adults   *United Kingdom*

Age	Mean number of teeth with:			Mean proportion of:		Base
	Fillings	**Amalgam fillings**	**Unsound amalgam fillings**	**Filled teeth that were filled with amalgam**	**Teeth filled with amalgam that were unsound**	
16-24	3.0	2.6	0.1	86	3	491
25-34	7.3	6.5	0.2	89	3	854
35-44	9.5	8.2	0.2	87	3	781
45-54	10.2	8.5	0.3	84	4	746
55-64	8.0	6.4	0.2	81	4	461
65 and over	7.0	5.4	0.2	77	4	484
All	7.6	6.4	0.2	84	3	3817

Table 3.2.13 **Distribution and mean number of artificial crowns**

Dentate adults   *United Kingdom*

Number of artificial crowns	Age						All
	16-24	**25-34**	**35-44**	**45-54**	**55-64**	**65 and over**	
	%	%	%	%	%	%	%
0	93	76	59	51	52	62	66
1	3	13	17	15	15	14	13
2	2	5	9	10	8	7	7
3	0	2	5	6	8	6	4
4	0	1	2	6	5	2	3
5	1	1	2	3	3	3	2
6 or more	0	2	5	8	8	6	5
Mean	0.1	0.6	1.1	1.6	1.6	1.3	1.0
Base	491	854	781	746	461	484	3817

Table 3.2.14 **Artificial crowns by characteristics of dentate adults**

Dentate adults   *United Kingdom*

	Percentage with artificial crowns	Base
All	34	*3817*
Age		
16-24	7	*491*
25-34	24	*854*
35-44	41	*781*
45-54	49	*746*
55-64	48	*461*
65 and over	38	*484*
Gender		
Men	31	*1745*
Women	37	*2072*
Country		
England	34	*2186*
Wales	36	*502*
Scotland	35	*668*
Northern Ireland	33	*461*
English region		
North	31	*617*
Midlands	32	*495*
South	35	*1074*
Social class of head of household		
I, II, IIINM	38	*1926*
IIIM	30	*1025*
IV,V	28	*625*
Usual reason for dental attendance		
Regular check-up	40	*2400*
Occasional check-up	28	*408*
Only with trouble	23	*1003*

Table 3.2.15 **Distribution and mean number of teeth with root surface fillings by age**

Dentate adults   *United Kingdom*

Number of teeth with root surface fillings	Age						All
	16-24	25-34	35-44	45-54	55-64	65 and over	
	%	%	%	%	%	%	%
None	100	96	92	80	65	57	85
1-5	0	4	7	18	32	37	14
6-11	0	0	1	1	3	5	2
12 or more	0	0	0	0	0	1	0
Mean	0	0.1	0.2	0.5	0.9	1.3	0.4
Base	*491*	*854*	*781*	*746*	*461*	*484*	*3817*

Table 3.2.16 **Distribution and mean number of teeth with root surface fillings by gender**

Dentate adults   *United Kingdom*

Number of teeth with root surface fillings	Gender		
	Men	Women	All
	%	%	%
None	85	84	85
1-5	13	14	14
6-11	1	2	2
12 or more	0	0	0
Mean	0.4	0.4	0.4
Base	1745	2072	3817

Table 3.2.17 **Distribution and mean number of teeth with root surface fillings by English region and country**

Dentate adults

Number of teeth with root surface fillings	English region			Country				
	North	Midlands	South	England	Wales	Scotland	Northern Ireland	United Kingdom
	%	%	%	%	%	%	%	%
None	85	90	81	84	86	89	91	85
1-5	14	9	17	14	12	11	8	14
6-11	1	1	2	2	2	0	1	2
12 or more	0	0	0	0	0	0	0	0
Mean	0.4	0.3	0.5	0.4	0.4	0.2	0.2	0.4
Base	617	495	1074	2186	502	668	461	3817

Table 3.2.18 **Distribution and mean number of teeth with root surface fillings by usual reason for dental attendance**

Dentate adults   *United Kingdom*

Number of teeth with root surface fillings	Usual reason for dental attendance			
	Regular check-up	Occasional check-up	Only with trouble	All
	%	%	%	%
None	81	92	89	85
1-5	17	7	10	14
6-11	2	1	0	2
12 or more	0	0	0	0
Mean	0.5	0.2	0.2	0.4
Base	2400	408	1003	3817

Table 3.2.19 **Distribution and mean number of all teeth with any coronal restorations**

Dentate adults   *United Kingdom*

Number of teeth with any coronal restorations	Age						All
	16-24	25-34	35-44	45-54	55-64	65 and over	
	%	%	%	%	%	%	%
None	29	4	2	2	6	8	8
1-5	52	29	14	10	20	30	25
6-11	18	46	40	34	34	31	35
12 or more	2	22	43	54	40	30	32
Mean number of restored teeth	3.1	7.9	10.6	11.8	9.6	8.3	8.6
Mean number of teeth	27.9	28.1	26.7	24.0	19.9	17.3	24.8
Mean proportion of restored teeth	11	28	40	49	48	48	35
Base	*491*	*854*	*781*	*746*	*461*	*484*	*3817*

Table 3.2.20 **Distribution and mean number of all teeth with any coronal or root restorations**

Dentate adults   *United Kingdom*

Number of teeth with any coronal or root restorations	Age						All
	16-24	25-34	35-44	45-54	55-64	65 and over	
	%	%	%	%	%	%	%
None	29	4	2	2	6	7	8
1-5	52	28	14	10	20	30	25
6-11	18	46	40	33	32	30	34
12 or more	2	22	44	56	43	32	33
Mean number of restored teeth	3.1	7.9	10.7	12.0	9.9	8.6	8.8
Mean number of teeth	27.9	28.1	26.7	24.0	19.9	17.3	24.8
Mean proportion of restored teeth	11	28	40	50	50	49	35
Base	*491*	*854*	*781*	*746*	*461*	*484*	*3817*

3.3 The condition of supporting structures

Summary

- The depth of any periodontal pocketing and the extent of any loss of attachment were measured to assess the condition of the structures which support natural teeth. The presence of visible plaque and calculus were also recorded as these are factors related to the development of periodontal disease.

- Nearly three-quarters (72%) of dentate adults had some visible plaque on their teeth. An average of 8.3 teeth had some plaque, representing a third of all teeth.

- The reported frequency of teeth cleaning was associated with the prevalence of visible plaque: 69% of dentate adults who reported cleaning their teeth at least twice a day had visible plaque compared with 79% of those who cleaned their teeth once a day and 87% of those who cleaned their teeth less than once a day (including those who never cleaned their teeth). Over half of teeth which were never cleaned or were cleaned less than once a day had visible plaque, compared with less than a third of teeth cleaned at least twice a day.

- Seventy-three per cent of dentate adults had some calculus and on average 23% of teeth had some calculus deposits.

- Fifty-four per cent of dentate adults had some periodontal pocketing of 4mm or more and 5% had deep pocketing (of 6mm or more); 43% had some loss of attachment of 4mm or more and 8% had loss of attachment of 6mm or more.

- The prevalence of pocketing, loss of attachment and calculus increased with age. For example, the proportion of dentate adults with loss of attachment of 4mm or more increased from 14% among those aged 16 to 24 years to 85% of those aged 65 and over.

- On average, older dentate adults also had a higher proportion of their teeth affected by periodontal conditions. For example, nearly a quarter (23%) of the teeth of those aged 65 and over were affected by pocketing of 4mm or more compared with 5% of those aged 16 to 24 years.

- Dentate men were more likely than dentate women to have some periodontal pocketing of 4mm or more (57% and 51% respectively) and loss of attachment of 4mm or more (46% and 40%). They were also more likely to have some teeth with visible plaque (76% and 68%) and to have teeth with calculus deposits (76% and 70%).

3.3.1 Introduction

Periodontal diseases comprise a range of conditions characterised by inflammation of the gums and loss of the tissues which support the natural teeth, including their supporting bone. The biological processes involved are complex with several factors potentially contributing to individual risk, and not all are fully understood at present[1]. The diseases can cause a variety of symptoms, including bleeding gums and looseness of the teeth, but are usually painless until the disease process has reached an advanced stage. However, if left untreated, tooth loss may result . The conditions affect different people to different degrees and even within an individual mouth can affect different teeth, or even individual surfaces of the same tooth, differently. This makes both the measurement and reporting of data from the population problematic.

As the understanding of periodontal diseases has developed, so the diagnostic criteria have changed for each national dental health survey and consequently detailed comparisons between the surveys have been difficult. The criteria adopted for this survey are an attempt to address current thinking on measuring periodontal diseases, and yet retain some comparability with the previous criteria. The features recorded were the presence of visible plaque (an indication of oral cleanliness and an essential factor in the progress of periodontal diseases), calculus (which is also related to oral cleanliness), periodontal pocketing and loss of attachment (see Appendix A for more detailed descriptions of these conditions). The latter two measurements are indicators of how much of the supporting tissues have been lost to disease and so give an indication of the impact of the condition on the mouth. Loss of attachment is included for the first time in this survey. It is a more robust measure of the experience of periodontal diseases than pocketing alone, particularly for older adults. Pocketing has been recorded separately since it is an important prognostic indicator and was the major measure used in the 1988 survey.

The presence of periodontal disease as measured by the survey is related to the number of teeth people have. Thus, those with few remaining natural teeth have, by definition, less chance of having disease in the supporting structures surrounding the teeth. While it is important to know how many people are affected, it is also useful to relate this prevalence to the number of teeth present. In this chapter the data relating to the different conditions are presented in a number of ways:

- the proportion of dentate adults with at least one tooth affected
- the mean number of teeth affected

- the proportion of teeth affected averaged over the population (calculated by dividing the mean number of affected teeth by the mean number of teeth for the given population)
- the distribution of the affected teeth around the mouth given for individual teeth and for teeth grouped into sextants (a sextant is one of six segments in the mouth, three in the upper jaw and three in the lower).

The measures used present particular difficulties for survey examiners[2,3]. Plaque and calculus are often difficult to see against a similarly coloured tooth surface. Measuring periodontal pockets and loss of attachment requires good lighting and a good view, not always easy under the field conditions of the dental examination. Consequently, there can be wide variation between examiners on all of these measures. However, for the reasons explained in Appendix B it was impossible to undertake a full calibration on this part of the examination. There was a relatively small number of examiners in each individual region and country so variations in the results for adults from different regions and countries should be interpreted with particular caution for all the measures reported.

All dentate adults who agreed to a dental examination were asked a series of screening questions (see Appendix C) to identify those whose health might be put at risk by the examination for calculus, pocketing and loss of attachment. Of the 3817 adults examined, 300 (8%) were excluded on these grounds. A further 10 adults who completed the other parts of the examination declined the examination of their gums.

3.3.2 Visible plaque

Dental plaque is the bacterial material which collects on the teeth or on other solid oral structures, such as dentures[1]. The bacteria within dental plaque have been implicated in both dental caries and periodontal disease. Most plaque is relatively easily removed by the individual, by cleaning the teeth with a tooth brush and other cleaning aids and good plaque control is an essential component in the prevention and management of the periodontal diseases, as well as being important in the prevention of caries. The presence of plaque, and by implication the effectiveness of hygiene, is therefore an important measure of oral health and the potential for oral health in the population.

Even moderate amounts of plaque can be difficult to see on the tooth with the naked eye (the criterion used in the examination), so where plaque is recorded it generally indicates a substantial accumulation. Visible deposits of plaque take some time to develop (at least 24 hours) and as most people brush their teeth once or twice a day, for plaque to be recorded in this survey it is likely to be at sites where tooth cleaning has been ineffective on a fairly consistent basis.

Respondents were asked if they wished to clean their teeth prior to the oral examination. Few (6%) chose to do so, although some respondents may have cleaned their teeth before the examiner arrived. The prevalence of visible plaque in those who cleaned their teeth immediately before the examination was no different from the prevalence for other dentate adults.

Table 3.3.1 shows that 72% of dentate adults had visible plaque on at least one tooth. An average of 8.3 teeth had some plaque representing a third of all teeth. The absence of recorded plaque in 28% of the dentate population does not indicate that there was no plaque present in these cases, but simply that there was not enough for it to be visible to the naked eye. Nevertheless, the absence of visible plaque suggests that oral hygiene is at least reasonably effective.

Although there was little significant difference with respect to age in the proportion of dentate adults with some plaque on their teeth, there were differences in the mean number of teeth with plaque by age group. Compared with older dentate adults those in the younger age groups had, on average, more teeth with visible plaque than older adults. However younger adults also had more teeth, on average, and therefore the number of teeth affected by plaque accounted for a smaller proportion of their teeth. For example, dentate adults aged 16 to 24 had on average 27.9 teeth of which an average of 9.5 had some visible plaque, representing a third (34%) of all teeth. In contrast, dentate adults aged 65 and over had an average of 17.3 teeth of which an average of 7.7 had some plaque, representing 44% of teeth.

Men were more likely to have plaque than women (76% compared with 68%) and they also had a higher proportion of affected teeth (38% compared with 29%). National and regional differences in the prevalence of plaque appear to be quite marked ranging from 52% in Wales to 75% in England. The proportion of dentate adults with some plaque varied from 70% to 80% within the different regions of England. The proportion of teeth affected by plaque varied according to country and English region in a similar way (from an average of 18% in Wales to 39% in the North of England). However, because of uncertainties about calibration and the regional distribution of examiners, regional differences should be interpreted with caution (see Appendix B).

Dentate adults in non-manual social classes were less likely than those in manual social classes to have mouths containing visible plaque; 70% of dentate adults living in non-manual households had some plaque on their teeth compared with 75% in skilled manual households and 78% in partly skilled or unskilled households.

Table 3.3.1

Dentate adults who reported that they attended the dentist for regular check-ups were less likely to have some plaque than other respondents: 68% had visible plaque compared with 72% of dentate adults who only went occasionally and 80% of dentate adults who only went when they had trouble with their teeth. On average, 43% of the teeth of dentate adults who only went to the dentist when in trouble had visible plaque compared with 32% of the teeth of those who went for an occasional check-up and 29% of the teeth of those who went for a regular check-up. This indicates differences between the groups in the efficiency with which they clean their own teeth rather than the impact of professional cleaning by a dentist.

The reported frequency of teeth cleaning established during the survey interview was associated with the prevalence of visible plaque: 69% of dentate adults who cleaned their teeth at least twice a day had visible plaque compared with 79% of those who cleaned their teeth once a day and 87% of those who cleaned their teeth less than once a day or who reported never cleaning their teeth. There was a similar pattern of results for the proportion of teeth affected; over half of teeth which were never cleaned or were cleaned less than once a day had visible plaque compared with less than a third of teeth cleaned at least twice a day or more. Such an association is not at all unexpected. However, the finding that 69% of people who clean their teeth twice a day or more still had plaque and that 30% of their teeth were affected by visible plaque suggests that while many people are making the effort to clean their teeth, the cleaning is not always very effective.

Figure 3.3.1, Table 3.3.2

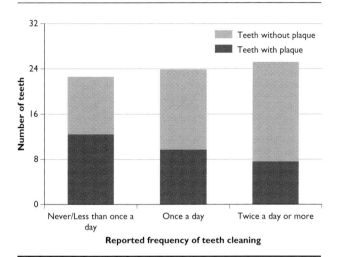

Fig 3.3.1 **Mean number of teeth with and without visible plaque by reported frequency of teeth cleaning**

3.3.3 Calculus

Dental calculus is the calcified or calcifying deposits which form on the teeth and dentures. While it is not in itself an indicator of periodontal disease, it is regarded as an important risk factor since it prevents effective cleaning and plaque removal. Once formed, calculus is extremely difficult to brush away and usually requires professional removal (scaling). The management of most periodontal disease begins with a combination of improved oral hygiene and meticulous removal of calculus from the teeth. The presence of calculus, particularly in the presence of periodontal disease, indicates a potential treatment need.

Seventy-three per cent of dentate adults had calculus present on at least one tooth which was similar to the prevalence of visible plaque among denate adults. On average, 23% of the teeth of dentate adults had some calculus deposits which was lower than the comparative figure for teeth with visible plaque (33%). The proportion of people with some calculus varied with age from 61% among those aged 16 to 24 years to 83% among those aged 65 years and over. The teeth of dentate adults aged 16 to 24 years were less than half as likely to have calculus as the teeth of those aged 65 and over (on average, 15% of teeth were affected by calculus in the former group compared with 33% of teeth in the latter).

Associations between demographic, regional and social class factors and prevalence of calculus were broadly similar to those for prevalence of visible plaque. Dentate women (70%) were less likely than dentate men (76%) to have some calculus. Wales, Scotland and Northern Ireland had a lower proportion of dentate adults with some calculus (61%, 62% and 67% respectively) than in England (75%), although, once

again regional differences should be interpreted with caution (see Appendix B). Dentate adults from non-manual households were less likely than those from skilled manual and partly skilled or unskilled households to have calculus present on at least one tooth (non-manual households, 71%; skilled manual households, 75%; partly skilled or unskilled households, 76%).

Table 3.3.3

As with visible plaque, there was a strong association between the prevalence of the condition and the usual reason for dental attendance. A smaller proportion of dentate adults who reported regular dental check ups had calculus (68%) compared with those who went occasionally (75%) and those who attended only when in trouble (82%).

The reported frequency of teeth cleaning was associated with the proportion of people with calculus, albeit less strongly than with visible plaque. The results for the proportion of teeth with calculus show this relationship more clearly: on average 21% of teeth which were cleaned at least twice a day had calculus compared with 29% of teeth cleaned once a day and 38% of teeth cleaned less than once a day or never.

The prevalence of calculus also varied by reported time since last dental visit. Sixty-eight per cent of those adults who reported that their last dental visit had been within the past year had calculus compared with 84% of adults who reported that their last dental visit had been between one and five years previously and 95% of those whose last visit had been 5 to 10 years previously. The same trend was evident when the proportion of teeth affected was reported; on average 19% of teeth among people who had visited in the last year had calculus compared with 51% of teeth among those who had last visited 5 to 10 years previously. The trend did not carry over to people who had not been to the dentist for ten years or more; 84% of them and 37% of their teeth were affected, both lower than the 5 to 10 years group. This may reflect that some people who have not needed to attend the dentist for over ten years have relatively problem free mouths.

Table 3.3.4

Calculus was not evenly distributed across different sites in the mouth. The average proportion of teeth with calculus in each sextant varied from 6% of teeth in the upper central sextant (upper incisors and canines) to 51% of teeth in the lower central sextant (lower incisors and canines). The mean number of teeth with calculus per sextant was less in the upper sextants than in the lower sextants showing calculus to be more widespread in the lower jaw. This illustrates a pattern of calculus occurrence around the mouth which will be familiar to any clinical dentist.

Figure 3.3.2, Table 3.3.5

Fig 3.3.2   Proportion of sextants containing teeth with calculus

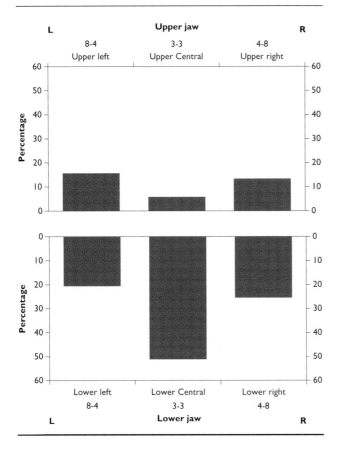

3.3.4 Periodontal pocketing

The progressive loss of the supporting structures of the teeth, which can ultimately lead to looseness of the tooth and then tooth loss if untreated, is the most important manifestation of periodontal disease. This can be quantified by measuring how much of the attachment between the tooth and the supporting tissues has been lost. There are two ways of doing this; one is to measure the depth of the *pockets* which form between the inflamed gum and the tooth when the attachment to the tooth is lost. The second measure, known as *loss of attachment*, involves measuring the distance between where the attachment is and where it "should" be (the neck of the tooth). Both are useful measures and both are reported here.

In the 1988 report, pocketing of 4mm or more but less than 6mm was referred to as *shallow pocketing*, while pocketing of 6mm or more was referred to as *deep pocketing*. The same terms are used in this report; in addition, the term *any pocketing* is used to refer to all pocketing of 4mm or more.

The prevalence of periodontal disease by age will be affected by the smaller number of natural teeth in older age groups, especially where teeth affected by advanced disease have been the ones removed. Loss of teeth is a confounding factor

in any analysis of the relationship between disease and socio-demographic variables and should be born in mind when considering the data which follow.

Fifty-four per cent of dentate adults had pocketing of 4mm or more on at least one tooth, while 5% had some deep pocketing of 6mm or more. One third (34%) of dentate adults aged 16 to 24 years and nearly half (47%) of those aged 25 to 34 had some pocketing. These proportions increased with age up to 67% of those aged 65 and over with pockets of 4mm or greater.

Dentate men were more likely than dentate women to have pocketing of 4mm or more (57% of dentate men were affected compared with 51% of dentate women). Pocketing of 4mm or more was less prevalent in Wales, Scotland, Northern Ireland and North of England (a half or fewer of dentate adults in each region had the condition) than in the Midlands (58%) and South (57%) of England, but such regional differences should be interpreted with caution.

Figure 3.3.3, Table 3.3.6

Fig 3.3.3   Proportion of dentate adults with periodontal pockets of 4mm or more by age

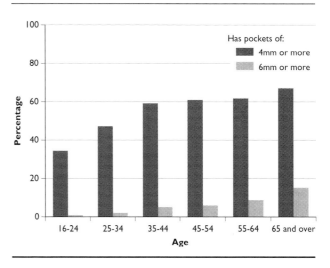

Dentate adults who regularly went for a check up were less likely to have pocketing (52%) than those who reported only attending with trouble (57%). The proportion of dentate adults with pocketing of 4mm or more also varied with the time since the last dental visit (68% of those who had not visited the dentist for over 5 years compared with 52% to 53% of those who had visited the dentist more recently) and with the reported frequency of teeth cleaning (52% of those who brushed twice a day or more compared with 59% of those who brushed less often than this).

Table 3.3.7

Deep pocketing (of 6mm or more) affected a much smaller proportion of the dentate population (5%) and varied from only 1% of the youngest age group (aged 16 to 24 years) up to 15% of those aged 65 years and over. These findings may well underestimate the real prevalence of the disease among adults since the data only refer to dentate adults, and even among dentate adults many of the most vulnerable teeth may already have been lost to periodontal disease. Differences in the proportion of the population affected by deep pocketing according to gender, geography and social class were marginal, suggesting that periodontal disease is not socially or culturally influenced in any major way. Similarly, differences according to the usual reason for dental attendance, the time since the most recent dental visit and the reported frequency of teeth cleaning were small and not statistically significant.

Tables 3.3.6-7

On average, 12% of all teeth of dentate adults were found to have some pocketing of 4mm or more. The average proportion of teeth with any pocketing of 4mm or more, like the overall proportion of people affected, increased with age. Dentate adults aged 16 to 24 years had, on average, 5% of their teeth affected by pocketing compared with an average of 23% of teeth among those aged 65 and over. The patterns in the average number of teeth and the proportion of teeth affected by pocketing with respect to other socio-demographic and behavioural characteristics were similar to those found for the prevalence of pocketing among these groups.

Table 3.3.8-9

When the analysis according to age and gender is restricted to people who had some pocketing, a clearer picture of the pattern of periodontal diseases emerges. Among people with pocketing of 4mm or more (54% of the sample), on average around 6 teeth were affected by pocketing, representing 23% of their natural teeth. There was little variation according either to age or gender but the group with the highest proportion of teeth affected were adults aged 55 years and over, who had an average of 30% affected teeth.

Table 3.3.10

When the analysis was restricted further to people with some pockets of 6mm or more, a mean of 2.6 teeth were affected by such deep pockets in these individuals, representing an average of 12% of their teeth. Men and older adults (aged 55 and over) tended to have more teeth and a higher proportion of teeth affected although small base numbers meant these differences were not significant. Individuals such as these, who have some deep pockets, may also be expected to have some shallower pockets as well. This group of people had a mean of 11.6 teeth with pockets of 4mm or more, representing 54% of all standing teeth. Although the number of teeth affected was lower in the people aged 55 and over (9.4 compared with 12.9 in the 35 to 44 year olds) the proportion of teeth affected remained fairly constant for the two age and gender groups reported (51% to 57%). Thus among people with deep pockets (5% of the sample, see Table 3.3.6) over half of their teeth are affected by some pocketing of at least 4mm. When left untreated, most periodontal diseases associated with deeper pockets are progressive and pocketing of 4mm or more in around half of the standing teeth could be seen as a significant threat to the dentition of these individuals.

Table 3.3.11

When the frequency of pocketing on individual teeth was examined, lower molars were the teeth most likely to be affected by any pockets of 4mm or more (18% to 28% depending on the tooth involved). Other teeth were less often affected, ranging from 5% to 19%. The proportion of any individual teeth affected by deep pocketing was very low and again no clear pattern of deep pocketing around the mouth emerged.

Table 3.3.12

3.3.5 Periodontal loss of attachment

Loss of attachment was measured in the 1998 survey to supplement the measure of pocketing. This was the first time loss of attachment was measured in a national adult dental health survey and it has been introduced in view of the increasing number of older dentate adults, as it is in many ways a more realistic measure of the impact of past disease. The same two threshold measurements have been used for loss of attachment as were used for pocketing, that is 4mm or more for *any* loss of attachment and 6mm or more for more *extensive* loss of attachment.

Forty-three per cent of dentate adults had loss of attachment of 4mm or more around at least one tooth. This increased with age from 14% among dentate adults aged 16 to 24 years to 85% among those aged 65 years and over. The proportion of dentate adults with extensive loss of attachment (of 6mm or more) also increased with age, particularly among those aged 45 years and over. None of those aged 16 to 24 years had any extensive loss of attachment but nearly one-third (31%) of dentate adults aged 65 and over had loss of attachment of 6mm or more on at least one tooth. Dentate men were more likely than dentate women to have some loss of attachment around at least one tooth (46% compared with 40%). Social class of the head of household did not significantly affect the likelihood of any loss of attachment of 4mm or more.

Figure 3.3.4, Table 3.3.13

Fig 3.3.4   Proportion of dentate adults with loss of attachment of 4mm or more by age

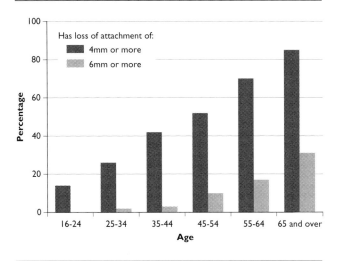

Dentate adults who had not visited their dentist for more than 5 years were twice as likely as those who had visited more recently to have extensive loss of attachment (of 6mm or more) around at least one tooth (14% compared with 7% respectively). People who brushed their teeth twice a day or more were less likely than those who brushed less often than that to have some extensive loss of attachment (7% compared with 11% and 13%). There was no significant difference with respect to the usual reason for dental attendance in the proportion of dentate adults with some extensive loss attachment. These findings appear to suggest at least some relationship between some dental behaviour and loss of attachment.

Table 3.3.14

On average, 10% of all teeth of dentate adults were affected by loss of attachment of 4mm or more. This varied markedly with age, from an average of 2% of the teeth of dentate adults aged 16 to 24 years to 30% of the teeth among those aged 65 years and over. The patterns in the average number of teeth and the proportion of teeth affected by loss of attachment with respect to other socio-demographic and behavioural characteristics were similar to those found for the prevalence of loss of attachment among these groups.

Tables 3.3.15–16

When analysis of loss of attachment is restricted to people who had some teeth with at least 4mm loss of attachment (43% of the sample, see Table 3.3.13), the mean number of teeth affected was 5.5 and was highest among men and those aged 55 years and over (5.8 in both groups). Among this affected group, an average of 24% of standing teeth had 4mm or more loss of attachment and this proportion increased with age as the mean number of teeth reduced. When analysis

was restricted further, just to the relatively small number of people with 6mm or more loss of attachment (8% of the dentate population) a mean of 9.8 teeth had loss of attachment of 4mm or more and 2.9 teeth had 6mm or more. The proportion of teeth affected among this group was similar to that found for the same analysis of pocketing data; teeth with 4mm or more attachment loss represented 54% of their teeth and those with 6mm or more attachment loss represented 16% of their teeth overall.

Tables 3.3.17–18

The pattern of loss of attachment around the mouth was slightly different from the pattern shown for pocketing. Upper and lower molars were equally likely to be affected by some loss of attachment (10 to 15%) while the proportion of other teeth affected varied from 6% to 11%.

Tables 3.3.19

Table 3.3.20 shows the pattern by sextants (where the mouth is divided into 6 segments, three in the upper jaw and three in the lower). The two central sextants (upper incisors and canines and lower incisors and canines) were less likely than the left and right posterior sextants (premolars and molars) to contain teeth with any pocketing or loss of attachment of 4mm or more.

Tables 3.3.20

References

1. Lindhe J, *Textbook of Clinical Periodontology (2nd Edition)*, Munksgaard, Copenhagen 1989

2. Mojon P, Chung J-p, Favre P, Budtz-Jörgensen E, *Examiner agreement on periodontal indices during dental surveys of elders*, Journal of Clinical Periodontology 1996 ; **23** : 56-59

3. Fleiss JL, Mann J, Paik M, Goultchin J, Chilton NW, *A study of inter- and intra- examiner reliability of pocket depth and attachment level*, Journal of Periodontal Research 1991 ; **26** : 122-128

Table 3.3.1   **Visible plaque by characteristics of dentate adults**

Dentate adults   *United Kingdom*

Characteristics of dentate adults	Proportion of dentate adults with visible plaque	Mean number of teeth with visible plaque	Mean proportion of teeth with visible plaque	Mean number of teeth	Base
All	72	8.3	33	24.8	3817
Age					
16-24	72	9.5	34	27.9	491
25-34	70	8.5	30	28.1	854
35-44	72	8.6	32	26.7	781
45-54	69	7.6	32	24.0	746
55-64	75	7.3	37	19.9	461
65 and over	78	7.7	44	17.3	484
Gender					
Men	76	9.4	38	25.0	1745
Women	68	7.1	29	24.6	2072
Country					
England	75	8.9	36	24.9	2186
Wales	52	4.3	18	24.2	502
Scotland	60	4.8	20	23.8	668
Northern Ireland	66	5.4	22	24.5	461
English Region					
North	80	9.7	39	24.6	617
Midlands	70	6.3	26	24.6	495
South	74	9.6	38	25.3	1074
Social class of head of household					
I, II, IIINM	70	7.7	30	25.4	1259
IIIM	75	8.8	36	24.2	706
IV,V	78	9.4	40	23.6	434

Table 3.3.2 Visible plaque by reported dental behaviour

Dentate adults   *United Kingdom*

Reported dental behaviour	Proportion of dentate adults with visible plaque	Mean number of teeth with visible plaque	Mean proportion of teeth with visible plaque	Mean number of teeth	Base
All	72	8.3	33	24.8	*3817*
Usual reason for dental attendance					
Regular check up	68	7.3	29	24.9	*2400*
Occasional check up	72	8.4	32	26.3	*408*
Only with trouble	80	10.3	43	24.0	*1003*
Frequency of teeth cleaning					
Never/less than once a day	87	12.4	55	22.6	*181*
Once a day	79	9.7	41	23.9	*801*
Twice a day or more	69	7.6	30	25.2	*2826*

Table 3.3.3 **Calculus by characteristics of dentate adults**

Dentate adults†   *United Kingdom*

Characteristics of dentate adults	Proportion of dentate adults with calculus	Mean number of teeth with calculus	Mean proportion of teeth with calculus	Mean number of teeth	Base
All	73	5.8	23.2	25.0	3507
Age					
16-24	61	4.3	15	27.9	471
25-34	71	5.8	21	28.1	828
35-44	74	6.7	25	26.8	735
45-54	77	6.2	26	24.1	689
55-64	77	5.5	28	19.8	400
65 and over	83	5.6	33	17.1	384
Gender					
Men	76	6.5	26	25.2	1614
Women	70	5.0	20	24.9	1893
Country					
England	75	6.0	24	25.2	2007
Wales	61	4.7	19	24.5	459
Scotland	62	3.9	16	24.0	614
Northern Ireland	67	4.6	19	24.8	427
English Region					
North	80	6.5	26	24.8	572
Midlands	76	6.2	25	25.0	452
South	72	5.7	22	25.4	983
Social class of head of household					
I, II, IIINM	71	5.4	21	25.6	1755
IIIM	75	6.1	25	24.5	950
IV,V	76	6.5	27	23.9	577

† 310 dentate adults were excluded from the periodontal examination. See Section 3.3.1.

Table 3.3.4 **Calculus by reported dental behaviour**

Dentate adults†   *United Kingdom*

Reported dental behaviour	Proportion of dentate adults with calculus	Mean number of teeth with calculus	Mean proportion of teeth with calculus	Mean number of teeth	Base
All	73	5.8	23	24.8	3507
Usual reason for dental attendance					
Regular check up	68	4.8	19	25.1	2188
Occasional check up	75	5.6	21	26.5	386
Only with trouble	82	7.7	32	24.3	927
Frequency of teeth cleaning					
Never/less than once a day	78	8.7	38	22.6	169
Once a day	79	7.0	29	23.9	741
Twice a day or more	70	5.2	21	25.2	2588
Time since last dental visit					
Less than 1 year	68	4.9	19.4	25.2	2620
Between 1 and 5 years	84	7.2	28.5	25.3	608
Over 5 up to 10 years	95	12.0	51.9	23.1	61
Over 10 years	84	8.7	36.9	23.6	212

† 310 dentate adults were excluded from the periodontal examination. See Section 3.3.1.

Table 3.3.5 **Teeth with calculus in different sextants of the mouth**

Dentate adults†   *United Kingdom*

	Sextant					
	Upper right 8-4	Upper central 3-3	Upper left 4-8	Lower right 8-4	Lower central 3-3	Lower left 4-8
Proportion of subjects with calculus in sextant	24	11	27	38	67	31
Mean number of teeth with calculus in sextant	0.5	0.3	0.5	0.9	3.0	0.7
Mean number of teeth in sextant	3.7	5.6	3.6	3.6	5.9	3.6
Mean proportion of teeth with calculus	13	6	16	26	51	21
Base	3249	3330	3262	3389	3490	3401

† 310 dentate adults were excluded from the periodontal examination. See section 3.3.1.
Base excludes those with no teeth in the relevant sextant.

Table 3.3.6 **Degree of pocketing by characteristics of dentate adults**

Dentate adults†   *United Kingdom*

Characteristics of dentate adults	No pocketing of 4mm or more	Pocketing 4mm or more	Pocketing 6mm or more	Base
		percentage		
All	46	54	5	3507
Age				
16-24	66	34	1	471
25-34	53	47	2	828
35-44	41	59	5	735
45-54	39	61	6	689
55-64	38	62	9	400
65 and over	33	67	15	384
Gender				
Men	43	57	6	1614
Women	49	51	5	1893
Country				
England	45	55	6	2007
Wales	53	47	4	459
Scotland	53	47	5	614
Northern Ireland	52	48	4	427
English region				
North	51	50	4	572
Midlands	42	58	3	452
South	43	57	8	983
Social class of head of household				
I, II, IIINM	48	52	6	1755
IIINM	44	56	5	950
IV,V	44	57	7	577

† 310 dentate adults were excluded from the periodontal examination. See Section 3.3.1.

Table 3.3.7 **Degree of pocketing by reported dental behaviour**

Dentate adults†   *United Kingdom*

Reported dental behaviour	No pocketing of 4mm or more	Pocketing 4mm or more	Pocketing 6mm or more	Base
		percentage		
All	46	54	5	*3507*
Usual reason for dental attendance				
Regular check up	48	52	5	*2188*
Occasional check up	45	56	4	*386*
Only with trouble	43	57	6	*927*
Time since last dental visit				
Less than 1 year	48	52	5	*2620*
Between 1 and 5 years	47	53	5	*608*
Over 5 years	32	68	8	*273*
Frequency of teeth cleaning				
Never/less than once a day	38	62	7	*169*
Once a day	42	58	6	*741*
Twice a day or more	48	52	5	*2588*

† 310 dentate adults were excluded from the periodontal examination. See Section 3.3.1.

Table 3.3.8   **The amount and degree of pocketing by characteristics of dentate adults**

Dentate adults[†]   *United Kingdom*

Characteristics of dentate adults	No pocketing of 4mm or more		Pocketing 4mm or more		Pocketing 6mm or more		Mean number of teeth	Base
	Mean number of teeth affected	Mean proportion of teeth affected	Mean number of teeth affected	Mean proportion of teeth affected	Mean number of teeth affected	Mean proportion of teeth affected		
All	21.7	87	3.1	12	0.1	0	25.0	3507
Age								
16-24	26.2	94	1.4	5	0.0	0	27.9	471
25-34	24.9	89	3.0	11	0.1	0	28.1	828
35-44	23.0	86	3.7	14	0.1	0	26.8	735
45-54	20.4	85	3.6	15	0.1	1	24.1	689
55-64	16.3	82	3.3	17	0.2	2	19.8	400
65 and over	13.1	77	3.9	23	0.4	2	17.1	384
Gender								
Men	21.5	85	3.4	13	0.2	1	25.2	1614
Women	22.0	88	2.8	11	0.1	0	24.9	1893
Country								
England	21.7	86	3.2	13	0.2	1	25.2	2007
Wales	22.2	91	2.2	9	0.1	0	24.5	459
Scotland	21.4	89	2.5	10	0.1	0	24.0	614
Northern Ireland	22.7	92	1.9	8	0.1	0	24.8	427
English Region								
North	21.8	88	2.9	12	0.1	0	24.8	572
Midlands	21.4	86	3.3	13	0.1	0	25.0	452
South	21.8	86	3.4	13	0.2	1	25.4	983
Social class of head of household								
I, II, IIINM	22.5	88	2.9	11	0.1	1	25.6	1755
IIINM	21.0	86	3.3	13	0.2	1	24.5	950
IV,V	20.1	84	3.6	15	0.2	1	23.9	577

† *310 dentate adults were excluded from the periodontal examination. See Section 3.3.1.*

Table 3.3.9   **The amount and degree of pocketing by reported dental behaviour**

Dentate adults†   *United Kingdom*

Reported dental behaviour	No pocketing of 4mm or more		Pocketing 4mm or greater		Pocketing 6mm or greater		Mean number of teeth	Base
	Mean number of teeth affected	Mean proportion of teeth affected	Mean number of teeth affected	Mean proportion of teeth affected	Mean number of teeth affected	Mean proportion of teeth affected		
All	21.7	87	3.1	12	0.1	0	25.0	3507
Usual reason for								
dental attendance								
Regular check up	22.2	88	2.8	11	0.1	0	25.1	2188
Occasional check up	23.5	89	2.8	10	0.1	0	26.5	386
Only with trouble	20.1	83	3.8	16	0.2	1	24.3	927
Time since								
last dental visit								
Less than 1 year	22.2	88	2.8	11	0.2	1	25.2	2620
Between 1 and 5 years	21.7	86	3.3	13	0.2	1	25.3	608
Over 5 years	18.2	79	4.7	20	0.2	1	23.1	273
Frequency of								
teeth cleaning								
Never/less than once a day	18.3	81	4.0	18	0.2	1	22.6	169
Once a day	20.1	84	3.8	16	0.2	1	23.9	741
Twice a day or more	22.4	89	2.8	11	0.1	0	25.2	1973

† 310 dentate adults were excluded from the periodontal examination. See Section 3.3.1.

Table 3.3.10 **The amount and degree of pocketing among those with pocketing of 4mm or more by age and gender**

Dentate adults with pocketing of 4mm or more   *United Kingdom*

	Pocketing of 4mm or more		Pocketing of 6mm or more		Mean number of teeth	Base
	Mean number of teeth affected	Mean proportion of teeth affected	Mean number of teeth affected	Mean proportion of teeth affected		
All	5.8	23	0.3	I	24.7	1837
Age						
16-34	5.6	20	0.2	I	28.4	509
35-54	6.0	24	0.2	I	25.6	846
55 and over	5.5	30	0.5	3	18.5	482
Gender						
Men	6.0	24	0.3	I	24.7	901
Women	5.5	22	0.2	I	24.6	936

Table 3.3.11 **The amount and degree of pocketing among those with pocketing of 6mm or more by age and gender**

Dentate adults with pocketing of 6mm or more   *United Kingdom*

	Pocketing of 4mm or more		Pocketing of 6mm or more		Mean number of teeth	Base
	Mean number of teeth affected	Mean proportion of teeth affected	Mean number of teeth affected	Mean proportion of teeth affected		
All	11.6	54	2.6	12	21.5	186
Age						
16-34	*	*	*	*	*	23
35-54	12.9	52	2.0	8	24.8	82
55 and over	9.4	55	2.6	15	17.0	81
Gender						
Men	11.7	57	3.0	14	20.6	94
Women	11.5	51	2.1	9	22.8	92

Table 3.3.12 **The proportion of teeth with different pocket depths**

Dentate adults†   *United Kingdom*

Level of pocketing	Upper Jaw															
	Left								Right							
	8	7	6	5	4	3	2	1	1	2	3	4	5	6	7	8
	%	%	%	%	%	%	%	%	%	%	%	%	%	%	%	%
No pocketing of 4mm or more	76	81	81	86	86	90	91	93	91	90	91	90	90	87	85	80
Pockets of 4mm or more	19	19	18	14	14	10	9	7	8	10	9	10	10	12	15	16
Pockets of 6mm or more	1	1	1	1	0	0	0	0	0	.1	0	1	0	1	1	1
Level of pocketing not measured	5	0	1	0	0	0	0	0	0	0	0	0	1	0	0	5
Base	1357	2726	2352	2592	2594	3168	3008	3132	3158	3018	3188	2532	2656	2351	2738	1429

Level of pocketing	Lower Jaw															
	Left								Right							
	8	7	6	5	4	3	2	1	1	2	3	4	5	6	7	8
	%	%	%	%	%	%	%	%	%	%	%	%	%	%	%	%
No pocketing of 4mm or more	65	80	81	85	88	92	94	94	93	93	91	88	84	78	76	64
Pockets of 4mm or more	27	20	18	14	12	8	5	6	6	7	8	12	15	21	23	28
Pockets of 6mm or more	1	1	1	1	0	1	0	0	0	0	0	1	1	1	1	1
Level of pocketing not measured	8	0	0	0	0	0	0	1	1	1	0	0	0	1	0	8
Base	1337	2649	2163	2816	3038	3436	3413	3368	3382	3389	3432	3029	2739	2214	2656	1334

† 310 dentate adults were excluded from the periodontal examination. See Section 3.3.1
Base excludes missing teeth

Table 3.3.13 **Degree of loss of attachment (LOA) by characteristics of dentate adults**

Dentate adults[†]   *United Kingdom*

Characteristics of dentate adults	No LOA of 4mm or more	LOA 4mm or more	LOA 6mm or more	Base
	percentage			
All	57	43	8	*3507*
Age				
16-24	86	14	0	*471*
25-34	74	26	2	*828*
35-44	58	42	3	*735*
45-54	48	52	10	*689*
55-64	30	70	17	*400*
65 and over	15	85	31	*384*
Gender				
Men	54	46	9	*1614*
Women	61	40	7	*1893*
Country				
England	56	44	8	*2007*
Wales	66	34	7	*459*
Scotland	56	44	8	*614*
Northern Ireland	61	39	6	*427*
English region				
North	55	45	7	*572*
Midlands	55	45	6	*452*
South	58	42	9	*983*
Social class of head of household				
I, II, IIINM	58	42	7	*1755*
IIINM	56	44	9	*950*
IV,V	53	47	11	*577*

† 310 dentate adults ere excluded from the periodontal examination. See Section 3.3.1.

Table 3.3.14 **Degree of loss of attachment (LOA) by reported dental behaviour**

Dentate adults†   *United Kingdom*

Reported dental behaviour	No LOA of 4mm or more	LOA 4mm or more	LOA 6mm or more	Base
		percentage		
All	57	43	8	3507
Usual reason for dental attendance				
Regular check up	57	43	7	2188
Occasional check up	61	39	5	386
Only with trouble	56	44	10	927
Time since last dental visit				
Less than 1 year	57	43	7	2620
Between 1 and 5 years	59	41	7	608
Over 5 years	50	50	14	273
Frequency of teeth cleaning				
Never/less than once a day	49	51	13	169
Once a day	51	49	11	741
Twice a day or more	59	41	7	2588

† 310 dentate adults were excluded from the periodontal examination. See Section 3.3.1.

Table 3.3.15 **The amount and degree of loss of attachment (LOA) by characteristics of dentate adults**

Dentate adults†   *United Kingdom*

Characteristics of dentate adults	No LOA of 4mm or more		LOA 4mm or more		LOA 6mm or more		Mean number of teeth	Base
	Mean number of teeth affected	Mean proportion of teeth affected	Mean number of teeth affected	Mean proportion of teeth affected	Mean number of teeth affected	Mean proportion of teeth affected		
All	22.5	90	2.4	10	0.0	0	25.0	3507
Age								
16-24	27.2	97	0.5	2	0.1	0	27.9	471
25-34	26.5	94	1.4	5	0.1	0	28.1	828
35-44	24.5	91	2.1	8	0.3	1	26.8	735
45-54	20.8	86	3.2	13	0.6	2	24.1	689
55-64	15.8	80	3.9	20	0.8	4	19.8	400
65 and over	11.9	70	5.1	30	0.2	1	17.1	384
Gender								
Men	22.2	88	2.7	11	0.3	1	25.2	1614
Women	22.7	91	2.0	8	0.2	1	24.9	1893
Country								
England	22.5	89	2.4	10	0.2	1	25.2	2007
Wales	22.9	93	1.5	6	0.2	1	24.5	459
Scotland	21.5	90	2.4	10	0.2	1	24.0	614
Northern Ireland	23.0	93	1.6	6	0.1	0	24.8	427
English Region								
North	22.3	90	2.3	9	0.2	1	24.8	572
Midlands	22.4	90	2.3	9	0.1	0	25.0	452
South	22.7	89	2.6	10	0.3	1	25.4	983
Social class of head of household								
I, II, IIINM	23.1	90	2.4	9	0.2	1	25.6	1755
IIINM	21.8	89	2.5	10	0.3	1	24.5	950
IV,V	20.9	87	2.7	11	0.3	1	23.9	577

† 310 dentate adults were excluded from the periodontal examination. See Section 3.3.1.

Table 3.3.16 **The amount and degree of loss of attachment (LOA) by reported dental behaviour**

Dentate adults[†]   *United Kingdom*

Reported dental behaviour	No LOA of 4mm or more		LOA 4mm or more		LOA 6mm or more		Mean number of teeth	Base
	Mean number of teeth affected	Mean proportion of teeth affected	Mean number of teeth affected	Mean proportion of teeth affected	Mean number of teeth affected	Mean proportion of teeth affected		
All	22.5	90	2.4	10	0.2	1	25.0	3507
Usual reason for dental attendance								
Regular check up	22.7	90	2.3	9	0.2	1	25.1	2188
Occasional check up	24.4	92	1.9	7	0.1	0	26.5	386
Only with trouble	21.2	87	2.7	11	0.4	2	24.3	927
Time since last dental visit								
Less than 1 year	22.8	90	2.3	9	0.2	1	25.2	2620
Between 1 and 5 years	22.8	90	2.3	9	0.3	1	25.3	608
Over 5 years	19.6	85	3.3	14	0.6	3	23.1	273
Frequency of teeth cleaning								
Never/less than once a day	18.9	84	3.4	15	0.3	1	22.6	169
Once a day	21.0	88	2.9	12	0.3	1	23.9	741
Twice a day or more	23.1	92	2.1	8	0.2	1	25.2	1973

* 310 dentate adults were excluded from the periodontal examination. See Section 3.3.1

Table 3.3.17 **The amount and degree of loss of attachment (LOA) among those with loss of attachment of 4mm or more by age and gender**

Dentate adults with loss of attachment of 4mm or more   *United Kingdom*

	LOA of 4mm or more		LOA of 6mm or more		Mean number of teeth	Base
	Mean number of teeth affected	Mean proportion of teeth affected	Mean number of teeth affected	Mean proportion of teeth affected		
All	5.5	24	0.5	2	22.8	1523
Age						
16-34	4.9	17	0.2	1	28.2	265
35-54	5.6	22	0.3	1	24.9	671
55 and over	5.8	33	0.9	5	17.7	587
Gender						
Men	5.8	25	0.6	3	23.0	783
Women	5.2	23	0.4	2	22.6	740

Table 3.3.18 **The amount and degree of loss of attachment (LOA) among those with loss of attachment of 6mm or more by age and gender**

Dentate adults with loss of attachment of 6mm or more   *United Kingdom*

	LOA of 4mm or more		LOA of 6mm or more		Mean number of teeth	Base
	Mean number of teeth affected	Mean proportion of teeth affected	Mean number of teeth affected	Mean proportion of teeth affected		
All	9.8	54	2.9	16	18.0	282
Age						
16-34	*	*	*	*	*	12
35-54	11.6	51	2.3	10	22.5	99
55 and over	8.5	56	3.0	20	15.2	171
Gender						
Men	9.4	54	3.1	18	17.6	162
Women	10.3	55	2.5	14	18.7	120

Table 3.3.19 **The proportion of teeth with loss of attachment**

Dentate adults† *United Kingdom*

Level of loss of attachment	Upper Jaw															
	Left								Right							
	8	7	6	5	4	3	2	1	1	2	3	4	5	6	7	8
	%	%	%	%	%	%	%	%	%	%	%	%	%	%	%	%
No loss of attachment of 4mm or more	83	86	85	88	89	91	93	94	93	93	92	90	90	87	87	84
Loss of attachment of 4mm or more	12	14	14	11	11	8	7	6	6	7	8	9	10	13	12	11
Loss of attachment 6mm or more	1	1	2	1	0	1	1	1	1	1	1	0	0	1	1	0
Loss of attachment not measured	5	0	1	0	0	0	0	0	0	0	0	1	1	0	1	5
Base	1357	2726	2352	2592	2594	3168	3008	3132	3158	3018	3188	2532	2656	2351	2738	1429

Level of loss of attachment	Lower Jaw															
	Left								Right							
	8	7	6	5	4	3	2	1	1	2	3	4	5	6	7	8
	%	%	%	%	%	%	%	%	%	%	%	%	%	%	%	%
No loss of attachment of 4mm or more	80	87	89	89	90	92	93	92	92	92	92	91	90	88	86	77
Loss of attachment of 4mm or more	12	12	10	11	9	7	7	7	8	7	8	9	10	11	13	15
Loss of attachment 6mm or more	0	1	1	1	1	1	1	1	1	1	1	1	1	1	1	0
Loss of attachment not measured	8	0	0	0	0	0	0	1	1	1	0	0	0	1	0	8
Base	1337	2649	2163	2816	3038	3436	3413	3368	3382	3389	3432	3029	2739	2214	2656	1334

† 310 dentate adults were excluded from the periodontal examination. See Section 3.3.1.
Base excludes missing teeth

Table 3.3.20 **Periodontal conditions around the sextants of the mouth**

Dentate adults†   *United Kingdom*

Periodontal condition	Sextant					
	Upper left 8-4	Upper central 3-3	Upper right 4-8	Lower left 8-4	Lower central 3-3	Lower right 4-8
	%	%	%	%	%	%
LOA 4mm or more	38	26	32	38	20	38
LOA 6mm or more	3	2	2	2	3	2
Pocketing 4mm or more	31	21	24	32	15	33
Pocketing 6mm or more	2	1	2	2	1	2
Base	3262	3330	3249	3401	3490	3390

† 310 dentate adults were excluded from the periodontal examination. See Section 3.3.1.
Base excludes those with no teeth in the relevant sextant.

Part 4

Social and behavioural characteristics and oral health

4.1 Social and behavioural characteristics and oral health

Summary

- Multivariate modelling techniques were used to analyse the inter-relationships between adults' socio-demographic characteristics, their dental behaviour and a range of clinical measures of oral health.
- Age was found to be the most significant variable in explaining the variation in the majority of measures of oral health. For example, adults aged 75 years and over were 144 times more likely to be edentate than adults aged 16 to 44 years and dentate adults aged 45 to 54 years were over 60 times more likely to have 12 or more restored (otherwise sound) teeth compared with those aged 16 to 24 years.
- The effect of age on whether or not respondents had any decayed or unsound teeth or any unrestorable teeth was not as large as for other measures of oral health.
- Other socio-demographic characteristics were also found to have independent effects on oral health. Social class of head of household or educational attainment or both were also found to be independently related to all the measures of oral health used. Region, gender, marital status and economic status were found to have an effect in relation to some of the measures investigated. The effects of all these socio-demographic factors were fairly small compared with the effects of age.
- One or more aspects of dental behaviour were found to have an effect for all the clinical measures of the oral health investigated except periodontal loss of attachment of 4mm or more. The specific behaviours that were significant varied according to which particular aspect of oral health was being analysed.
- The usual reason for dental attendance was the factor most strongly associated with dentate adults having unrestorable teeth; dentate adults who only went to the dentist with trouble were over five times more likely to have unrestorable teeth than those who went for regular check-ups.
- The reported frequency of tooth cleaning and whether the respondent had fissure sealants placed on their teeth were independently associated with the number of sound and untreated teeth among younger dentate adults (aged 16 to 34).

4.1.1 Introduction

The previous chapters (in Parts 2 and 3) have described the variations in a number of clinical measures of oral health with regard to age, gender, social class, geographical area (English region or country in the United Kingdom) and usual reason for dental attendance. The relationships between each of these characteristics and oral health were investigated independently of any other factors. However complex inter-relationships exist between socio-demographic characteristics, dental behaviour and oral health. Age was strongly associated with many of these clinical measures of oral health. Over time, an adult accumulates experience of dental disease and restorative treatment that affects their current oral health. People's oral health also varied according to their other socio-demographic characteristics, but these too can vary with age, especially social class and educational attainment. Part 6 shows that dental behaviour, such as attendance at the dentist or frequency with which adults clean their teeth, also vary with age and other socio-demographic characteristics. The combination of these characteristics and dental behaviour may have an impact on adults' experience of dental disease and restorative treatment and could also have a cumulative effect on their oral health over their lifetime.

This chapter will report on the interaction between age, other socio-demographic characteristics and dental behaviour on the following range of clinical measures of oral disease and treatment used in previous chapters:

- whether or not adults had any natural teeth
- the number of teeth dentate adults had
- the initial experience of dental caries among young dentate adults (aged 16 to 34 years)
- dentate adults with some decayed or unsound teeth
- dentate adults with some unrestorable teeth
- dentate adults with 12 or more restored (otherwise sound teeth)
- dentate adults with artificial crowns and
- dentate adults with periodontal loss of attachment of 4mm or more.

4.1.2 Analysis techniques

Various different methods of analysis are available for investigating the relationships between a number of different variables. This chapter presents data on the clinical measures of oral health analysed using age standardisation and two multivariate modelling techniques (see Appendix F for more details on the techniques used).

Age standardisation was used before any modelling was carried out to investigate whether the measures of oral health varied with respect to social class and educational attainment once age had been taken account. As previously mentioned a strong relationship exists between age and many measures of dental health and some social characteristics, especially social class and educational attainment. It is therefore important to take age into account when investigating the relationship between dental health and these characteristics. Age-standardised values for each clinical measure were calculated and showed that each of the oral health measures varied with regard to social class and educational attainment independently of age. Age standardisation was carried out as a precursor to the multivariate modelling so the results of the age standardisation for each measure of oral health are not discussed separately in this chapter, but tables showing the age standardised values are presented together with the tables showing the results of the modelling for that measure.

Multivariate modelling was carried out to identify which socio-demographic and behavioural factors (independent variables) were independently associated with the clinical measures of dental health (dependent variables) once all the other factors in the models had been taken into account. Two different modelling techniques were used to analyse the data. Multiple regression was used to analyse the variations of a continuous dependent variable, such as the number of teeth present in the mouth, and logistic regression was used to analyse dependent variables which had only two possible outcomes, for example whether or not the respondent had any artificial crowns (see Appendix F).

In general, the same socio-demographic and behavioural factors were used in each of the regression models. A number of socio-demographic characteristics were collected during the interview, but not all them were entered into the models. Analysis which is not reported here in detail showed that variations in adult dental health with respect to vehicle ownership, household income and social class were all very similar. This is because certain social characteristics are related to one other; for example adults living in households with low incomes were less likely to have a car and were more likely to live in households headed by someone with a manual occupation. Social class was used in the modelling rather than income or vehicle ownership to maintain consistency with the rest of the report. The socio-demographic factors used in the modelling were:

- age
- gender
- social class of head of household
- educational attainment

- country or English region (the three English regions were entered into the models separately, so overall values for England were not compared with the oral health of those from the other countries)
- marital status and
- economic status.

For the models relating to the oral health of dentate adults the following behavioural factors were also entered:

- the usual reason for dental attendance
- the reported frequency of teeth cleaning and
- whether or not respondents used any dental hygiene products other than an ordinary toothbrush and toothpaste.

The results of any analysis using modelling techniques are always limited by the number and characteristics of the factors entered into the model. The results account for a part but not all of the variation in the dependent variable. It is likely that there are other factors, not in the model, which could affect a person's oral health. Such factors could not always be measured by the survey, for technical reasons or because it would not have been cost-effective to do so. For example, the design and sample size means that any effect of fluoridation of water on adults' oral health could not be analysed. The amount of variation which is explained by the model can be described using the R^2 statistic. The closer this statistic is to 1, the better fitting the model. The tables show the R^2 statistic for each of the models used in the analysis.

4.1.3 Adults who were edentate

Overall 13% of adults within the United Kingdom had lost all of their natural teeth. Chapter 2.1 showed the proportion of adults who had lost all their natural teeth varied by age, gender, geographical area and social class. Logistic regression was carried out to identify factors independently associated with adults losing all their natural teeth. The analysis was confined to the effects of socio-demographic characteristics as the questions on dental behaviour were different for dentate and edentate adults. Very few adults under the age of 45 had lost their teeth, so they were treated as one category in this model.

After taking account of the other factors in the model, age had the strongest association with becoming edentate, as one would expect given the information from previous chapters. The odds of being edentate increased dramatically with age. The odds of being edentate were almost eleven times higher in the 45 to 54 years age group than those aged 16 to 44 (the reference category). Adults aged 75 and over were 144 times

more likely to be edentate compared with adults aged between 16 and 44.

Once all other factors in the model had been taken into account, educational attainment had the next largest effect after age on the odds of becoming edentate. Compared with adults who had qualifications at degree level or above the odds of being edentate were almost nine times higher for those adults with no qualifications and almost four times higher for adults with qualifications below degree level (OR= 8.79 and 3.95 respectively).

The analysis in Chapter 2.1 identified that adults from the South of England were the least likely to be edentate. The logistic regression also identified region as being independently associated with being edentate. Adults from the other areas within Great Britain had higher odds of being edentate compared with adults from the South of England. Adults from Scotland had the largest increased odds (OR=3.46). However, once all other factors had been taken into account the odds of adults from Northern Ireland being edentate were not significantly different from adults living in the South of England.

The odds of being edentate for adults from manual households were over twice those of adults from non-manual households. (OR = 2.14 for adults from skilled manual households and 2.21 for adults from partly skilled and unskilled households).

Adults' marital status also had an independent association with becoming edentate. All other factors in the model being equal, adults who were single or widowed have slightly higher odds of becoming edentate compared with adults who were married or cohabiting (OR=1.45 for single adults and 1.62 for those who were widowed).

Table 2.1.4 showed that women aged 65 and over were more likely than men to have lost all their natural teeth. However, in the modelling when other factors were taken into account, gender was not found to be significant. The gender differences observed in the earlier bivariate analysis may, in part, be explained by other factors, particularly marital status. In general, women tend to live longer than men and are therefore more likely to be widowed. Women also account for a larger proportion of all adults in the older age groups. Most widowed adults aged 65 and over are women. The results from the logisitic regression showed that all other things being equal, adults who were widowed (and therefore more likely to be women) had higher odds of being edentate.

Figure 4.1.1, Tables 4.1.1–2

The modelling showed that all the relationships investigated separately in Chapter 2.1 were found to have an independent association with total tooth loss with the exception of gender. These relationships probably reflect a broad and complex set of social and behavioural interactions rather than simple differences in the risk of disease or treatment choices that ultimately resulted in an adult losing all of their natural teeth.

4.1.4 The number of natural teeth present in the mouth

The number of natural teeth retained in the mouth is an important aspect of oral health. People with a small number of teeth are at risk of becoming edentate in the future. Moreover, a minimum number of teeth are required for adequate oral function without the need for partial dentures (see Chapter 2.3).

Multiple regression modelling was carried out to find out which combinations of socio-demographic and behavioural factors independently affected the number of teeth among dentate adults. Table 4.1.4 show the coefficients for all the variables entered into the regression model. The coefficients show the predicted difference from the overall average number of teeth for each factor once all other factors in the model have been taken into account. The coefficients can be used to estimate the number of teeth of dentate adults with particular characteristics, by adding the coefficients for each relevant characteristic to the constant calculated by model. (More information on the modelling technique can be found in Appendix F). For example, a 32-year-old woman from a non-manual background, living in Scotland, who was married, had a degree and who brushed her teeth twice a day would have a predicted number of 27.9 teeth:

22.40	(constant)
+ 3.79	(being aged 25 to 34)
+ 0	(women)
- 0.85	(Scotland)
+ 0.64	(non-manual households)
+ 0.99	(degree)
+ 0.40	(married)
+ 0.54	(cleans teeth twice a day)
= 27.9	

Age accounted for the largest differences in the predicted number of teeth. Taking into account all other factors in the model, being aged 16 to 24 years and 25 and 34 years increased the predicted number of teeth by almost four, whereas being aged 65 and over decreased the predicted number of teeth by almost six.

Other socio-demographic and behavioural characteristics contributed to the variation in the number of teeth, but to a much lesser degree than age. The characteristics that significantly increased the predicted number of teeth were:

- having qualifications at degree level or above (0.99);
- living in the Midlands or the South of England (0.65 and 1.00 respectively);
- living in a non-manual household (0.64);
- being male (0.42);
- being married or cohabiting (0.40);
- going to the dentist for a regular check-up (0.54);
- cleaning one's teeth twice a day or more (0.54).

Conversely, adults who had no qualifications, who lived in Scotland, who were economically inactive, who lived in partly skilled or unskilled households, who only went to the dentist when they had trouble with their teeth or who cleaned their teeth less than once a day were all likely to have fewer teeth when all other factors were taken into consideration.

Whether or not respondents used dental hygiene products other than a toothbrush and toothpaste was entered into the model but not found to to have an independent effect on the number of teeth among dentate adults.

Tables 4.1.3–4

4.1.5 Initial experience of dental caries among young dentate adults

Describing the characteristics related to the risk of developing dental disease in the relatively recent past will give an indication as to which factors contribute to the risk of developing disease and suggest which sections of the population would most benefit from preventive measures.

The following analysis shows the social and behavioural characteristics associated with the number of sound and untreated teeth among adults aged under 35. These adults have relatively low experience of dental disease and restorative treatment and have, on the whole, benefited from contemporary concepts of dental disease management and restoration techniques. They also have, on average, a large number of natural teeth. The analysis was restricted to these younger adults in order to exclude some of the confounding historical effects of dental treatment on the ability to measure disease accurately.

The number of sound and untreated teeth shows how many teeth remain unaffected by dental caries and can therefore be used to indicate the relative susceptibility to caries. During the survey examination no distinction was made between unerupted teeth (unerupted wisdom teeth will have been quite common in younger respondents) and teeth that were lost through extraction. Nothing is known about why the teeth were lost and the condition of the teeth when they were removed. As a consequence the number of sound and untreated teeth as determined during the home dental examination is not a perfect indicator of caries susceptibility as healthy teeth that have been extracted for orthodontic purposes or remain unerupted cannot be taken into account.

Table 4.1.5 shows the distribution of sound and untreated teeth for young adults. Three-quarters (74%) of all dentate adults aged under 35 have 18 or more sound and untreated teeth. This group had, on average, 20.8 sound and untreated teeth. Even among these younger adults differences in the number of sound and untreated teeth with regard to age were apparent. On average, adults aged 16 to 24 years had 23.4 sound and untreated teeth compared with an average of 19.1 among adults aged 25 to 34 years.

Table 4.1.5

A multiple regression was carried out to find out which combination of socio-demographic and behavioural factors were independently associated with the number of sound and untreated teeth among young dentate adults. Table 4.1.6 shows the coefficients from the regression model for the variables entered into the model. Age accounted for the biggest differences in the predicted number of sound and untreated teeth. All other things being equal, being aged between 16 and 19 years increased the predicted number of sound and untreated teeth by almost two and half (2.38); being aged between 30 and 34 years decreased the predicted number by almost three (-2.96).

After age, region had the next largest independent effect on the predicted number of sound and untreated teeth. Living in the Midlands or the South of England increased the predicted number of sound and untreated teeth (by 2.20 and 1.48 respectively) whereas living in Scotland or Northern Ireland decreased the number by over two (-2.15 and -2.38 respectively). Living in either the North of England or Wales did not alter the predicted number of sound and untreated teeth. This pattern, of the South of England and the Midlands having lower disease risk, the North of England and Wales having moderate disease risk and Scotland and Northern Ireland having the greatest risk strongly suggests differences in lifestyle or environmental factors according to region. Many of the potential behavioural factors are taken into account in the model, but other regional differences such as diet, the use of fluoridated toothpaste or fluoridation of water supplies were not measured in this survey and so could not be taken into account in the model.

Of the other socio-demographic characteristics entered into the model: social class, educational attainment and marital status were found to have an association with the number of sound and untreated teeth of young dentate adults. The effects of these variables, and therefore the coefficients, were smaller than those found for either age or region. Having qualifications at degree level or above or being from a non-manual background increased the predicted number of sound and untreated teeth. Taking all other factors into account, young adults who were divorced or separated had fewer sound and untreated teeth and those married or cohabiting had more sound and untreated teeth. Gender and economic status were entered into the model but were found not to be statistically significant using this model.

Behavioural characteristics were entered into the model as well as socio-demographic ones. Once all other factors had been taken into account, the frequency with which adults cleaned their teeth affected the predicted number of teeth. Adults who cleaned their teeth less than once a day had fewer sound and untreated teeth (coefficient of -1.75), whereas cleaning teeth once a day or more had a positive effect and increased the predicted number of sound and untreated teeth by almost one.

Whether or not the respondent had fissure sealants was entered as an extra factor into the model for the number of sound and untreated teeth. Fissure sealants are placed on teeth to prevent the onset of dental decay. All other things in the model being equal, young dentate adults with sealants had a predicted number of sound and untreated teeth 1.32 higher than those who did not have any sealants. This finding suggests that, at population level, sealants play an effective role in preventing disease at this age.

The usual reason for attending the dentist and whether or not people used other dental hygiene products were entered into the model but were not found to have an independent effect on the number of sound and untreated teeth among young dentate adults.

Figure 4.1.2, Table 4.1.6

4.1.6 Dentate adults with decayed or unsound teeth

It was not possible to treat the number of decayed or unsound teeth as a continuous variable as a large proportion (45%) of dentate adults had no decayed or unsound teeth and, in general, those adults with decayed or unsound teeth had a small number of such teeth. Thus logistic regression modelling was used to investigate the socio-demographic and behavioural characteristics independently associated with whether or not dentate adults had some decayed or unsound teeth. This gives an indication of the current treatment needs among dentate adults irrespective of the amount of treatment they have had in the past. Adults with decayed or unsound teeth may have teeth with previously untreated decay, teeth that have been decayed, restored and have suffered further decay, or teeth with damaged restorations that may or may not need repair. Although this measure gives a strong indication of current treatment need it may not necessarily be a precise indicator of the risk of developing new disease, as the presence of a decayed or unsound tooth will partly reflect the tendency of an individual to seek regular dental treatment.

Many aspects of oral health show marked variations with age, but this is not the case with the prevalence of decayed and unsound teeth. This finding was confirmed by the logistic regression modelling. Once all other factors in the model had been taken into account, only two age groups, adults aged 25 to 34 and 45 to 54, had significantly higher odds (OR=1.74 and 1.52 respectively) of having decayed or unsound teeth than adults aged 16 to 24 (the reference group).

Other social and personal characteristics were independently associated with having at least one decayed or unsound tooth. The groups of people that had increased odds of having decayed or unsound teeth were:

● men (OR=1.23);
● adults from partly skilled or unskilled households (OR=1.36) compared with adults from non-manual households;
● adults with no qualifications (OR=1.41) compared with adults who had qualifications at degree level or above;
● adults from the North of England (OR=1.86), Scotland (OR=1.40) compared with adults from the South of England;
● adults who were divorced or separated (OR=1.34) or widowed (OR=1.47) compared with adults who were married or cohabiting.

Of the behavioural factors entered into the logistic regression only one was independently associated with having decayed or unsound teeth: the usual reason for attending the dentist. The odds of having decayed or unsound teeth for adults who only went to the dentist when they had trouble with their teeth were nearly twice (OR=1.83) the odds of those who went to the dentist for a regular check-up. This confirmed that having decayed or unsound teeth reflected an individual's tendency to seek dental treatment.

Factors not found to be significant by the model were economic status, reported frequency of teeth cleaning and the use of other dental hygiene products other than a toothbrush and toothpaste.

Tables 4.1.7–8

4.1.7 Dentate adults with unrestorable teeth

Over time teeth can become so decayed that, if left untreated, they become unrestorable. Eight per cent of all dentate adults had at least one tooth deemed by the examiner to be unrestorable. Whether or not a dentate adult has any unrestorable teeth is likely to be more useful as an indicator of oral neglect than of decay risk.

Logistic regression was used to investigate which factors were independently associated with having unrestorable teeth. The factor that had the strongest association with having unrestorable teeth was the usual reason for going to the dentist. Dentate adults who only went to the dentist when they had trouble with their teeth had odds five times higher than those who went to the dentist for a regular check-up.

Unlike having decayed or unsound teeth which generally did not show much variation with age, the odds of having unrestorable teeth increased with age even when other factors had been taken into account. The increased odds of having unrestorable teeth, by comparison with dentate adults aged 16 to 24 years, varied from under two for dentate adults aged 25 to 34 years and 35 to 44 years (OR=1.90 and 1.79 respectively) to 3.25 for those aged 65 and over.

The previous section showed that men had higher odds of having any decayed or unsound teeth compared to women and this section shows they also had higher odds of having some unrestorable teeth (OR=1.71). However the differences by geographical area were different from those with respect to decayed and unsound teeth. The only region to show any statistically significant difference to adults from the South of England in the odds of having unrestorable teeth was the Midlands, where adults were less likely to have them (OR=0.54). Other characteristics that increased the likelihood of having unrestorable teeth were having no qualifications (OR=1.82) and living in a household headed by someone with a manual occupation (OR=1.50 for skilled manual households and OR=1.84 for partly skilled and unskilled households).

The other behavioural characteristic that affected the odds of having unrestorable teeth was whether or not the respondent used dental hygiene products other than a toothbrush and toothpaste, such as dental floss or mouthwash. All other things being equal, if the respondent did not use other dental hygiene products they had slightly higher odds of having unrestorable teeth than those who used these products (OR=1.35).

Factors not found to be significant by the model were economic status, marital status and reported frequency of teeth cleaning.

Figure 4.1.3, Tables 4.1.7–8

4.1.8 Restorative treatment among dentate adults

If a tooth has suffered from dental caries it may be restored to prevent further decay. However, over time, the tooth may suffer further decay and be subject to more restorative work. Some teeth that have been subject to years of restorative treatment may eventually be lost. The extent and type of restorations found will reflect treatment provided over the respondent's lifetime. Chapter 3.2 describes in detail the variations in different amount and types of restorative treatment. Logistic regression was used to analyse the social and behavioural characteristics associated with the receipt of restorative dental care among dentate adults using two measures: having 12 or more restored (otherwise sound) teeth and having artificial crowns. These measures may also give insight into some of the factors which act as barriers to the receipt of effective dental care.

Age had the largest effect on increasing the odds of a dentate adult having 12 or more restored (otherwise sound) teeth. Dentate adults aged 45 to 54 were the most likely to have 12 or more restored (otherwise sound) teeth. The odds of dentate adults in this age group were over 60 times those of dentate adults aged 16 to 24. The odds of having 12 or more restored (otherwise sound) teeth decreased above the age of 55, but older adults were still much more likely to have 12 or more restored (otherwise sound) teeth than the youngest dentate adults. Dentate adults aged 65 and over were over 25 times more likely to have 12 or more restored (otherwise sound) teeth than those aged 16 to 24.

Once other factors including age had been taken into account, the usual reason for attending the dentist had the strongest association with the odds of having 12 or more restored (otherwise sound) teeth. Dentate adults who went to the dentist for check-ups, either occasionally or regularly, were more likely to have 12 or more restored (otherwise sound) teeth than dentate adults who only went to the dentist when they were having trouble with their teeth. If the respondent stated they went for an occasional check-up the odds of them

having 12 or more restored (otherwise sound) teeth were twice (OR=2.07) those of people who only went to the dentist when in trouble. The odds were even higher for people who went for a regular check-up (OR=2.74).

Social class and educational attainment were also found to be independently associated with the odds of having 12 or more restored (otherwise sound) teeth. All other factors in the model being equal, the odds of having 12 or more were higher for those living in non-manual households (OR=1.83) or skilled manual households (OR=1.34) compared with those who lived in households with a partly-skilled or unskilled head of household. Dentate adults with qualification at degree level or above were more likely to have 12 or more restored (otherwise sound) teeth (OR= 1.40) than dentate adults with no qualifications (the reference category).

Dentate adults living in the North had lower odds of having 12 or more restored (otherwise sound) teeth (OR=0.72) compared with dentate adults in the South of England (the reference category). The odds of dentate adults from Northern Ireland having 12 or more restored (otherwise sound) teeth were almost twice (OR=1.89) those of dentate adults from the South of England.

Other socio-demographic characteristics were also found to be independently associated with having 12 or more restored (otherwise sound) teeth. Once other factors in the model had been controlled for, dentate adults who were unemployed had lower odds of having 12 or more restored (otherwise sound) teeth (OR=0.49) and dentate adults in part-time employment had higher odds (OR=1.24) than dentate adults in full-time employment (the reference category). Dentate adults who were divorced or separated had lower odds of having 12 or more restored (otherwise sound) teeth than those who were married or cohabiting.

Using other dental hygiene products (in addition to a toothbrush and toothpaste) was independently associated with the odds of having 12 or more restored (otherwise sound) teeth. All other things in the model being equal, dentate adults who used dental hygiene products had slightly higher odds (OR=1.27) of having 12 or more restored (otherwise sound) teeth compared with dentate adults who did not use them (the reference category). An explanation for this may be that dentate adults who have suffered decay and have had it treated may use dental hygiene products other than a toothbrush and toothpaste in an attempt to avoid further decay occuring.

Factors not found to be significant by the model were gender and reported frequency of teeth cleaning.

Logistic regression modelling was also used to identify the factors that were independently associated with having artificial crowns. Of all the factors entered into the model only four were found to be significant. Again, age had the largest effect in increasing the odds of having artificial crowns; dentate adults aged 45 to 54 years and 55 to 64 years had odds almost fifteen times higher than those for dentate adults aged 16 to 24. The other factors that increased the odds of having artificial crowns were having qualifications at degree level or above; going to the dentist for a regular check-up; and using dental hygiene products other than a toothbrush and toothpaste. Although there were regional differences in the odds of having 12 or more restored (otherwise sound) teeth, region was not found to be independently associated with having artificial crowns.

Tables 4.1.9–10

4.1.9 Dentate adults with periodontal disease

The condition of the gums and supporting structures are important aspects of oral health, as severe periodontal disease can ultimately lead to the loss of otherwise healthy teeth. Chapter 3.3 has described in detail the variations in periodontal conditions. This section describes the results of the logistic regression modelling carried out to identify the behavioural and social factors independently associated with having some periodontal loss of attachment of 4mm or more.

In general periodontal conditions, including loss of attachment, tend to affect people as they get older. The results of the modelling showed that age was the factor most strongly associated with having loss of attachment of 4mm or more. The odds of dentate adults aged 65 or more having loss of attachment were 33 times greater than the odds for dentate adults aged 16 to 24. Educational attainment, gender and region were the other factors that were associated with dentate adults having loss of attachment, although their effects were much smaller than effects due to age. All other things being equal, dentate adults with no qualifications or qualifications below degree level were more likely than dentate adults with qualifications at degree level or above to have loss of attachment of 4mm or more (OR = 1.74 and 1.35 respectively). Men were one and a half times more likely to have this condition than women. Chapter 3.3 showed that men were more likely to have plaque than women and as the bacteria within dental plaque have been implicated in periodontal disease this may in part account for men having higher odds of having loss of attachment.

Dentate adults living in two areas were found to have significantly different odds of having some loss of attachment compared with dentate adults from the South of England; dentate adults from the North of England has slightly higher odds (OR=1.30) whereas dentate adults from Wales had slightly lower odds (OR=0.65). These geographical variations should be interpreted with caution as they may reflect differences in the way in which individual dental examiners assessed the periodontal condition of adults' mouths rather than actual regional differences in the prevalence of periodontal disease.

Once all the other factors had been taken into consideration, none of the behavioural factors entered into the model were found to be independently associated with dentate adults having some periodontal loss of attachment of 4mm or more. Social class of head of household, economic status and marital status were also not found to be significant by the model.

Tables 4.1.11–12

Table 4.1.1 Edentate adults by social class of head of household, education and age with age-standardised ratios

All adults   *United Kingdom*

Characteristic	Age							All	Age standardisation		
	16-24	25-34	35-44	45-54	55-64	65-75	75 and over		Expected %	Ratio[††]	Standard error
					percentage edentate						
Social class of head of household											
I, II, IIINM	0	0	0	3	7	27	43	8	12	64 *	3.8
IIIM	0	1	2	9	28	40	61	15	12	121 *	5.9
IV, V	0	0	2	10	35	49	70	22	16	141 *	6.6
Educational attainment											
No qualifications	0	2	3	14	30	46	66	32	25	129 *	3.3
Qualifications below degree	0	0	1	5	14	28	41	6	8	73 *	4.4
Qualifications at degree level or above	0	0	0	2	4	4	-	1	7	18 *	6.4
All[†]	0	0	1	6	20	36	58	13			
					Base						
Social class of head of household											
I, II, IIINM	279	521	530	499	345	338	220	2732			
IIIM	176	339	305	322	249	220	118	1729			
IV, V	133	205	181	192	196	170	144	1221			
Educational attainment											
No qualifications	63	136	211	289	375	422	354	1850			
Qualifications below degree	574	795	716	637	368	279	154	3523			
Qualifications at degree level or above	64	226	187	180	85	53	24	819			
All[†]	702	1158	1114	1106	833	758	533	6204			

† Includes those for whom social class or education level were not known.

†† The standardised ratio is given as the observed rate divided by the expected rate and the ratio multiplied by 100. A 2-tailed test was used to show whether the ratio was significantly different from 100.

* Significant at 95% level.

- Percentage not shown as base is under 30.

Table 4.1.2   **Likelihood of an adult having lost all their natural teeth (based on odds ratios from logistic regression)**

All adults   *United Kingdom*

Characteristic	Odds ratios		95% confidence intervals	
			Lower	Upper
Age				
16-44	1.00		..	..
45-54	10.77	***	6.18	18.78
55-64	34.05	***	20.07	57.77
65-74	71.95	***	42.36	122.21
75 and over	144.05	***	83.28	249.15
Gender				
Men	-		-	-
Women	-		-	-
English region or country				
North	2.02	***	1.55	2.62
Midlands	1.59	***	1.22	2.07
South	1.00		..	..
Wales	1.92	**	1.25	2.96
Scotland	3.46	***	2.44	4.89
Northern Ireland	1.47		0.77	2.81
Social class of head of household				
I, II, IIINM	1.00		..	..
IIIM	2.14	***	1.68	2.72
IV, V	2.21	***	1.71	2.86
Education attainment				
At degree level or above	1.00		..	..
Below degree level	3.95	***	1.97	7.91
No qualifications	8.79	***	4.38	17.63
Economic status				
Employed full time	-		-	-
Employed part time	-		-	-
ILO unemployed	-		-	-
Economically inactive	-		-	-
Marital status				
Single	1.45	*	1.00	2.11
Married/cohabiting	1.00		..	..
Divorced/separated	1.43		0.99	2.07
Widowed	1.62	***	1.24	2.11
Nagelkerke R^2	0.50			

* *Significant at 95% level, ** significant at 99% level, *** significant at 99.9% level.*

- *Factors were available to be entered into the model but were not found to be significant, therefore odds ratios were not calculated.*

Table 4.1.3 **Dentate adults with 21 or more teeth by social class of head of household, educational attainment and age with age-standardised ratios**

Dentate adults   *United Kingdom*

Characteristic	Age						All	Age standardisation		
	16-24	25-34	35-44	45-54	55-64	65 and over		Expected %	Ratio[††]	Standard error
percentage with 21 or more teeth										
Social class of head of household										
I, II, IIINM	100	99	98	89	68	47	86	82	105 *	0.8
IIIM	100	98	94	72	52	33	79	83	96 *	1.3
IV, V	98	99	85	75	37	25	73	83	91 *	1.8
Educational attainment										
No qualifications	100	94	86	64	42	29	59	69	85 *	2.1
Qualifications below degree	100	99	95	84	62	44	88	87	102 *	0.6
Qualifications at degree										
level or above	100	100	98	93	79	59	93	85	109 *	1.4
All[†]	100	98	94	82	57	38	83			
Base										
Social class of head of household										
I, II, IIINM	*215*	*402*	*395*	*394*	*247*	*273*	*1926*			
IIIM	*116*	*254*	*214*	*204*	*115*	*122*	*1025*			
IV, V	*90*	*148*	*121*	*107*	*85*	*74*	*625*			
Educational attainment										
No qualifications	*43*	*87*	*134*	*148*	*173*	*221*	*806*			
Qualifications below degree	*395*	*585*	*502*	*453*	*226*	*199*	*2360*			
Qualifications at degree										
level or above	*53*	*181*	*145*	*145*	*62*	*63*	*649*			
All[†]	*491*	*854*	*781*	*746*	*461*	*484*	*3817*			

† *Includes those for whom social class or education level were not known.*

†† *The standardised ratio is given as the observed rate divided by the expected rate and the ratio multiplied by 100. A 2-tailed test was used to show whether the ratio was significantly different from 100.*

* *Significant at 95% level.*

Table 4.1.4 **Factors independently affecting the number of teeth based on coefficients from multiple regression**

Dentate adults *United Kingdom*

Characteristic	Coefficient	95% confidence intervals	
		Lower	Upper
Age			
16-24	3.90 ***	3.44	4.36
25-34	3.79 ***	3.46	4.11
35-44	2.39 ***	2.06	2.72
45-54	-0.30	-0.65	0.05
55-64	-3.95 ***	-4.35	-3.54
65 and over	-5.82 ***	-6.31	-5.33
Gender			
Men	0.42 *	0.07	0.77
Women	0.00	..	..
English region or country			
North	-0.31	-0.66	0.03
Midlands	0.65 **	0.28	1.02
South	0.97 ***	0.67	1.26
Wales	-0.05	-0.66	0.55
Scotland	-0.85 **	-1.35	-0.36
Northern Ireland	-0.39	-1.20	0.41
Social class of head of household			
I, II, IIINM	0.64 ***	0.41	0.86
IIIM	-0.20	-0.43	0.04
IV, V	-0.44 **	-0.72	-0.16
Education attainment			
At degree level or above	0.99 ***	0.69	1.30
Below degree level	0.13	-0.08	0.35
No qualifications	-1.13 ***	-1.43	-0.82
Economic status			
Employed full time	0.14	-0.18	0.47
Employed part time	0.07	-0.28	0.43
ILO unemployed	0.23	-0.46	0.93
Economically inactive	-0.45 *	-0.82	-0.07
Marital status			
Single	0.14	-0.28	0.55
Married/cohabiting	0.40 **	0.10	0.69
Divorced/separated	0.01	-0.47	0.49
Widowed	-0.54	-1.17	0.09
Reason for dental attendance			
Regular check-up	0.54 ***	0.31	0.77
Occasional check-up	0.16	-0.17	-0.49
Only with trouble	-0.70 ***	-0.96	-0.44
Frequency of teeth cleaning			
Twice a day or more	0.54 **	0.22	0.87
Once a day	0.31	-0.03	0.66
Less than once a day	-0.86 **	-1.40	-0.31
Use other dental hygiene products			
Yes	0.04	-0.27	0.36
No	0.00	..	..
Constant	22.40	21.88	22.92
Adjusted R^2	0.44		

* *Significant at 95% level,* ** *significant at 99% level,* *** *significant at 99.9% level.*

Table 4.1.5 **Distribution of the number of sound and untreated teeth by age**

Dentate adults aged 16-34 *United Kingdom*

Number of sound and untreated teeth	Age		All
	16 – 24	25 – 34	
	%	%	%
0	0	0	0
1 – 5	1	1	1
6 – 11	2	9	6
12 – 17	8	25	18
18 – 23	33	42	38
24 or more	57	23	36
Mean	23.4	19.1	20.8
Base	*491*	*854*	*1345*

Table 4.1.6   **Factors independently affecting the number of sound and untreated teeth (based on coefficients from multiple regression)**

Dentate adults aged 16-34   *United Kingdom*

Characteristic	Coefficient	95% confidence intervals	
		Lower	**Upper**
Age			
16-19	2.38 ***	1.72	3.05
20-24	1.82 ***	1.29	2.36
25-29	-1.25 ***	-1.76	-0.74
30-34	-2.96 ***	-3.47	-2.44
Gender			
Men	0.33	-0.28	0.93
Women	0.00	..	..
English region or country			
North	-0.07	-0.68	0.54
Midlands	2.20 ***	1.53	2.87
South	1.48 ***	0.94	2.01
Wales	0.92	-0.20	2.04
Scotland	-2.15 ***	-3.04	-1.26
Northern Ireland	-2.38 ***	-3.71	-1.06
Social class of head of household			
I, II, IIINM	0.77 ***	0.37	1.17
IIIM	-0.75 **	-1.18	-0.33
IV, V	-0.02	-0.49	0.46
Education attainment			
At degree level or above	0.68 *	0.08	1.29
Below degree level	-0.30	-0.75	0.16
No qualifications	-0.39	-1.10	0.33
Economic status			
Employed full time	-0.20	-0.76	0.36
Employed part time	-0.24	-0.87	0.38
ILO unemployed	0.65	-0.44	1.73
Economically inactive	-0.21	-0.87	0.46
Marital status†			
Single	0.54	-0.11	1.18
Married/cohabiting	0.86 **	0.27	1.45
Divorced/separated	-1.40 **	-2.41	-0.38
Reason for dental attendance			
Regular check-up	-0.20	-0.60	0.20
Occasional check-up	0.38	-0.16	0.91
Only with trouble	-0.18	-0.61	0.25
Frequency of teeth cleaning			
Twice a day or more	0.85 **	0.25	1.46
Once a day	0.90 **	0.28	1.57
Less than once a day	-1.75 **	-2.81	-0.69
Use other dental hygiene products			
Yes	-0.12	-0.70	0.45
No	0.00	..	..
Has fissure sealants			
Yes	1.32 **	0.45	2.19
No	0.00	..	..
Constant	19.04	18.09	19.99
Adjusted R^2	0.24		

*   *Significant at 95% level, ** significant at 99% level,*
*** *significant at 99.9% level.*
† *One respondent who was widowed was excluded from the analysis.*

Table 4.1.7 **Decayed or unsound teeth and unrestorable teeth by social class of head of household, educational attainment age with age-standardised ratios**

Dentate adults   *United Kingdom*

Characteristic	Age						All	Age standardisation		
	16-24	25-34	35-44	45-54	55-64	65 and over		Expected %	Ratio[††]	Standard error
percentage with decayed or unsound teeth										
Social class of head of household										
I, II, IIINM	44	54	43	55	53	49	50	54	92 *	1.8
IIIM	56	64	58	58	44	57	57	54	105	2.9
IV, V	58	70	62	55	68	59	62	54	115 *	3.9
Educational attainment										
No qualifications	63	73	61	65	62	57	62	54	114 *	3.2
Qualifications below degree	51	60	50	54	49	54	54	55	98	1.3
Qualifications at degree level or above	38	53	45	55	52	38	49	55	88 *	4.0
All[†]	51	60	51	57	54	54	55			
percentage with unrestorable teeth										
Social class of head of household										
I, II, IIINM	2	3	4	7	5	10	5	8	62 *	6.2
IIIM	14	7	9	10	7	17	10	8	126 *	11.4
IV, V	3	17	13	9	28	10	13	8	169 *	17.4
Educational attainment										
No qualifications	15	16	14	10	19	12	14	9	153 *	12.7
Qualifications below degree	5	7	6	8	5	13	7	8	90	5.5
Qualifications at degree level or above	1	3	4	6	5	7	4	8	54 *	11.8
All[†]	6	7	7	8	10	12	8			
Base										
Social class of head of household										
I, II, IIINM	215	402	395	394	247	273	1926			
IIIM	116	254	214	204	115	122	1025			
IV, V	90	148	121	107	85	74	625			
Educational attainment										
No qualifications	43	87	134	148	173	221	806			
Qualifications below degree	395	585	502	453	226	199	2360			
Qualifications at degree level or above	53	181	145	145	62	63	649			
All[†]	491	854	781	746	461	484	3817			

† *Includes those for whom social class or education level were not known.*
†† *The standardised ratio is given as the observed rate divided by the expected rate and the ratio multiplied by 100. A 2-tailed test was used to show whether the ratio was significantly different from 100.*
* *Significant at 95% level.*

Table 4.1.8 **Likelihood of a dentate adult having decayed or unsound teeth, or having unrestorable teeth (based on odds ratios from logistic regression)**

Dentate adults   *United Kingdom*

Characteristic	Decayed or unsound teeth			Unrestorable teeth		
	Odds ratios	95% confidence intervals		Odds ratios	95% confidence intervals	
		Lower	Upper		Lower	Upper
Age						
16-24	1.00	..	..	1.00	..	..
25-34	1.74 ***	1.34	2.26	1.90 *	1.12	3.23
35-44	1.18	0.89	1.56	1.79 *	1.03	3.13
45-54	1.52 **	1.13	2.03	2.74 ***	1.58	4.74
55-64	1.30	0.95	1.80	2.83 ***	1.58	5.06
65 and over	1.19	0.85	1.66	3.25 ***	1.83	5.77
Gender						
Men	1.23 **	1.06	1.41	1.71 ***	1.27	2.29
Women	1.00	..	..	1.00	..	..
English region or country						
North	1.86 ***	1.55	2.23	1.01	0.72	1.42
Midlands	1.01	0.83	1.22	0.54 **	0.36	0.82
South	1.00	..	..	1.00	..	..
Wales	1.01	0.73	1.39	0.62	0.30	1.28
Scotland	1.40 *	1.07	1.83	0.65	0.37	1.14
Northern Ireland	1.48	0.95	2.31	0.93	0.42	2.08
Social class of head of household						
I, II, IIINM	1.00	..	..	1.00	..	..
IIIM	1.18	1.00	1.40	1.50 *	1.07	2.10
IV, V	1.36 **	1.11	1.67	1.84 **	1.27	2.65
Education attainment						
At degree level or above	1.00	..	..	1.00	..	..
Below degree level	1.13	0.93	1.38	1.19	0.73	1.94
No qualifications	1.41 **	1.09	1.82	1.82 *	1.05	3.14
Economic status						
Employed full time	-	-	-	-	-	-
Employed part time	-	-	-	-	-	-
ILO unemployed	-	-	-	-	-	-
Economically inactive	-	-	-	-	-	-
Marital status						
Single	1.13	0.91	1.39	-	-	-
Married/cohabiting	1.00	..	..	-	-	-
Divorced/separated	1.34 *	1.03	1.76	-	-	-
Widowed	1.47 *	1.01	2.15	-	-	-
Reason for dental attendance						
Regular check-up	1.00	..	..	1.00	..	..
Occasional check-up	1.26	1.00	1.58	1.18	0.64	2.17
Only with trouble	1.83 ***	1.55	2.16	5.41 ***	3.98	7.35
Frequency of teeth cleaning						
More than twice a day	-	-	-	-	-	-
Twice a day	-	-	-	-	-	-
Once a day	-	-	-	-	-	-
Less than once a day	-	-	-	-	-	-
Use other dental hygiene products						
Yes	-	-	-	1.35 *	1.02	1.80
No	-	-	-	1.00	..	..
Nagelkerke R^2	0.08			0.19		

* *Significant at 95% level, ** significant at 99% level, *** significant at 99.9% level.*

\- *Factors were available to be entered into the model but were not found to be significant, therefore odds ratios were not calculated.*

Table 4.1.9 **Twelve or more restored (otherwise sound) teeth and artificial crowns by social class of head of household, educational attainment and age with age-standardised ratios**

Dentate adults   *United Kingdom*

Characteristic	Age						All	Age standardisation		
	16-24	25-34	35-44	45-54	55-64	65 and over		Expected %	Ratio††	Standard error
percentage with 12 or more restored (otherwise sound) teeth										
Social class of head of household										
I, II, IIINM	1	16	42	54	44	33	32	27	118 **	3.0
IIIM	2	19	27	37	35	18	23	27	86 **	4.8
IV, V	2	14	26	35	14	12	17	25	67 **	6.4
Educational attainment										
No qualifications	0	18	29	30	28	16	22	30	73 **	5.0
Qualifications below degree	2	17	35	48	38	33	26	25	105 **	2.5
Qualifications at degree level or above	1	14	43	58	49	44	34	28	121 **	6.8
All†	2	16	36	46	36	26	27			
percentage with artificial crowns										
Social class of head of household										
I, II, IIINM	4	24	43	55	56	50	38	34	110 *	2.6
IIIM	9	24	34	45	41	27	30	34	89 *	4.2
IV, V	9	29	38	36	36	24	28	32	87 *	5.9
Educational attainment										
No qualifications	10	27	37	35	38	27	31	39	80 *	4.2
Qualifications below degree	6	25	43	50	52	47	33	32	106 *	2.1
Qualifications at degree level or above	6	20	39	60	59	61	38	35	110	5.7
All†	7	24	41	49	48	38	34			
Base										
Social class of head of household										
I, II, IIINM	*215*	*402*	*395*	*394*	*247*	*273*	*1926*			
IIIM	*116*	*254*	*214*	*204*	*115*	*122*	*1025*			
IV, V	*90*	*148*	*121*	*107*	*85*	*74*	*625*			
Educational attainment										
No qualifications	*43*	*87*	*134*	*148*	*173*	*221*	*806*			
Qualifications below degree	*395*	*585*	*502*	*453*	*226*	*199*	*2360*			
Qualifications at degree level or above	*53*	*181*	*145*	*145*	*62*	*63*	*649*			
All†	*491*	*854*	*781*	*746*	*461*	*484*	*3817*			

† Includes those for whom social class or education level were not known.

†† The standardised ratio is given as the observed rate divided by the expected rate and the ratio multiplied by 100. A 2-tailed test was used to show whether the ratio was significantly different from 100.

* Significant at 95% level.

Table 4.1.10 **Likelihood of a dentate adult having twelve or more restored (otherwise sound) teeth or having artificial crowns (based on odds ratios from logistic regression)**

Dentate adults   *United Kingdom*

Characteristic	12 or more restored (otherwise sound) teeth			Artificial crowns		
	Odds ratios	95% confidence intervals		Odds ratios	95% confidence intervals	
		Lower	Upper		Lower	Upper
Age						
16-24	1.00	..	..	1.00	..	..
25-34	13.56 ***	6.50	28.30	5.11 ***	3.44	7.60
35-44	36.18 ***	17.31	75.62	9.93 ***	6.70	14.71
45-54	61.37 ***	29.19	129.05	14.75 ***	9.93	21.92
55-64	39.47 ***	18.52	84.10	14.83 ***	9.77	22.52
65 and over	25.24 ***	11.61	54.88	11.15 ***	7.32	16.99
Gender						
Men	-	-	-	-	-	-
Women	-	-	-	-	-	-
English region or country						
North	0.72 **	0.57	0.89	-	-	-
Midlands	0.83	0.66	1.05	-	-	-
South	1.00	..	..	-	-	-
Wales	0.91	0.61	1.34	-	-	-
Scotland	1.35	0.99	1.84	-	-	-
Northern Ireland	1.89 *	1.12	3.18	-	-	-
Social class of head of household						
I, II, IIINM	1.83 ***	1.40	2.38	1.21	0.96	1.52
IIIM	1.34 *	1.01	1.78	0.94	0.74	1.19
IV, V	1.00	..	..	1.00	..	..
Education attainment						
At degree level or above	1.40 *	1.11	1.77	1.36 **	1.11	1.67
Below degree level	1.32	0.98	1.78	1.26	0.96	1.65
No qualifications	1.00	..	..	1.00	..	..
Economic status						
Employed full time	1.00	..	..	-	-	-
Employed part time	1.24 *	1.01	1.53	-	-	-
ILO unemployed	0.49 *	0.24	0.97	-	-	-
Economically inactive	1.02	0.78	1.32	-	-	-
Marital status						
Single	1.16	0.88	1.53	-	-	-
Married/cohabiting	1.00	..	..	-	-	-
Divorced/separated	0.66 **	0.48	0.90	-	-	-
Widowed	0.87	0.57	1.33	-	-	-
Reason for dental attendance						
Regular check-up	2.74 ***	2.19	3.43	1.66 ***	1.38	2.00
Occasional check-up	2.07 ***	1.49	2.87	1.18	0.89	1.58
Only with trouble	1.00	..	..	1.00	..	..
Frequency of teeth cleaning						
More than twice a day	-	-	-	1.07	0.70	1.63
Twice a day	-	-	-	0.70	0.46	1.09
Once a day	-	-	-	1.08	0.69	1.70
Less than once a day	-	-	-	1.00	..	..
Use other dental hygiene products						
Yes	1.27 **	1.07	1.51	1.18 *	1.09	1.38
No	1.00	..	..	1.00	..	..
Nagelkerke R^2	0.26			0.18		

* Significant at 95% level, ** significant at 99% level, *** significant at 99.9% level.

\- Factors were available to be entered into the model but were not found to be significant, therefore odds ratios were not calculated.

Table 4.1.11 **Periodontal loss of attachment of 4mm or more by social class of head of household, educational attainment and age with age-standardised ratios**

Dentate adults   *United Kingdom*

Characteristic	Age						All	Age standardisation		
	16-24	25-34	35-44	45-54	55-64	65 and over		Expected %	Ratio††	Standard error
percentage with loss of attachment of 4mm or more										
Social class of head of household										
I, II, IIINM	17	20	39	54	67	86	42	44	97	2.1
IIIM	12	32	40	54	69	84	44	44	102	3.4
IV, V	12	31	56	52	85	83	47	42	112 *	4.7
Educational attainment										
No qualifications	14	38	50	52	67	87	59	57	105	3.1
Qualifications below degree	15	27	42	51	73	84	39	39	101	1.7
Qualifications at degree level or above	10	17	33	55	65	86	37	42	88 *	4.8
All†	14	26	42	52	70	85	43			
Base										
Social class of head of household										
I, II, IIINM	207	387	375	360	215	211	1755			
IIIM	113	249	199	191	98	100	950			
IV, V	86	143	111	100	74	63	577			
Educational attainment										
No qualifications	42	85	119	134	151	177	708			
Qualifications below degree	379	566	479	420	194	159	2197			
Qualifications at degree level or above	50	176	137	135	55	47	600			
All†	471	828	735	689	400	384	3507			

† Includes those for whom social class or education level were not known.

†† The standardised ratio is given as the observed rate divided by the expected rate and the ratio multiplied by 100. A 2-tailed test was used to show whether the ratio was significantly different from 100.

* Significant at 95% level.

Table 4.1.12 **Likelihood of a dentate adult having periodontal loss of attachment of 4mm or more (based on odds ratio from logistic regression)**

Dentate adults   *United Kingdom*

Characteristic	Odds ratios	95% confidence intervals	
		Lower	Upper
Age			
16-24	1.00	..	..
25-34	2.12 ***	1.57	2.87
35-44	4.50 ***	3.33	6.07
45-54	7.28 ***	5.37	9.88
55-64	14.39 ***	10.16	20.38
65 and over	33.31 ***	22.34	49.64
Gender			
Men	1.48 ***	1.26	1.74
Women	1.00	..	..
English region or country			
North	1.30 *	1.06	1.59
Midlands	1.10	0.88	1.37
South	1.00	..	..
Wales	0.65 *	0.44	0.96
Scotland	1.27	0.94	1.72
Northern Ireland	1.14	0.69	1.88
Social class of head of household			
I, II, IIINM	-	-	-
IIIM	-	-	-
IV, V	-	-	-
Education attainment			
At degree level or above	1.00	..	..
Below degree level	1.35 *	1.08	1.68
No qualifications	1.74 **	1.33	2.28
Economic status			
Employed full time	-	-	-
Employed part time	-	-	-
ILO unemployed	-	-	-
Economically inactive	-	-	-
Marital status			
Single	-	-	-
Married/cohabiting	-	-	-
Divorced/separated	-	-	-
Widowed	-	-	-
Reason for dental attendance			
Regular check-up	-	-	-
Occasional check-up	-	-	-
Only with trouble	-	-	-
Frequency of teeth cleaning			
More than twice a day	-	-	-
Twice a day	-	-	-
Once a day	-	-	-
Less than once a day	-	-	-
Use other dental hygiene products			
Yes	-	-	-
No	-	-	-
Nagelkerke R^2	0.28		

* Significant at 95% level, ** significant at 99% level, *** significant at 99.9% level.

- Factors were available to be entered into the model but were not found to be significant, therefore odds ratios were not calculated.

Part 5

Trends in tooth loss and the condition of natural teeth

5.1   Trends in total tooth loss

Summary

- Since 1978, when the first United Kingdom Adult Dental Health Survey was carried out, there has been an improvement in the proportion of people retaining their natural teeth: the proportion who had lost their natural teeth decreased from 30% in 1978 to 21% in 1988 and to 13% in 1998.
- The first Adult Dental Health Survey, in 1968, covered only England and Wales. The combined data for England and Wales can, therefore, be compared over thirty years: in 1968, 37% of adults were edentate compared with 29% in 1978, 20% in 1988 and 12% in 1998.
- Since 1978, there have been improvements in the proportion of edentate adults in all age groups, particularly in the middle and older age groups.
- The incidence of total tooth loss in the decade 1988 to 1998 was small at all ages, and showed a reduction from the values reported in 1988. In the younger age groups (those aged under 45 years) total tooth loss is virtually a thing of the past.
- There have been improvements in the retention of natural teeth among men and women, and among adults from different social classes. The rates of improvement between 1978 and 1998 have varied slightly and differences in the proportion of adults with no natural teeth still exist between these groups in 1998. For example, the prevalence of total tooth loss among adults from skilled manual backgrounds in 1998 is similar to that seen among adults from non-manual backgrounds in 1988 (15% and 14% respectively).
- All four countries of the United Kingdom showed marked improvements in the retention of natural teeth over the previous two decades, although there are still differences in the proportion of edentate adults within each country in 1998.
- Predictions for the proportions of edentate adults of different ages in 1998 were given in the 1988 Adult Dental Health Survey report. These predictions proved to be accurate, with the proportions of edentate adults generally falling between the highest and lowest 1988 forecasts.
- Predictions of future total tooth loss have been made based on the data from the 1998 survey (using similar methods as were used in 1988). Assuming no increase in the incidence in total tooth loss, the proportion of adults with no natural teeth can be expected to reduce to single figure percentages in all age groups up to the age of 74 by 2028.

5.1.1 Introduction

National surveys of adult dental health in the United Kingdom were first undertaken in 1968, on that occasion covering only England and Wales. In 1972 an equivalent survey was conducted in Scotland, and then from 1978 the surveys covered the whole of the United Kingdom. The sampling and examination protocols have been very similar in all of the surveys so the data from each survey are comparable. The 1998 survey, as the fourth in the series, allows a unique opportunity to track the trends in oral health and disease for up to 30 years.

5.1.2 Changes in total tooth loss in England and Wales since 1968, and in the United Kingdom since 1978

Table 5.1.1 shows the comparative figures for 1968, 1978 and 1988 along with the 1998 figures. In 1968 only England and Wales were surveyed, so to allow comparison the 1978 figures for England and Wales are given as well. In Chapter 2.1 a threshold of 95% of the population with some natural teeth was used as a marker to measure the changing oral health of different groups within the population. The highlighted figures show which age groups achieved this threshold in each survey. The proportion dentate has increased by between 7% and 9% in each of the intervening decades, starting at 63% for England and Wales in 1968, and increasing to 87% for the United Kingdom in 1998. Not only does the table illustrate this steady increase in the overall total for the proportion dentate, but the age related trends can also be seen clearly. In 1968, 7% of 25 to 34 year olds in England and Wales were edentate, increasing with age so that 41% of 45 to 54 year olds and 64% of 55 to 64 year olds had no natural teeth. By 1998 being edentate before the age of 45 was very rare indeed (1%), and it was only among those aged 75 and over where there was a majority with no remaining natural teeth.

Figure 5.1.1, Table 5.1.1

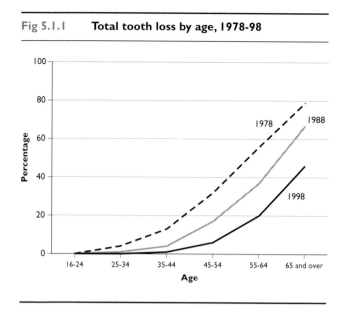

Fig 5.1.1   **Total tooth loss by age, 1978-98**

Table 5.1.2 shows data for the United Kingdom, by age, gender, country and social class of head of household, in 1978, 1988 and 1998. Overall there has been a 17 percentage point reduction in the proportion of adults with no natural teeth in the last 20 years. An improvement was seen in all age groups. Two other trends are apparent. The improvements in oral health have been demonstrated consistently among both men and women. The last decade has seen a reduction from 16% to 10% in men, but a greater reduction of ten percentage points from 25% to 15% for women, suggesting that the differences between the two genders may be starting to even out.

The overall reduction in the proportion of adults without teeth in England has been eight percentage points in both of the last two decades, from 28% in 1978 to 20% in 1988 and to 12% in 1998. The rate of improvement in Scotland has been more rapid than for England. There has been a reduction of 21 percentage points in the proportion edentate in 20 years. Wales had seen the biggest reduction of the four countries in the proportion edentate in the decade from 1978 to 1988, from 37% down to 22%. In the decade from 1988 to 1998 the reduction was less, only another 5 points down to 17%. Northern Ireland has also shared in the reduction in the proportion dentate with a decrease from 18% to 12% in the decade between 1988 and 1998 (the 1978 survey did not provide data for Northern Ireland but Chapter 7.4 gives results from a separate survey conducted in 1979). More detailed analysis of the change in levels of total tooth loss within the countries of the United Kingdom can be found in the respective chapters about each country (Chapters 7.1 to 7.4).

While all social classes have demonstrated marked improvements since 1978, in the last decade the greatest improvements have been among people from manual backgrounds. The proportion edentate among people from unskilled manual backgrounds decreased from 32% to 22% in the last decade, a reduction of ten percentage points compared with only six points in the decade from 1978 to 1988. People from skilled manual backgrounds saw a reduction from 24% to 15%; a nine-point reduction compared to only five percentage points in the previous decade. People from non-manual backgrounds on the other hand have shown a slightly slower rate of change, with a six percentage point drop from 14% edentate in 1988 to 8% in 1998, the decade before saw an eight point drop. However, substantial differences according to social class remain. In 1998, people from skilled manual backgrounds demonstrated a prevalence of total tooth loss equivalent to the non-manual population in 1988, and people from unskilled manual backgrounds in 1998 with the same prevalence of total tooth loss as shown by people from non-manual backgrounds in 1978 (22%).

Figure 5.1.2, Table 5.1.2

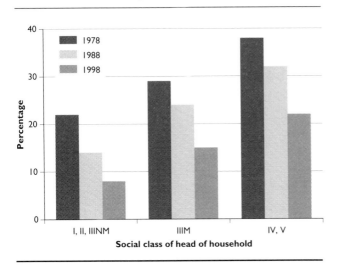

Fig 5.1.2   **Total tooth loss by social class of head of household, 1978-98**

In England and Wales these statistics can be compared for an extra decade, back to 1968. Table 5.1.3 shows the proportion of the adult population with no natural teeth in England and Wales since 1968, by age, gender and social class, allowing social and gender variations in the overall pattern to be measured over the period. This illustrates a clear continuation of the improvement that was observed between the survey in 1968 and the one conducted in 1988. In 1968 in England and Wales, 37% of the population were without natural teeth. This proportion has reduced in each subsequent decade, by 8 percentage points in the decade to 1978, by 9 percentage points in the next decade and by another 8 percentage points in the decade to 1998 resulting in only 12% without any natural teeth in 1998. There has been no sign that the trend has diminished over the last decade. In

the younger groups, those aged under 45, being edentate is almost a thing of the past, and the edentate population is increasingly concentrated in the older age groups.

In 1968 there was a difference of seven percentage points between men and women, with 33% of men and 40% of women edentate. Over the subsequent two decades both men and women showed an improvement in the proportion who were dentate, but between 1968 and 1988 men improved slightly more quickly than women, so that by 1988 there was a difference of 9 percentage points. In the last decade, men have continued to improve on this, but the proportion edentate has reduced by 6 percentage points, rather less than in previous decades, suggesting a slowing down in the trend. Women on the other hand showed an 11 percentage point decrease in the proportion edentate, much more than the 7 to 8 points of the previous two decades, and a more rapid decrease than demonstrated by the men over the last ten years.

All social classes appear to have shared the increase in the rate of retention of natural teeth in England and Wales. The last decade saw people from non-manual backgrounds maintain their steady improvement which can be traced back to 1968, with a 7 percentage point decrease in the proportion edentate, very similar to previous decades. People from unskilled manual backgrounds showed a 9 percentage point decrease in the proportion edentate since 1988, higher than the previous decade, but the same as the period from 1968 to 1978. It was people from skilled manual backgrounds who demonstrated an apparent acceleration of the trends since 1968, with an decrease of 10 percentage points, in the last decade, in the proportion with no natural teeth, which represents a marked rise on the 4 point decrease in the previous decade.

Table 5.1.3

5.1.3 Comparison between the 1988 estimates and the 1998 findings

In the report of the 1988 survey of Adult Dental Health a series of predictions were made to establish the likely limits of the rate of change of the proportion edentate in future years. Two estimates were made. One, based on the continuation of the incidence of total tooth loss over the previous decade, was considered the most "pessimistic" estimate. The other, based on there being no further loss of teeth from the position in 1988, was considered the most "optimistic" estimate. Although use of the terms "optimistic" and "pessimistic" pre-suppose that retention of natural teeth is an advantage in all situations, which may not be the case, they have been retained in this report because retention of a natural dentition is generally associated with the positive aspects of general and oral health. The "pessimistic" estimate would be accurate if the decennial incidence of total tooth loss had reached the minimum possible level and no further improvement was to occur. The "optimistic" estimate represented the other limit of what was possible. It assumed an immediate halt in the incidence of total tooth loss in 1988. As total tooth loss over the previous decade was running at 3% or more, this abrupt halt was inconceivable, but when combined with the most "pessimistic" estimate, allowed the boundaries to be set, within which predictions of future levels of total tooth loss could be made. If the incidence continued to fall below the 1988 to 1998 levels, the result in future decades would lie somewhere between the most optimistic and the most pessimistic estimates.

It is now possible to see how accurate these predictions were in the decade 1988 to 1998. Table 5.1.4 shows the proportion of edentate adults by age group in 1998 predicted by the two models. These are presented alongside the figures reported in the 1998 survey. Up until the age of 65, and for the overall total, the 'actual' figures for 1998 fall almost exactly half-way between the most optimistic and most pessimistic estimates, indicating a continued reduction in the incidence of total tooth loss at younger ages. In the smaller number of adults aged 65 and over the proportion of the population with no remaining natural teeth was found to be slightly lower even than the optimistic estimate from the 1988 figures, suggesting a net gain in the proportion of people with natural teeth. However, it should be noted that the results from these two different surveys will have been subject to sampling errors; the 1998 figures for the proportion edentate did in fact fall within the boundaries of the 95% confidence intervals of the 1988 estimates of about +/- 4% for the two oldest age groups. Demographic variations may also have contributed to this, apparently unlikely, scenario. For example, if good oral health is associated with other aspects of good health, more deaths among the edentate than the dentate population in these age groups may be expected. This would result in a relative increase in the dentate population with age which may be partially responsible for the very low apparent incidence of total tooth loss at these ages.

Figure 5.1.3, Table 5.1.4

Fig 5.1.3
Comparison of 1998 results with estimates of future total tooth loss based on 1988 data

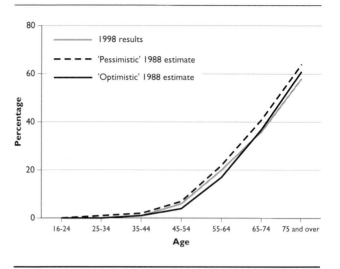

5.1.4 The incidence of total tooth loss in the decade 1988-98

By calculating the incidence of total tooth loss between 1988 and 1998 the limits within which future trends are likely to operate can be recalculated to estimate the impact that current trends will have on total tooth loss over the first few decades of the new millennium.

Although completely different samples were involved in 1988 and 1998, almost identical sampling strategies were used and both were nationally representative. Consequently, one way in which it is possible to estimate the incidence of total tooth loss over the period is by comparing the prevalence figures for 1998 with previous surveys in the United Kingdom in 1988 and 1978, and with 1968 in England and Wales. Using this way of calculating the incidence of total tooth loss, the trends since 1988 are a little different from the previous two ten-year periods.

In 1998, up until the age of 44 the incidence of new cases of total tooth loss appeared to be negligible. There were a very few individuals who were edentate at this age, and unless there is a dramatic reversal in the pattern observed over the past thirty years, only a very small number of people will become edentate in these age groups in future decades. However, using this method, the ten years between 1988 and 1998 did see a continued incidence of new cases of complete tooth loss for those aged 45 to 64 years in 1998. Over this ten year period 2% to 3% of these age groups of the population became edentate for the first time. This is in contrast to the estimated incidence of 4% to 5% for the ten years from 1978 to 1988, and 7% over the ten years before that (1968 to 1978). The incidence among the oldest group was very low indeed; for those aged 65 to 74 years in 1998 there appeared to be no new cases of total tooth loss at all using this method to calculate incidence. However, for the reasons already discussed above (sampling error and the association between oral and general health), the comparison of prevalence data is probably a less reliable way of calculating incidence data for older age groups. Data from the 1988 survey showed a similarly small increase over the decade from 1978 to 1988, suggesting that similar factors were involved.

Table 5.1.5

An alternative strategy for calculating the incidence of total tooth loss, and one which is free from the problems of comparing prevalence values in different surveys, is to ask the edentate respondents in the survey when they lost the last of their natural teeth. From these data it is possible to calculate incidence by age group and the decade in which the loss occurred. This method has its own limitations in that retrospective data of any kind are subject to memory errors. However, using this approach there again appears to have been a consistent reduction in the incidence of total tooth loss over the ten years between 1988 and 1998 compared with the previous decade. Among participants under the age of 65 in 1998 the incidence of total tooth loss using this method was similar to that shown in Table 5.1.5, with rates of 0 to 3% for the ten year period, compared with 0 to 6% for the previous decade. In the older age groups (those who were 65 or over in 1998) there was little difference between the decade from 1988 to 1998 and the one before, with incidence levels of 3% to 4% in the most recent decade. However, there are no data available from the 1978 to 1988 period with which the very oldest group in 1998 (aged 75 to 84 years) can be compared.

Table 5.1.6

In the younger age groups the scope for improvement is now very limited indeed; with only negligible numbers of new cases of total tooth loss occurring in 1998 not much further improvement is possible. There is some scope for a further reduction in the incidence of total tooth loss in the middle and older age categories. Among those aged between 35 and 64 years there is some indication that the incidence fell further in the last decade, and further minor improvements are possible. There is no clear evidence of any reduction in incidence for those aged 65 years and over in 1998.

The prevalence of total tooth loss has fallen dramatically as the historic effect of early tooth loss in older populations is lost, but there are also clear signs that the incidence of total tooth loss is continuing to diminish, particularly for those

under the age of 65. Certainly as those aged under 35 years in 1998 (who were shown in 1988 to have much lower levels of disease than their predecessors) continue to get older, the incidence of total tooth loss in this group should be expected to remain at the very low levels observed in 1998.

The incidence of total tooth loss in the oldest age groups is more difficult to predict. Elderly people may be at additional risk of total tooth loss for two reasons. The first is because of increased vulnerability to dental disease for systemic, social, functional and behavioural reasons (such as dry mouth resulting from medication, altered dietary intake and physical difficulties in maintaining oral hygiene). The second is the cumulative effect of dental disease and loss of individual teeth. Effective dental prevention may have delayed the onset of disease and treatment may have prevented tooth loss from occurring, but prevention may not work indefinitely and the effects of disease and treatment are cumulative, so this process cannot go on forever. The question which then arises, but which cannot yet be answered, is whether the prevention and treatment of dental disease is sufficient to delay tooth loss beyond the normal life span. Consequently, it is difficult to make as confident a prediction of future trends in those aged 65 and over as it is for the younger groups.

5.1.5 Predictions of future total tooth loss

The trend towards the retention of natural teeth throughout life, and the resulting steady reduction in both the prevalence and incidence of total tooth loss among the United Kingdom population observed between 1968 and 1988, clearly continued between 1988 and 1998. This section will examine whether the trends observed over the last 30 years are likely to continue for the next thirty years, and what impact they will have on the dental status of the United Kingdom population over that period.

Edentate respondents in the survey were asked when they lost the last of their natural teeth. In the 1988 survey report these questionnaire data were used to calculate the incidence of total tooth loss over the previous ten years, and then project this forward to calculate levels of total tooth loss in future decades. The 1998 data have been used to revise and update these predictions made using the 1988 data. With data from national surveys available for thirty years, predictions have been made here for the next 30 years, through to 2028. Beyond this there are too many uncertainties to be able to make predictions with any confidence. Questionnaire data were used in preference to comparisons of prevalence because, on balance, the minor disadvantages of the former (the possibility of problems with accurate recall of events over a ten year period) were outweighed by the more substantial problems caused by the chance variations which affect the comparisons of prevalence from two different samples.

Table 5.1.7 shows the predicted values for the proportion of the population who will be edentate from 2008 to 2028, assuming that the current incidence continues to affect each subsequent age cohort as they pass through the age bands, and also assuming that population age and demographic structure remain the same. To do this, the oldest group (generally reported as 75 years and over elsewhere in this chapter) was broken down into 75 to 84 year olds and those aged 85 years and over so that the trends in the oldest groups of the population could be calculated more accurately. This table represents the most "pessimistic" (that is, the smallest) estimate of the reductions in the proportion of the population with no natural teeth. The figures are markedly lower than the equivalent values in 1988, reflecting the further reduction in incidence in younger age groups over the last decade. Using this calculation, total tooth loss, predicted until 2028 will not be eliminated; there will continue to be low levels in the population (around 5% by the year 2018), concentrated in the elderly.

Figure 5.1.4, Table 5.1.7

Fig 5.1.4   **Predictions for future levels of total tooth loss**

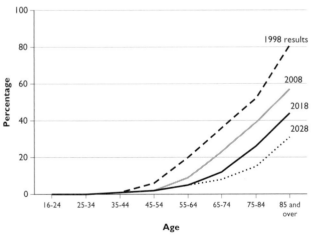

Table 5.1.8 shows the most "optimistic" predictions for total tooth loss; the estimate based on the greatest possible reductions in total tooth loss assuming that no further cases occurred after 1998. By 2028 total tooth loss would be eliminated in the population under the age of 65 years and almost eliminated in the population under the age of 75 years (1% of 65 to 74 year olds would be affected). Only adults aged 85 years and over would have an appreciable proportion of the population without natural teeth (20%). While this is

a very unlikely scenario, it does indicate the limit of what is possible in terms of eliminating total tooth loss, working on the basis of the current status of the population of the United Kingdom.

Table 5.1.8

The incidence of total tooth loss has seen further reduction, so that the incidence measured from 1998 data is now closer than in 1988 to the 'ideal' of no new cases. This means that the 'pessimistic' and 'optimistic' estimates of the prevalence of total tooth loss are closer than they were in the 1988 report.

Table 5.1.9

Over the past thirty years the population in the South of England has consistently set the trend for the retention of natural teeth, and has generally been about 10 years ahead of the rest of the United Kingdom. The sample for this region is large enough to make the calculation of the incidence of total tooth loss over the past ten years feasible. This may give some indication of whether further reductions in the incidence of total tooth loss in the United Kingdom as a whole can be expected over the next ten years. Based on the results of the questionnaire, the incidence was not significantly lower in the South than in the United Kingdom as a whole. These data suggest that, if the South is an indicator for what may be expected over the next decade, any reduction in the incidence of total tooth loss is likely to be very small.

Table 5.1.10

There are good clinical reasons why a continuing, although perhaps very low, incidence of total tooth loss may be expected to occur in the near future. Some forms of dental disease, although rare, are destructive and very difficult to treat. This specifically applies to some types of periodontal disease. More importantly though, the effects of the more common dental diseases, caries and periodontal disease, are cumulative over a lifetime. With life expectancy increasing, there may always be some people for whom the life of their dentition is shorter than their own life span. The proportion of the United Kingdom population who have no natural teeth is low in 1998 and is predicted to get much lower. However, it is likely that there will continue to be a very small proportion of the population without natural teeth, particularly among older adults.

Table 5.1.1 Dental status by age, 1968-98

All adults

Age		England and Wales				United Kingdom					
		1968		1978		1978		1988		1998	
		Dentate	Edentate	Dentate	Edentate	Dentate	Edentate	Dentate	Edentate	Dentate	Edentate
16-24	%	**99**	1	**100**	0	**100**	0	**100**	0	**100**	0
25-34	%	93	7	**97**	3	**96**	4	**99**	1	**100**	0
35-44	%	78	22	88	12	87	13	**96**	4	**99**	1
45-54	%	59	41	71	29	68	32	83	17	94	6
55-64	%	36	64	52	48	44	56	63	37	80	20
65-74	%	21	79	26	74	} 21	} 79	43	57	64	36
75 and over	%	12	88	13	87			20	80	42	58
All	%	63	37	71	29	70	30	79	21	·87	13

For bases see Tables Appendix
1988 Table 2.3

Table 5.1.2 Total tooth loss by age, gender, country and social class of head of household, 1978-98

All adults   *United Kingdom*

	1978	1988	1998
	percentage edentate		
All	30	21	13
Age			
16-24	0	0	0
25-34	4	1	0
35-44	13	4	1
45-54	32	17	6
55-64	56	37	20
65 and over	79	67	46
Gender			
Men	25	16	10
Women	33	25	15
Country			
England	28	20	12
Wales	37	22	17
Scotland	39	26	18
Northern Ireland	..	18	12
Social class of head of household			
I, II, IIINM	22	14	8
IIIM	29	24	15
IV,V	38	32	22

For bases see Tables Appendix
1988 Table 3.5

Table 5.1.3 Total tooth loss by age, gender and social class of head of household, 1968-98

All adults   *England and Wales*

	1968	1978	1988	1998
	percentage edentate			
All	37	29	20	12
Age				
16-24	1	0	0	0
25-34	7	3	1	0
35-44	22	12	3	1
45-54	41	29	15	6
55-64	64	48	36	18
65-74	79	74	56	34
75 and over	88	87	80	57
Gender				
Men	33	24	16	10
Women	40	32	25	14
Social class of head of household				
I,II,IIINM	27	21	14	7
IIIM	34	28	24	14
IV,V	46	37	31	22

For bases see Tables Appendix
1988 Table 3.1

Table 5.1.4   **Most 'optimistic' and 'pessimistic' estimates for 1998 from 1988 data compared with 1998 data**

All adults   *United Kingdom*

Age	Estimates for 1998 from 1988 data		1998 Results	Base (1998)
	'Optimistic'	'Pessimistic'		
	percentage edentate			
16-24	0	0	0	702
25-34	0	1	0	1158
35-44	1	2	1	1114
45-54	4	7	6	1106
55-64	17	22	20	833
65-74	37	41	36	758
75 and over	61	64	58	533
All	12	14	13	6204

For bases see Tables Appendix
1988 Table 2.6, 2.7

Table 5.1.5   **The incidence of total tooth loss between the surveys, 1968 to 1998, based on prevalence data**

All adults

Age	England and Wales			United Kingdom					
	percentage edentate 1968	estimated incidence 68-78	percentage edentate 1978	percentage edentate 1978	estimated incidence 68-78	percentage edentate 1988	estimated incidence 88-98	percentage edentate 1998	
16-24	1		0	0		0		0	
		2			1		0		
25-34	7		3	4		1		0	
		5			0		0		
35-44	22		12	13		4		1	
		7			4		2		
45-54	41		29	32		17		6	
		7			5		3		
55-64	64		48	56		37		20	
		10			1		†		
65-74			74			57		36	

†   *As estimated incidence decreases there is a greater chance that the sampling error of the difference between the two samples is equal to or greater than the estimated incidence. This happens between the figures for 55-64 in 1988 and 65-74 in 1998. See text.*
For bases see Tables Appendix
1988 Table 2.4

Table 5.1.6   **The incidence of total tooth loss
estimated from the interview**

All adults   *United Kingdom*

	Incidence of total tooth loss	
	1978-1988	**1988-1998**
Between age groups:		
16-24 and 25-34	0%	0%
25-34 and 35-44	1%	1%
35-44 and 45-54	4%	1%
45-54 and 55-64	6%	3%
55-64 and 65-74	4%	3%
65-74 and 75-84	-	4%

*For bases see Tables Appendix
1988 Table 2.5*

Table 5.1.7   **Predictions of future levels of total tooth loss**

All adults   *United Kingdom*

Age	**Percentage edentate in 1998**	**Assumed future incidence of total tooth loss†**	**Predicted future levels for:**			**Base (1998)**
			2008	**2018**	**2028**	
			estimated percentage edentate			
16-24	0		0	0	0	702
		0.2				
25-34	0.5		0	0	0	1158
		0.5				
35-44	0.9		1	1	1	1114
		1.4				
45-54	6.4		2	2	2	1106
		2.7				
55-64	19.7		9	5	5	833
		2.8				
65-74	35.9		23	12	8	758
		3.5				
75-84	52.5		39	26	15	432
		4.7				
85 and over	80.9		57	44	31	101
All	12.6		8	5	4	6204

† *Based on the assumption that future incidence of total tooth loss will be the same as 1988-1998 as estimated from the 1998 interview and shown in Table 5.1.6
1988 Table 2.6*

Table 5.1.8 **Predictions of future levels of tooth loss if no-one else became edentate**

All adults   *United Kingdom*

Age	Total tooth loss in 1998	Predicted future levels for:			Base (1998)
		2008	2018	2028	
		estimated percentage edentate			
16-24	0	0	0	0	702
25-34	0	0	0	0	1158
35-44	1	0	0	0	1114
45-54	6	1	0	0	1106
55-64	20	6	1	0	833
65-74	36	20	6	1	758
75-84	52	36	20	6	432
85 and over	81	52	36	20	101
All	13	7	3	1	6204

1988 Table 2.7

Table 5.1.9 **Comparison of estimates of total tooth loss based on 1988 and 1998 data**

All adults   *United Kingdom*

	'Optimistic'		'Pessimistic'	
	1988	1998	1988	1998
	estimated percentage edentate			
1998	12	-	14	-
2008	6	7	10	8
2018	2	3	7	5
2028	1	1	6	4
2038	0	0	6	3
2048	-	0	-	3

1988 Table 2.8

Table 5.1.10 **Incidence of total tooth loss[†] between 1988 and 1998 in the South of England and United Kingdom**

All adults

Age	Incidence of total tooth loss	
	United Kingdom	South of England
Between age groups:		
16-24 and 25-34	0%	0%
25-34 and 35-44	0%	0%
35-44 and 45-54	1%	1%
45-54 and 55-64	3%	1%
55-64 and 65-74	3%	3%
65-74 and 75-84	4%	5%
		Base
16-24	*702*	*177*
25-34	*1158*	*339*
35-44	*1114*	*302*
45-54	*1106*	*273*
55-64	*833*	*214*
65-74	*758*	*179*
75-84	*432*	*102*

† As estimated from the interview

5.2 Trends in the condition of natural teeth

Summary

- The results in this Chapter are based on the same criteria as those used in the previous surveys. Results presented here differ from those presented in Part 3 and elsewhere in this report which are based on the new criteria for 1998 which included an assessment of visual caries (see Chapter 3.1).

Changes in the number of teeth in dentate adults
- The improvement in the mean number of teeth observed in the United Kingdom among dentate adults between 1978 (23.0) and 1988 (24.2) has continued to 1998 (24.8).
- Larger increases have occurred in the older age groups, with those aged 55 and over having an average of 1.8 more teeth in 1998 than 1988. The comparable increase for the 16 to 24 age group was 0.3 teeth.
- There has been an increase in the proportion of dentate adults with 21 or more teeth over the twenty years since 1978; from 73% in 1978 to 80% in 1988 and to 83% in 1998.
- The largest improvement was in the 65 to 74 year age group. In 1998 almost half (46%) of the dentate adults in this age group had 21 or more teeth compared with a quarter (25%) in 1988.
- If the equivalent ten-year age cohorts from the three separate surveys are compared they generally show that dentate adults have lost very few teeth over the past twenty years since 1978.
- Most of the improvement was found among those who reported attending the dentist only with trouble, there was little change among those who attended for regular check-ups.

Changes in sound and untreated teeth
- In 1998, dentate adults in the UK had, on average, 15.7 sound and untreated teeth compared with 14.8 in 1988 and 13.0 in 1978.
- Between 1988 and 1998, the number of sound and untreated teeth present increased in 16 to 44 year olds, as had the proportion of the population with 18 or more sound and untreated teeth.
- Among dentate adults aged 55 and over, there had been little change between 1988 and 1998 in either the number of sound and untreated teeth or the proportion of the population with 18 or more such teeth. For older dentate adults these figures are affected by the increase in both the numbers of older dentate adults and the average number of teeth they have. In 1998, among older people, more teeth were being preserved than hitherto, albeit not in a "sound untreated" state.
- Dentate adults who reported attending for regular check-ups and those who reported attending only with trouble showed the same pattern of increase in numbers of sound teeth and the proportion with 18 or more sound teeth.
- Most of the improvements over the decade between 1988 and 1998 in the proportion of sound and untreated teeth in 16 to 24 year olds occurred in the first and second molars. The increases were apparent irrespective of the reported usual reason for dental attendance and ranged from 16 to 36 percentage points.

Changes in decayed or unsound teeth

● The average number of decayed or unsound teeth decreased between 1978 and 1988, from 1.9 to 1.1, but did not change significantly between 1988 and 1998 (1.0). These results were repeated across all age groups, and by gender, social class and usual reason for dental attendance.

Changes in the condition of individual teeth

● Considerably higher proportions of third molar teeth were present in 1998 than in 1988. This may reflect changes in attitudes to removal of these teeth among patients and dentists, but may also be influenced by reductions in dental disease.

5.2.1 Introduction

This chapter considers the change over time in the numbers of teeth and in their condition among dentate adults from 1978 to 1998 in the UK and from 1968 to 1998 in England and Wales. The final part of the chapter looks at the change from 1988 to 1998 in the condition of individual teeth including restorative treatment. The changing experience of restorative treatment among dentate adults is covered in Chapter 5.3.

Two important points need to be considered with reference to the data presented in this and the following chapter. The first concerns the difference between the 1998 survey and the earlier surveys in the criteria used for determining dental caries, which is discussed in detail in Section 5.2.6. This change affects the results for each of the different conditions of the natural teeth except for whether or not they are present. Thus, results presented in this and the following chapter, which are based on the criteria of the previous studies, will differ from the equivalent figures presented in Part 3. The second point to take into account is the effect of the changing age distribution on the overall results which is discussed in the following section.

5.2.2 Changes in the age structure of the dentate population

The changes presented in this and the following chapters must be viewed in the context of the increase in the dentate population. As people keep their teeth for longer the age distribution of the dentate population has changed with an increase in the proportion of older people represented. The last table in this chapter presents the age distribution of the dentate population measured by the three surveys since 1978 and shows that the proportion of dentate adults aged 55 or more has increased from 15% in 1978, to 19% in 1988 and to 24% in 1998 while the proportion of those aged 16 to 24 has decreased from 23% in 1978 to 16% in 1998. This change in age distribution will affect comparisons over time, based on the overall dentate population, where the measures being compared vary with age. This is particularly true for measures such as the overall average number of natural teeth, the number of sound and untreated teeth, and the number of restored teeth each of which are strongly associated with age. The effect of the changing age distribution will be noted where comparisons of these data are presented.

Table 5.2.38

5.2.3 Changes in the number of teeth among dentate adults

The previous reports of the surveys of adult dental health have shown data on the numbers of missing teeth among dentate adults. As dental health has improved and more teeth retained, the focus has changed to the numbers of teeth which are present. The variation in the numbers of teeth found among dentate adults in 1998 is discussed in Chapter 2.2. Two measures were used; the mean numbers of teeth present and the proportion of dentate adults with 21 or more teeth. In order to continue this analysis for the investigation of change over time, data from the previous surveys have been converted from numbers of teeth missing to the number of teeth present. Full distributions of the numbers of teeth are shown and a further measure is discussed: the proportion of dentate adults with 27 or more teeth (this is equivalent to fewer than 6 missing teeth).

The average number of teeth increased from 23.0 in 1978 to 24.2 in 1988 and 24.8 in 1998. The size of change varied with age, being highest among those aged 35 or more. For example, dentate adults aged 45 to 54 and those aged 55 and over were found to have 1.8 more teeth in 1998 compared with 1988 while the equivalent figure for those aged 16 to 24 was 0.3 teeth.

Tables 5.2.1–2

The smaller overall increase in the average number of teeth between 1988 and 1998 compared with that between 1978 and 1988 is mainly the result of the changing age distribution discussed in 5.2.2. If there had been no change in the age distribution of the dentate population since 1988 the increase in the average number of teeth between 1988 and 1998 would have been about one tooth[1] (rather than the increase of 0.6 teeth shown). Thus, the overall improvement in dental health as measured by the increase in the number of older people who retain any teeth has led to an apparent slowing of improvement in the average numbers of teeth retained among dentate people. Figure 5.2.1 illustrates that, by including those who have lost all their teeth, the overall rate of increase in numbers of teeth has in fact remained fairly stable.

Figure 5.2.1, Table 5.2.3

Fig 5.2.1   **Distribution of the number of teeth among all adults, 1978-98**

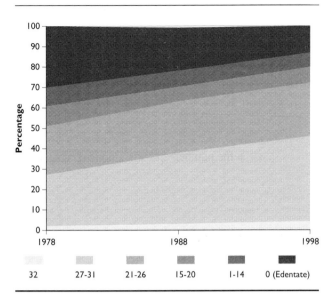

| 32 | 27-31 | 21-26 | 15-20 | 1-14 | 0 (Edentate) |

The proportion of dentate adults with 27 or more teeth (fewer than 6 missing) has increased across the age ranges. Figure 5.2.2 shows this proportion by age for each decennial survey together with the changes for particular age cohorts tracked across the 20 year period. Care must be taken with the interpretation of these results, particularly for the younger age groups as third molar teeth continue to erupt into the mouth at this age increasing the number of standing teeth. However, this figure indicates clinically important improvements seen across the age cohorts in the retention of teeth and shows the rate of tooth loss for each cohort.

Figure 5.2.2

Fig 5.2.2   **Proportion of dentate adults with 27 or more teeth, 1978-98**

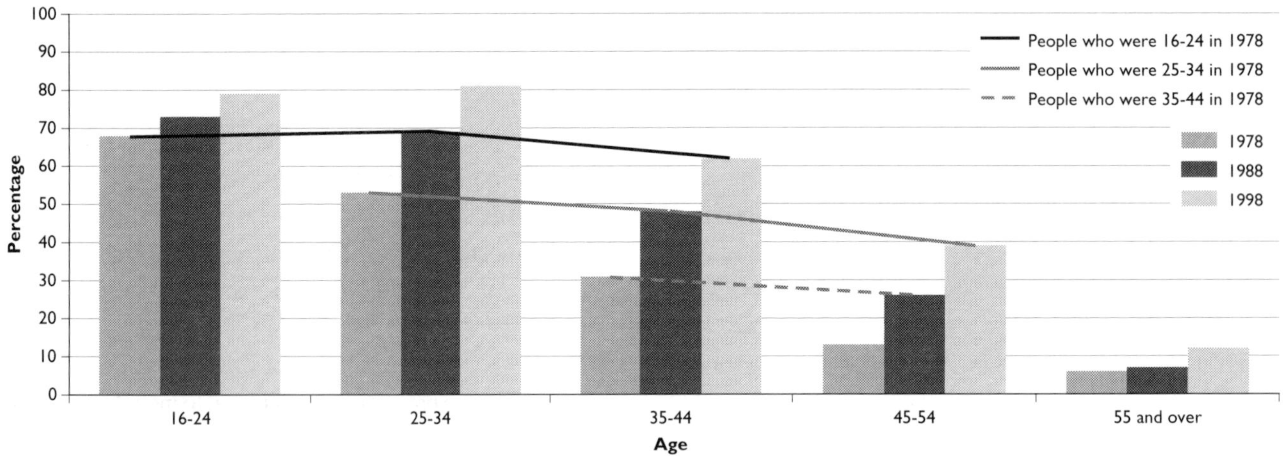

Comparison of the proportion of dentate adults with 21 or more teeth in 1998 with those for 1978 and 1988 showed an improvement over the twenty years; from 73% in 1978 to 80% in 1988 and to 83% in 1998. The largest improvement was in the 65 to 74 year age group. In 1998 almost half (46%) of the dentate adults in this age group had 21 or more teeth compared with a quarter in 1988. If the equivalent ten-year age cohorts from the separate surveys are compared they generally show that few dentate adults had lost enough teeth over the past twenty years to take them below the threshold of having 21 or more teeth. For example, in 1978, 89% of dentate 25 to 34 year olds had 21 or more teeth, compared with 86% of dentate adults aged 35 to 44 in 1988 and 82% of dentate adults aged 45 to 54 in 1998. The exception is the cohort aged 55 to 64 in 1998 who had suffered significantly more tooth loss over the previous ten years than the other age cohorts. In 1998, 57% of dentate adults in this age group had 21 or more natural teeth compared with 72% of the equivalent cohort (those aged 45 to 54) in 1988.

Figure 5.2.3, Table 5.2.4

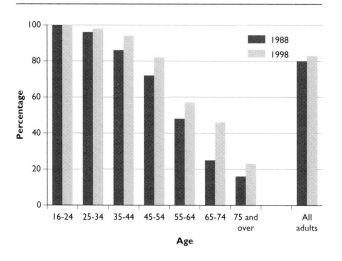

Fig 5.2.3 **Dentate adults with 21 or more teeth by age, 1988-98**

The rate of change was very similar for men and women. All social classes showed an increase in the average numbers of teeth since 1978. However, among those from unskilled manual backgrounds, there was no significant increase in the average numbers of teeth between 1988 and 1998; 23.4 to 23.6 compared with 24.8 in 1988 to 25.4 in 1998 among those from a non-manual background. It was shown in Chapter 5.1 that the reduction in total tooth loss was greatest among those from unskilled manual households which means that this social class group will have experienced the largest changes to the underlying age structure of their dentate population. This may affect the interpretation of the mean numbers of missing teeth (see section 5.2.2). The larger

increase of dentate adults in this group may partly explain the difference between the non-manual and unskilled manual groups in the proportion with very few (1-14) teeth; a significantly smaller proportion of dentate adults from non-manual backgrounds had 1 to 14 teeth in 1998 compared with 1988 (5% compared with 9%) while there was no change among those from unskilled non-manual backgrounds.

Table 5.2.5–6

In terms of the usual reason for dental attendance, all improvement in the numbers of teeth among dentate adults has been in the group that reported attending the dentist only with trouble; the 1998 distribution of missing teeth among those who reported having regular check-ups was virtually identical to the 1988 results. The average number of teeth among those who attend only with trouble increased from 21.1 in 1978 to 22.4 in 1988 and to 24.0 in 1998. It must, however, be appreciated that there may be clinically significant differences in disease levels, susceptibility and attitudes to dental care between the groups identified by their reported usual reason for dental attendance.

Table 5.2.7

Results for England and Wales are available from four surveys over a 30 year period since 1968 and the observations above also apply. On these data, the different patterns of improvement among those with different reported usual reasons for attendance were more clear, with those who attend only with trouble consistently showing a greater survey-to-survey improvement. Thus, over the past thirty years, the average number of teeth among dentate adults who reported attendance only with trouble increased by 4.6 (19.6 to 24.2) compared with an increase of only 0.8 teeth among those who report regular check-ups, resulting, for the first time in this series of surveys, in similar average numbers of teeth in 1998 among these two groups.

Tables 5.2.8–10

5.2.4 Future expectations for retaining 21 or more teeth

Future expectations for retaining 21 or more teeth have been calculated based on the overall adult population (that is including those who were edentate). Table 5.2.4 showed the proportion of dentate adults with 21 or more teeth and Table 5.2.11 shows the proportion recalculated to include all adults. In 1998 the overall proportion of UK adults with 21 or more natural teeth was 72%, a figure which has shown a consistent increase since the 51% reported in 1978. The change has occurred in all age groups, but has been particularly pronounced in the older age groups. In 1978 only 10% of all people aged 55 and over had 21 or more teeth. In 1998, 45%

of 55 to 64 year olds and 29% of 65 to 74 year olds had at least 21 teeth; while the group aged 75 and over (those who were aged 55 and over in 1978) still had 10% with 21 or more natural teeth.

If the rate of change observed over the last decade continues over the next 3 decades the expectation would be of a continued increase in the proportion of the United Kingdom population who have 21 or more natural teeth, reaching 90% of all 16 to 74 year olds by 2018 and nearly three quarters (74%) of 65 to 74 year olds by 2028. This prediction assumes no slow down in the rate of tooth loss. However, the loss of individual teeth depends on the loss of tooth tissue and periodontal tissue, accumulated over a lifetime through disease and treatment. For the age groups under the age of 35 in 1998, the experience of decay has been less than for their predecessors, and the treatment techniques for early disease have been generally less destructive. The loss of teeth resulting from the burden of accumulated disease by late middle age might be expected to be a little less, or a little later in life, than it has been for their predecessors. The 15 percentage point drop-out from the 21 or more teeth group for those aged 45 to 54 in 1988, shown in Table 5.2.12, may be over pessimistic for future cohorts. Even higher levels of retention of a functional dentition could reasonably be hoped for in decades to come.

Tables 5.2.11–12

5.2.5 Changes in the proportion of people with natural teeth and dentures

The close relationship between the numbers of teeth people have and having dentures is discussed in Chapter 2.2. This section looks briefly at the overall change in the provision of dentures in conjunction with natural teeth over the past twenty years.

The proportion of the adult population who had dentures with natural teeth has decreased from 21% in 1978 to 16% in 1998. This decrease was consistent across all age groups for people aged less than 55. The wearing of dentures with natural teeth among 16 to 34 year olds has almost been eliminated, while among adults aged 35 to 44 the proportion has halved over the past twenty years; in 1978 26% of those aged 35 to 44 had dentures compared with 10% in 1998. The figures for the two oldest age groups are largely a reflection of past experience. The retention of at least some natural teeth among these older people has increased but their overall experience of tooth loss is considerably higher than those in the younger age groups. The proportion of denture wearers among those aged 55 to 64 has remained relatively stable since 1978 with no significant change between 1988 and 1998, while among

those aged 65 and over the proportion has increased from 14% in 1978 to 23% in 1988 and to 31% in 1998.

Table 5.2.13–14

5.2.6 The diagnostic criteria used for comparison with the results from the previous surveys

Before reporting changes in the condition of the teeth over time, it is important to consider how and why the diagnostic criteria have changed between the 1988 and 1998 surveys and how valid comparisons can still be drawn from 1998 data when they are presented in a way which is compatible with the conventions used in the earlier surveys.

In the 1998 survey the criteria for measuring decay (dental caries) were explicitly changed from those used in the earlier surveys in response to increased understanding about the progress of dental caries and developments in dental epidemiological methods. For the first time, visual dentine caries (that is, decay in the dentine which has not yet cavitated) was included in the measure of decay used in these surveys. In previous Adult Dental Health Surveys, a tooth was recorded as decayed only if cavitated caries was present. In the 1988 survey and earlier, teeth with untreated visual dentinal caries were recorded as sound and untreated, but in 1998 they were classified as decayed using a new code that allowed calculation of results with or without the inclusion of the condition. This ensured that an appropriate estimate of the 1998 prevalence of dentinal caries could be made, while the data could still be made comparable to the earlier surveys in order to examine trends in the condition of the teeth. Chapter 3.1 discusses the results of all these changes in more detail[2].

All 1998 figures presented in the following sections and in Chapter 5.3 are based on the criteria used by the earlier surveys, which exclude visual caries.

5.2.7 Changes in sound and untreated teeth

In 1998, dentate adults in the UK had, on average, 15.7 sound and untreated teeth compared with 14.8 in 1988 and 13.0 in 1978. The increase in the average number of sound and untreated teeth was seen across all age groups with the exception of those aged 55 and over. For example, 25 to 34 year olds had, on average 16.0 sound and untreated teeth in 1988 and 19.7 in 1998, an increase of 3.7 teeth on average.

These variations are also seen in the proportion of dentate adults with 18 or more sound and untreated teeth. Between

1978 and 1998, the proportion of 16 to 44 year olds with 18 or more sound and untreated teeth doubled. For example, in 1978, 50% of 16 to 24 year olds had 18 or more such teeth compared with 92% in 1998. Among the oldest age group (55 and over), however, the proportion with 18 or more such teeth has decreased from 9% in 1978 to 5% in 1998. This apparent decrease in the dental health of older dentate adults must be viewed in the context of the increase in the numbers of dentate adults in this age group (see Section 5.2.2) and in the increase in the average number of teeth among these dentate adults. Thus in 1998, among older people, more teeth are being preserved than before, although not in a sound and untreated state.

Figure 5.2.4, Tables 5.2.15–16

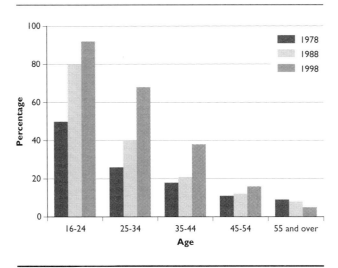

Fig 5.2.4   **Proportion of dentate adults with 18 or more sound and untreated teeth by age, 1978-98**

Dentate men and women showed a similar increase in the number of sound and untreated teeth (0.9) from 1988 to 1998. By 1998, the differences observed in 1978 between the mean numbers of sound and untreated teeth among dentate adults from manual and non-manual households had almost disappeared, as had the previously observed differences in the proportions of dentate adults in these groups who had 18 or more sound and untreated teeth. All the social class groupings showed improvements between 1978 and 1998 but among those in the unskilled manual group the increase between 1988 and 1998 was not statistically significant.

Tables 5.2.17–18

There was no difference between the rates of improvement between 1978 and 1998 in either the mean number of sound and untreated teeth or the proportion with 18 or more such teeth between dentate adults who reported attending the dentist regularly for check-ups and those who only went when

they had trouble with their teeth. However, the mean number of sound teeth for those giving a check-up as the reason for dental attendance was still lower than among those who attend only with trouble. Comparison by age of the proportion with 18 or more sound and untreated teeth by usual reason for attendance between 1988 and 1998 showed similar variation within age groups to that seen overall.

Tables 5.2.19–20

Results over a 30 year period for England and Wales confirmed the trends described in this section for sound and untreated teeth among dentate adults by age, gender, social class and the usual reason for dental attendance.

Tables 5.2 21–23

5.2.8 Changes in sound and untreated teeth among 16 to 24 year olds

One of the key age groups to monitor in order to understand changes in dental health over the last decade is the 16 to 24 year old group for whom the influence of the previous provision of restorative care is minimised in comparison with the older groups. Figure 5.2.5 shows for 16 to 24 year olds, by individual tooth type, the proportion of teeth which were sound and untreated in 1988 and 1998. It is evident that most of the improvement occurred in the first and second molar teeth (the 6's and 7's). In order to explore the significance of these changes and the possible influence of reported usual pattern of dental attendance, Figure 5.2.5 also shows the changes for those in this age group who reported attending for regular check-ups, occasional check-ups and only with trouble. The overall patterns were similar for these three groups, although the increase in the numbers of sound and untreated upper incisors (UL2, UL1, UR1, UR2) were slightly greater for those reporting regular check-ups. The improvements over the decade in the proportion of sound and untreated first and second molars were similar across the groups and varied between 16% and 36%. Further analysis determined that significant differences between the proportions of sound and untreated teeth were present between year of examination and tooth type but not between reported attendance groups. Thus, for the first and second molar teeth in 16 to 24 year olds, dental health had improved significantly irrespective of reported usual reason for dental attendance.

Figure 5.2.5, Table 5.2.24

Fig 5.2.5   **Sound and untreated teeth by usual reason for dental attendance among dentate adults aged 16-24, 1988-98**

Dentate adults aged 16-24

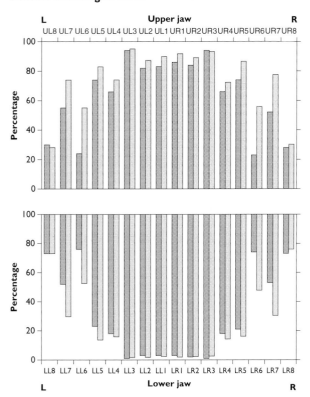

Dentate adults aged 16-24: Regular check-ups

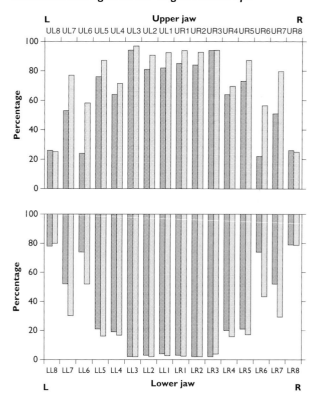

Dentate adults aged 16-24: Occasional check-ups

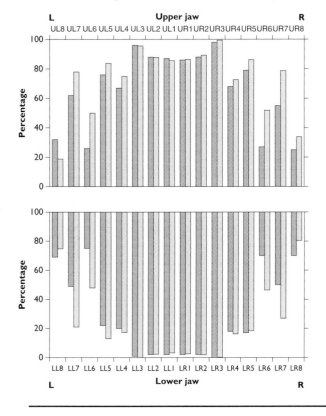

Dentate adults aged 16-24: Only attends with trouble

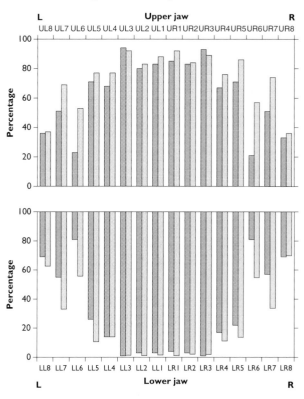

The condition of the natural teeth   1988   1998

5.2.9 Future expectations for the numbers of sound and untreated teeth

Previous sections have calculated, for total tooth loss and the proportion retaining 21 or more teeth, the likely levels for the future based on current findings (Sections 5.1.6 and 5.2.4). Predictions relating to tooth loss are relatively straightforward in that the assumptions can be simply stated. However, predictions relating to the condition of the natural teeth are more complex.

Direct extrapolation from the "conventional" measures of dental and oral health that have been used in this series of UK Adult Dental Health Surveys[3] is becoming more difficult following the significant shifts in the patterns of tooth loss (see Chapter 5.1) and changes in the understanding, assessment and clinical management of periodontal diseases and dental caries.

Extrapolation is made more complex because of the increasing diversity between young and old groups within the population and, between subsets within any age cohort. Age is not the only factor to be taken into consideration, over the last few decades social changes have taken place which have affected different social groups differentially.

It is also possible that the real improvements in caries prevalence seen in the early teenage years may be accompanied, for a substantial subset of young adults, by "delayed progression". In this situation decay is not eradicated but merely progresses undetected at a very slow rate, typically being detected in late teenage or early adulthood once significant and irreversible destruction of tooth tissue has taken place. Thus, difficulties in making accurate predictions are increasing.

Given these provisos, Table 5.2.25 shows estimates for the proportion of dentate adults aged 16 to 64 who are likely to have 18 or more sound and untreated teeth in 2008. The estimates are made on the following assumptions:

- the proportion of 16 to 24 year olds who have 18 or more sound and untreated teeth will remain at the same level (92%) over the next ten years
- there will continue to be a marked decrease in the proportion with 18 or more sound and untreated teeth between the two youngest age groups
- the increase in the numbers of older dentate adults will not affect the distribution.

On this basis, the largest increase in the proportion of dentate adults aged 16 to 64 with 18 or more sound and untreated teeth will be among those aged 35 to 44 while the overall figure for those aged 16 to 64 could increase by around 10 percentage points. However, the predicted increase in the proportion of older people retaining their natural teeth may serve to reduce the overall level of these improvements since the proportion of sound and untreated teeth declines with age.

Table 5.2.25

5.2.10 Changes in decayed or unsound teeth

As described in Section 5.2.6, in order to compare the 1998 data directly with those from the 1978 and 1988 surveys, visual dentine caries has been excluded from the measure of tooth decay.

Overall, the average number of decayed or unsound teeth in dentate adults decreased between 1978 and 1988, from 1.9 to 1.1, but did not change significantly between 1988 and 1998. Similarly there was no significant difference between 1988 and 1998 in the proportion of dentate adults with no active decay (56% and 58%). These findings were repeated across the age groups.

Figure 5.2.6 Tables 5.2.26–27

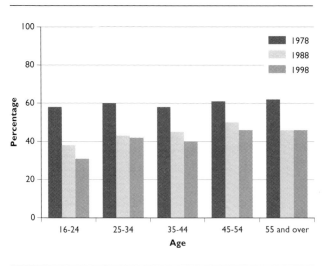

Fig 5.2.6 **Proportion of dentate adults with some decayed or unsound teeth by age, 1978-98**

Analysis by gender, social class, and the usual reason for dental attendance all showed a decrease in the levels of decayed or unsound teeth between 1978 and 1988 and little change between 1988 and 1998.

Tables 5.2.28–30

The figures over the thirty year period for England and Wales from 1968 to 1998 showed that the largest decrease in the level of decayed and unsound teeth had come between 1978 and 1988, from 1.9 teeth on average to 0.9 teeth, although there had been a significant decrease between 1968 and 1978, from 2.2 teeth to 1.9. These findings were repeated across the age groups and for those who reported only attending with trouble. Among those who reported attending for regular check-ups the only decrease was between 1978 and 1988 from 1.1 decayed and unsound teeth to 0.6 such teeth.

Tables 5.2.31–33

5.2.11 A summary of the condition of the natural teeth

Table 5.2.34 presents a summary of the mean number of teeth in each of the conditions discussed so far and for restored (otherwise sound) teeth which are discussed in Chapter 5.3. The table shows how the distribution of teeth in each condition is shifting away from missing teeth towards sound and untreated teeth while there has been less change in the overall average numbers of decayed or unsound teeth and none in the overall average number of restored (otherwise sound) teeth. This table also highlights the increasing diversity between the younger and older groups referred to earlier. Thus, in 1998, the older age groups were keeping more of their teeth than in previous decades but were more likely to have had them restored. The younger adults not only had more teeth, but also had more teeth with no experience of disease.

Table 5.2.34

5.2.12 Changes in the condition of individual teeth

In order to give an overview of the changes in the distribution of different tooth conditions around the mouth between 1988 and 1998, Figure 5.2.7 compares the proportions of individual teeth in each condition in 1988 and 1998 for all dentate adults, young dentate adults (aged 16 to 24) and older dentate adults (aged 45 to 54). The figure shows stacked bar charts for each tooth site, the proportion which were sound and untreated, restored (otherwise sound) – either filled or artificially crowned –, decayed or unsound, and missing. The differences between the decades discussed earlier can be seen, in particular the greater proportion of sound and untreated teeth and smaller proportion of missing teeth in 1998, even for the decay susceptible first molars (the "6s"). The second part shows a similar presentation for the 16 to 24 year old group and shows similar changes since 1988. The increases in sound and untreated teeth are matched by a decreased proportion of restored (otherwise sound) teeth in this age band. The third part of Figure 5.2.7 shows data presented in the same way but for the 45 to 54 year old age group. In this group the improvements in sound and untreated teeth are generally accompanied by significant increases in the proportion of restored (otherwise sound) teeth around the mouth. It is also noteworthy that the proportions of third molar teeth present are considerably higher in 1998 than in 1988. This may reflect changes in attitudes to removal of these teeth among patients and dentists, but may also be influenced by reductions in dental disease.

Figure 5.2.7, Tables 5.2.35–37

Notes and references

1.　This figure was calculated by using a form of direct age standardisation. In this method the mean values for each age group in 1998 were applied to the 1988 age distribution and an overall value was calculated.

2.　A further change to note is that in the earlier surveys examiners were supplied with a blunted 0.7mm diameter dental probe and asked to use the probe to investigate lesions "if doubt exists" when considering a carious cavity, while examiners in 1998 used a 0.5 mm CPITN ball ended probe to clean fissures only but not as a diagnostic aid.

3.　See Chapter 1.1

Fig 5.2.7   Comparison of tooth conditions around the mouth 1988-98

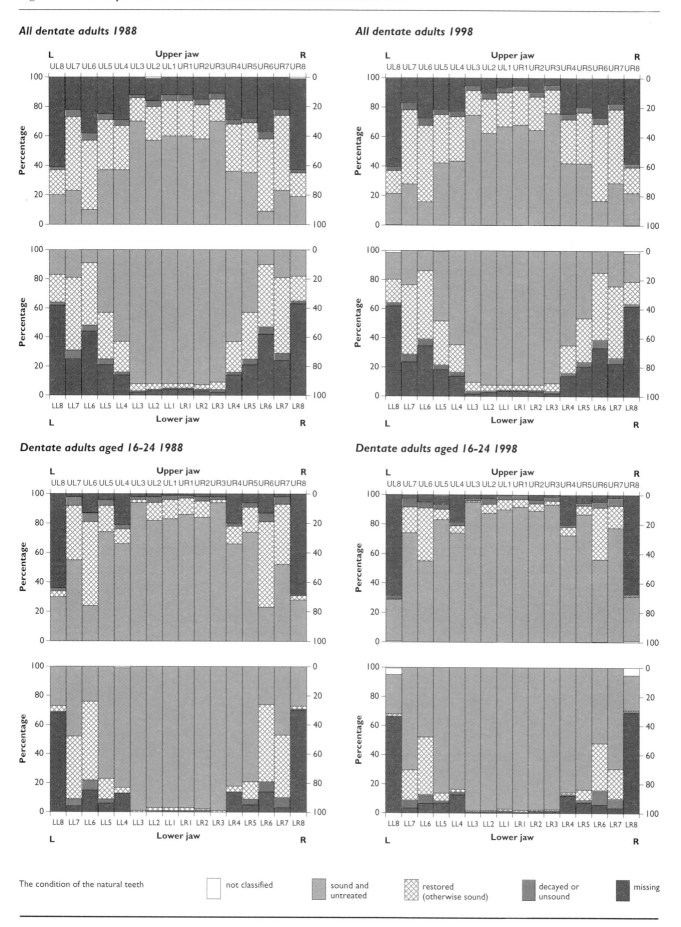

All dentate adults 1988

All dentate adults 1998

Dentate adults aged 16-24 1988

Dentate adults aged 16-24 1998

The condition of the natural teeth   ☐ not classified   ▨ sound and untreated   ⊠ restored (otherwise sound)   ▨ decayed or unsound   ■ missing

Fig 5.2.7   **Comparison of tooth conditions around the mouth 1988-98** (continued)

Dentate adults aged 45-54 1988   *Dentate adults aged 45-54 1998*

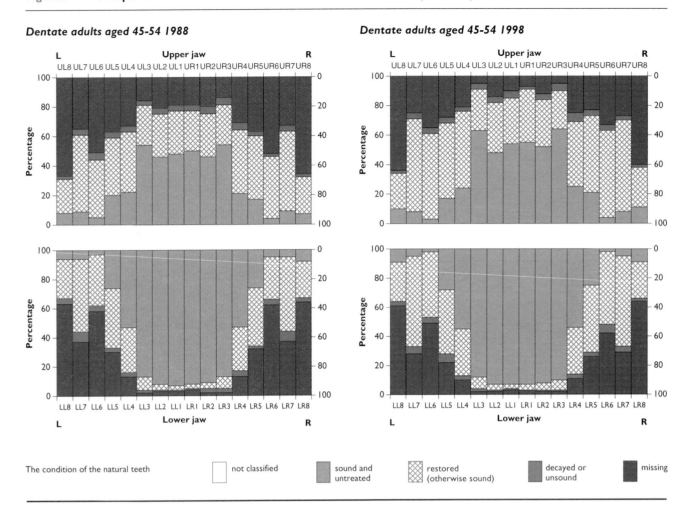

The condition of the natural teeth   ☐ not classified   ▨ sound and untreated   ⊠ restored (otherwise sound)   ▦ decayed or unsound   ■ missing

Table 5.2.1 **Distribution of the numbers of teeth, 1978-98**

Dentate adults   *United Kingdom*

Number of teeth	1978	1988	1998
	%	%	%
1-14	13	10	8
15-20	14	9	9
21-26	34	32	30
27-31	36	44	48
32	3	4	5
Mean	23.0	24.2	24.8
Base	3495	4331	3817

1988 Table 5.13

Table 5.2.2 **Distribution of the numbers of teeth by age, 1978-98**

Dentate adults   *United Kingdom*

Number of teeth	Age														
	16-24			25-34			35-44			45-54			55 and over		
	1978	1988	1998	1978	1988	1998	1978	1988	1998	1978	1988	1998	1978	1988	1998
	%	%	%	%	%	%	%	%	%	%	%	%	%	%	%
1-14	1	0	0	4	2	0	12	5	2	23	12	7	42	47	27
15-20	2	0	0	7	2	1	13	9	4	27	16	11	29	25	26
21-26	29	26	21	36	28	18	44	38	32	37	46	43	23	22	35
27-31	63	68	71	48	60	70	29	46	55	13	25	38	6	7	12
32	5	5	8	5	9	11	2	2	7	0	1	1	0	0	0
Mean	27.1	27.6	27.9	25.6	27.1	28.1	22.8	25.1	26.7	19.4	22.2	24.0	15.8	16.8	18.6

For bases see Tables Appendix.
1988 Table 5.13

Table 5.2.3   **Distribution of the numbers of teeth for all adults, 1978-98**

All adults   *United Kingdom*

Number of teeth	1978	1988	1998
	%	%	%
0 (Edentate)	30	21	13
1-14	9	8	7
15-20	10	7	8
21-26	24	25	26
27-31	25	35	42
32	2	3	4

The distribution of the numbers of teeth for all adults was obtained by recalculating the percentage of dentate adults (based on the survey dental examination) to include the proportion of edentate adults (based on the interview data). As these figures come from different samples the bases for the resulting percentage are not shown.

Table 5.2.4   **Proportion with 21 or more teeth, 1978-98**

Dentate adults   *United Kingdom*

Age	1978	1988	1998
	percentage with 21 or more teeth		
16-24	97	100	100
25-34	89	96	98
35-44	75	86	94
45-54	50	72	82
55-64		48	57
65-74	29	25	46
75 and over		16	23
All	73	80	83

For bases see Tables Appendix
1988 Tables 4.1, 4.3

Table 5.2.5   **Distribution of the numbers of teeth by gender, 1978-98**

Dentate adults   *United Kingdom*

Number of teeth	Gender					
	Men			Women		
	1978	1988	1998	1978	1988	1998
	%	%	%	%	%	%
1-14	14	10	8	13	10	8
15-20	13	9	9	14	10	9
21-26	34	31	28	34	34	32
27-31	36	45	48	37	43	47
32	3	4	7	2	3	4
Mean	22.9	24.4	25.0	23.2	24.1	24.6

For bases see Tables Appendix
1988 Table 5.14

Table 5.2.6   **Distribution of the numbers of teeth by social class of head of household, 1978-98**

Dentate adults   _United Kingdom_

Number of teeth	Social class of head of household								
	I, II, IIINM			IIIM			IV,V		
	1978	1988	1998	1978	1988	1998	1978	1988	1998
	%	%	%	%	%	%	%	%	%
1-14	12	9	6	13	12	10	16	13	13
15-20	12	9	8	16	9	10	15	11	12
21-26	35	30	30	33	35	32	35	34	29
27-31	38	49	50	35	40	44	32	39	42
32	3	4	6	3	4	4	2	3	4
Mean	23.6	24.8	25.4	22.8	23.1	24.2	22.4	23.4	23.6

For bases see Tables Appendix
1988 Table 5.15

Table 5.2.7   **Distribution of the numbers of teeth by usual reason for dental attendance, 1978-98**

Dentate adults   _United Kingdom_

Number of teeth	Usual reason for dental attendance*					
	Regular check-up			Only with trouble		
	1978	1988	1998	1978	1988	1998
	%	%	%	%	%	%
1-14	7	6	6	21	19	13
15-20	11	10	9	17	11	10
21-26	38	35	34	32	30	24
27-31	42	47	46	28	37	47
32	2	3	4	2	4	6
Mean	24.4	24.9	24.9	21.1	22.4	24.0

*Excludes people who only visit the dentist for an occasional check-up
For bases see Tables Appendix
1988 Table 5.16

Table 5.2.8   **Distribution of the numbers of teeth by gender, 1968-98**

Dentate adults   *England and Wales*

Number of teeth	Gender											
	Men				Women				All			
	1968	1978	1988	1998	1968	1978	1988	1998	1968	1978	1988	1998
	%	%	%	%	%	%	%	%	%	%	%	%
1-14	15	14	10	8	18	12	10	8	17	13	10	8
15-20	17	13	9	9	15	13	9	9	16	13	9	9
21-26	35	33	30	28	35	34	34	31	35	33	32	30
27-31	31	37	46	49	30	38	44	48	30	38	45	48
32	2	3	5	7	2	3	3	4	2	3	4	6
Mean	22.2	23.1	24.5	25.1	21.7	23.3	24.2	24.7	21.9	23.2	24.4	24.9

For bases see Tables Appendix
1988 Table 5.7

Table 5.2.9   **Distribution of the numbers of teeth by age, 1968-98**

Dentate adults   *England and Wales*

Number of teeth	Age																			
	16-24				25-34				35-44				45-54				55 and over			
	1968	1978	1988	1998	1968	1978	1988	1998	1968	1978	1988	1998	1968	1978	1988	1998	1968	1978	1988	1998
	%	%	%	%	%	%	%	%	%	%	%	%	%	%	%	%	%	%	%	%
1-14	1	1	0	0	7	3	2	0	14	12	5	2	34	23	11	6	47	41	38	26
15-20	3	1	0	0	10	7	2	1	19	11	8	4	27	27	15	10	32	29	25	26
21-26	33	27	25	20	41	35	26	17	44	43	37	32	30	36	46	43	17	23	30	35
27-31	60	66	69	72	39	50	61	71	22	32	48	56	8	14	26	40	4	6	7	12
32	3	5	6	8	3	5	9	11	1	2	3	8	1	0	1	1	0	1	0	0
Mean	26.8	27.3	27.7	27.9	24.5	26.0	27.3	28.2	21.5	23.2	25.4	26.9	17.5	19.7	22.5	24.3	14.7	16.0	16.9	18.7

For bases see Tables Appendix
1988 Table 5.6

Table 5.2.10 **Distribution of the numbers of teeth by usual reason for dental attendance, 1968-98**

Dentate adults   England and Wales

Number of teeth	Usual reason for dental attendance*							
	Regular check-up				Only with trouble			
	1968	1978	1988	1998	1968	1978	1988	1998
	%	%	%	%	%	%	%	%
1-14	7	7	6	6	26	20	18	12
15-20	13	10	10	9	19	16	10	10
21-26	38	37	34	34	34	31	30	23
27-31	40	43	48	46	20	30	38	48
32	2	3	3	5	1	3	4	7
Mean	24.2	24.6	25.0	25.0	19.6	21.4	22.6	24.2

*Excludes people who only visit the dentist for an occasional check-up
For bases see Tables Appendix
1988 Table 5.8

Table 5.2.11 **Proportion with 21 or more natural teeth, 1978-98**

United Kingdom

Age	1978	1988	1998
		percentages	
16-24	97	100	100
25-34	85	95	98
35-44	65	83	93
45-54	34	60	77
55-64		30	45
65-74	} 10	} 8	29
75 and over			10
All	51	63	72

The percentage of all adults with 21 or more teeth was calculated by multiplying the percentage of adults who are dentate (taken from the interview sample) by the percentage of dentate adults with 21 or more teeth (taken from the examination sample). As these figures come from different samples the base for the resulting percentage is not shown.
1988 Table 4.3

Table 5.2.12 Future levels of adults having 21 or more natural teeth

All adults   *United Kingdom*

Age	Percentage with 21 or more natural teeth 1998	Assumed future rate of loss*	2008	2018	2028	Base for 1998 figures
			estimated percentage with 21 or more natural teeth			
16-24	100		100	100	100	702
		2				
25-34	98		98	98	98	1158
		2				
35-44	93		96	96	96	1114
		6				
45-54	77		87	90	90	1106
		15				
55-64	45		62	72	75	833
		1				
65-74	29		44	71	74	758
16-74	78		85	90	90	5671

* Based on the assumption that the future decrease in the proportion of dentate adults with 21 or more natural teeth will be the same as between 1988 - 1998.
1988 Table 4.4

Table 5.2.13 The reliance on natural teeth and dentures, 1978-1998

All adults   *United Kingdom*

Dental status	1978	1988	1998
	%	%	%
Natural teeth only	49	60	72
Natural teeth and dentures	21	19	16
No natural teeth (edentate)	30	21	13
Base	5967	6825	6204

1988 Table 16.13

Table 5.2.14 The reliance on natural teeth and dentures by age, 1978-98

All adults   *United Kingdom*

Dental status	Age																	
	16-24			25-34			35-44			45-54			55-64			65 and over		
	1978	1988	1998	1978	1988	1998	1978	1988	1998	1978	1988	1998	1978	1988	1998	1978	1988	1998
	%	%	%	%	%	%	%	%	%	%	%	%	%	%	%	%	%	%
Natural teeth only	94	97	99	79	89	97	61	76	89	33	52	75	15	30	46	7	9	23
Natural teeth and dentures	6	3	1	17	10	3	26	20	10	34	31	18	29	33	34	14	23	31
No natural teeth (edentate)	0	0	0	4	1	0	13	4	1	32	17	6	56	37	20	79	67	46

For bases see Tables Appendix
1988 Table 16.13

Table 5.2.15 Distribution of the numbers of sound and untreated teeth, 1978-98

Dentate adults   *United Kingdom*

Number of sound and untreated teeth	1978	1988	1998
	%	%	%
0	1	1	1
1-5	10	8	7
6-11	32	27	24
12-17	32	29	26
18 or more	25	35	42
Mean	13.0	14.8	15.7
Base	3495	4331	3817

1998 data based on 1988 criteria (see Section 5.2.6)
1988 Table 6.16

Table 5.2.16 **Distribution of the numbers of sound and untreated teeth by age, 1978-98**

Dentate adults   *United Kingdom*

Number of sound and untreated teeth	Age														
	16-24			25-34			35-44			45-54			55 and over		
	1978	1988	1998	1978	1988	1998	1978	1988	1998	1978	1988	1998	1978	1988	1998
	%	%	%	%	%	%	%	%	%	%	%	%	%	%	%
0	0	0	0	1	1	0	1	0	0	2	1	0	5	1	3
1-5	2	1	0	6	3	1	10	6	5	13	11	9	25	21	18
6-11	15	2	2	28	20	7	41	37	18	49	41	39	40	45	47
12-17	33	17	6	39	37	24	30	36	38	25	35	35	21	24	26
18 or more	50	80	92	26	40	68	18	21	38	11	12	16	9	8	5
Mean	17.1	21.2	24.0	13.9	16.0	19.7	12.1	13.0	15.7	10.4	11.5	11.9	9.0	9.4	9.2

1998 data based on 1988 criteria (see Section 5.2.6)
For bases see Tables Appendix
1988 Table 6.16

Table 5.2.17 **Distribution of the numbers of sound and untreated teeth by gender, 1978-98**

Dentate adults   *United Kingdom*

Number of sound and untreated teeth	Gender					
	Men			Women		
	1978	1988	1998	1978	1988	1998
	%	%	%	%	%	%
0	2	1	1	1	1	1
1-5	11	8	7	9	7	8
6-11	29	26	22	36	28	26
12-17	30	28	27	32	31	26
18 or more	28	38	44	22	32	39
Mean	13.4	15.2	16.1	12.7	14.3	15.2

1998 data based on 1988 criteria (see Section 5.2.6)
For bases see Tables Appendix
1988 Table 6.17

Table 5.2.18 **Distribution of the numbers of sound and untreated teeth by social class of head of household, 1978-98**

Dentate adults   *United Kingdom*

Number of sound and untreated teeth	Social class of head of household								
	I, II, IIINM			IIIM			IV, V		
	1978	1988	1998	1978	1988	1998	1978	1988	1998
	%	%	%	%	%	%	%	%	%
0	1	1	1	1	0	1	2	1	1
1-5	11	8	7	9	8	7	8	6	7
6-11	37	28	24	31	27	25	28	28	24
12-17	32	31	27	31	29	28	31	25	25
18 or more	19	33	41	28	36	40	31	41	42
Mean	12.1	14.4	15.7	13.4	14.7	15.1	14.0	15.4	15.7

1998 data based on 1988 criteria (see Section 5.2.6)
For bases see Tables Appendix
1988 Table 6.18

Table 5.2.19 **Distribution of the numbers of sound and untreated teeth by usual reason for dental attendance, 1978-98**

Dentate adults   *United Kingdom*

Number of sound and untreated teeth	Usual reason for dental attendance*					
	Regular check-up			Only with trouble		
	1978	1988	1998	1978	1988	1998
	%	%	%	%	%	%
0	1	0	1	2	1	1
1-5	9	7	7	12	10	8
6-11	39	31	27	27	35	18
12-17	34	32	28	28	26	24
18 or more	17	29	36	31	38	48
Mean	11.9	14.0	14.9	13.7	14.8	16.5

*Excludes people who only visit the dentist for an occasional check-up
1998 data based on 1988 criteria (see Section 5.2.6)
For bases see Tables Appendix
1988 Table 6.19

Table 5.2.20 **18 or more sound and untreated teeth by usual reason for dental attendance and age, 1988-98**

Dentate adults   *United Kingdom*

| Age | Usual reason for dental attendance† | | | | | |
| | Regular check-up | | Only with trouble | | All | |
	1988	1998	1988	1998	1988	1998
			percentage with 18 or more sound and untreated teeth			
16-24	79	94	79	91	80	92
25-34	30	68	47	68	40	68
35-44	16	34	30	46	21	38
45-54	10	12	15	20	12	16
55-64	5	4	17	6	10	5
65 and over	3	4	5	6	5	5
All	29	36	38	48	35	42

†Excludes people who only visit the dentist for an occasional check-up
1998 data based on 1988 criteria (see Section 5.2.6)
For bases see Tables Appendix
1988 Table 6.8

Table 5.2.21 **Distribution of the numbers of sound and untreated teeth by gender, 1968-98**

Dentate adults   *England and Wales*

| Number of sound and untreated teeth | Gender | | | | | | | | | | | |
| | Men | | | | Women | | | | All | | | |
	1968	1978	1988	1998	1968	1978	1988	1998	1968	1978	1988	1998
	%	%	%	%	%	%	%	%	%	%	%	%
0	1	1	0	1	1	1	1	1	1	1	1	1
1-5	11	10	8	6	9	9	7	8	10	9	7	7
6-11	27	28	25	22	36	35	28	25	31	32	26	23
12-17	34	31	27	26	35	32	31	26	35	31	29	26
18 or more	27	30	40	45	19	23	34	40	23	27	37	43
Mean	13.3	13.6	15.5	16.4	12.3	12.9	14.5	15.4	12.8	13.2	15.0	15.9

1998 data based on 1988 criteria (see Section 5.2.6)
For bases see Tables Appendix
1988 Table 6.10

Table 5.2.22 Distribution of the numbers of sound and untreated teeth by age, 1968-98

Dentate adults   *England and Wales*

Number of sound and untreated teeth	Age																			
	16-24				25-34				35-44				45-54				55 and over			
	1968	1978	1988	1998	1968	1978	1988	1998	1968	1978	1988	1998	1968	1978	1988	1998	1968	1978	1988	1998
	%	%	%	%	%	%	%	%	%	%	%	%	%	%	%	%	%	%	%	%
0	0	0	0	0	0	1	1	0	1	0	0	0	2	2	1	0	4	4	1	3
1-5	2	2	0	0	6	6	2	0	9	9	6	4	15	12	10	8	28	25	22	18
6-11	15	14	2	2	26	26	18	6	35	40	36	16	45	49	41	38	38	38	45	48
12-17	39	31	15	4	39	39	37	23	36	31	36	39	28	26	35	36	20	22	25	26
18 or more	44	53	83	94	28	28	42	71	19	20	23	40	10	11	13	17	10	11	8	5
Mean	16.4	17.5	21.6	24.3	13.6	14.1	16.4	20.1	12.5	12.5	13.3	16.0	10.3	10.6	11.7	12.1	8.9	9.1	9.5	9.3

1998 data based on 1988 criteria (see Section 5.2.6)
For bases see Tables Appendix
1988 Table 5.6

Table 5.2.23 Distribution of the numbers of sound and untreated teeth by usual reason for dental attendance, 1968-98

Dentate adults   *England and Wales*

Number of sound and untreated teeth	Usual reason for dental attendance*							
	Regular check-up				Only with trouble			
	1968	1978	1988	1998	1968	1978	1988	1998
	%	%	%	%	%	%	%	%
0	1	1	0	1	1	2	1	1
1-5	8	9	7	7	13	11	9	8
6-11	38	37	31	26	26	26	24	18
12-17	40	35	32	28	29	28	26	24
18 or more	13	18	30	37	31	33	40	50
Mean	11.8	12.2	14.1	15.0	13.3	14.1	15.1	16.8

Excludes people who only visit the dentist for an occasional check-up
1998 data based on 1988 criteria (see Section 5.2.6)
For bases see Tables Appendix
1988 Table 6.11

Table 5.2.24 Sound and untreated teeth by usual reason for dental attendance, 1988-98

Dentate adults aged 16-24   *United Kingdom*

	Left								Right							
	8	7	6	5	4	3	2	1	1	2	3	4	5	6	7	8

percentage of teeth sound and untreated

	8	7	6	5	4	3	2	1	1	2	3	4	5	6	7	8
All																
Upper jaw																
1988	30	55	24	74	66	94	82	83	86	84	94	66	74	23	52	28
1998	28	74	55	83	74	95	87	90	92	89	93	72	87	56	78	30
Lower jaw																
1988	27	48	24	77	82	99	97	97	97	98	99	82	79	26	47	27
1998	27	70	48	86	84	98	98	98	98	98	97	86	84	52	70	24
Regular check-ups																
Upper jaw																
1988	26	53	24	76	64	94	81	82	85	84	94	64	73	22	51	26
1998	25	77	58	87	71	97	91	92	94	93	94	70	87	56	80	25
Lower jaw																
1988	22	48	26	79	81	98	97	96	97	98	98	80	79	26	48	21
1998	20	70	48	84	83	98	98	97	98	98	96	84	83	57	71	21
Occasional check-ups																
Upper jaw																
1988	32	62	26	76	67	96	88	87	86	88	98	68	79	27	55	25
1998	19	78	50	84	75	96	88	86	86	89	99	73	86	52	79	34
Lower jaw																
1988	31	51	25	78	80	99	98	98	98	98	100	82	83	30	50	30
1998	25	79	52	87	83	100	98	97	97	98	100	84	82	54	73	20
Only with trouble																
Upper jaw																
1988	36	51	23	71	68	94	80	83	85	83	93	67	71	21	51	33
1998	37	69	53	77	77	92	83	88	92	84	89	76	86	57	74	36
Lower jaw																
1988	31	45	19	74	86	99	97	97	96	97	99	83	78	19	43	31
1998	37	67	44	89	86	99	99	98	99	98	98	89	86	45	66	30

1998 data based on 1988 criteria (see Section 5.2.6)
1988 Tables 10.1, 10.11, 10.12, 10.13

Table 5.2.25 **Future level of dentate adults with 18 or more sound and untreated teeth**

Dentate adults   *United Kingdom*

Age	Percentage with 18 or more sound and untreated teeth in 1988	Percentage with 18 or more sound and untreated teeth in 1998	Assumed future rate of change	Estimates for 2008	Base for 1998 figures
		percentage			
16-24	80	92		92	491
			12		
25-34	40	68		80	854
			2		
35-44	21	38		66	781
			5		
45-54	12	16		33	746
			7		
55-64	-	5		9	461
16-64	-	47		58	3333

Table 5.2.26 **Distribution of the number of decayed or unsound teeth, 1978-98**

Dentate adults   *United Kingdom*

Number of decayed or unsound teeth	1978	1988	1998
	%	%	%
0	40	56	58
1-5	50	41	39
6 or more	10	3	3
Mean	1.9	1.1	1.0
Base	3495	4331	3817

1998 data based on 1988 criteria (see Section 5.2.6)
1988 Table 8.22

Table 5.2.27 Distribution of the numbers of decayed or unsound teeth by age, 1978-98

Dentate adults   *United Kingdom*

Number of decayed or unsound teeth	Age														
	16-24			25-34			35-44			45-54			55 and over		
	1978	1988	1998	1978	1988	1998	1978	1988	1998	1978	1988	1998	1978	1988	1998
	%	%	%	%	%	%	%	%	%	%	%	%	%	%	%
0	42	62	69	40	57	58	42	55	60	39	50	54	38	54	54
1-5	48	35	27	50	40	39	49	42	37	54	48	43	53	42	44
6 or more	10	3	4	10	4	2	9	3	3	7	3	3	9	4	2
Mean	2.0	0.9	0.8	1.9	1.1	1.0	1.8	1.0	0.9	1.9	1.1	0.9	2.0	1.1	1.0

1998 data based on 1988 criteria (see Section 5.2.6)
For bases see Tables Appendix
1988 Table 8.22

Table 5.2.28 Distribution of the numbers of decayed or unsound teeth by gender, 1978-98

Dentate adults   *United Kingdom*

Number of decayed or unsound teeth	Gender					
	Men			Women		
	1978	1988	1998	1978	1988	1998
	%	%	%	%	%	%
0	37	52	55	44	60	62
1-5	52	42	41	50	37	37
6 or more	11	4	4	6	2	2
Mean	2.2	1.2	1.1	1.6	0.9	0.8

1998 data based on 1988 criteria (see Section 5.2.6)
For bases see Tables Appendix
1988 Table 8.23

Table 5.2.29 **Distribution of the numbers of decayed or unsound teeth by social class of head of household, 1978-98**

Dentate adults   *United Kingdom*

Number of decayed or unsound teeth	Social class of head of household								
	I, II, IIINM			IIIM			IV,V		
	1978	1988	1998	1978	1988	1998	1978	1988	1998
	%	%	%	%	%	%	%	%	%
0	44	62	63	40	51	55	34	47	52
1-5	50	36	36	50	45	42	54	47	43
6 or more	6	2	1	10	4	4	12	6	5
Mean	1.6	0.8	0.7	2.1	1.2	1.1	2.4	1.5	1.2

1998 data based on 1988 criteria (see Section 5.2.6)
For bases see Tables Appendix
1988 Table 8.24

Table 5.2.30 **Distribution of the numbers of decayed or unsound teeth by usual reason for dental attendance, 1978-98**

Dentate adults   *United Kingdom*

Number of decayed or unsound teeth	Usual reason for dental attendance†					
	Regular check-up			Only with trouble		
	1978	1988	1998	1978	1988	1998
	%	%	%	%	%	%
0	52	66	64	27	41	44
1-5	45	34	34	49	52	50
6 or more	3	1	1	24	8	6
Mean	1.1	0.6	0.7	3.0	1.7	1.6

†Excludes people who only visit the dentist for an occasional check-up
1998 data based on 1988 criteria (see Section 5.2.6)
For bases see Tables Appendix
1988 Table 8.25

Table 5.2.31 **Distribution of the numbers of decayed or unsound teeth by gender, 1968-98**

Dentate adults   *England and Wales*

Number of decayed or unsound teeth	Gender											
	Men				Women				All			
	1968	1978	1988	1998	1968	1978	1988	1998	1968	1978	1988	1998
	%	%	%	%	%	%	%	%	%	%	%	%
0	33	37	53	55	40	44	61	61	36	41	57	58
1-5	53	52	43	41	52	49	37	37	52	50	40	39
6 or more	14	11	4	4	8	7	2	1	12	9	3	3
Mean	2.6	2.2	1.2	1.1	1.8	1.6	0.8	0.7	2.2	1.9	1.0	0.9

1998 data based on 1988 criteria (see Section 5.2.6)
For bases see Tables Appendix
1988 Table 8.16

Table 5.2.32 **Distribution of the numbers of decayed or unsound teeth by age, 1968-98**

Dentate adults   *England and Wales*

Number of decayed or unsound teeth	Age																			
	16-24				25-34				35-44				45-54				55 and over			
	1968	1978	1988	1998	1968	1978	1988	1998	1968	1978	1988	1998	1968	1978	1988	1998	1968	1978	1988	1998
	%	%	%	%	%	%	%	%	%	%	%	%	%	%	%	%	%	%	%	%
0	38	42	63	70	37	40	58	58	33	43	56	61	36	37	50	54	38	38	55	52
1-5	50	48	34	26	51	50	38	40	57	49	42	36	53	55	48	43	49	53	41	46
6 or more	12	10	2	4	12	10	4	2	10	8	3	3	11	8	3	3	13	9	4	3
Mean	2.1	2.0	0.8	0.8	2.3	1.9	1.1	0.9	2.2	1.8	1.0	0.9	2.2	1.9	1.1	0.9	2.2	2.0	1.1	1.0

1998 data based on 1988 criteria (see Section 5.2.6)
For bases see Tables Appendix
1988 Table 8.15

Table 5.2.33 **Distribution of the numbers of decayed or unsound teeth by usual reason for dental attendance, 1968-98**

Dentate adults   *England and Wales*

Number of decayed or unsound teeth	Usual reason for dental attendance†							
	Regular check-up				Only with trouble			
	1968	1978	1988	1998	1968	1978	1988	1998
	%	%	%	%	%	%	%	%
0	50	51	66	64	23	28	42	44
1-5	47	46	34	35	58	55	50	50
6 or more	3	3	1	1	19	17	7	6
Mean	1.1	1.1	0.6	0.6	3.2	2.9	1.7	1.5

†Excludes people who only visit the dentist for an occasional check-up
1998 data based on 1988 criteria (see Section 5.2.6)
For bases see Tables Appendix
1988 Table 8.17

Table 5.2.34 **The mean number of teeth in each condition, 1978-98**

Dentate adults *United Kingdom*

| | Condition of teeth | | | | | | | | | | | |
| | Missing | | | Decayed or unsound | | | Restored, otherwise sound | | | Sound & untreated | | |
	1978	1988	1998	1978	1988	1998	1978	1988	1998	1978	1988	1998
							mean number of teeth					
All	9.0	7.8	7.2	1.9	1.1	1.0	8.1	8.4	8.1	13.0	14.8	15.7
Age												
16-24	4.9	4.4	4.1	2.0	0.9	0.8	8.0	5.5	2.9	17.1	21.2	24.0
25-34	6.4	4.9	3.9	1.9	1.1	1.0	9.8	10.0	7.4	13.9	16.0	19.7
35-44	9.2	6.9	5.3	1.8	1.0	0.9	8.9	11.1	10.1	12.1	13.0	15.7
45-54	12.6	9.8	8.0	1.9	1.1	1.0	7.1	9.6	11.1	10.4	11.5	11.9
55-64	)	13.4	12.1	)	1.1	1.0	)	7.1	9.0	)	10.4	9.8
65-74	) 16.2	16.9	13.7	) 2.0	1.0	0.9	) 4.8	5.7	8.2	) 9.0	8.4	9.1
75 and over	)	19.3	16.6	)	1.5	1.1	)	3.7	6.5	)	7.4	7.6

1998 data based on 1988 criteria (see Section 5.2.6)
For bases see Tables Appendix

Table 5.2.35 Condition of the individual teeth, 1988-98

Dentate adults   *United Kingdom*

	Upper jaw															
	Left								Right							
	8	7	6	5	4	3	2	1	1	2	3	4	5	6	7	8
	%	%	%	%	%	%	%	%	%	%	%	%	%	%	%	%
1988																
Sound and untreated	20	23	10	37	37	70	57	60	60	58	70	36	35	9	23	19
Restored, otherwise sound	17	50	47	34	30	16	23	24	24	23	15	32	34	49	51	16
Decayed or unsound	2	5	5	4	4	2	4	4	4	4	4	3	3	5	4	2
Missing	61	22	38	25	29	12	15	12	12	15	11	29	28	37	22	62
1998																
Sound and untreated	21	28	16	42	43	74	62	67	68	65	76	42	42	17	29	22
Restored, otherwise sound	15	50	52	33	30	17	23	23	24	23	16	30	35	52	50	17
Decayed or unsound	3	5	5	3	4	3	4	3	3	3	3	3	3	4	4	2
Missing	60	17	28	22	23	6	10	7	6	10	5	25	20	27	18	58
Not classified	0	0	0	0	0	0	0	0	0	0	0	0	0	0	0	0

	Lower jaw															
	Left								Right							
	8	7	6	5	4	3	2	1	1	2	3	4	5	6	7	8
	%	%	%	%	%	%	%	%	%	%	%	%	%	%	%	%
1988																
Sound and untreated	17	19	9	43	63	92	92	92	92	93	91	63	43	10	19	18
Restored, otherwise sound	19	50	43	32	21	5	4	3	3	3	5	21	32	43	52	17
Decayed or unsound	2	6	4	4	2	1	1	1	1	1	2	2	4	5	5	2
Missing	62	25	44	21	14	2	3	4	4	3	2	14	21	42	24	63
1998																
Sound and untreated	18	24	14	48	65	91	93	92	92	92	91	65	46	15	24	19
Restored, otherwise sound	16	48	47	30	19	6	4	3	3	4	5	19	30	46	49	15
Decayed or unsound	2	5	5	4	3	2	1	1	1	1	2	2	3	6	4	2
Missing	62	24	34	18	14	2	3	4	4	3	2	14	20	33	22	62
Not classified	2	0	0	0	0	0	0	0	0	0	0	0	0	0	0	2

1998 data based on 1988 criteria (see Section 5.2.6)
1988 Table 10.6

Table 5.2.36 **Condition of the individual teeth, 1988-98**

Dentate adults aged 16-24   *United Kingdom*

Upper jaw

	Left								Right							
	8	**7**	**6**	**5**	**4**	**3**	**2**	**1**	**1**	**2**	**3**	**4**	**5**	**6**	**7**	**8**
	%	%	%	%	%	%	%	%	%	%	%	%	%	%	%	%
1988																
Sound and untreated	30	55	24	74	66	94	82	83	86	84	94	66	74	23	52	28
Restored, otherwise sound	4	37	57	18	10	2	12	13	11	11	2	12	17	58	41	3
Decayed or unsound	2	6	6	4	3	2	4	3	2	3	2	2	3	6	5	1
Missing	64	2	13	4	21	2	2	1	1	2	2	20	6	13	2	68
1998																
Sound and untreated	28	74	55	83	74	95	87	90	92	89	93	72	87	56	78	30
Restored, otherwise sound	1	18	36	7	5	1	6	7	5	5	3	6	6	35	15	1
Decayed or unsound	2	6	4	3	2	1	4	2	3	2	3	1	2	4	5	1
Missing	68	2	5	6	19	3	3	1	0	4	2	20	5	5	2	67
Not classified	1	0	0	0	0	0	0	0	0	0	0	0	0	0	0	1

Lower jaw

	Left								Right							
	8	**7**	**6**	**5**	**4**	**3**	**2**	**1**	**1**	**2**	**3**	**4**	**5**	**6**	**7**	**8**
	%	%	%	%	%	%	%	%	%	%	%	%	%	%	%	%
1988																
Sound and untreated	27	48	24	77	82	99	97	97	97	98	99	82	79	26	47	27
Restored, otherwise sound	4	43	54	14	4	1	2	2	2	1	1	4	12	53	43	2
Decayed or unsound	1	5	7	3	0	0	1	0	0	1	0	0	4	7	7	1
Missing	68	4	15	6	13	0	0	1	1	0	0	14	5	14	3	70
1998																
Sound and untreated	27	70	48	86	84	98	98	98	98	98	97	86	84	52	70	24
Restored, otherwise sound	1	21	40	5	2	1	1	1	1	1	1	1	7	32	21	1
Decayed or unsound	1	6	6	2	2	1	1	1	1	0	1	1	2	10	6	1
Missing	66	3	6	7	12	0	0	1	0	1	1	12	7	6	3	69
Not classified	4	0	0	0	0	0	0	0	0	0	0	0	0	0	0	5

1998 data based on 1988 criteria (see Section 5.2.6)
1988 Table 10.1

Table 5.2.37 **Condition of the individual teeth, 1988-98**

Dentate adults aged 45-54 *United Kingdom*

Upper jaw

	Left								**Right**							
	8	7	6	5	4	3	2	1	1	2	3	4	5	6	7	8
	%	%	%	%	%	%	%	%	%	%	%	%	%	%	%	%
1988																
Sound and untreated	8	9	5	20	22	54	46	48	50	46	54	21	17	4	9	7
Restored, otherwise sound	23	52	39	39	41	27	29	29	27	29	27	43	43	42	54	25
Decayed or unsound	2	4	5	4	4	3	4	4	4	5	5	5	3	2	4	2
Missing	67	35	51	37	33	16	21	19	19	20	14	31	37	52	33	66
1998																
Sound and untreated	10	8	3	17	24	63	48	54	55	52	64	25	21	4	8	11
Restored, otherwise sound	24	63	58	51	52	28	34	31	36	32	26	44	52	59	62	27
Decayed or unsound	2	4	4	4	3	4	4	5	2	4	5	6	4	4	3	2
Missing	64	24	35	28	21	5	14	9	8	13	5	25	24	34	26	60
Not classified	0	0	0	0	0	0	0	0	0	0	0	0	0	0	0	0

Lower

	Left								**Right**							
	8	7	6	5	4	3	2	1	1	2	3	4	5	6	7	8
	%	%	%	%	%	%	%	%	%	%	%	%	%	%	%	%
1988																
Sound and untreated	6	6	3	26	53	87	92	92	93	91	87	53	26	5	5	8
Restored, otherwise sound	27	50	35	41	31	9	4	3	3	4	8	30	40	29	51	25
Decayed or unsound	4	7	4	3	3	2	1	1	1	3	3	4	2	4	7	3
Missing	63	37	58	30	13	2	3	3	4	2	2	13	32	62	37	64
1998																
Sound and untreated	9	4	2	28	54	88	93	93	92	93	89	54	25	3	5	9
Restored, otherwise sound	27	62	45	44	32	8	4	3	4	5	7	32	46	50	62	25
Decayed or unsound	3	5	4	6	3	2	1	1	0	1	1	3	3	6	4	2
Missing	61	28	49	22	10	2	2	3	3	2	2	11	26	42	29	64
Not classified	0	0	0	0	0	0	0	0	0	0	0	0	0	0	0	1

1998 data based on 1988 criteria (see Section 5.2.6)
1988 Table 10.4

Table 5.2.38 **Change$ in the age structure of the dentate population, 1978-98**

Dentate adults   *United Kingdom*

Age	Based on interview sample			Based on examination sample		
	1978	1988	1998	1978	1988	1998
	%	%	%	%	%	%
16-24	23	24	16	23	24	16
25-34	26	23	22	26	23	23
35-44	21	21	21	21	21	20
45-54	15	15	18	15	14	18
55-64	10	11	12	} 14	10	11
65-74	} 5	6	8		} 8	} 12
75 and over		2	4			
Base	*4082*	*5280*	*5281*	*3495*	*4331*	*3817*

5.3   Trends in restorative treatment

Summary

- Overall, the average number of teeth which were restored (otherwise sound) has varied little from 1978 to 1998. In 1998, dentate adults had, on average 8.1 such teeth which was the same as that found in 1978 and similar to the level in 1988 (8.4).

- There was a marked disparity between younger and older dentate adults. Since 1988, the average number of restored (otherwise sound) teeth in dentate adults had decreased up to the age of 44 and increased in adults over this age. For example, among dentate adults aged 16 to 24 the average number of restored (otherwise sound) teeth decreased from 8.0 in 1978 to 2.9 in 1998, whereas among those age 55 and over it increased from 4.8 to 8.3.

- Among the youngest age group there was a sharp rise in the proportion with no restored (otherwise sound) teeth from 9% in 1978 to 13% in 1988 and to 30% in 1998, while, for those aged 55 and over, the proportion with no restored (otherwise sound) teeth has decreased from 26% in 1978 to 10% in 1998.

- There was a similar disparity between the restorative experience of those who reported regular dental check-ups and those who attended only with trouble. Among the former group the proportion with no restored (otherwise sound) teeth had remained at a low level (5% in 1998), while among the latter group the proportion with no restored (otherwise sound) teeth almost halved between 1978 and 1998 from 31% to 16%.

- Overall, from 1988 to 1998, there was an increase in the proportion of dentate adults with artificial crowns, from just over a quarter in 1988 (26%) to just over a third in 1998 (34%). There were increases in all but the two youngest age groups and the changes were greater among those aged over 45.

- The distribution of artificial crowns among dentate adults shifted upward between 1988 and 1998, with an increase in those with 3 or more crowns and a decrease in those with no crowns

- The increase in the provision of artificial crowns was similar for both men and women, among those with and without dentures in addition to their natural dentition, for all social classes and for each reported pattern of dental attendance.

5.3.1 Introduction

Restored teeth includes those which are filled and those with artificial crowns. In the previous surveys of Adult Dental Health[1], data on both these types of restorations have mainly been presented together and all these teeth have been described as 'filled'. As artificial crowns have become more common, it has become useful, for analysis, to identify these separately and Chapter 3.2 presents data for 1998 relating to filled teeth and to teeth with artificial crowns. For comparison with previous surveys the first section of this current chapter presents data on restored teeth (that is, filled teeth together with artificial crowns – previously called 'filled') and the second section looks at artificial crowns. It should be noted that all the results in this chapter are based on the same criteria as those used in the previous surveys, and this excludes visual caries. Thus, results presented here differ from those based on the new criteria for 1998 presented in Part 3 and elsewhere in this report (see Section 5.2.6).

5.3.2 Changes in restored (otherwise sound) teeth

Overall, the average number of teeth which were restored (otherwise sound) has varied little from 1978 to 1998. In 1998, dentate adults had, on average 8.1 such teeth which was the same as that found in 1978 and similar to the level in 1988 (8.4). The proportion with 12 or more restored (otherwise sound) teeth decreased slightly between 1988 and 1998 from 31% to 28%.

Table 5.3.1

There was a marked disparity in the average number of restored (otherwise sound) teeth between the younger (more dentally healthy) and the older (apparently more dentally unhealthy) age groups. It should be remembered, however, that in the previous decades many of the teeth in the older cohorts would have been extracted.

Dentate adults aged less than 45 showed a decrease in the number of restored (otherwise sound) teeth between 1988 and 1998 while those aged 45 and over showed an increase. Among the youngest age group (aged 16 to 24 years) there was a sharp rise in the proportion with no restored (otherwise sound) teeth from 13% in 1988 to 30% in 1998, while, for those aged 55 and over, the proportion with no restored (otherwise sound) teeth decreased from 18% in 1988 to 10% in 1998. All but the two oldest age groups showed lower proportions with 12 or more restored (otherwise sound) teeth in 1998 compared with 1988. The apparent deterioration of dental health among older dentate adults can be explained by the observation that this group has retained more teeth than had their equivalent age groups in 1978 and 1988.

Table 5.3.2

The changing pattern of restorations by age is clearly shown by Figure 5.3.1 which includes data for 1968 from the survey of England and Wales[1]. By 1978 those aged 25 to 34 were most likely to have 12 or more restored (otherwise sound) teeth. The cohort effect can be seen as this peak moves through the age groups for 1988 and 1998 so that by 1998 it was those aged 45 to 54 who were most likely to have 12 or more restored (otherwise sound) teeth.

Figure 5.3.1

Fig 5.3.1   **Proportion of dentate adults with 12 or more restored (otherwise sound) teeth**

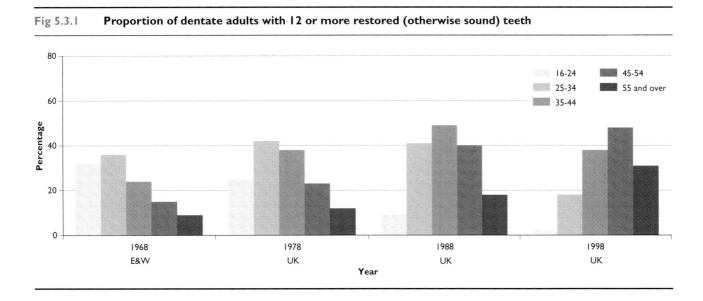

Dentate men and women both conformed to the pattern described above for the whole population, but there was variation with respect to social class of head of household. Among dentate adults from non-manual backgrounds the average number of restored (otherwise sound) teeth decreased from 9.9 in 1978 to 8.9 in 1998, while those from manual backgrounds showed an increase in the average number of restored (otherwise sound) teeth. No social class showed a statistically significant change between 1988 and 1998.

Tables 5.3.3–4

There was a similar disparity between the restorative experience of those who reported regular dental check-ups and those who attended only with trouble. Among the former group the proportion with 12 or more restored (otherwise sound) teeth had decreased from 48% in 1978 to 35% in 1998 and the proportion with no restored (otherwise sound) teeth had remained at a low level. However among those who attended only with trouble there had been little change in the proportion with 12 or more restored (otherwise sound) teeth but the proportion with no restored (otherwise sound) teeth almost halved between 1978 and 1998 from 31% to 16%.

Further analysis of the proportion with 12 or more teeth by usual reason for dental attendance and age highlighted these disparities. For example, among those who reported attending for regular check-ups, this proportion decreased from 35% in 1978 to 1% in 1998 among the youngest age group, while the opposite could be seen for those aged 45 to 54 who attended only with trouble (increasing from 5% in 1978 to 33% in 1998).

Tables 5.3.5–6

The comparable results over 30 years for England and Wales repeat the patterns described above.

Tables 5.3.7–9

5.3.3 Future levels of restored teeth

The difficulties of predicting the condition of teeth are discussed in Section 5.2.9. Given the provisions discussed there, Table 5.3.10 shows estimates for the proportion of dentate adults aged 16 to 64 who are likely to have 12 or more restored (otherwise sound) teeth in 2008. The estimation is made on the following assumptions:

● the proportion of 16 to 24 year olds who have 12 or more restored (otherwise sound) teeth will remain at the same level (2%) over the next ten years

● the change in the proportion between successive age groups over the last ten years will continue at the same levels

● the increase in the numbers of older dentate adults will not affect the distribution.

On this basis, the figures indicate that the proportion of dentate adults with 12 or more restored (otherwise sound) teeth will decrease among those aged 16 to 54 and increase among those age 55 to 64 while the overall figure for those aged 16 to 64 could decrease by around 7 percentage points from 28% to 21%. However, the predicted increase in the older dentate population may serve to reduce this difference since the proportion of restored (otherwise sound) teeth increases with age.

Table 5.3.10

5.3.4 Changes in teeth with artificial crowns, 1988-98

From 1988 to 1998 there was an increase in the proportion of dentate adults with artificial crowns, from just over a quarter in 1988 (26%) to just over a third in 1998 (34%). There were increases in all but the two youngest age groups and the changes were larger among those aged over 45. For example, in 1988 just over a quarter (26%) of adults aged 55 to 64 had some artificial crowns compared with almost half (48%) in 1998.

Table 5.3.11

The distribution of artificial crowns among dentate adults shifted upward between 1988 and 1998, with an increase in those with 3 or more crowns and a reduction in the proportion with no crowns. This trend was apparent only among those aged 35 and over. Further analysis with respect to whether or not dentate adults had dentures in addition to natural dentition showed similar variations for both groups in the changing provision of artificial crowns and similar variations with age.

Tables 5.3.12–14

Both dentate men and women reflected the levels of overall change. There was some variation with social class with a slightly larger increase in the proportion of those from an unskilled manual background having artificial crowns but the difference between these two groups was still large; in 1998 28% of those from unskilled manual backgrounds had some artificial crowns compared with 38% among those from non-manual backgrounds. There was also little variation with usual reason for dental attendance; the proportion with no artificial crowns decreased among both those who attended for check-ups and those who attended only with trouble, while the proportion with 3 or more crowns increased for both groups.

Figure 5.3.2, Tables 5.3.15–17

Fig 5.3.2 **Proportion of dentate adults with at least one artificial crown by social class of head of household**

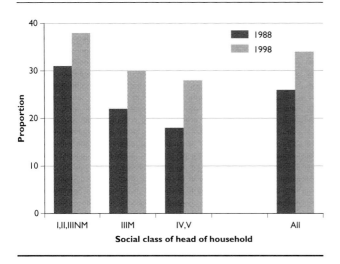

5.3.5 Changes in surfaces with amalgam fillings

The different types of fillings and their condition were discussed in Chapter 3.2. The last table in this section presents the change in the numbers of tooth surfaces with amalgam fillings by age. The overall mean number of surfaces with amalgam fillings decreased from 14.2 in 1988 to 12.9 in 1998. There was no change in the proportion with no surfaces with amalgam fillings but the proportion with 30 or more had decreased from 12% to 8%. The variation with age was broadly similar to that seen with respect to restored (otherwise sound) teeth; among those aged 65 and over the proportion with large numbers of surfaces with amalgam fillings increased while among the youngest age group the proportion with no such filled surfaces increased.

Table 5.3.18

Notes and references
1.   See Chapter 1.1

5.3: Trends in restorative treatment

Table 5.3.1 **Distribution of the number of restored (otherwise sound) teeth, 1978-98**

Dentate adults   *United Kingdom*

Number of restored (otherwise sound) teeth	1978	1988	1998
	%	%	%
0	14	10	9
1-5	23	25	27
6-11	33	35	36
12 or more	30	31	28
Mean	8.1	8.4	8.1
Base	3495	4331	3817

1998 data based on 1988 criteria (see Section 5.2.4)
1988 Table 7.17

Table 5.3.2 **Distribution of the number of restored (otherwise sound) teeth by age, 1978-98**

Dentate adults   *United Kingdom*

Number of restored (otherwise sound) teeth	Age														
	16-24			25-34			35-44			45-54			55 and over		
	1978	1988	1998	1978	1988	1998	1978	1988	1998	1978	1988	1998	1978	1988	1998
	%	%	%	%	%	%	%	%	%	%	%	%	%	%	%
0	9	13	30	10	5	4	15	5	2	16	8	2	26	18	10
1-5	26	39	52	16	17	32	17	14	18	27	19	12	36	33	27
6-11	40	39	17	32	37	46	30	31	42	34	33	38	26	30	33
12 or more	25	9	2	42	41	18	38	49	38	23	40	48	12	18	31
Mean	8.0	5.5	2.9	9.8	10.0	7.4	8.9	11.0	10.1	7.1	9.6	11.1	4.8	6.3	8.3

1998 data based on 1988 criteria (see Section 5.2.4)
For bases see Tables Appendix
1988 Table 5.13, 7.17

Table 5.3.3 **Distribution of the number of restored (otherwise sound) teeth by gender, 1978-98**

Dentate adults   *United Kingdom*

Number of restored (otherwise sound) teeth	Gender					
	Men			Women		
	1978	1988	1998	1978	1988	1998
	%	%	%	%	%	%
0	17	12	10	11	8	7
1-5	26	27	30	20	23	25
6-11	33	34	35	34	36	37
12 or more	24	28	25	35	34	31
Mean	7.3	7.9	7.6	8.9	8.9	8.6

1998 data based on 1988 criteria (see Section 5.2.4)
For bases see Tables Appendix
1988 Table 7.18

Table 5.3.4 **Distribution of the number of restored (otherwise sound) teeth by social class of head of household, 1978-98**

Dentate adults *United Kingdom*

Number of restored (otherwise sound) teeth	Social class of head of household								
	I, II, IIINM			IIIM			IV, V		
	1978	1988	1998	1978	1988	1998	1978	1988	1998
	%	%	%	%	%	%	%	%	%
0	6	7	8	17	10	7	25	17	12
1-5	18	21	23	26	28	31	28	30	33
6-11	35	34	35	32	36	37	29	36	38
12 or more	41	39	34	25	25	25	18	18	17
Mean	9.9	9.5	8.9	7.3	7.7	7.9	5.9	6.5	6.6

1998 data based on 1988 criteria (see Section 5.2.4)
For bases see Tables Appendix
1988 Table 7.19

Table 5.3.5 **Distribution of the number of restored (otherwise sound) teeth by usual reason for dental attendance, 1978-98**

Dentate adults *United Kingdom*

Number of restored (otherwise sound) teeth	Usual reason for dental attendance[†]					
	Regular check-up			Only with trouble		
	1978	1988	1998	1978	1988	1998
	%	%	%	%	%	%
0	1	4	5	31	18	16
1-5	12	18	22	35	35	39
6-11	39	35	39	23	32	31
12 or more	48	43	35	11	15	14
Mean	11.4	10.4	9.3	4.4	5.8	5.8

† Excludes people who only visit the dentist for an occasional check-up
1998 data based on 1988 criteria (see Section 5.2.4)
For bases see Tables Appendix
1988 Table 7.20

Table 5.3.6   **12 or more restored (otherwise sound) teeth by age and usual reason for dental attendance, 1978-1998**

Dentate adults   *United Kingdom*

Age	Usual reason for dental attendance†					
	Regular check-up			**Only with trouble**		
	1978	**1988**	**1998**	**1978**	**1988**	**1998**
	percentage with 12 or more restored (otherwise sound) teeth					
16-24	35	11	1	15	6	2
25-34	63	53	20	18	25	14
35-44	59	62	44	12	23	21
45-54	42	53	55	5	21	33
55-64	27	41	48	3	5	9
65 and over	27	21	34	3	2	5
All	48	43	35	11	15	14

† Excludes those who only visit the dentist for an occasional check-up
1998 data based on 1988 criteria (see Section 5.2.4)
For bases see Tables Appendix
1988 Table 7.8, 1978 Table B33

Table 5.3.7   **Distribution of the number of restored (otherwise sound) teeth by gender, 1968-98**

Dentate adults   *England and Wales*

Number of restored (otherwise sound) teeth	Gender								All			
	Men				**Women**							
	1968	**1978**	**1988**	**1998**	**1968**	**1978**	**1988**	**1998**	**1968**	**1978**	**1988**	**1998**
	%	%	%	%	%	%	%	%	%	%	%	%
0	26	17	12	10	20	11	8	7	23	14	10	9
1-5	26	26	27	30	22	21	23	25	24	24	25	28
6-11	26	35	34	35	30	36	36	37	28	35	35	36
12 or more	22	22	28	24	28	32	33	30	25	27	31	27
Mean	6.2	7.0	7.9	7.5	7.5	8.5	8.8	8.5	6.8	7.8	8.3	8.0

1998 data based on 1988 criteria (see Section 5.2.4)
For bases see Tables Appendix
1988 Table 7.11

Table 5.3.8 **Distribution of the number of restored (otherwise sound) teeth by age, 1968-98**

Dentate adults   *England and Wales*

Number of restored (otherwise sound) teeth	Age																			
	16-24				25-34				35-44				45-54				55 and over			
	1968	1978	1988	1998	1968	1978	1988	1998	1968	1978	1988	1998	1968	1978	1988	1998	1968	1978	1988	1998
	%	%	%	%	%	%	%	%	%	%	%	%	%	%	%	%	%	%	%	%
0	14	9	14	31	16	10	5	4	24	15	5	2	27	15	8	2	43	26	17	9
1-5	20	27	39	53	19	15	18	34	23	18	14	19	35	29	19	12	29	38	34	27
6-11	34	41	39	15	29	35	37	46	29	33	31	43	23	35	33	39	19	27	30	33
12 or more	32	23	7	1	36	40	40	16	24	34	49	36	15	21	41	48	9	9	19	31
Mean	8.2	7.7	5.3	2.7	8.6	9.6	9.9	7.1	6.7	8.6	11.1	9.9	5.0	6.8	9.7	11.2	3.5	4.6	6.2	8.4

1998 data based on 1988 criteria (see Section 5.2.4)
For bases see Tables Appendix
1988 Table 7.10

Table 5.3.9 **Distribution of the number of restored (otherwise sound) teeth by usual reason for dental attendance, 1968-98**

Dentate adults   *England and Wales*

Number of restored (otherwise sound) teeth	Usual reason for dental attendance[†]							
	Regular check-up				Only with trouble			
	1968	1978	1988	1998	1968	1978	1988	1998
	%	%	%	%	%	%	%	%
0	2	1	4	5	43	31	18	16
1-5	12	13	18	22	34	36	35	40
6-11	37	42	35	39	18	24	32	31
12 or more	49	44	42	34	5	9	15	13
Mean	11.1	10.8	10.2	9.2	3.0	4.3	5.8	5.7

† Excludes people who only visit the dentist for an occasional check-up.
1998 data based on 1988 criteria (see Section 5.2.4)
For bases see Tables Appendix
1988 Table 7.12

Table 5.3.10 **Future level of dentate adults with 12 or more restored (otherwise sound) teeth**

Dentate adults   United Kingdom

Age	Percentage with 12 or more restored teeth in 1988	Percentage with 12 or more restored teeth in 1998	Assumed future rate of change	Estimates for 2008	Base for 1998 figures
			percentage		
16-24	9	2		2	491
			9		
25-34	41	18		11	854
			-3		
35-44	49	38		15	781
			-1		
45-54	40	48		37	746
			-3		
55-64	-	37		45	461
All aged 16-64	-	28		21	3333

1988 Table 7.17

Table 5.3.11 **Artificial crowns by age, 1988-98**

Dentate adults   United Kingdom

Age	Percentage with artifical crowns	
	1988	1998
16-24	11	7
25-34	29	24
35-44	35	41
45-54	33	49
55-64	26	48
65-74	21	39
75 and over	18	36
All	26	34
Base	4331	3817

1988 Table 9.1

Table 5.3.12 **The distribution of artificially crowned teeth by age, 1988-98**

Dentate adults *United Kingdom*

Number of teeth with artificial crowns	Age														All	
	16-24		25-34		35-44		45-54		55-64		65-74		75 and over			
	1988	1998	1988	1998	1988	1998	1988	1998	1988	1998	1988	1998	1988	1998	1988	1998
	%	%	%	%	%	%	%	%	%	%	%	%	%	%	%	%
0	89	93	71	76	65	59	67	51	74	52	79	61	82	64	74	66
I	6	3	13	13	14	17	13	15	10	15	8	12	8	16	II	13
2	4	2	8	5	10	9	8	10	7	8	6	7	2	8	7	7
3 or more	I	I	8	6	II	15	12	23	9	25	7	20	8	12	8	14

For bases see Tables Appendix
1988 Table 9.1

Table 5.3.13 **The distribution of artificially crowned teeth by dental status, 1988-98**

Dentate adults *United Kingdom*

Number of teeth with artificial crowns	Dental status				All	
	Natural teeth and dentures		Natural teeth only			
	1988	1998	1988	1998	1988	1998
	%	%	%	%	%	%
0	74	65	74	66	74	66
I	9	12	12	13	II	13
2	7	8	7	7	7	7
3 or more	10	15	7	14	8	14
Base	685†	740	2286†	3077	4331	3817

† Weighted base
1988 Table 9.6

Table 5.3.14 **The distribution of artificially crowned teeth by age and dental status, 1988-98**

Dentate adults   *United Kingdom*

Age and number of teeth with artificial crowns	Dental status				All	
	Natural teeth and dentures		Natural teeth only			
	1988	1998	1988	1998	1988	1998
	%	%	%	%	%	%
16-34						
0	61	66	81	83	80	83
1	12	17	9	9	9	9
2	11	4	6	4	6	4
3 or more	16	12	4	4	5	4
35-54						
0	68	55	64	55	65	55
1	11	15	15	16	14	16
2	9	12	9	10	9	10
3 or more	12	18	12	19	12	19
55 and over						
0	81	69	69	45	77	57
1	6	11	5	18	9	14
2	5	6	8	9	6	8
3 or more	8	14	8	28	8	21

For bases see Tables Appendix
1988 Table 9.7

Table 5.3.15 **The distribution of artificially crowned teeth by gender, 1988-98**

Dentate adults   *United Kingdom*

Number of teeth with artificial crowns	Gender				All	
	Men		Women			
	1988	1998	1988	1998	1988	1998
	%	%	%	%	%	%
0	77	69	71	63	74	66
1	10	13	12	14	11	13
2	6	8	8	6	7	7
3 or more	7	11	9	17	8	14

For bases see Tables Appendix
1988 Table 9.3

Table 5.3.16 **The distribution of artificially crowned teeth by social class of head of household, 1988-98**

Dentate adults   *United Kingdom*

Number of teeth with artificial crowns	Social class of head of household						All†	
	I, II, IIINM		IIIM		IV,V			
	1988	1998	1988	1998	1988	1998	1988	1998
	%	%	%	%	%	%	%	%
0	69	62	78	70	82	72	74	66
1	13	14	10	13	7	13	11	13
2	8	8	6	7	5	6	7	7
3 or more	9	16	6	11	7	10	8	14

†Includes those for whom social class was not known and Armed Forces
For bases see Tables Appendix
1988 Table 9.4

Table 5.3.17 **The distribution of artificially crowned teeth by usual reason for dental attendance, 1988-98**

Dentate adults   *United Kingdom*

Number of teeth with artificial crowns	Usual reason for dental attendance						All	
	Regular check-up		Occasional check-up		Only with trouble			
	1988	1998	1988	1998	1988	1998	1988	1998
	%	%	%	%	%	%	%	%
0	65	60	81	72	84	77	74	66
1	15	14	7	10	7	11	11	13
2	9	9	8	4	4	4	7	7
3 or more	11	16	4	14	5	8	8	14

For bases see Tables Appendix
1988 Table 9.5

Table 5.3.18 **The number of sound amalgam filled surfaces by age, 1988-98**

Dentate adults   *United Kingdom*

Number of sound amalgam filled surfaces	Age															All	
	16-24		25-34		35-44		45-54		55-64		65-74		75 and over			
	1988	1998	1988	1998	1988	1998	1988	1998	1988	1998	1988	1998	1988	1998	1988	1998
	%	%	%	%	%	%	%	%	%	%	%	%	%	%	%	%
None	15	37	6	6	6	4	8	4	18	11	19	14	38	15	11	12
1-9	49	51	22	42	20	24	25	18	33	26	44	29	39	42	31	33
10-19	27	10	36	34	26	37	26	38	22	36	26	30	19	27	28	31
20-29	8	1	22	14	25	24	24	25	19	18	9	16	4	13	18	17
30 or more	1	1	14	4	23	11	16	15	8	9	2	10	0	3	12	8
Mean[†]	-	3.9	-	11.6	-	16.3	-	18.2	-	14.0	-	13.0	-	10.2	14.2	12.9

† *Data not available by age from the 1988 Adult Dental Health Survey report*
For bases see Tables Appendix
1988 Table 14.4

Part 6

Dental attitudes and reported behaviour

6.1   Usual reason for dental attendance

Summary

- Twice as many dentate adults in the United Kingdom in 1998 reported attending for regular dental check-ups (59%) than reported only attending when they had trouble with their teeth (30%).

- Women were much more likely than men to report seeking regular dental check-ups; around two-thirds of women (66%) compared with just over half of men (52%).

- Younger men were one of the groups least likely to seek regular check-ups. Only 42% of men aged 16 to 24 and 44% of men aged 25 to 34 did so. This is similar to the average proportion of the population as a whole reporting attending for regular check-ups in the 1968 survey of Adult Dental Health in England & Wales (40%) which shows that reported attendance behaviour among young males in the United Kingdom is 20 to 30 years behind that of the population average.

- Over half of dentate adults (53%) said they went to a dentist at about the same frequency as five years previously, 20% claimed to go more often and 27% said they went less often than before.

- Almost half of young adults aged 16 to 24 (48%) said they now went to a dentist less often than 5 years previously.

- Over three-quarters (76%) of those who said they sought regular dental check-ups said they visited 10 or more times over the last 5 years compared with 6% of those who said they only attended when they had trouble with their teeth.

- Among those who reported usually attending only with trouble 42% said they had visited a dentist for a check up in the preceding 5 years and of those who said they sought regular check-ups 49% said they made between 1 to 4 visits in the last 5 years in response to a specific dental problem.

- Just under 40% of those who said they attended more frequently than 5 years ago said it was because they were more aware of the need to look after their teeth while a similar proportion said it was because they were having more trouble with their teeth.

6.1.1 Introduction

Since the first survey of Adult Dental Health in 1968 participants have been asked "In general do you go to a dentist for a regular check-up, an occasional check-up or only when you're having trouble with your teeth?" The question is concerned with establishing a person's usual motive for attendance in relation to whether they need to experience symptoms before they would visit a dentist or whether they would attend for an examination screening for dental disease in the absence of symptoms. It has been made clear in previous reports that seeking regular dental check-ups is not intended to be interpreted as attendance at a particular frequency and it has been shown that observed frequency of attendance is largely unrelated to a person's stated motive for attending.[1] In order to clarify this, data relating to frequency of attendance and reasons for attendance over the preceding 5 years are also examined in this chapter.

6.1.2 Usual reason for dental attendance in the United Kingdom

Table 6.1.1 gives a breakdown by country and by English region of the reason given by dentate adults for visiting a dentist. Over half of the UK dentate population in 1998 (59%) reported that they usually attended a dentist for a regular check-up. Northern Ireland and Scotland had the lowest percentages of dentate adults who reported going for regular dental check-ups 51% and 55% respectively compared with 59% in Wales and 60% in England. Within England there was no significant difference between the regions in the proportion of dentate adults who reported going for regular dental check-ups; 58% in the North, 63% in the Midlands and 60% in the South.

Table 6.1.1

Differences between the countries of the United Kingdom may be a reflection of their social class structures. This is shown in Table 6.1.2 where the proportion of dentate adults who report going for regular dental check-ups in each of three social class groups is compared across the countries of the United Kingdom. Overall, there were differences between social class groups; those from a non-manual background were more likely to attend for regular check-ups than those from unskilled manual backgrounds; 65% and 49% respectively. However, there were also differences within the same social class groups between countries suggesting that differences between those stating they had regular dental check-ups reflect more than just the social class structure of each country. For example, among those from unskilled manual backgrounds, 57% of dentate adults in Wales reported

attendance for regular check-ups compared with 34% of the same group in Northern Ireland.

Table 6.1.2

Table 6.1.3 shows the usual reason given for dental attendance by age group. The proportion of dentate adults who reported attending for dental check-ups increased with age up to 64 years old, from 48% of those aged 16 to 24 years to 68% of those aged 55 to 74 years. Age-related patterns in dental health among dentate adults have to be considered against changes that occur solely from the loss of people who become edentate which disproportionately affects the older age groups. However, in this case the percentage change in attendance pattern with age among the younger age groups greatly exceeds that of the percentage lost from each age group through people becoming edentate.

Women were much more likely than men to report seeking regular dental check-ups; around two-thirds of women (66%) compared with just over half of men (52%). Over half of the women in all age groups claimed to attend for regular dental check-ups. However, it can be seen that although attendance for regular dental check-ups seems quite well established in the dentate population as a whole it is less usual in young men. Only 42% of men aged 16 to 24 and 44% of men aged 25 to 34 claimed to seek regular dental check-ups in 1998. This is similar to the average proportion of the population as a whole reporting attending for regular check-ups in the 1968 survey of Adult Dental Health in England & Wales (40%) or in the 1978 survey in Scotland (38%); this shows that reported attendance behaviour among young males in the United Kingdom is 20 to 30 years behind that of the population average.

Figure 6.1.1, Table 6.1.3

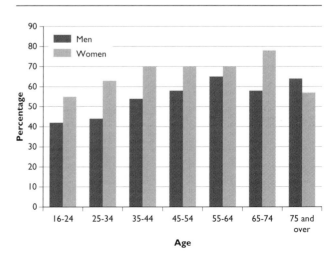

Fig 6.1.1 **Proportion of dentate adults who reported attending for a regular check-up by age and gender**

Table 6.1.4 shows the relationship of gender, social class of head of household and age to the usual reason given for attending a dentist. Men from unskilled manual working backgrounds were less likely to claim to visit for regular dental check-ups (42%) than women (55%) from similar backgrounds and from men with a non-manual background (58%). Among men and women aged 16 to 24, however, there was no difference between those from manual and non-manual backgrounds in their usual reason for attending a dentist.

Table 6.1.4

Figures for the reported usual reason for attendance and employment status of the head of household among dentate adults are shown in Table 6.1.5. There are several factors that may be related to attendance and employment. The group who were most likely to report regular dental check-ups were those in part-time employment (68%) or retired (63%). Far fewer of those in the unemployed category (49%) claimed to seek regular check-ups, although this was not significantly lower than those in full time employment (56%). Table 6.1.6 shows that the figures for women reflected the overall pattern of variation. This was not the case for men among whom there was no difference between those in full time and those in part time employment in the proportion reporting regular dental check-ups (52%) while retired men were the most likely to report regular dental check-ups (62%).

Tables 6.1.5–6

6.1.3 Reported changes in usual reason for dental attendance

Dentate adults who reported attending for regular check-ups in 1998 were asked if there had ever been a time when they had not been for regular check-ups. Conversely, those who said they were not currently regular attenders were asked if there had been a time when they did go for regular check-ups. Just under a third (29%) of dentate adults claimed to be lifelong regular attenders, 30% said they now sought regular dental check-ups but had not always done so in the past and another 30% said they used to attend for regular dental check-ups but now attended only for occasional check-ups or when they had trouble with their teeth.

Many young adults aged 16 to 24 said that they had been regular attenders but were not now (45%). This suggests that many in this age group may be changing their attendance behaviour, as they become adults. However, it is also the case that more of those in the youngest age group said they had always sought regular check-ups (37%) and were continuing to do so than those in other age groups.

Table 6.1.7

People may lapse from regular attendance and then change back to regular check-ups as they get older. Dentate adults who had changed their reason for attending were asked their reasons for the pattern they had in the past. Those who had lapsed from seeking regular dental check-ups were most likely to say that they had no choice but to attend regularly in the past (53%). This perhaps suggests that compulsion is not a particularly effective method of building a habit of regular attendance. Other frequently mentioned reasons for past regular attendance were that they wanted to look after their teeth (18%), treatment was free (16%) and that their dentist sent a reminder (11%). Those who had become regular attenders were asked why they were not regular attenders in the past. Over a third (36%) of those who did not go for regular check-ups in the past said this was due to apathy or laziness, fewer (27%) mentioned fear as a reason for not going to a dentist regularly in the past, 18% said it was because they had no trouble with their teeth and cost was mentioned by 10%. The overall picture this gives is that many people change their usual reason for attendance because there is no clear value to them in seeking regular dental check-ups. Many appear to go if coerced, but without compulsion they admit to feeling apathetic about arranging a regular dental check-up.

Table 6.1.8

6.1.4 Usual reason for dental attendance compared with frequency of dental attendance

It has been made clear in previous reports that seeking regular dental check-ups is not necessarily related to attending at a particular frequency. However, the interview also included questions relating to the number of visits made in the past 5 years and the reason for those visits.

Table 6.1.9 shows reported attendance frequency over 5 years and the reason for the visits according to the person's usual reason for going to a dentist. Over three-quarters (76%) of those who said they sought regular dental check-ups said they visited 10 or more times over the last 5 years which on average is once every 6 months. Most of those who said they only attended when they had trouble with their teeth visited less than once a year; 31% had not attended at all during the preceding 5 years and 48% attended 4 or fewer times. Most of those who claimed to seek occasional dental check-ups said they went to a dentist at least once in the 5 year period but fewer than 10 times. Just under half of this group (44%) had attended at least 5 times but fewer than 10 (that is an average of annually), while 37% had attended 1 to 4 times (less than annually).

Reported frequency of attendance therefore does generally fit with the reason given for going to a dentist by those who report attending for check-ups; most people who attend for regular check-ups say they have been to a dentist at least 10 times in 5 years whereas most of those who claim only to seek occasional check-ups say they have visited once a year or less often in the last 5 years. However, comparison of the reasons for attendance in the last 5 years showed a less clear cut relationship. Among those who reported usually attending only with trouble 42% said they had visited a dentist for a check-up in the preceding 5 years and of those who said they sought regular check-ups only 40% said that they had made no visits over the last 5 years in response to a specific dental problem, while 49% said they made between 1 to 4 visits in the last 5 years in response to a specific dental problem.

Figure 6.1.2, Table 6.1.9

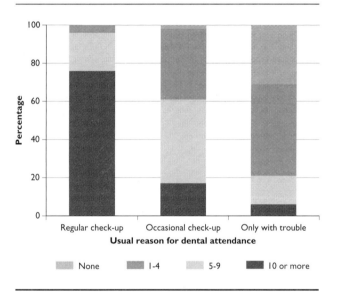

Fig 6.1.2   **Number of visits in the last 5 years by usual reason for dental attendance**

Table 6.1.10 presents the reason given for the most recent dental visit according to the reason stated for usually attending a dentist. Around a fifth (19%) of those who reported usually seeking dental check-ups last attended because they had some problem with their teeth as did just under a third (31%) of those who reported that they sought occasional check-ups. However, here again the unexpected finding was that 30% of people who said they usually only go to a dentist if they have some dental trouble said their most recent visit was for a dental check-up. Some of these people may have changed their attendance pattern in the last 5 years but the finding does tend to suggest that it may be difficult for some people to interpret their dental attendance behaviour comprehensively using the categories that are traditionally used in these surveys.

Table 6.1.10

Table 6.1.11 shows the time people thought had elapsed since their last dental visit. Overall, more than half of the dentate population (56%) said they had visited a dentist in the last 6 months and 11% said their last visit was over 5 years ago. The table also shows that a small group of people who say they go for regular dental check-ups (3%) last did so over a year ago.

Table 6.1.11

6.1.5 Change in frequency of attendance compared with 5 years previously

People were also asked whether they went to the dentist more frequently or less frequently now than 5 years previously (Table 6.1.12). Over half of dentate adults (53%) said they went to a dentist at about the same frequency as 5 years previously, 20% claimed to go more often and 27% said they now went less often. The finding that a greater proportion of people said they currently went less often than they used to would seem at odds with other findings in this chapter that show a greater proportion of dentate adults seek regular check-ups than in the past. Looking at the proportions of people attending "less often" and "more often" by age showed that almost half of 16 to 24 year olds said they attended less often than 5 years previously (48%) compared with between 20% and 27% of those in other age groups. In referring to their attendance 5 years previously these young adults would have been considering when they were aged between 11 and 19. So for many the reported decline in frequency of attendance in adulthood was made in relation to when they were under 16. Among dentate adults aged 25 to 64 similar proportions changed either to a less or to a more frequent pattern of attendance.

Figure 6.1.3, Table 6.1.12

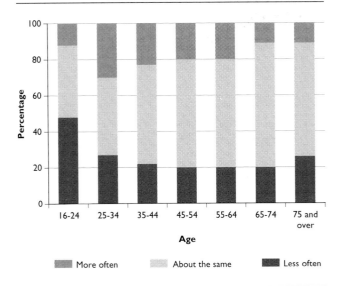

Fig 6.1.3   **Reported change in frequency of attendance compared with 5 years previously by age**

Legend: More often / About the same / Less often

Table 6.1.13 shows the relationship between the reported change to or from regular dental check-ups and the reported change in frequency of attending over the past 5 years. A fifth of those who said they had always attended for regular dental check-ups said that their frequency of attendance had changed; 12% said they now went more often than 5 years before and 9% said they now went less often. Among those who reported that they had never attended for regular check-ups, a greater proportion reported a change in frequency of attendance compared with 5 years ago; 33% said they went less often and 12% said they went more often. All these variations suggest that the reported reason for attending ("regular check-up, occasional check-up and only when having trouble with teeth") is not seen as being directly equivalent to frequency of attendance by many people as many considered the frequency at which they visit a dentist had changed but that their underlying motive for attending had not. The results also indicate that those who had never attended regularly were almost three times as likely to report a decrease in frequency of attendance over the past 5 years as to report an increase (33% compared with 12%).

Table 6.1.13

Over a quarter of dentate adults (27%) said they attended less often in 1998 than 5 years previously, and 20% said they attended more frequently in 1998 than in the past (Table 6.1.12). Table 6.1.14 examines the reasons given by people for changing their frequency of visiting. The reason most commonly given by those who attend less often than in the past was that they now had less trouble with their teeth or gums (29%). A quarter of those who attended less frequently said they did so because of cost (28%). Fewer (15%) reported

they could not be bothered or had lost the habit of attending frequently and a similar percentage reported fear (14%) as reasons for attending less often. Those who attended more often felt this was due to being more dentally aware (39%) or having more trouble with their teeth or gums than in the past (38%).

Table 6.1.14

6.1.6. Usual reason for dental attendance in relation to other attitudes to dentistry

Table 6.1.15 compares four attitudes towards dental health with the reported reason for usually attending a dentist. A much larger proportion of those who use dental services only when they have some trouble (72%) thought that they would need treatment if they went to a dentist tomorrow than those who sought regular dental check-ups (29%). This may give an insight into the motivation for dental attendance among those who only attend when they have trouble with their teeth. Clearly the view among most of these attenders is that they have some form of current treatment need, yet this has not motivated them to arrange an appointment. Therefore, there are likely to be other motivating factors involved, one of which may be the experience of oral pain. Those who attend only when having trouble with their teeth were also much more likely to opt for the extraction of an aching tooth (37%) than those who said they sought check-ups (regular check-ups; 13%, occasional check-ups 19%). However, despite this, many of those claiming only to attend when they have trouble with their teeth (52%) say they would find the need for full dentures very upsetting. More of those who reported only to attend when they had some trouble with their teeth reported that they were more likely to expect to need full dentures (16%) than those who sought either regular or occasional check-ups (9% and 10%).

Table 6.1.15

Notes and references

1.   Eddie S. Frequency of attendance in the General Dental Service in Scotland a comparison with claimed attendance. *Br Dent J* 1984; 157: 267-270.

Table 6.1.1   **Usual reason for dental attendance by English region and country**

Dentate adults

Usual reason for attendance	English region			Country				United Kingdom
	North	Midlands	South	England	Wales	Scotland	Northern Ireland	
	%	%	%	%	%	%	%	%
Regular check-up	58	63	60	60	59	55	51	59
Occasional check-up	12	7	11	10	12	12	14	11
Only with trouble	31	30	29	30	29	33	35	30
Base	823	731	1456	3010	682	953	636	5281

1988 Table 20.1 (part)

Table 6.1.2   **Attending for regular check-up by country and social class of head of household**

Dentate adults

Social class of head of household	Country				All
	England	Wales	Scotland	Northern Ireland	
	percentage attending for regular check-ups				
I, II, IIINM	65	69	61	59	65
IIIM	58	49	53	50	57
IV,V	50	57	41	34	49
	Base				
I, II, IIINM	1611	343	517	261	2732
IIIM	948	266	307	208	1729
IV,V	628	165	271	157	1221

Table 6.1.3 **Attending for a regular check-up by age and gender**

Dentate adults *United Kingdom*

Age	Gender		All
	Men	**Women**	
	percentage attending for regular check-ups		
16-24	42	55	48
25-34	44	63	53
35-44	54	70	62
45-54	58	70	64
55-64	65	70	68
65-74	58	78	68
75 and over	64	57	60
All ages	52	66	59

For bases see Tables Appendix

Table 6.1.4 **Attending for regular check-ups by social class of head of household, gender and age**

Dentate adults *United Kingdom*

Social class of head of household	Age							All
	16-24	**25-34**	**35-44**	**45-54**	**55-64**	**65-74**	**75 and over**	
	percentage attending for regular check-ups							
Men								
I, II, IIINM	41	48	61	64	74	65	68	58
IIIM	49	42	51	55	58	46	65	50
IV,V	38	35	40	41	53	58	48	42
Women								
I, II, IIINM	54	68	74	77	79	82	58	71
IIIM	60	63	70	69	66	75	63	66
IV,V	53	45	64	56	53	71	53	55

For bases see Tables Appendix

Table 6.1.5   Usual reason for dental attendance by economic status

Dentate adults   *United Kingdom*

Usual reason for dental attendance	Economic status					All†
	Employed full-time	Employed part-time	Unemployed	Retired	Other inactive	
	%	%	%	%	%	
Regular check-up	56	68	49	63	55	59
Occasional check-up	13	10	10	4	12	11
Only with trouble	32	21	41	33	33	30
Base	2461	1260	169	634	747	5281

† Includes those for whom economic status was not known
1988 Table 20.8

Table 6.1.6   Usual reason for dental attendance by economic status and gender

Dentate adults   *United Kingdom*

Usual reason for dental attendance	Economic status										All†	
	Employed full-time		Employed part-time		Unemployed		Retired		Other inactive			
	Men	Women	Men	Women	Men	Women	Men	Women	Men	Women	Men	Women
	%	%	%	%	%	%	%	%	%	%	%	%
Regular check-up	52	63	52	71	42	54	62	70	47	60	52	66
Occasional check-up	11	13	16	10	9	13	6	3	13	12	11	11
Only with trouble	37	24	31	19	49	33	33	27	40	28	37	23
Base	1561	900	140	1120	103	66	343	291	251	496	2406	2875

† Includes those for whom economic status was not known
1988 Table 20.8

Table 6.1.7   Reported change in usual reason for attendance

Dentate adults   *United Kingdom*

Dental attendance	Age							All
	16-24	25-34	35-44	45-54	55-64	65-74	75 and over	
	%	%	%	%	%	%	%	%
Always regular	37	23	28	30	30	30	29	29
Regular now but not always	11	30	34	34	38	38	30	30
Was regular but not now	45	40	28	26	19	16	14	30
Never been regular	7	7	10	10	13	16	26	10
Base	701	1151	1099	1016	645	456	213	5281

1988 Table 20.22

Table 6.1.8 **Reasons for previous attendance at the dentist**

Dentate adults who reported a change in their dental attendance   *United Kingdom*

Reason for previous attendance behaviour

Used to attend regularly;	%
occasional check-ups or only with trouble now	
Had no choice	53
Wanted to look after teeth	18
Treatment was free	16
Dentist sent reminder card	11
Needed, wanted or was receiving treatment	9
Was taking the family	6
Had more time or fewer commitments	5
Was child taken by parents or seeing school dentist	2
Other	8
Used not to attend regularly;	%
regular check-ups now	
Apathy/laziness	36
Scared of dentist	27
Have no trouble with teeth	18
Too expensive	10
Difficult to get to dentist/make the journey	8
Lack of time/would go if surgery open different hours	8
Moved to new area, moved a lot, was abroad or in Forces	6
Afraid dentist will find trouble or cause damage	3
Not happy with treatment/bad experience	3
Was child then and not taken by parents	3
Difficult to get appointment	2
Other	6

	Base
Used to attend regularly;	
occasional check-up or only for trouble now	1526
Used not to attend regularly;	
regular check-ups now	1542

Percentages may add to more than 100% because respondents may have given more than one answer

Table 6.1.9   **Number and type of visits in the last 5 years by usual reason for dental attendance**

Dentate adults   *United Kingdom*

Number and type of visits in last 5 years	Usual reason for attendance			All†
	Regular check-up	**Occasional check-up**	**Only with trouble**	
Total number of visits	%	%	%	%
None	0	2	31	10
1 - 4 (less than annually)	4	37	48	21
5 - 9 (at least annually but less than 6 monthly)	20	44	15	21
10 or more (6 monthly or more often)	76	17	6	49
Number of visits for dental check-ups	%	%	%	%
None	1	6	58	18
1 - 4 (less than annually)	7	51	34	20
5 - 9 (at least annually but less than 6 monthly)	23	37	6	19
10 or more (6 monthly or more often)	70	7	2	42
Number of visits for trouble with teeth	%	%	%	%
None	40	44	44	42
1 - 4 (less than annually)	49	47	48	48
5 - 9 (at least annually but less than 6 monthly)	8	6	5	7
10 or more (6 monthly or more often)	4	2	3	3
Base	*3121*	*575*	*1572*	*5281*

† *Includes those for whom the usual reason for dental attendance was not known*

Table 6.1.10 **Usual reason for dental attendance by reason for most recent visit**

Dentate adults   *United Kingdom*

Reason for most recent visit	Usual reason for attendance			All†
	Regular check-up	**Occasional check-up**	**Only with trouble**	
	%	%	%	%
Trouble with teeth	19	31	65	34
Check-up	79	66	30	63
Other	2	3	5	3
Base	*3121*	*575*	*1572*	*5281*

† *Includes those for whom the usual reason for dental attendance was not known*
1988 Table 24.28

Table 6.1.11 **Reported time since last visit to the dentist by usual reason for attending**

Dentate adults *United Kingdom*

Reported time since last visit to dentist	Usual reason for attendance			All†
	Regular check-up	Occasional check-up	Only with trouble	
	%	%	%	%
Up to 6 months ago	82	32	14	56
Over 6 months, up to 1 year ago	15	26	9	15
Over 1 year, up to 2 years ago	2	24	12	8
Over 2 years, up to 5 years ago	1	14	32	12
Over 5 years, up to 10 years ago	0	3	17	6
Over 10 years, up to 20 years ago	0	0	9	3
Over 20 years ago	0	0	5	2
Never been to dentist	0	0	1	0
Base	3121	575	1572	5281

† Includes those for whom the usual reason for dental attendance was not known

Table 6.1.12 **Reported change in frequency of dental attendance compared with 5 years ago**

Dentate adults *United Kingdom*

Reported change in frequency compared with 5 years ago	Age							All
	16-24	25-34	35-44	45-54	55-64	65-74	75 and over	
	%	%	%	%	%	%	%	%
More often	12	30	23	20	20	11	11	20
About the same	40	43	55	60	60	69	63	53
Less often	48	27	22	20	20	20	26	27
Base	701	1151	1099	1016	645	456	213	5281

1988 Table 20.17

Table 6.1.13 **Reported change in frequency of attendance over past 5 years by (any) reported change in reason for attending the dentist**

Dentate adults   *United Kingdom*

Reported change in frequency since 5 years ago	Reported change in usual reason for attending				All†
	Always been regular	Regular now but not always	Was regular but not now	Never been regular	
	%	%	%	%	%
More often	12	40	12	12	20
About the same	80	52	28	56	53
Less often	9	8	60	33	27
Base	1575	1542	1524	620	5281

† *Includes those for whom the usual reason for dental attendance was not known*

Table 6.1.14 **Reason for change in frequency of attendance compared with 5 years previously**

Dentate adults who reported a change in frequency of attendance   *United Kingdom*

Reason for change in frequency of attendance

	%
Attends less often now than 5 years ago	
Have no or less trouble with teeth or gums	29
Too expensive	28
Can't be bothered/got out of habit	15
Afraid of dentists or injections	14
Lack of time, would go if surgery open at different hours	12
No regular dentist/dentist retired	6
Difficult to get to dentist/make the journey	6
Have fewer teeth now	5
Not happy with treatment/bad experience	3
Dentist no longer sends reminders	2
Other	11

	%
Attends more often now than 5 years ago	
More dentally aware/want to keep my teeth	39
Have (more) trouble with teeth or gums	38
Have good dentist now	11
More incentive to go	10
Less frightened of dentist or treatment now	7
Taking children, family pressure	5
Dentist sends reminders or will strike off NHS if doesn't go	4
Other	12

	Base
Attended less often now than 5 years ago	1368
Attended more often now than 5 years ago	1073

Percentages add to more than 100% because respondents may have given more than one answer

Table 6.1.15 **Attitudes towards other aspects of dental health by usual reason for attendance**

United Kingdom

	Usual reason for attendance			All
	Regular check-up	Occasional check-up	Only with trouble	
	%	%	%	%
Whether thinks would need treatment if went to dentist tomorrow				
Yes	29	48	72	44
No	68	46	24	53
Don't know	3	5	5	4
Whether would have aching back tooth extracted or filled				
Extracted	13	19	37	21
Filled	87	81	63	79
Base (dentate adults)	*3121*	*575*	*1572*	*5281*
	%	%	%	%
How upsetting would find loss of all natural teeth and having full dentures				
Very upsetting	67	59	52	61
A little/not at all upsetting	33	41	48	39
Expectation of need for full dentures				
Expects to need full dentures	9	10	16	11
Expect to keep some natural teeth	84	82	75	81
Don't know	7	9	8	8
Base (adults with natural teeth only)	*2512*	*485*	*1180*	*4188*

6.2   Opinions about dental visits

Summary

- Dentate adults were shown a series of statements related to going to the dentist and asked whether their own feelings were reflected by these statements.

- Dentate adults were most likely to identify with the following statements: *I'd like to know more about what the dentist is going to do and why* (71%); *I'd like to be given an estimate without commitment* (70%); *I'd like to be able to drop in at the dentist without an appointment* (69%); and *I'm nervous of some kinds of dental treatment* (64%).

- The organisational aspects of going to the dentist seemed to be of most importance to people, both in terms of the proportion of dentate adults whose feelings were reflected by the statements in some way and those most frequently selected as most important. Statements relating to cost were the next most important. This was followed by feelings of anxiety about going to the dentist. The statements concerned with the long-term value of going to the dentist were least important to people.

- The youngest dentate adults (aged between 16 and 24 years) were among those least likely to feel the same way as the statements relating to cost, and reflect their relatively limited experience of paying for dental treatment. For example, 38% of dentate 16 to 24 year olds felt that NHS dental treatment was expensive compared more than 50% of those aged 25 years or over.

- Dentate adults who had private treatment were also more likely to feel they would like to pay for their dental treatment by instalments (53%) and would like to be given an estimate without commitment (75%) than those who had their treatment through the NHS (47% and 69% respectively).

- Overall, 64% of dentate adults identified with being nervous of some kinds of dental treatment., The groups least likely to identify with being nervous were men (55%); and those aged 65 years and over (51%).

- Dentate adults from manual backgrounds were more likely than those from non-manual backgrounds to have feelings that were reflected by the organisational, cost and long-term value statements. There was little significant difference among dentate adults from different social classes in whether they felt nervous or anxious about going to the dentist.

- Generally, the proportion of dentate adults who went to the dentist for a regular check-up was lower among those who 'definitely felt like' an individual statement compared with those whose feelings were not reflected by that statement.

- Two long-term value statements *I don't see any point in visiting the dentist unless I need to* and *It will cost me less in the long run if I only go to the dentist when I have trouble with my teeth* showed the largest variations in dental attendance. For example, 20% of dentate adults who definitely felt that they did not see any point in visiting the dentist unless they had to went to the dentist for a regular check-up compared with 80% of those who did not feel like that.

6.2.1 Introduction

Many people do not find the idea or experience of going to the dentist very pleasant. People's opinions towards the dentist may influence how often they go to the dentist. In some instances their feelings may be so strong that they avoid going to the dentist altogether[1].

During the interview dentate adults were shown a series of statements related to going to the dentist (see Figure 6.2.1). For each statement respondents had to decide whether they 'definitely felt like that', 'felt like that to some extent' or 'did not feel like that'. The statements were on cards which were shuffled between each interview to randomise the order in which respondents looked at the statements. If the respondent stated that they 'definitely felt like that' for more than one statement they were asked to rank these statements in order of importance.

The statements related to different aspects of going to the dentist, such as fear or cost. The relationships between the responses to each statement were examined[2] to identify which statements could be grouped together. Four factors were identified which were used to group the statements:

- organisational aspects of going to the dentist
- cost
- fear
- the long-term value of dental treatment.

In the 1988 Adult Dental Health Survey respondents were also asked about their feelings towards going to the dentist. Comparison between how dentate adults felt towards going to the dentist in 1988 and 1998 are reported in Chapter 6.9.

6.2.2 Feelings towards going to the dentist

Figure 6.2.1 shows the proportions of dentate adults whose feelings towards going to the dentist were reflected by each of the statements.

Four of the eleven statements had a high proportion (64% or more) of dentate adults who felt, either definitely or to some extent, the same way as the statement. Two of these related to organisational aspects of going to the dentist: *I'd like to know more about what the dentist is going to do and why* (71%) and *I'd like to be able to drop in at the dentist without an appointment* (69%). A third statement related to cost – *I'd like to be given an estimate without commitment* (70%). These three statements also had the highest proportions of dentate adults who 'definitely felt like that'; ranging from

40% to 50%. A slightly lower proportion of dentate adults felt the same way as a fourth statement, which related to fear – *I'm nervous of some kinds of dental treatment* (64%).

The two statements that had the lowest proportions of dentate adults identifying with them both related to the long-term value of going to the dentist: *I don't see any point in visiting the dentist unless I need to* (36%) and *It will cost me less in the long run if I only go to the dentist when I have trouble with my teeth* (35%). About half of dentate adults felt, either definitely or to some extent, the same way as the remaining five statements.

Figure 6.2.1, Table 6.2.1

Ten per cent of dentate adults did not consider any of the statements to be ones they 'definitely felt like'; 13% definitely felt like only one of the statements; and the remaining 78% definitely felt like two or more statements. This latter group were asked to order the statements according how important they felt these were to themselves. For those who definitely felt like only one statement this ranking was automatic. There was not a large variation in the proportion of dentate adults selecting each of the statements as most important; ranging from 4% to 15%. The statements most frequently selected as most important were:

- I'd like to know more about what the dentist is going to do and why (15%)
- I'd like to be able to drop in at the dentist without an appointment (13%)
- I find NHS dental treatment expensive (13%)
- I always feel anxious about going to the dentist (11%)

Two of these four statements were about the organisational aspects of going to the dentist (*I'd like to know more about what the dentist is going to do and why* and *I'd like to be able to drop in at the dentist without an appointment*) and were the same statements that reflected the feelings of the highest proportions of dentate adults. The other two statements related to cost and fear and had, respectively, only 51% and 49% of dentate adults whose feelings were reflected by the statements. As mentioned above, there were other cost and fear statements that had higher proportions of dentate adults who felt like that, but these were not so frequently ranked as most important.

The statements relating to long-term value were among those that were least likely to have been selected as the one respondents felt was most important to them; the proportion of dentate adults who selected each individual long-term value statements was between 3 and 5%.

Fig 6.2.1   **Feelings towards going to the dentist**

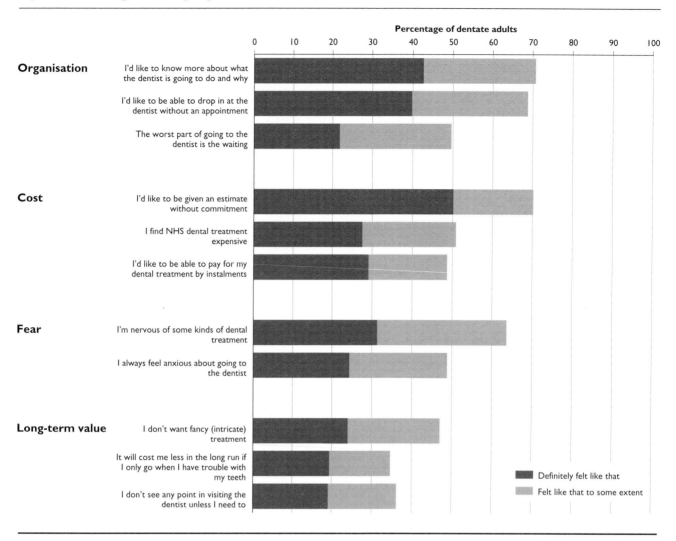

Table 6.2.2 also shows what type of statement dentate adults selected as most important: 34% of adults selected an organisational statement, 27% selected a cost statement, 19% selected a fear statement; and 12% selected a value statement.

Table 6.2.2

In summary, it appears that the organisational aspects of going to the dentist were of most importance to people, both in terms of the proportion of dentate adults whose feelings were reflected by the statements in some way and the statements most frequently selected as most important. The statements relating to cost were the next most important. This was followed by feelings of anxiety about going to the dentist. The statements concerned with the long-term value of going to the dentist were least important to people.

6.2.3 The feelings of people with different characteristics and experience

There are many things that can influence people's current attitudes towards going to the dentist such as their experiences

or their socio-demographic characteristics. This section describes the variations in the proportion of dentate adults whose feelings were reflected by the statements, either definitely or to some extent, with respect to age, gender and social class of head of household. The statements relating to the costs of going to dentist were also analysed by whether the respondent had their last dental treatment through the NHS or whether it was private and whether or not respondents thought they would be currently entitled to free NHS dental treatment.

The analysis describes the variations in the proportion of dentate adults whose feelings were reflected by the statements and the combined proportions of those who definitely felt like that or felt like that to some extent are reported. In general, the variations in the proportion of dentate adults who 'definitely felt like' each statement were similar to these combined proportions and are not reported separately although the data are shown in the tables. In some cases, the individual proportions for those who 'definitely felt like that' and those who 'felt like that to some extent' do not sum to the combined proportion due to rounding. For ease of presentation of the data, Tables 6.2.3 to 6.2.8 show the variations for all the statements by each characteristic in turn. However, the analysis considers the statements by type.

Organisational statements

It has already been stated that two of the three statements relating to the organisational aspects of going to dentist were among the statements that had the highest proportions of dentate adults whose feelings were reflected by the statements.

The variations with respect to age were different for the individual organisational statements, however the feelings of dentate adults aged 65 years and over were the least likely to be reflected by the statements. For example, 56% of dentate adults aged 65 years and over felt they would like to drop in at the dentist without an appointment compared with between 66% and 73% of the other age groups. Younger dentate adults were most likely to want to know what the dentist is going to do and why (77% of those aged 16 to 24 years and 76% of those aged 25 to 34 years).

A higher proportion of dentate women than dentate men felt that waiting was the worst part of going to the dentist; 52% of women compared with 47% of men. The two other organisational statements did not show significant differences with respect to gender.

Dentate adults who lived in households where the head of household had a manual occupation were more likely to agree with the organisational statements. For example, 72% of dentate adults who lived in manual households identified with being able to drop into the dentist without an appointment compared with 65% of those from non-manual households.

Figure 6.2.2, Tables 6.2.3–5

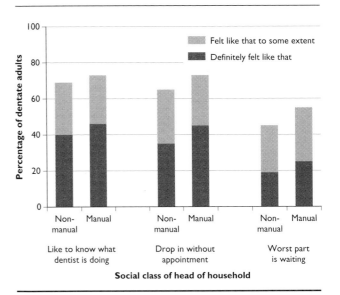

Fig 6.2.2   Statements relating to organisational aspects of going to the dentist by social class of head of household

Cost statements

The youngest dentate adults (aged between 16 and 24 years) were among those least likely to feel the same way as the cost statements. They were also most likely not to be able to give an opinion at all. For example, 38% of 16 to 24 year olds felt that NHS dental treatment was expensive but 45% did not know how they felt about the statement. If these figures are compared with those for dentate adults aged 25 years or over, more than 50% felt NHS dental treatment was expensive and 20% or less did not know how they felt. These feelings towards the costs of going to the dentist may in part be explained by the fact that many 16 to 24 year olds would have had little experience of paying for their dental treatment as they would not have been required to contribute to the cost of NHS dental treatment until their 18th birthday or until age 19 if they were a full-time student (see Chapter 6.3 for futher details).

The youngest and oldest age groups were the least likely to identify with wanting an estimate without commitment; 61% of dentate adults aged 16 to 24 years and 65% of those aged 65 years and over felt this way compared with over 70% of those in other age groups. The lower proportion for the youngest age group will reflect their relatively limited experience of paying for treatment. For dentate adults aged 25 years and over, the proportion who would like to be able to pay for their dental treatment by instalments varied with age from 60% adults aged 25 to 34 years to 27% of those aged 65 years and over. Just over half (51%) of 16 to 24 year olds felt this way. Feelings towards paying for dental treatments by instalments also varied by gender and social class of head of household. Dentate men were less likely than dentate women to feel that they would like to be able to pay for their dental treatment by instalments (46% compared with 51%); 46% of dentate adults from non-manual households felt that way compared with 52% of those from manual households. These variations may reflect different attitudes in general towards paying for items by instalments as well as particularly wanting to spread the costs of dental treatment. Dentate adults from manual backgrounds were also more likely to feel that NHS dental treatment was expensive (55%) compared with those from non-manual backgrounds (48%).

Figure 6.2.3

People's opinions on the financial aspects of going to the dentist are likely to be affected by the costs, if any, they have had to pay for their dental treatment in the past and also on their ability to pay those costs. The costs of going to the dentist can vary greatly depending on, for example, the treatment required; whether it was NHS or private treatment; whether the cost of the dental treatment had to be paid in full or whether they were eligible for (some) free treatment under the NHS. These in turn vary according to an individual's personal characteristics and circumstances. Detailed analyses on the costs and provision of dental treatment are given in Chapter 6.3, however the relationships between people's opinions on the costs of going to the dentist and their recent experience of NHS dental treatment are discussed below.

During the interview the respondents were asked whether their dental treatment at their most recent dental visit was through the NHS or private. People who had NHS treatment at their last dental visit were more likely than those who had private treatment to have felt that NHS dental treatment was expensive, 53% compared with 44% respectively. However, almost a third (32%) of adults who had received private treatment were unable to give their opinion on whether or not they felt that NHS dental treatment was expensive compared with just under a fifth (19%) of adults who had NHS treatment. This most probably reflects the fact that people who have had private treatment have had little recent experience of NHS treatment and its costs.

Dentate adults who had private treatment were also more likely to feel they would like to pay for their dental treatment by instalments (53%) and would like to be given an estimate without commitment (75%) than those who had their treatment through the NHS (47% and 69% respectively).

Figure 6.2.4

Dentate adults were also asked if they went to an NHS dentist within the next four weeks whether they thought they would be eligible for free dental treatment. As one might expect,

those who thought they would be eligible for free NHS treatment were less likely to having feelings that were reflected by the cost statements and more likely to be unable to give an opinion than those who thought they would have to pay. For example, 36% of dentate adults who thought they would be eligible for free NHS dental treatment thought such treatment was expensive compared with 56% of those who thought they would have to contribute.

Tables 6.2.3–7

Fig 6.2.3 **Statements relating to cost of going to the dentist by age**

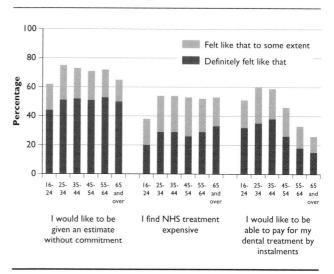

Fig 6.2.4 **Statements relating to cost of going to the dentist by type of dental service**

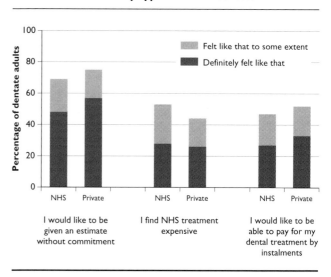

Fear statements

Overall, 64% of dentate adults identified with being nervous of some kinds of dental treatment. The groups least likely to feel nervous were men (55%); and those aged 65 years and over (51%).

These same groups were among those who were less likely to identify with being anxious about going to the dentist. Forty-three per cent of dentate men felt anxious about going to the dentist compared with 55% of dentate women. The youngest as well as the oldest dentate adults were the least likely to feel anxious. Forty per cent of dentate adults aged 16 to 24 years and 42% of those aged 65 years and over and felt anxious about going to the dentist compared with 49% or more of the other age groups.

There was little significant difference among dentate adults from different social classes in whether they felt nervous or anxious.

Figure 6.2.5, Tables 6.2.3–5

when they had trouble with their teeth, however there was no consistent trend across the age groups.

Figure 6.2.6

Dentate men were more likely than dentate women to feel the same way as the long-term value statements. For example, 42% of men identified with not seeing the point in visiting the dentist unless they had to compared with 31% of women.

As with the cost and organisational statements, dentate adults from manual backgrounds were more likely than those from non-manual backgrounds to have feelings that were reflected by the long-term value statements. For example, 30% of dentate adults from non-manual households felt they did not see any point in visiting the dentist unless they had to compared with over 43% of those from manual households.

Tables 6.2.3–5

Fig 6.2.5 **Statements relating to fear of going to the dentist by gender**

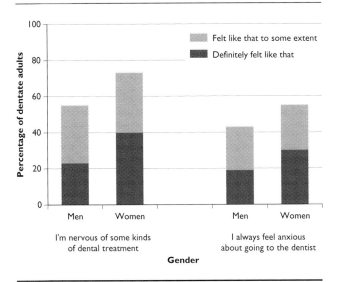

Fig 6.2.6 **Statements relating to the long-term value of dental treatment by age**

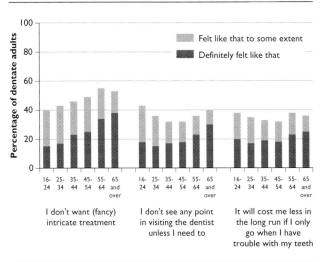

Long-term value statements

The statements referring to the long-term value aspects of dental treatment had the lowest proportions of dentate adults whose feelings were reflected by the statements.

People's opinions towards the long-term value of going to the dentist varied with age, although the individual statements showed different patterns with respect to age. Forty-seven percent of dentate adults felt they did not want fancy or intricate dental treatment. This proportion varied from 40% among 16 to 24 year olds to 53% or more in those aged 55 years and over. The proportion of dentate adults whose feelings were reflected by the statement *I don't see any point in visiting the dentist unless I need to* varied from 43% among dentate adults aged 16 to 24 years to 32% of those in the middle age ranges (aged between 35 and 54 years) to 40% of those aged 65 and over. Overall, 35% of dentate adults felt that it would cost less in the long run if they only went dentist

6.2.4 Opinions towards going to the dentist and usual reason for attending the dentist

People's attitudes and opinions can influence their behaviour. Chapter 6.1 chapter discussed dentate adults' reported usual reason for attending at the dentist. This section investigates whether there was any relationship between the feelings of dentate adults towards going to the dentist and their usual reason for dental attendance.

Figure 6.2.7 shows the usual reason for dental attendance for dentate adults who definitely felt like, felt like to some extent and did not feel like each statement. Generally, the proportion of dentate adults who went to the dentist for a regular check-up was lower among those who definitely felt like the statement compared with those whose feelings were not reflected by the statement. Conversely, dentate adults who did not feel the same way as the statement were less likely than those who 'definitely felt like that' to state they went to the dentist only when they had trouble with their teeth. Adults' feelings could therefore be acting as barriers towards their attendance at the dentist. However, some of the statements showed larger differences in the usual reason for dental attendance among those who 'definitely felt like that' and those who 'did not feel like that' and it would appear that some feelings could be stronger barriers than others. Moreover, the statements with the highest proportions of dentate adults who felt that way (for example, *I'd like to know more about what the dentist is going to do and why*) were not the statements that showed the greatest differences in dental attendance among those who 'definitely felt like that' and those who 'did not feel like that'.

Long-term value statements

The two statements that had the lowest proportion of dentate adults whose feelings were reflected by them, *I don't see any point in visiting the dentist unless I need to* and *It will cost me less in the long run if I only go to the dentist when I have trouble with my teeth*, were the two statements that showed the largest variations in dental attendance. These statements were also the only two where among dentate adults who definitely felt like that a higher proportion attended only with trouble than attended for regular check-ups. For all other statements, the reverse was true. For example, 20% of dentate adults who 'definitely felt that they did not see any point in visiting the dentist unless they had to' went to the dentist for a regular check-up compared with 80% of those who did not feel like that. Dentate adults who definitely felt it would cost them less in the long run if they only went to the dentist when they had trouble with their teeth were more likely to have reported attending the dentist only with trouble (54%) than those who did not feel this way (15%).

Although lower proportions of dentate adults had feelings which were reflected by these two long-term value statements than by other statements it appears that if adults did feel that way, their feelings could be acting as possible barrier to their attendance at the dentist.

Figure 6.2.7, Table 6.2.8

Cost statements

Dentate adults' feelings towards the costs of dental treatment were also related to their usual reason for dental attendance. There was a 25 percentage point difference in the proportions of dentate adults who went for a regular check-up between those who definitely felt NHS dental treatment was expensive (47%) and those who did not feel that way (72%); the largest

difference observed in statements other than those relating to long-term value. The same difference in the proportion of regular attenders was observed between those who definitely felt they would like to pay for their dental treatment by instalments and those who did not feel that way (45% and 70% respectively).

Table 6.2.8

Fear statements

A smaller proportion of dental adults who definitely felt anxious about going to the dentist or nervous of some kinds of dental treatment went to the dentist for a regular check-up compared with those who felt like that to some extent or did not feel like that. For example, 43% of people who definitely felt anxious went to the dentist for a regular check-up compared with 63% who were anxious to some extent and 66% of those who did not feel anxious. It appears those who are only slightly anxious or nervous could overcome their feelings and were still able to go to the dentist, whereas those who are very anxious or nervous were more likely to avoid going to the dentist unless they had to.

Table 6.2.8

Organisational statements

The differences in dental attendance of dentate adults whose feeling were reflected by the organisational statements and those who did not feel that way were among the smallest observed, but were still large enough to be significant. For example, 55% of adults who definitely thought they would like to know more about what the dentist is doing and why, went to the dentist for a regular check-up compared with 67% of those who did not feel like that.

Even though two of the organisational statements were those that had the highest proportion of dentate adults whose feelings were reflected by the statements (*I'd like to know more about what the dentist is doing and why* and *I'd like to be able to drop in at the dentist without an appointment*), these feelings did not appear to have a strong relationship with whether or not dentate adults attended the dentist regularly.

Table 6.2.8

Fig 6.2.7   Usual reason for dental attendance by feelings towards going to see the dentist

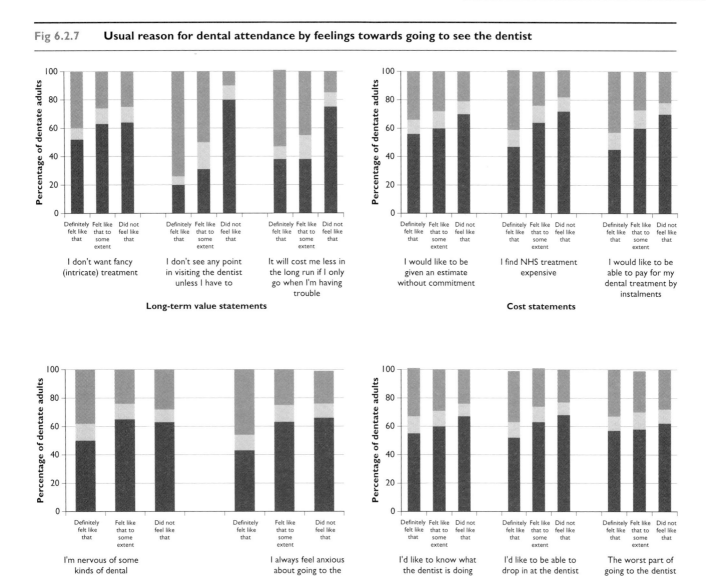

Notes and references

1.   For futher details on factors which may inhibit people going to the dentist, see Finch et al, *Barriers to the receipt of dental care,* Social and Community Planning Research, 1988.

2.   Principal components factor analysis was used to analyse the relationships between the responses to the individual statements. This technique analyses the correlations between a set of variables to identify any underlying dimensions or factors. The first stage is the creation of a correlation matrix, which is used to generate a factor loading matrix. The values in this matrix range from –1 to +1, and show the strength of association between each variable and the derived factors. The larger the factor loading (positive or negative), the greater the degree of association between the variables. The statements were grouped according whichever factor they had the strongest association with.

Table 6.2.1 Feelings towards going to the dentist

Dentate adults   *United Kingdom*

Statement

		%	
Organisation			
I'd like to know more about what the dentist is going to do and why	Definitely felt like that	43	} 71
	Felt like that to some extent	28	
	Did not feel like that	26	
	Don't know	4	
I'd like to be able to drop in at the dentist without an appointment	Definitely felt like that	40	} 69
	Felt like that to some extent	29	
	Did not feel like that	27	
	Don't know	4	
The worst part of going to the dentist is the waiting	Definitely felt like that	22	} 50
	Felt like that to some extent	28	
	Did not feel like that	47	
	Don't know	3	
Costs			
I'd like to be given an estimate without commitment	Definitely felt like that	50	} 70
	Felt like that to some extent	20	
	Did not feel like that	20	
	Don't know	10	
I find NHS dental treatment expensive	Definitely felt like that	28	} 51
	Felt like that to some extent	23	
	Did not feel like that	27	
	Don't know	22	
I'd like to be able to pay for my dental treatment by instalments	Definitely felt like that	29	} 49
	Felt like that to some extent	20	
	Did not feel like that	40	
	Don't know	11	
Fear			
I'm nervous of some kinds of dental treatment	Definitely felt like that	31	} 64
	Felt like that to some extent	32	
	Did not feel like that	34	
	Don't know	2	
I always feel anxious about going to the dentist	Definitely felt like that	24	} 49
	Felt like that to some extent	24	
	Did not feel like that	49	
	Don't know	2	
Long-term value			
I don't want fancy (intricate) treatment	Definitely felt like that	24	} 47
	Felt like that to some extent	23	
	Did not feel like that	38	
	Don't know	15	
I don't see any point in visiting the dentist unless I need to	Definitely felt like that	19	} 36
	Felt like that to some extent	17	
	Did not feel like that	61	
	Don't know	3	
It will cost me less in the long run if I only go when I have trouble with my teeth	Definitely felt like that	19	} 35
	Felt like that to some extent	15	
	Did not feel like that	56	
	Don't know	9	
Base		5281	

1988 Table A21.1

Table 6.2.2 **The statement dentate adults felt most strongly about**

Dentate adults *United Kingdom*

Statement	All
	%
Did not feel strongly about any statement	10

Organisation

I'd like to know more about what the dentist is going to do and why	15	
I'd like to be able to drop in at the dentist without an appointment	13	} 34
The worst part of going to the dentist is the waiting	6	

Cost

I'd like to be given an estimate without commitment	9	
I find NHS dental treatment expensive	13	} 27
I'd like to be able to pay for my dental treatment by instalments	4	

Fear

I'm nervous of some kinds of treatment	7	} 19
I always feel anxious about going to the dentist	11	

Long term value

I don't want fancy (intricate) treatment	4	
I don't see any point in visiting the dentist unless I need to	5	} 12
It will cost me less in the long run if I only go when I have trouble with my teeth	3	

Base	5281

1998 Table 21.14 (part)

Table 6.2.3 **Feelings towards going to the dentist by age**

Dentate adults   *United Kingdom*

Statement		Age						All
		16-24	25-34	35-44	45-54	55-64	65 and over	
		%	%	%	%	%	%	%
Organisation								
I'd like to know more about what the dentist is going to do and why	Definitely felt like that	44 } 77	44 } 76	44 } 71	40 } 66	44 } 69	39 } 61	43 } 71
	Felt like that to some extent	33	31	28	26	25	22	28
	Did not feel like that	19	21	26	30	28	33	26
	Don't know	4	3	2	4	3	6	4
I'd like to be able to drop in at the dentist without an appiontment	Definitely felt like that	37 } 70	41 } 71	44 } 73	39 } 70	40 } 66	35 } 56	40 } 69
	Felt like that to some extent	33	31	29	31	26	21	29
	Did not feel like that	24	25	23	26	31	40	27
	Don't know	6	4	4	4	3	5	4
The worst part of going to the dentist is the waiting	Definitely felt like that	19 } 50	21 } 47	21 } 50	25 } 57	24 } 49	22 } 44	22 } 50
	Felt like that to some extent	31	26	29	32	25	21	28
	Did not feel like that	47	50	47	41	48	52	47
	Don't know	3	3	3	2	3	5	3
Costs								
I'd like to be given an estimate without commitment	Definitely felt like that	44 } 61	51 } 75	52 } 73	51 } 71	53 } 72	50 } 65	50 } 70
	Felt like that to some extent	18	24	21	20	19	15	20
	Did not feel like that	15	18	19	22	21	28	20
	Don't know	23	7	8	7	7	7	10
I find NHS dental treatment expensive	Definitely felt like that	20 } 38	29 } 54	29 } 53	26 } 53	29 } 52	34 } 54	28 } 51
	Felt like that to some extent	18	25	25	27	23	20	23
	Did not feel like that	18	28	28	32	28	29	27
	Don't know	45	18	19	15	20	17	22
I'd like to be able to pay for my dental treatment by instalments	Definitely felt like that	32 } 51	36 } 60	38 } 58	26 } 46	18 } 33	15 } 27	29 } 49
	Felt like that to some extent	19	25	21	20	15	11	20
	Did not feel like that	22	31	34	46	58	64	40
	Don't know	27	8	8	8	9	9	11
Fear								
I'm nervous of some kinds of dental treatment	Definitely felt like that	32 } 62	33 } 69	33 } 67	32 } 64	33 } 64	22 } 51	31 } 64
	Felt like that to some extent	30	36	35	32	31	28	32
	Did not feel like that	36	29	31	34	35	46	34
	Don't know	2	3	2	2	2	3	2
I always feel anxious about going to the dentist	Definitely felt like that	18 } 40	26 } 51	26 } 53	26 } 54	28 } 49	23 } 42	24 } 49
	Felt like that to some extent	21	26	27	28	21	19	24
	Did not feel like that	58	47	45	44	50	55	49
	Don't know	2	2	2	2	2	4	2
Long-term value								
I don't want fancy (intricate) treatment	Definitely felt like that	15 } 40	17 } 43	23 } 46	24 } 50	34 } 55	38 } 53	24 } 47
	Felt like that to some extent	25	26	23	24	21	14	23
	Did not feel like that	40	39	39	38	33	35	38
	Don't know	20	18	14	13	12	12	15
I don't see any point in visiting the dentist unless I need have to	Definitely felt like that	18 } 43	15 } 36	17 } 32	18 } 32	23 } 36	30 } 40	19 } 36
	Felt like that to some extent	25	21	15	14	13	10	17
	Did not feel like that	53	61	65	65	61	56	61
	Don't know	4	2	2	2	3	4	3
It will cost me less in the long run if I only go when I have trouble with my teeth	Definitely felt like that	20 } 38	17 } 34	19 } 33	18 } 31	23 } 38	25 } 35	19 } 35
	Felt like that to some extent	18	18	14	14	15	11	15
	Did not feel like that	48	58	60	60	51	55	56
	Don't know	14	8	7	8	11	10	9
Base		*701*	*1151*	*1099*	*1016*	*645*	*669*	*5281*

Table 6.2.4 **Feelings towards going to the dentist by gender**

Dentate adults *United Kingdom*

Statement		Gender		All
		Men	**Women**	
		%	%	%
Organisation				
I'd like to know more about what the dentist is going to do and why	Definitely felt like that Felt like that to some extent Did not feel like that Don't know	41 } 71 29 25 4	44 } 71 27 26 3	43 } 71 28 26 4
I'd like to be able to drop in at the dentist without an appiontment	Definitely felt like that Felt like that to some extent Did not feel like that Don't know	42 } 70 29 25 5	38 } 67 30 29 4	40 } 69 29 27 4
The worst part of going to the dentist is the waiting	Definitely felt like that Felt like that to some extent Did not feel like that Don't know	20 } 47 28 50 3	24 } 52 28 45 3	22 } 50 28 47 3
Costs				
I'd like to be given an estimate without commitment	Definitely felt like that Felt like that to some extent Did not feel like that Don't know	50 } 70 20 20 10	50 } 70 20 20 10	50 } 70 20 20 10
I find NHS dental treatment expensive	Definitely felt like that Felt like that to some extent Did not feel like that Don't know	27 } 49 22 28 22	28 } 52 24 26 22	28 } 51 23 27 22
I'd like to be able to pay for my dental treatment by instalments	Definitely felt like that Felt like that to some extent Did not feel like that Don't know	27 } 46 19 42 12	32 } 51 20 38 11	29 } 49 20 40 11
Fear				
I'm nervous of some kinds of dental treatment	Definitely felt like that Felt like that to some extent Did not feel like that Don't know	23 } 55 32 42 3	40 } 72 33 26 2	31 } 64 32 34 2
I always feel anxious about going to the dentist	Definitely felt like that Felt like that to some extent Did not feel like that Don't know	19 } 43 24 55 2	30 } 55 25 43 2	24 } 49 24 49 2
Long term value				
I don't want fancy (intricate) treatment	Definitely felt like that Felt like that to some extent Did not feel like that Don't know	26 } 49 24 37 14	23 } 45 22 39 17	24 } 47 23 38 15
I don't see any point in visiting the dentist unless I need to	Definitely felt like that Felt like that to some extent Did not feel like that Don't know	22 } 42 19 55 3	16 } 31 15 67 2	19 } 36 17 61 3
It will cost me less in the long run if I only go when I have trouble with my teeth	Definitely felt like that Felt like that to some extent Did not feel like that Don't know	22 } 38 16 52 10	17 } 31 14 60 9	19 } 35 15 56 9
Base		*2406*	*2875*	*5281*

1988 Table A21.2

Table 6.2.5 Feelings towards going to the dentist by social class of head of household

Dentate adults   *United Kingdom*

Statement		Social class of head of household		All†
		Non-manual	**Manual**	
		%	%	%
Organisation				
I'd like to know more about what the dentist is going to do and why	Definitely felt like that Felt like that to some extent Did not feel like that Don't know	40 } 69 29 29 2	46 } 73 27 23 5	43 } 71 28 26 4
I'd like to be able to drop in at the dentist without an appointment	Definitely felt like that Felt like that to some extent Did not feel like that Don't know	35 } 65 30 32 3	45 } 72 28 22 5	40 } 68 29 27 41
The worst part of going to the dentist is the waiting	Definitely felt like that Felt like that to some extent Did not feel like that Don't know	19 } 45 26 52 3	25 } 55 30 42 3	22 } 50 28 47 3
Costs				
I'd like to be given an estimate without commitment	Definitely felt like that Felt like that to some extent Did not feel like that Don't know	48 } 69 21 23 8	53 } 72 19 18 11	50 } 70 20 20 10
I find NHS dental treatment expensive	Definitely felt like that Felt like that to some extent Did not feel like that Don't know	23 } 48 25 31 21	32 } 55 23 23 22	28 } 51 23 27 22
I'd like to be able to pay for my dental treatment by instalments	Definitely felt like that Felt like that to some extent Did not feel like that Don't know	25 } 46 21 45 10	33 } 52 19 36 12	29 } 49 20 40 11
Fear				
I'm nervous of some kinds of dental treatment	Definitely felt like that Felt like that to some extent Did not feel like that Don't know	30 } 65 35 34 1	33 } 62 30 34 3	31 } 64 32 34 2
I always feel anxious about going to the dentist	Definitely felt like that Felt like that to some extent Did not feel like that Don't know	21 } 47 26 52 1	29 } 51 22 46 3	24 } 49 24 49 2
Long term value				
I don't want fancy (intricate) treatment	Definitely felt like that Felt like that to some extent Did not feel like that Don't know	21 } 46 25 40 14	28 } 49 21 34 17	24 } 48 23 38 15
I don't see any point in visiting the dentist unless I need to	Definitely felt like that Felt like that to some extent Did not feel like that Don't know	14 } 30 16 68 2	25 } 43 18 54 4	19 } 36 17 61 3
It will cost me less in the long run if I only go when I have trouble with my teeth	Definitely felt like that Felt like that to some extent Did not feel like that Don't know	15 } 28 14 64 8	25 } 42 17 47 11	19 } 35 15 56 9
Base		2483	2346	5281

† *Includes those for whom social class of head of household was not known and armed forces*
1988 Table A21.3

Table 6.2.6 **Feelings towards the costs of going to the dentist by type of dental service**

Dentate adults   *United Kingdom*

Statement		Type of dental service			All†
		NHS	**Private**	**Other**	
		%	%	%	%
I'd like to be given an estimate without commitment	Definitely felt like that	48 } 69	57 } 75	52 } 68	50 } 70
	Felt like that to some extent	21	18	16	20
	Did not feel like that	20	18	22	20
	Don't know	10	7	10	10
I find NHS dental treatment expensive	Definitely felt like that	28 } 53	26 } 44	29 } 48	28 } 51
	Felt like that to some extent	25	18	19	23
	Did not feel like that	28	23	22	27
	Don't know	19	32	30	22
I'd like to be able to pay for my dental treatment by instalments	Definitely felt like that	27 } 47	33 } 53	40 } 58	29 } 49
	Felt like that to some extent	20	19	18	20
	Did not feel like that	40	41	35	40
	Don't know	13	6	8	11
Base		4087	870	284	5281

† Includes those whom type of dental service was not known

Table 6.2.7 **Feelings towards the costs of going to the dentist by whether respondents thought they were entitled to free NHS treatment**

Dentate adults   *United Kingdom*

Statement		Respondents thought they:			All
		Were entitled to free NHS treatment	**Would have to pay**	**Don't know**	
		%	%	%	%
I'd like to be given an estimate without commitment	Definitely felt like that	40 } 58	53 } 74	52 } 63	50 } 70
	Felt like that to some extent	17	21	11	20
	Did not feel like that	22	19	22	20
	Don't know	20	7	15	10
I find NHS dental treatment expensive	Definitely felt like that	21 } 36	30 } 56	18 } 34	28 } 51
	Felt like that to some extent	16	26	16	23
	Did not feel like that	27	27	25	27
	Don't know	37	17	42	22
I'd like to be able to pay for my dental treatment by instalments	Definitely felt like that	28 } 44	30 } 50	25 } 42	29 } 49
	Felt like that to some extent	17	21	17	20
	Did not feel like that	30	43	38	40
	Don't know	26	7	20	11
Base		1101	3957	213	5281

Table 6.2.8   Usual reason for dental attendance by feelings towards going to the dentist

Dentate adults   *United Kingdom*

Statement		Usual reason for dental attendance			Base
		Regular check-up	Occasional check-up	Only with trouble	
Organisation					
I'd like to know more about what the dentist is going to do and why	Definitely felt like that	% 55	12	34	2246
	Felt like that to some extent	% 60	11	29	1425
	Did not feel like that	% 67	9	24	1397
I'd like to be able to drop in at the dentist without an appointment	Definitely felt like that	% 52	11	36	2090
	Felt like that to some extent	% 63	11	27	1492
	Did not feel like that	% 68	9	23	1436
The worst part of going to the dentist is the waiting	Definitely felt like that	% 57	10	33	1216
	Felt like that to some extent	% 58	12	29	1415
	Did not feel like that	% 62	10	28	2453
Costs					
I'd like to be given an estimate without commitment	Definitely felt like that	% 56	10	34	2556
	Felt like that to some extent	% 60	12	28	1071
	Did not feel like that	% 70	9	21	1118
I find NHS dental treatment expensive	Definitely felt like that	% 47	12	42	1490
	Felt like that to some extent	% 64	12	24	1263
	Did not feel like that	% 72	10	19	1452
I'd like to be able to pay for my dental treatment by instalments	Definitely felt like that	% 45	12	43	1538
	Felt like that to some extent	% 60	13	27	993
	Did not feel like that	% 70	8	22	2155
Fear					
I'm nervous of some kinds of dental treatment	Definitely felt like that	% 50	12	38	1665
	Felt like that to some extent	% 65	11	24	1687
	Did not feel like that	% 63	9	28	1802
I always feel anxious about going to the dentist	Definitely felt like that	% 43	11	46	1329
	Felt like that to some extent	% 63	12	25	1273
	Did not feel like that	% 66	10	23	2554
Long-term value					
I don't want fancy (intricate) treatment	Definitely felt like that	% 52	8	40	1279
	Felt like that to some extent	% 63	11	26	1132
	Did not feel like that	% 64	11	25	2040
I don't see any point in visiting the dentist unless I need to	Definitely felt like that	% 20	6	74	1048
	Felt like that to some extent	% 31	19	50	840
	Did not feel like that	% 80	10	10	3230
It will cost me less in the long run if I only go when I have trouble with my teeth	Definitely felt like that	% 38	9	54	1021
	Felt like that to some extent	% 38	17	45	781
	Did not feel like that	% 75	10	15	2965
All		% 59	11	30	5281

6.3   Visiting the dentist

Summary

- Fourteen per cent of dentate adults said their most recent visit to the dentist was the first time at that particular practice and 86% said they had visited the practice before, with 61% reporting that they had attended for 5 years or more.

- Forty-five per cent of men and 41% of women in full time employment take time off work to attend for dental treatment. In most cases this was estimated to take less than two hours per visit.

- Most dentate adults (83%) attended a dentist who was within 5 miles of their home, work or place of study; 26% visited a dentist who was less than half a mile away.

- Seventy-one per cent of dentate adults had visited a dentist within the year prior to the survey, 19% had visited the dentist between one and five years ago and 10% had last visited the dentist over five years ago.

- Sixty per cent of dentate adults had no teeth filled or extracted during the most recent course of treatment prior to this survey. A third of those who only attend when they have some trouble with their teeth (34%) required some extractions during their last course of treatment compared with 4% of those who sought regular dental check-ups.

- Over three-quarters (77%) of dentate adults had their last course of the treatment provided under the NHS while 18% said they had private dental care and 2% had a mix of NHS and private care.

- Two-thirds (67%) of dentate adults who had visited the dentist in the previous two years paid a contribution to the costs of their last course of dental treatment. Those in the 16 to 24 year age group were the least likely to have paid something, 32% compared with between 64% and 90% of other age groups.

- Those who had private treatment were likely to have paid more for their dental treatment than those who had their treatment through the NHS. For example, 4% of those who had NHS treatment in the previous two years paid over £100 compared with 15% who had private treatment.

6.3.1 Introduction

As in previous Adult Dental Health surveys, adults in this survey were asked as part of the dental interview about various aspects of the dental practice that they had visited most recently. These questions included the reasons for their choosing the practice they last attended, and aspects associated with its location and accessibility. They were also asked about the treatment they had received at their most recent visit, including the type of treatment received and the cost, and whether they intended attending the same practice at their next visit. Where appropriate the experiences of dentate and edentate adults are discussed separately.

6.3.2 Previous attendance at a dental practice

Dentate adults were asked whether, on the most recent occasion they visited a dentist, they had been to that practice before, or whether they were first-time attenders to this dentist. Those who had been to the practice before were asked how long they had been attending. Fourteen per cent said their most recent visit to the dentist was the first time at that particular practice and 86% said they had visited the practice before, with 61% reporting that they had attended for 5 years or more. Dentate men were more likely to be first-time attenders than dentate women, 16% compared with 12%. Those aged 25 to 34 years were the most likely to be first time attenders, 18% of this age group were first-time attenders compared with between 11% and 14% of those in other age groups.

Dentate adults in Wales were less likely to be first-time attenders (10%) compared with those in England (14%) and Scotland (15%). There were no significant differences between the regions in England in the proportion of dentate adults who were first-time attenders.

Dentate adults from unskilled manual backgrounds were more likely to be first-time attenders (18%) than those from non-manual backgrounds (12%) and were correspondingly less likely to have been with the same practice for 5 years or longer (58% compared with 63%).

Dentate adults who reported attending the dentist for regular check-ups were the least likely to have been first time attenders at their last dental visit (5%). By contrast almost a third (32%) of those who said they usually only attend when they had some trouble with their teeth said their last dental visit was to a dentist they were visiting for the first time.

Table 6.3.1

6.3.3 Reasons for choosing a dental practice

Dentate adults were asked how they chose the dental practice they had visited most recently. For about a third of all dentate adults (34%) the choice had been made on recommendation. For over a quarter, 27%, the practice had been chosen because it was their family dentist, that is they had attended the practice since childhood or their decision was influenced by its being the practice attended by their spouse or children. A similar proportion, 27%, said that it was the nearest practice.

Table 6.3.2

Table 6.3.3 shows the characteristics of dentate adults mentioning each of the three most frequently mentioned reasons for choosing their dental practice. There were marked differences in the reasons for choosing a dental practice given by dentate adults in different age groups. The youngest age group, those aged 16 to 24 years, were most likely to mention that the practice they attended was chosen because it was their family dentist (61%) whereas less than 15% of those aged 55 years and over gave this as the reason for choosing their dentist. A larger proportion of these older adults, in comparison to the youngest age group, said the practice they visited had been recommended to them or that it was the nearest practice.

Dentate women were more likely than dentate men to have chosen the dental practice they most recently visited on the basis of recommendation (37% compared with 30%) and were less likely than men to have chosen it on the basis that it was their family dentist (25% compared with 30%).

There were some differences in the reasons for choosing a dental practice by country and, in England, by region. In comparison to the rest of the United Kingdom dentate adults in Wales and Northern Ireland were more likely to have chosen the family dentist (31% and 35%) whereas those in Scotland were more likely to have chosen a practice because it was the nearest to their home or place of work (37%). Dentate adults living in the North of England were more likely to have chosen a practice because it was the family dentist (33%) than those in the South (24%).

There was little difference in the proportion of dentate adults from different social classes who chose their dental practice because it was nearest to them or because it was the family dentist. However, those from non-manual backgrounds were more likely than others to have chosen their dental practice because it had been recommended to them; 37% of those from non-manual backgrounds gave this reason compared with 31% of those from skilled manual backgrounds and 30% of those from unskilled manual backgrounds.

Those who said they went for regular check-ups were more likely to have chosen their dental practice on the basis of recommendation (38%) than either those who went for occasional check-ups (32%) or only when they had some trouble with their teeth (27%).

Figure 6.3.1, Table 6.3.3

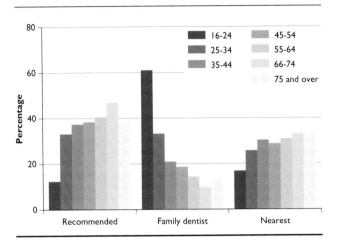

Fig 6.3.1 **Main reason given for choosing dental practice by age among dentate adults**

6.3.4 Using a dental practice near home or work

One factor to consider when deciding which dental practice to attend is its location; overall just over one in four dentate adults (27%) said proximity to home or place or work was a reason for their choice of the last dental practice they visited. Dentate adults who were working or studying were asked about the location of the dental practice they visited in relation to whether it was nearer to their place of work or study, or their home, or about the same distance from both. The majority, 60%, said that they attended a practice nearer to their home, 20% said it was nearer to their place of work or study and the remaining 20% that it was the same distance from both. Dentate men (63%) who were currently working or studying were more likely than dentate women (56%) to say the practice they attended was nearer to their home than their place of work.

There were no significant differences between those living in the different regions of England, but dentate adults in Scotland were most likely and those in Wales least likely to attend a dental practice which was nearer to their place of work or study (15% and 25% respectively).

Dentate adults from skilled manual backgrounds were less likely to attend a practice nearer to their home (56%) than adults from other backgrounds (62%).

Table 6.3.4

6.3.5 Time taken off work to visit a dental practice

Dentate adults who were working were asked if they usually took time off work to visit the dentist, and if they did, how much work time the visit usually took. Overall, over a third (36%) of dentate adults who were working usually took some time off work to visit the dentist. In most cases this was less than 2 hours.

The proportion of working dentate adults who took time off to visit the dentist varied with age; from 30% of those aged 16 to 24 years to 42% of those aged 65 and over. There was little variation in the proportion of those taking time off to visit the dentists in the different countries of the United Kingdom, however there were significant differences within the different regions of England. Those living in the South were more likely to take time of work (41%) than those living in the other regions (32%). Social class of head of household was also associated with whether dentate adults usually took time off work to visit the dentist; 41% of those from non-manual backgrounds took time off compared with 34% of those from skilled manual backgrounds and 29% of those from unskilled manual households.

Table 6.3.5

Working patterns of men and women tend to be different, with a higher proportion of women working part-time. As a consequence a higher proportion of dentate men who were working (44%) usually took time off work to visit the dentist compared with women (27%). However, Table 6.3.6 shows that among dentate men and women who worked full-time a similar proportion said they usually took time off work to visit the dentist (45% and 41% respectively).

Table 6.3.6

6.3.6 Distance travelled to the dental practice

Dentate adults were asked to estimate the distance to the dental practice they visited most recently from their home, or place of work or study, whichever was the nearest. One quarter (26%) of dentate adults said that their practice was no more than half a mile from their home or work or place of study. Overall, 43% travelled up to one mile and 60% up to two miles. Only 7% of dentate adults reported travelling more than 10 miles to reach their dental practice.

There was very little difference in the distances travelled to a dental practice between men and women or between age groups. There were however some differences between the constituent countries and between the regions in England.

Thus 10% of dentate adults in Scotland and 11% of those in Northern Ireland travelled more than 10 miles to their dental practice compared with 6% of those in Wales and 7% of those in England. Furthermore the sampling scheme for this survey excluded very remote areas of the United Kingdom such as the Scottish Islands which will tend to produce an underestimate of the proportion who are far away from their nearest practice. Within England a higher proportion of dentate adults from the North travelled less than half a mile to their dental practice; 33% compared with 18% of those from the Midlands. Dentate adults living in households without access to a car or van were more likely than those with access to a vehicle to have travelled no further than half a mile to their dentist (34% compared with 24%). This group were also less likely to travel more than 10 miles; 5% of those without access to a vehicle travelled more than 10 miles compared with 8% of those with access.

Figure 6.3.2, Table 6.3.7

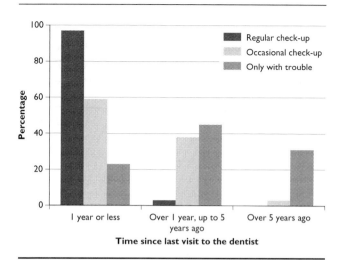

Fig 6.3.3   **Time since last visit to the dentist by usual reason for dental attendance among dentate adults**

Fifty per cent of those who had no teeth said they had last visited a dentist over 10 years previously. Fewer of the younger edentate said they had not attended in the last 10 years but this is perhaps more likely to reflect that they have experienced total tooth loss more recently than an actual difference in attendance behaviour since becoming edentate. The same process may underlie the finding that fewer edentate adults in Wales (48%), Scotland (46%) or Northern Ireland (47%) had visited a dentist in the previous 10 years than in England (51%).

Table 6.3.9

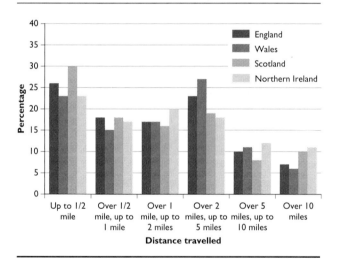

Fig 6.3.2   **Distance travelled to dental practice by country among dentate adults**

6.3.7   Time since the last visit to the dentist

Table 6.3.8 shows for dentate adults the relationship between their estimate of the length of time since they last visited the dentist and their socio-demographic characteristics. Around three-quarters of dentate adults aged between 35 and 74 years reported visiting the dentist within the previous year compared with around two-thirds of oldest and youngest age groups. Six per cent of dentate men and a similar proportion (7%) of those from unskilled manual backgrounds said they had not been to a dentist in the previous 10 years compared with 3% of dentate women and 3% of those from non-manual households. There were no significant differences in the time since the last visit to a dentist by country or English region.

Figure 6.3.3, Table 6.3.8

6.3.8   Treatment received at the most recent visit by dentate adults

Dentate adults were asked what treatment they had received at their most recent dental visit. The 'most recent visit' was taken to include all the visits that were part of the most recent completed course of treatment. Those who were in the middle of a course of treatment at the time of survey were excluded from the analysis. Most courses of treatment start with an examination and this part of the treatment is not identified or discussed separately. Some respondents reported that they had more than one type of treatment at their most recent visit. The proportion of dentate adults who received specific treatments are shown in Table 6.3.10 and Table 6.3.11 summarises the types of treatment received according to whether or not a filling or an extraction or both took place for various socio-demographic factors. Detailed information regarding specific treatments by gender, English region and country, social class of head of household, usual reason for dental attendance and length of time since the last dental visit can be found in Tables 6.3.12 to 6.3.16 but these are not discussed in the text.

The most common individual treatment was a scale and polish; over half (58%) of dentate adults received this treatment at their most recent visit to the dentist. Just under one third, 29%, had a tooth or teeth filled, and 27% had an x-ray taken. Fourteen per cent said that their most recent visit was to have a tooth extracted. Less than 10% mentioned other treatments such as the fitting of a crown (7%) or a partial denture (5%), treatment of an abscess (5%) or having a denture repaired (2%).

Table 6.3.10

The majority of dentate adults (60%) had neither fillings nor extractions at their most recent visit to the dentist. Dentate adults who had neither extractions nor fillings were mainly those who had an examination or a scale and polish, although a small proportion had more complex treatment such as an abscess treated, a partial denture made or a crown (re)fitted (data not shown).

There were variations in the experience of fillings or extractions at the most recent dental visit by age and gender. Dentate adults aged 16 to 24 years were the most likely to have no extractions or fillings at their most recent visit to the dentist. The proportion who had some extractions and no fillings increased with age; from 4% of those aged 16 to 24 years to 16% of those aged 65 and over. Dentate women were more likely than dentate men to have had no extractions and no fillings (65% compared with 56%) and other combinations of treatment tended to be more common among men than women.

Dentate adults in England and Wales were more likely than those in Scotland and Northern Ireland to have had neither extractions nor fillings during their most recent course of treatment. Dentate adults from Northern Ireland were the most likely to have had some fillings, 41%, and those in Scotland were the most likely to have had some extractions, 20%. There were no significant differences in the main types of treatment received between dentate adults living in different regions of England.

There was some variation in the treatment patterns experienced at the most recent dental visit by dentate adults from different social classes. Nearly two thirds (65%) of those from non-manual backgrounds had neither an extraction nor a filling at their most recent visit to the dentist compared with over half (55%) of dentate adults from manual backgrounds. Correspondingly those from non-manual backgrounds were less likely than others to have had any extractions; 10% of dentate adults from a non-manual background had an extraction compared with 17% of those from skilled manual backgrounds and 21% from unskilled manual backgrounds.

A higher proportion of those who went to the dentist for regular dental check-ups had neither an extraction nor a filling at their most recent visit to the dentist (71%) compared with those who usually attend only when having trouble with their teeth (40%). Furthermore, 98% of those whose last visit was for a check-up had no extractions compared with 64% of those who reported attending with trouble.

Figure 6.3.4, Tables 6.3 .11–16

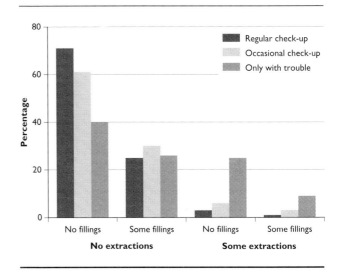

Fig 6.3.4 **Treatment received at most recent dental visit by usual reason for attendance among dentate adults**

6.3.9 Type of service used at the most recent visit

Since the previous Adult Dental Health Survey there have been changes in the payment system for dentists within the NHS and also in the charges to patients. In particular, dental check-ups have become subject to a patient charge, except to specified groups of patients[1]. Changes in regulations have also introduced the ability to mix private dental care with NHS care subject to specific rules. As was the case at the last survey, people who are not exempt from some or all patient charges pay a proportion of their NHS dental treatment costs. This requirement may confuse some adults, as it is possible that some consider this as being the same as having "private" treatment. Similarly some paying adult patients may think they are getting NHS treatment when some or all of their course of treatment was private. Notwithstanding this all adults were asked whether their treatment was "under the NHS, was it private, or was it something else?". Responses were not prompted by the interviewer, and were subsequently coded to one of 10 categories, including, National Health Service (NHS), private, NHS and private, through insurance or with a dental plan (which are different forms of arrangements for payment of private treatment).

Overall, just over three-quarters of dentate adults (77%) reported that their most recent course of treatment had been provided through the NHS while 18% said their last course of treatment was provided privately. A further 2% said their treatment had been a mix of NHS and private treatment.

Dentate adults who had dentures in combination with their natural teeth were more likely than those with natural teeth only to have had their last course of treatment through the NHS rather than privately (80% compared with 76%).

Table 6.3.17

Table 6.3.18 looks at type of service used during the last dental visit by the time since that visit. Those who visited the dentist over five years ago were most likely to have had NHS treatment at that time; 81% of those whose most recent visit had been between five and ten years ago and 86% of those whose most recent visit had been over ten years ago compared with 76% of those who visited the dentist within the last five years.

Table 6.3.18

Dentate adults aged 16 to 24 years were most likely to have been treated through the NHS at their most recent dental visit (86% compared with between 73% and 78% of those in the older age groups) and least likely to have been treated privately (10% compared with between 15% and 21% of other age groups). This is probably influenced by the fact that, for many of this group, their most recent course of treatment would have been while they were under 18 (or under 19 if they were in full-time education) and they therefore would have been exempt from contributing towards the cost of any NHS treatment. Apart from the difference between the youngest and other age groups there was no clear association between the type of service and age. There was also no significant difference between dentate men and women in the type of service provided at their most recent course of treatment.

Dentate adults living in England were more likely than those living in Scotland or Northern Ireland to report having been treated as a private patient; 19% of dentate adults in England were treated privately compared with 14% in Scotland and 10% in Northern Ireland. In Wales 17% of dentate adults said they had been treated privately and further 4% said they had a mix of NHS and private treatment. Within England, dentate adults living in the South were more likely than those living elsewhere to report having being treated as a private patient at their most recent visit; 24% said they had been treated privately compared with 18% in the Midlands and 9% in the North.

Dentate adults from non-manual backgrounds were more likely than those from manual backgrounds to say their most recent course of treatment was provided privately. Of those from non-manual backgrounds 21% reported having private treatment, compared with 15% of those from skilled manual and 15% of those from unskilled manual backgrounds. There were no significant differences in the type of service used at the most recent course of treatment associated with the usual reason for dental attendance.

Figure 6.3.5, Table 6.3.19

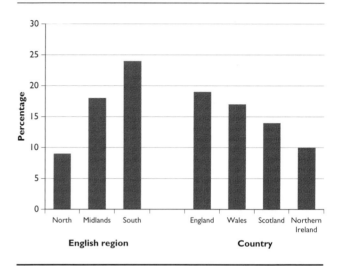

Fig 6.3.5   **Dentate adults who had private dental treatment at their most recent visit to the dentist by English region and country**

Among edentate adults who had visited a dentist in the previous 10 years the proportions using different types of service were similar to those for dentate adults; 78% of the edentate group had been treated under the NHS, 19% privately and 2% said their treatment was a mix of NHS and private service. Generally there were few statistically significant differences in the type of service used by the edentate group associated with the various socio-demographic characteristics considered, due in part at least to the relatively small size of many of the sub-groups.

Table 6.3.20

6.3.10 Payment for treatment

Overall, 61% of dentate adults had paid something towards the cost of their dental treatment, 30% said they had not paid anything and 10% did not know or could not remember. The proportion who reported paying something decreased as the time since the most recent visit increased; over two-thirds (69%) of dentate adults who had visited the dentist in the last year said they paid something towards the cost of their treatment compared with half (50%) of those whose last visit was between one and five years ago and less than a third of those who last visited the dentist over five years ago. There was a corresponding increase in the proportions who said they had not paid anything for their treatment and in the proportions who did not know or could not remember. Subsequent analysis has been restricted to those who had completed their most recent course of dental treatment within the two years prior to the survey.

Table 6.3.21

Among dentate adults who completed their most recent course of treatment within the last two years, 67% had paid something towards the cost of their treatment. There were significant differences among dentate adults in different age groups as to whether they had paid anything for their treatment, which reflects the system of payment for NHS treatment. Thus a third (32%) of those aged 16 to 24 years reported paying something compared with around two-thirds (64%) of those aged 25 to 34 years. The proportion paying something continued to increase steadily with age up to those aged 65 to 74 years, 90% of whom reported paying something towards their most recent course of treatment. In the oldest age group, those aged 75 and over, 73% said they had paid something.

A smaller proportion of dentate women (64%) contributed to the cost of their dental treatment compared with dentate men (71%). A higher proportion of dentate adults from non-manual backgrounds (74%) paid something compared with those from manual backgrounds (66% of those from skilled manual backgrounds and 56% from unskilled manual backgrounds paid something towards their dental treatment).

The proportion of dentate adults paying towards treatment was highest for those living in England, (68%) and lowest for those in Wales (61%) and Northern Ireland (61%). Within the regions of England there were also differences; 70% of dentate adults in the South contributed to their costs of treatment compared with 68% in the Midlands and 64% in the North.

Seventy per cent of dentate adults who attended the dentist for regular check-ups paid something towards their dental treatment compared with 56% of those who said they only attended when they had trouble with their teeth.

Almost nine out of ten of those who had private treatment (87%) or a mix of NHS and private treatment (89%) paid something towards the cost of their dental treatment compared with 64% of those who had their treatment through the NHS. Thirty percent of those treated under a dental plan said they had paid something towards their most recent course of treatment.

Table 6.3.22

Over a quarter (27%) of dentate adults who had visited the dentist in the previous two years said they did not pay anything towards their most recent completed course of treatment and almost half (45%) paid under £30. Almost a quarter (23%) paid £30 or more towards their dental treatment; 9% paid between £30 and £49, 8% paid between £50 and £99 and 6% paid £100 or more. Not surprisingly there were large differences in the cost depending on whether the dental treatment had been carried out through the NHS or privately. Thirty per cent of dentate adults who had NHS dental treatment did not pay anything towards their treatment compared with 8% of those who had private treatment. Those who had private treatment were likely to have paid more for their dental treatment than those who had their treatment through the NHS. For example, among those who had NHS treatment, 25% paid between £10 and £19 and 4% paid over £100 whereas 14% of those who had private treatment paid between £10 and £19 and 15% paid over £100. Among dentate adults whose treatment was covered by a dental plan over half (53%) said they did not pay anything for their last course of treatment, 16% said they paid less than £10 and 17% did not know or could not remember how much they paid.

Figure 6.3.6, Table 6.3.23

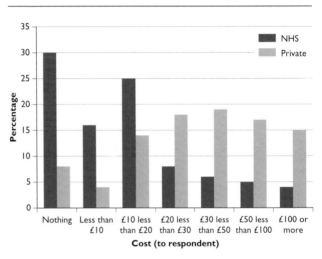

Fig 6.3.6 **Costs of dental treatment by type of dental service among dentate adults**

Table 6.3.24 shows how much was paid for the most recent course of treatment by the usual reason for dental attendance. Since there was such a wide variation in cost associated with whether or not the treatment was provided under the NHS, the table is restricted to those being treated wholly under the NHS. Overall nearly half, 45%, of those who only attend when they have trouble with their teeth said they had not paid anything for their most recent NHS treatment compared with 27% of regular and 38% of occasional attenders.

However regular attenders were more likely to have paid lower amounts than either occasional attenders or those attending only with trouble. For example, 47% of regular attenders paid less than £20 for their NHS treatment, compared with 27% of occasional attenders, and 12% of those attending only with trouble. Twenty-one per cent of those attending with trouble paid £50 or more for their NHS treatment compared with 10% of those who attended for occasional check-ups and 7% of those who had regular check-ups.

Table 6.3.24

Dentate adults who had paid something towards the cost of their dental treatment were asked whether the cost was more, less or about what they expected. Among those who had visited the dentist within the previous two years 63% said the cost of their most recent completed course of treatment had been about what they expected, 19% said the costs had been more than expected and 12% less than expected. Five per cent said they had no idea what the cost was likely to have been. Dentate adults' expectations of the costs of their treatment were related to their usual reason for dental attendance. Those who went to the dentist for an occasional check-up or only when they had trouble with their teeth were more likely than those who went for a regular check-up to have expected the costs of their dental treatment to be lower than they were; 38% of those who only went to the dentist with trouble and 28% of those who went for an occasional check-up said the costs where more than they had expected compared with 16% of those who went for a regular check-up.

Table 6.3.25

Those who had private dental treatment were more likely than those who had NHS treatment to have said that the costs of their treatment were more than they expected (28% compared with 16%). Conversely, a higher proportion of those who had NHS treatment compared with those who had private treatment said the costs were less than expected (14% and 6% respectively). As one would expect, whether respondents said the costs were more or less than they expected was related to the amount they had paid. For example, the proportion of those who said the costs were more than they expected varied from 6% of those who paid less than £10 to 25% of those who paid between £30 and £49 and to 52% of those who paid £100 or more.

Tables 6.3.26–27

The survey could not assess respondents' eligibility for free NHS dental treatment, however dentate adults were asked whether, if they visited a dentist in the next four weeks, they thought they would be entitled to free treatment. Overall,

21% of dentate adults thought they would be entitled to free treatment. Although this information was based on the respondent's own understanding of the eligibility criteria, and some people may have been mistaken in determining their eligibility, the variations in the proportion of dentate adults who thought they would be entitled to free treatment reflected both the eligibility rules[1] and the variations in the proportions of those who had not paid towards their treatment at their most recent visit.

The youngest dentate adults (aged 16 to 24 years) were the most likely to think they would be entitled to free NHS treatment; 40% compared with 23% or less in the other age groups. A higher proportion of dentate women than dentate men thought they would be entitled to free NHS treatment (23% compared with 16%).

Comparing the constituent countries of the United Kingdom, dentate adults in Northern Ireland were the most likely and those in England the least likely to think they would be entitled to free NHS treatment, 26% and 19% respectively. Within England, dentate adults in the South were less likely to think they would be entitled to free NHS treatment (17%) than those in the Midlands (22%) or the North (21%).

The proportion of dentate adults who thought they would be entitled to free NHS treatment also varied by social class of head of household; 14% of those from non-manual backgrounds thought they could have free treatment compared with 22% of those from skilled manual backgrounds and 30% of those from unskilled manual backgrounds.

There were no significant differences between those who attended regularly, occasionally and only with trouble in the proportions who thought they would be exempt from charges for treatment.

Table 6.3.28

6.3.11 Change of dental practice

Dentate adults were asked whether they intended going back to the practice they most recently attended. Overall, a very high proportion of dentate adults (85%) said they intended to return to the same practice for future dental treatment and 15% said they would change their dental practice. There was some variation by age group, with those in the younger age groups being more likely to want to change dental practice than older dentate adults. For example, 18% of those aged 16 to 24 years and 21% of those aged 25 to 34 years said they intended to change dental practice compared with 14% of 35 to 54 year olds.

Differences in intention to change dental practice were particularly marked between dentate adults who attend a dentist for different reasons; 5% of those who said they went for regular dental check-ups intended to change practice compared with 37% of those who only attend when they have trouble with their teeth. Intention to change practice also varied according to the length of time since the most recent visit to the dentist. Only 5% of those who visited the dentist within the last year intended to change practice compared with 54% of those who last visit was between five and ten years ago and 68% of those who last visited the dentist over 10 years previously.

Table 6.3.29

Those who said they would change their practice were asked their reasons and their unprompted answers were classified as shown in Table 6.3.30. The most frequently mentioned reason for changing practice was that the present practice was no longer convenient (43%) followed by dissatisfaction with the treatment received (20%). A further 13% said that their dentist had moved, retired or died and 11% felt they were no longer in the system or on their current practice patient list. Fewer people who intended to change dentist, were doing so because their dentist no longer provided NHS care (5%) and 4% were intending to go to another practice because of the cost of treatment. A small proportion (2%) said it was because they found it difficult to get an appointment at their present practice.

Table 6.3.30

The relationship between the four most frequently mentioned reasons for changing dental practice (no longer convenient, dissatisfied with dental treatment, dentist has moved, retired or died, and no longer in system or on practice list) and socio-demographic characteristics is shown in Table 6.3.31.

The proportion of dentate adults who said the practice was no longer convenient was lower among older adults; 55% of 16 to 24 year olds gave this as a reason for intending to change dentist compared with 28% of those aged 65 and over. Convenience was also a factor in the intention to change among those from non-manual backgrounds; nearly half, 49%, of those from non-manual backgrounds said their present practice was no longer convenient compared with just over a third of those from skilled manual and unskilled manual backgrounds (37% and 36% respectively).

Dentate adults who attended for regular check-ups were, not surprisingly, less likely to say they were changing practice because it was no longer convenient (32%) compared with those who attended occasionally (48%) or only when they were having trouble (46%).

There were no significant differences in the proportions who said they would change practice because they were dissatisfied with the treatment they had received and the various socio-demographic and behavioural characteristics, although this may partly be due to the relatively small size of some of the sub-groups. The proportion of dentate adults whose dentist was no longer available, because he or she had moved, died or retired, was strongly associated with the age of the respondent and varied from 5% of those aged 16 to 24 years to 34% of those aged 65 years and over.

Table 6.3.31

Notes and references

1. At the time of the 1998 Adult Dental Health Survey all NHS treatment was free for those who were under 18, under 19 and in full-time education, pregnant women and women with children under one year old. Some adults on low incomes were also entitled to free NHS dental treatment if they received certain Social Security benefits and others received partial exemption depending on their disposable income. Certain treatments, such as repair of a denture or arrest of haemorrhage, were free of charge to all patients.

Table 6.3.1 **How long respondents had been attending their present dental practice by characteristics of dentate adults**

Dentate adults†   *United Kingdom*

Characteristics		How long respondent had been attending their present dental practice				Base
		First-time attender	Attending for:			
			Less than 5 years	5 years or more	Don't know/ can't remember	
All	%	14	22	61	3	5268
Age						
16-24	%	13	21	63	3	698
25-34	%	18	30	50	3	1149
35-44	%	14	24	60	3	1095
45-54	%	13	18	66	3	1015
55-64	%	14	14	69	3	645
65-74	%	11	17	68	5	454
75 and over	%	13	12	68	7	212
Gender						
Men	%	16	21	60	3	2398
Women	%	12	22	63	3	2870
Country						
England	%	14	22	61	3	3000
Wales	%	10	20	63	7	680
Scotland	%	15	20	60	5	952
Northern Ireland	%	13	17	67	3	636
English region						
North	%	12	17	70	2	822
Midlands	%	14	21	61	5	727
South	%	15	25	57	2	1451
Social class of head of household						
I, II, IIINM	%	12	22	63	3	2482
IIIM	%	16	20	62	3	1425
IV,V	%	18	20	58	5	911
Usual reason for dental attendance						
Regular check-up	%	5	24	68	3	3121
Occasional check-up	%	13	24	59	3	574
Only with trouble	%	32	16	49	4	1564

† Excludes those who have never been to the dentist.
1988 Tables 25.1-5

Table 6.3.2 **Reasons for choosing present dental practice**

Dentate adults†   *United Kingdom*

Reasons for choosing present dental practice	%
Recommended	34
Family dentist	27
Nearest	27
By chance	4
No choice	3
Emergency dentist	2
Female dentist	0
Cannot remember	2
Other reasons	10
Base	*5268*

Percentages may add to more than 100% as respondents could give more than one answer.
† Excludes those who have never been to the dentist
1998 Table 25.7

Table 6.3.3 **Main reasons for choosing present dental practice by characteristics of dentate adults**

Dentate adults†   *United Kingdom*

Characteristic		Most frequently mentioned reasons for choosing present dental practice			Base††
		Recommended	Family dentist	Nearest	
All	%	34	27	27	5268
Age					
16-24	%	12	61	17	698
25-34	%	33	33	26	1149
35-44	%	37	21	30	1095
45-54	%	38	19	29	1015
55-64	%	41	14	31	645
65-74	%	47	10	33	454
75 and over	%	38	13	34	212
Gender					
Men	%	30	30	28	2398
Women	%	37	25	27	2870
Country					
England	%	34	27	27	3000
Wales	%	31	31	27	680
Scotland	%	30	25	37	952
Northern Ireland	%	31	35	23	636
English region					
North	%	31	33	29	822
Midlands	%	35	28	28	727
South	%	36	24	25	1451
Social class of head of household					
I, II, IIINM	%	37	26	26	2482
IIIM	%	31	29	28	1425
IV,V	%	30	28	30	911
Usual reason for dental attendance					
Regular check-up	%	38	27	26	3121
Occasional check-up	%	32	30	31	574
Only with trouble	%	27	28	29	1564

† Excludes those who have never been to the dentist.
†† Base also includes those who gave other answers.
1988 Tables 25.8-12

Table 6.3.4 **Location of dental practice by characteristics of dentate adults**

Dentate adults currently working or studying†   *United Kingdom*

		Dental practice was:			Base
		Nearer to home	Nearer to place of work or study	Same distance from both	
All	%	60	20	20	3554
Age					
16-24	%	64	21	15	608
25-34	%	58	22	20	901
35-44	%	64	18	18	894
45-54	%	57	21	23	796
55-64	%	55	18	27	318
65-74	%	73	11	16	34
75 and over	%	*	*	*	3
Gender					
Men	%	63	19	18	1768
Women	%	56	22	22	1786
Country					
England	%	60	20	20	2070
Wales	%	68	15	17	434
Scotland	%	53	25	22	645
Northern Ireland	%	57	21	22	405
English region					
North	%	63	18	18	533
Midlands	%	62	21	17	498
South	%	58	20	22	1039
Social class of head of household					
I, II, IIINM	%	62	19	19	1745
IIIM	%	56	23	21	962
IV,V	%	62	19	19	561
Usual reason for dental attendance					
Regular check-up	%	59	21	20	2118
Occasional check-up	%	60	20	20	452
Only with trouble	%	62	19	20	983

† Excludes those who have never been to the dentist.
1988 Tables 25.37-39

Table 6.3.5   **Usual time taken off work to attend the dentist by characteristics of dentate adults**

Dentate adults currently working†   *United Kingdom*

Characteristic		Usual time taken off work				Base
		Does not take time off work	Less than I hour	I hour - less than 2 hours	2 hours or more	
All	%	64	16	14	6	3272
Age						
16-24	%	70	13	11	6	400
25-34	%	64	14	14	8	859
35-44	%	62	17	15	6	874
45-54	%	62	16	16	6	789
55-64	%	59	21	14	5	318
65 and over	%	58	14	22	5	32
Gender						
Men	%	56	19	17	8	1656
Women	%	73	11	11	4	1616
Country						
England	%	63	15	15	7	1943
Wales	%	68	14	12	6	395
Scotland	%	66	19	10	5	580
Northern Ireland	%	63	16	16	6	354
English region						
North	%	68	14	12	6	494
Midlands	%	68	17	12	3	475
South	%	59	15	17	8	974
Social class of head of household						
I, II, IIINM	%	59	19	16	7	1624
IIIM	%	66	14	13	6	909
IV,V	%	71	11	12	6	522
Usual reason for dental attendance						
Regular check-up	%	63	18	14	5	1949
Occasional check-up	%	59	16	15	9	399
Only with trouble	%	67	10	14	8	923

† Excludes those who have never been to the dentist.
1988 Tables 25.30-32, 25.34, 25.36

Table 6.3.6 **Time taken off work to attend the dentist by whether working full or part-time and gender**

Dentate adults currently working†   *United Kingdom*

Time taken off work	Whether working full or part time				All	
	Working full-time		Working part-time			
	Men	Women	Men	Women	Men	Women
	%	%	%	%	%	%
Does not take time off work	55	59	79	91	56	73
Less than 1 hour	20	17	10	4	19	11
1 hour - less than 2 hours	17	18	8	3	17	11
2 hours or more	8	7	3	2	8	4
Base	1552	891	103	723	1656	1616

† Excludes those who have never been to the dentist.
1988 Table 25.33

Table 6.3.7 **Estimated distance to dental practice from home/work/college[†] by characteristics of dentate adults**

Dentate adults††   *United Kingdom*

Characteristic		Estimated distance from home/work/college						Base
		Up to ½ mile	Over ½ mile, up to 1 mile	Over 1 mile, up to 2 miles	Over 2 miles, up to 5 miles	Over 5 miles, up to 10 miles	Over 10 miles	
All	%	26	17	17	23	10	7	5268
Age								
16-24	%	28	19	15	21	8	9	698
25-34	%	26	14	18	23	10	9	1149
35-44	%	27	18	18	22	9	5	1095
45-54	%	25	17	17	25	10	6	1015
55-64	%	24	21	19	19	12	6	645
65-74	%	25	17	18	25	9	6	454
75 and over	%	24	20	20	24	6	7	212
Gender								
Men	%	26	18	17	23	9	7	2398
Women	%	26	17	17	23	10	7	2870
Country								
England	%	26	18	17	23	10	7	3000
Wales	%	23	15	17	27	11	6	680
Scotland	%	30	18	16	19	8	10	952
Northern Ireland	%	23	17	20	18	12	11	636
English region								
North	%	33	18	16	17	9	7	822
Midlands	%	18	19	21	25	11	7	727
South	%	26	17	16	25	9	6	1451
Social class of head of household								
I, II, IIINM	%	24	17	16	24	11	9	2482
IIIM	%	28	20	18	20	9	4	1425
IV, V	%	28	16	19	25	7	6	911
Usual reason for dental attendance								
Regular check-up	%	25	17	16	24	10	7	3121
Occasional check-up	%	28	20	17	18	10	6	574
Only with trouble	%	27	17	19	22	9	7	1564
Household has access to a car or van								
Has car or van	%	24	17	17	24	10	8	3732
No car or van	%	34	19	19	16	7	5	867

† Whichever was the nearest
†† Exclude those who have never been to the dentist
1988 Tables 25.41-46

Table 6.3.8   Time since last visit to the dentist by characteristics of dentate adults

Dentate adults†   *United Kingdom*

Characteristic		Time since last visit to the dentist				Base
		1 year or less	Over 1 year, up to 5 years ago	Over 5 years, up to 10 years ago	Over 10 years ago	
All	%	71	19	6	4	5268
Age						
16-24	%	65	27	6	1	698
25-34	%	67	24	7	3	1149
35-44	%	73	17	5	4	1095
45-54	%	75	15	5	5	1015
55-64	%	75	13	6	6	645
65-74	%	74	14	4	8	454
75 and over	%	66	16	5	13	212
Gender						
Men	%	65	21	7	6	2398
Women	%	77	17	4	3	2870
Country						
England	%	71	19	6	4	3000
Wales	%	72	19	5	4	680
Scotland	%	71	21	4	4	952
Northern Ireland	%	69	21	7	3	636
English region						
North	%	71	20	5	4	822
Midlands	%	71	19	5	5	727
South	%	71	18	6	4	1451
Social class of head of household						
I, II, IIINM	%	75	17	5	3	2482
IIIM	%	68	21	6	5	1425
IV,V	%	63	22	8	7	911
Usual reason for dental attendance						
Regular check-up	%	97	3	0	0	3121
Occasional check-up	%	59	38	3	0	574
Only with trouble	%	23	45	18	14	1564

† Excludes those who have never been to the dentist.
1988 Tables 24.3, 24.6, 24.9, 24.14

Table 6.3.9 **Time since last visit to the dentist by characteristics of edentate adults**

Edentate adults†   *United Kingdom*

Characteristics		Time since last visit to the dentist				Base
		1 year or less	Over 1 year, up to 5 years ago	Over 5 years, up to 10 years ago	Over 10 years ago	
All	%	10	21	19	50	914
Age						
16-24	%	*	*	*	*	1
25-34	%	*	*	*	*	7
35-44	%	*	*	*	*	15
45-54	%	20	23	16	41	89
55-64	%	11	32	23	35	188
65-74	%	10	19	24	48	298
75 and over	%	8	15	14	63	316
Gender						
Men	%	10	18	21	52	356
Women	%	11	22	18	49	558
Country						
England	%	10	19	18	51	422
Wales	%	17	23	13	48	147
Scotland	%	8	30	17	46	249
Northern Ireland	%	11	22	21	47	96
English region						
North	%	11	20	20	49	150
Midlands	%	12	22	22	44	125
South	%	8	16	18	58	147
Social class of head of household						
I, II, IIINM	%	12	20	17	50	247
IIIM	%	10	22	25	43	296
IV,V	%	7	21	14	57	303

† *Excludes those who have never been to the dentist.*
1988 Tables 24.4, 24.7, 24.10, 24.15

Table 6.3.10 Treatment received at the most recent visit to the dentist by age

Dentate adults†   *United Kingdom*

Treatment received	Age							All
	16-24	25-34	35-44	45-54	55-64	65-74	75 and over	
	%	%	%	%	%	%	%	%
Scale and polish	44	57	62	65	66	54	50	58
Teeth filled	27	33	32	30	27	25	23	29
X-rays	24	29	30	33	24	22	14	27
Teeth extracted	8	13	12	15	17	20	20	14
Crown (re)fitted	2	8	7	10	7	5	6	7
Abscess treated	4	6	5	5	3	3	2	5
Partial denture fitted	0	1	4	5	12	12	14	5
Partial denture repaired	0	0	1	2	4	5	5	2
Base	648	1060	1015	937	590	431	204	4885

† Excludes those who have never been to the dentist or were in the middle of a course of treatment.
Percentages may add to more than 100% as respondents could give more than one answer.

Table 6.3.11 **Summary of treatment received at most recent course of treatment by characteristics of dentate adults**

Dentate adults†   *United Kingdom*

Characteristic		Summary of treatment received				Base
		No extractions and:		Some extractions and:		
		no fillings	some fillings	no fillings	some fillings	
All	%	60	26	10	4	4885
Age						
16-24	%	69	23	4	4	648
25-34	%	59	28	8	5	1060
35-44	%	59	29	10	3	1015
45-54	%	58	27	12	3	937
55-64	%	60	23	14	3	590
65-74	%	59	21	16	4	431
75 and over	%	60	19	16	4	204
Gender						
Men	%	56	28	12	5	2232
Women	%	65	24	9	3	2653
Country						
England	%	62	25	10	4	2802
Wales	%	62	24	12	3	628
Scotland	%	50	31	15	5	874
Northern Ireland	%	49	37	10	4	581
English region						
North	%	62	25	9	4	758
Midlands	%	63	22	12	3	675
South	%	61	26	9	3	1369
Social class of head of household						
I, II, IIINM	%	65	25	7	3	2294
IIIM	%	55	28	12	5	1318
IV,V	%	55	24	17	4	849
Usual reason for dental attendance						
Regular check-up	%	71	25	3	1	2856
Occasional check-up	%	61	30	6	3	535
Only with trouble	%	40	26	25	9	1485
Reason for most recent visit to a dentist						
Trouble with teeth	%	28	36	27	9	1727
Check-up	%	77	21	1	1	3021
Other reason	%	67	16	15	2	129

† Excludes those who have never been to the dentist or were in the middle of a course of treatment.
1988 Tables 24.34-39

Table 6.3.12 **Treatment received at the most recent visit to the dentist by gender**

Dentate adults†   *United Kingdom*

Treatment received	Gender		All
	Men	Women	
	%	%	%
Scale and polish	57	59	58
Teeth filled	32	27	29
X-rays	29	25	27
Teeth extracted	16	11	14
Crown (re)fitted	7	7	7
Abscess treated	5	4	5
Partial denture fitted	6	4	5
Partial denture repaired	2	2	2
Base	2232	2653	4885

† Excludes those who have never been to the dentist or were in the middle of a course of treatment.
Percentages may add to more than 100% as respondents could give more than one answer.

Table 6.3.13 **Treatment received at the most recent visit to the dentist by English region and country**

Dentate adults†

Treatment received	English Region			Country				United Kingdom
	North	Midlands	South	England	Wales	Scotland	Northern Ireland	
	%	%	%	%	%	%	%	%
Scale and polish	57	59	60	59	52	54	62	58
Teeth filled	29	26	30	29	26	36	41	29
X-rays	25	24	32	28	20	22	25	27
Teeth extracted	14	15	12	13	14	20	14	14
Crown (re)fitted	6	6	7	7	6	7	8	7
Abscess treated	5	4	4	4	4	6	5	5
Partial denture fitted	6	6	4	5	5	6	6	5
Partial denture repaired	2	2	1	2	3	2	4	2
Base	758	675	1369	2802	628	874	581	4885

† Excludes those who have never been to the dentist or were in the middle of a course of treatment.
Percentages may add to more than 100% as respondents could give more than one answer.
1988 Table 24.32

Table 6.3.14 **Treatment received at the most recent visit to the dentist by social class of head of household**

Dentate adults†   *United Kingdom*

Treatment received	Social class of head of household			All
	I, II, IIINM	IIIM	IV,V	
	%	%	%	%
Scale and polish	60	56	53	58
Teeth filled	28	33	28	29
X-rays	27	27	27	27
Teeth extracted	10	17	21	14
Crown (re)fitted	6	7	7	7
Abscess treated	3	7	4	5
Partial denture fitted	3	6	7	5
Partial denture repaired	2	2	2	2
Base	2294	1318	849	4885

† Excludes those who have never been to the dentist or were in the middle of a course of treatment.
Percentages may add to more than 100% as respondents could give more than one answer.

Table 6.3.15 **Treatment received at the most recent visit to the dentist by usual reason for dental attendance**

Dentate adults†   *United Kingdom*

Treatment received	Usual reason for dental attendance			All
	Regular check-up	Occasional check-up	Only with trouble	
	%	%	%	%
Scale and polish	63	62	47	58
Teeth filled	26	33	35	29
X-rays	22	28	37	27
Teeth extracted	4	9	34	14
Crown (re)fitted	6	7	9	7
Abscess treated	2	5	9	5
Partial denture fitted	2	3	10	5
Partial denture repaired	2	2	2	2
Base	2856	535	1485	4885

† Excludes those who have never been to the dentist or were in the middle of a course of treatment.
Percentages may add to more than 100% as respondents could give more than one answer.

Table 6.3.16 **Treatment received at the most recent visit to the dentist by time since last dental visit**

Dentate adults†   *United Kingdom*

Treatment received	Time since last dental visit				All
	I year ago or less	More than I year - 5 years ago	More than 5 years - 10 years ago	More than 10 years ago	
	%	%	%	%	%
Scale and polish	61	53	52	35	58
Teeth filled	28	34	31	28	30
X-rays	26	32	33	24	27
Teeth extracted	7	24	28	47	14
Crown (re)fitted	6	8	9	4	7
Abscess treated	4	8	6	4	5
Partial denture fitted	3	8	10	16	5
Partial denture repaired	1	2	3	2	2
Base	3396	978	134	372	4885

† Excludes those who have never been to the dentist or were in the middle of a course of treatment.
Percentages may add to more than 100% as respondents could give more than one answer.
1988 Table 24.33

Table 6.3.17 **Type of dental service used at most recent course of treatment by dental status**

Dentate adults†   *United Kingdom*

Type of service	Dental status		All
	Natural teeth only	Natural teeth and dentures	
	%	%	%
NHS	76	80	77
Private	19	15	18
NHS and private	2	3	2
Community Dental Service	0	0	0
Armed Forces	1	0	1
Dental Hospital/Hospital	0	0	0
Workplace dentist	0	0	0
Through insurance	0	1	1
With a Dental Plan	1	1	1
Other	1	0	1
Base	3861	1024	4885

† Excludes those who have never been to the dentist or in the middle of a course of treatment.
1988 Table 24.45

Table 6.3.18 **Time since last dental visit by type of dental service**

Dentate adults†   *United Kindgom*

Time since last dental visit		Type of dental service			Base
		NHS	**Private**	**Other**	
1 year or less	%	76	19	5	3398
Over 1 year, up to 5 years ago	%	76	20	4	978
Over 5 years, up to 10 years ago	%	81	11	8	135
Over 10 years ago	%	86	6	8	374
All	%	77	18	5	4885

† Excludes those who have never been to the dentist or were in the middle of a course of treatment.

Table 6.3.19 **Type of dental service by characteristics of dentate adults**

Dentate adults†   *United Kingdom*

Characteristics		Type of service					Base
		NHS	**Private**	**NHS and private**	**Dental Plan**	**Other**	
All	%	77	18	2	1	3	4885
Age							
16-24	%	86	10	1	1	2	648
25-34	%	76	19	2	0	3	1060
35-44	%	74	19	2	2	3	1015
45-54	%	73	21	2	1	2	937
55-64	%	74	21	1	2	2	590
65-74	%	76	20	1	1	2	431
75 and over	%	77	15	6	0	2	204
Gender							
Men	%	75	19	1	1	4	2232
Women	%	78	18	2	1	1	2653
Country							
England	%	76	19	2	1	3	2802
Wales	%	77	17	1	3	2	628
Scotland	%	81	14	1	2	3	874
Northern Ireland	%	85	10	4	0	1	581
English region							
North	%	87	9	0	1	3	758
Midlands	%	78	18	2	0	2	675
South	%	70	24	2	1	3	1369
Social class of head of household							
I, II, IIINM	%	73	21	2	1	2	2294
IIIM	%	80	15	2	1	2	1318
IV,V	%	81	15	1	1	3	849
Usual reason for dental attendance							
Regular check-up	%	76	19	2	2	2	2856
Occasional check-up	%	76	20	2	0	1	535
Only with trouble	%	79	16	1	0	4	1485

† Excludes those who have never been to the dentist or were in the middle of a course of treatment.
1988 Table 24.47

Table 6.3.20 **Type of dental service by characteristics of edentate adults**

Edentate adults who had visited the dentist in the last ten years†   *United Kingdom*

Characteristics		Type of service				Base
		NHS	**Private**	**NHS and private**	**Other**	
All	%	78	19	2	1	*503*
Gender						
Men	%	83	16	0	1	*189*
Women	%	76	21	2	1	*314*
Country						
England	%	78	19	2	1	*215*
Wales	%	71	26	2	0	*87*
Scotland	%	81	17	1	1	*148*
Northern Ireland	%	81	19	0	0	*53*
English region						
North	%	87	11	1	0	*80*
Midlands	%	80	19	1	0	*63*
South	%	71	24	2	3	*72*
Social class of head of household						
I, II, IIINM	%	79	18	2	1	*142*
IIIM	%	75	22	1	2	*157*
IV,V	%	81	17	2	0	*169*

† *Excludes those who were in the middle of a course of treatment.*

Table 6.3.21 **Whether respondents paid anything towards the cost of their dental treatment by time since last visit to the dentist**

Dentate adults†   *United Kingdom*

Time since last visit to dentist		Whether respondents paid anything			Base
		Paid something	**Did not pay anything**	**Don't know/ can't remember**	
1 year or less	%	69	26	5	3398
Over 1 year, up to 5 years ago	%	50	33	16	978
Over 5 years, up to 10 years ago	%	36	42	22	280
Over 10 years ago	%	22	51	28	229
All	%	61	30	10	4885

† *Excludes those who have never been to the dentist or were in the middle of a course of treatment.*
1988 Table 24.45

Table 6.3.22 **Whether respondents paid anything towards the cost of their dental treatment by characteristics of dentate adults**

Dentate adults who visited the dentist in the last two years†   *United Kingdom*

Characteristics		Whether respondents paid anything			Base
		Paid something	Did not pay anything	Don't know/ can't remember	
All	%	67	27	6	3798
Age					
16-24	%	32	63	5	502
25-34	%	64	30	6	804
35-44	%	73	21	6	814
45-54	%	76	18	6	740
55-64	%	78	16	6	462
65-74	%	90	7	3	332
75 and over	%	73	18	9	144
Gender					
Men	%	71	24	6	1600
Women	%	64	30	6	2198
Country					
England	%	68	26	6	2169
Wales	%	61	30	9	496
Scotland	%	63	30	7	677
Northern Ireland	%	61	31	8	456
English region					
North	%	64	29	7	590
Midlands	%	68	26	5	506
South	%	70	25	5	1073
Social class of head of household					
I, II, IIINM	%	74	20	6	1898
IIIM	%	66	28	6	975
IV,V	%	56	40	4	591
Usual reason for dental attendance					
Regular check-up	%	70	25	5	2838
Occasional check-up	%	62	29	9	440
Only with trouble	%	56	38	6	520
Type of dental service					
NHS	%	64	30	6	2930
Private	%	87	8	6	657
NHS and private	%	89	9	2	69
Dental plan	%	30	53	17	63
Other	%	15	80	5	61

† Excludes those who were in the middle of a course of treatment.

Table 6.3.23 **Amount paid towards cost of dental treatment by type of dental service**

Dentate adults who visited the dentist in the last two years†   *United Kingdom*

Amount paid	Type of service					All
	NHS	**Private**	**NHS and private**	**Dental Plan**	**Other**	
	%	%	%	%	%	%
Nothing	30	8	9	53	80	27
Less than £10	16	4	8	16	2	13
£10 less than £20	25	14	16	10	9	22
£20 less than £30	8	18	11	4	0	10
£30 less than £50	6	19	15	0	2	9
£50 less than £100	5	17	13	1	0	8
£100 or more	4	15	26	0	1	6
Don't know/can't remember	6	6	2	17	5	6
Base	*2747*	*618*	*68*	*49*	*56*	*3798*

† Excludes those who were in the middle of a course of treatment.

Table 6.3.24 **Amount paid towards cost of dental treatment by usual reason for dental attendance**

Dentate adults who visited the dentist in the last two years and treatment was through the NHS†   *United Kingdom*

Amount paid	Usual reason for dental attendance			All
	Regular check-up	**Occasional check-up**	**Only with trouble**	
	%	%	%	%
Nothing	27	38	45	30
Less than £10	18	9	5	16
£10 less than £20	29	18	7	25
£20 less than £30	8	9	6	8
£30 less than £50	6	8	9	6
£50 less than £100	4	6	12	5
£100 or more	3	4	9	4
Don't know/can't remember	5	8	7	6
Base	*2195*	*337*	*398*	*2930*

† Excludes those who were in the middle of a course of treatment.

Table 6.3.25 **Whether costs of dental treatment were what respondents expected by usual reason for dental attendance**

Dentate adults who paid for their dental treatment†   *United Kingdom*

Costs were:	Usual reason for dental attendance			All
	Regular check-up	Occasional check-up	Only with trouble	
	%	%	%	%
More than expected	16	28	38	19
About what expected	67	53	44	63
Less than expected	12	12	13	12
Did not know what to expect	5	7	5	5
Base	1958	278	288	2524

† Includes those who have been to the dentist in the past two years and were not in the middle of a course of treatment.

Table 6.3.26 **Whether costs of dental treatment were what respondents expected by type of dental service**

Dentate adults who paid for their dental treatment†   United Kingdom

Costs were:	Type of dental service			All
	NHS	Private	Other	
	%	%	%	%
More than expected	16	28	30	19
About what expected	65	61	47	63
Less than expected	14	6	15	12
Did not know what to expect	5	5	9	5
Base	1958	278	288	2524

† Includes those who have been to the dentist in the past two years and were not in the middle of a course of treatment.

Table 6.3.27 **Amount paid towards the cost of dental treatment by whether those costs were as expected**

Dentate adults who paid for their dental treatment†   *United Kingdom*

Amount paid		Costs were:				Base
		More than expected	About what was expected	Less than expected	Did not know what to expect	
Less than £10	%	6	73	18	3	481
£10 less than £20	%	11	72	13	4	841
£20 less than £30	%	17	67	11	5	371
£30 less than £50	%	25	58	10	7	323
£50 less than £100	%	36	45	10	9	282
£100 or more	%	52	38	7	4	220
All	%	19	63	12	5	2524

† Includes those who have been to the dentist in the past two years and were not in the middle of a course of treatment

Table 6.3.28 **Whether respondents thought they would be entitled to free NHS treatment[†] by characteristics of dentate adults**

Dentate adults   *United Kingdom*

Characteristics	Proportion who thought they would be entitled to free NHS treatment	Base
All	21	5281
Age		
16-24	40	701
25-34	23	1151
35-44	15	1099
45-54	12	1016
55-64	12	645
65-74	12	456
75 and over	22	213
Gender		
Men	16	2406
Women	23	2875
Country		
England	19	3010
Wales	24	682
Scotland	22	953
Northern Ireland	26	636
English region		
North	21	823
Midlands	22	731
South	17	1456
Social class of head of household		
I, II, IIINM	14	2483
IIIM	22	1431
IV,V	30	915
Usual reason for dental attendance		
Regular check-up	19	3121
Occasional check-up	24	575
Only with trouble	20	1572

[†] *Respondents were asked if they had dental treatment under the NHS in the next 4 weeks whether they thought they would be entitled to free treatment or would have to pay something towards the cost.*
1988 Table 24.64

Table 6.3.29 **Whether respondents intended to change their dental practice by characteristics of dentate adults**

Dentate adults[†]   *United Kingdom*

Characteristics	Proportion intending to change dental practice	Base
All	15	5268
Age		
16-24	18	698
25-34	21	1149
35-44	14	1095
45-54	13	1015
55-64	11	645
65-74	11	454
75 and over	16	212
Gender		
Men	18	2398
Women	13	2870
Country		
England	16	3000
Wales	12	680
Scotland	12	952
Northern Ireland	7	636
English region		
North	13	822
Midlands	15	727
South	18	1451
Social class of head of household		
I, II, IIINM	15	2482
IIIM	15	1425
IV,V	18	911
Usual reason for dental attendance		
Regular check-up	5	3121
Occasional check-up	14	574
Only with trouble	37	1564
Time since last visit to dentist		
1 year ago or less	5	2572
More than 1 year - 5 years ago	31	681
More than 5 years - 10 years ago	54	84
More than 10 years ago	68	217

[†] *Excludes those who have never been to the dentist.*
1988 Table 25.15

Table 6.3.30 **Reasons given for wanting to change to a different dental practice**

Dentate adults intending to change dental practice   *United Kingdom*

Reasons given	%
No longer convenient	43
Dissatisfied with dental treatment	20
Dentist has moved, or died or retired	13
No longer in system - on practice patient list	11
Never go regularly to dentist	8
Last visit was not to usual practice	5
Present dentist no longer offers NHS treatment	5
Cost	4
Difficult to get an appointment	2
Heard of better dentist	1
Other reasons	5
Base	*702*

Percentages may add to more than 100% as respondents could give more than one answer.
1988 Table 25.16

Table 6.3.31 **Most frequently mentioned reasons for wanting to change dental practice by characteristics of dentate adults**

Dentate adults wanting to change dental practice   *United Kingdom*

Characteristic		Reasons given				Base†
		No longer convenient	Dissatisfied with dental treatment	Dentist has moved/died/ retired	No longer in system/on practice list	
All	%	43	20	13	11	702
Age						
16-24	%	55	21	5	11	*111*
25-34	%	53	22	10	6	*193*
35-44	%	36	19	15	14	*130*
45-54	%	39	21	17	16	*117*
55-64	%	26	22	14	15	*64*
65-74	%	29	24	24	18	*50*
75 and over	%	28	12	34	7	*36*
Gender						
Men	%	46	18	12	11	*366*
Women	%	40	24	15	12	*336*
Country						
England	%	43	21	13	12	*466*
Wales	%	38	17	20	7	*80*
Scotland	%	46	20	14	12	*114*
Northern Ireland	%	51	19	17	3	*42*
English region						
North	%	40	19	16	15	*99*
Midlands	%	38	26	13	10	*111*
South	%	46	19	11	11	*256*
Social class of head of household						
I, II, IIINM	%	49	19	12	9	*325*
IIIM	%	37	23	15	12	*181*
IV,V	%	36	24	12	15	*144*
Usual reason for dental attendance						
Regular check-up	%	32	28	16	13	*127*
Occasional check-up	%	48	22	14	2	*70*
Only with trouble	%	46	18	13	12	*497*
Time since last visit to dentist						
1 year ago or less	%	26	35	12	8	*168*
More than 1 year - 5 years ago	%	50	23	11	9	*262*
More than 5 years - 10 years ago	%	56	11	11	14	*75*
More than 10 years ago	%	39	6	21	18	*197*

Percentages do not add to 100% as only the most frequently mentioned reasons are listed.
† Base also includes those who gave other answers.
1988 Tables

6.4   Treatment preferences of dentate adults

Summary

- More than three quarters of dentate adults (79%) would prefer to have an aching back tooth filled in preference to extraction.
- When considering the choice of a crown or extraction for a back tooth about two thirds (68%) would prefer the tooth to be crowned while for a front tooth (92%) of respondents said they would prefer to have a crown rather than an extraction.
- There were no differences between men and women in their preference for the extraction of a tooth rather than having it crowned or filled.
- A higher proportion of those from manual backgrounds preferred extraction for either a back or front tooth than of those from non-manual backgrounds.
- A quarter (26%) of those who reported attending for regular dental check-ups said they would prefer to have a back tooth extracted rather than crowned compared with 44% of those who reported attending only when having trouble with their teeth.

6.4.1 Introduction

Questions on treatment preferences have been a part of the Adult Dental Health surveys since 1968 and seem to represent a significant element in people's attitudes to the way they value their dental condition. The two original items concerned whether the respondent would have an aching front, or an aching back tooth, extracted or filled. The question asked:

If you went to the dentist with an aching back tooth would you prefer the dentist to take it out or fill it (supposing it could be filled)?

Latterly the option of having a front tooth extracted or filled was removed and two new items introduced which asked:

If the dentist said a front tooth would have to be extracted (taken out) or crowned, what would you prefer?

If the dentist said a back tooth would have to be extracted (taken out) or crowned, what would you prefer?

These new items do not specifically mention pain but do indicate that a choice has to be made.

6.4.2 Treatment preferences of dentate adults

Overall, more than three quarters of dentate adults (79%) would prefer to have an aching back tooth filled in preference to extraction. When considering the choice of a crown or extraction for a back tooth, the proportion preferring extraction was higher (32%), nevertheless about two thirds (68%) would prefer the tooth to be crowned. For a front tooth, dentate adults were considerably more likely to express a preference for treatment to retain and restore the tooth; the majority (92%) of respondents said they would prefer to have a crown rather than have the tooth taken out.

Treatment preferences for an aching back tooth varied among adults in different age groups but there was no discernible trend in treatment preferences with age but there was an increased preference for extraction rather than filling among those aged 65 and over. In choosing between a crown and extraction for a back tooth, the proportion of adults preferring a crown rather than extraction decreased from 71% of 45 to 54 year olds to 45% of those aged 75 and over. A higher proportion of adults in all age groups preferred a front tooth to be crowned, but this also decreased with age from 95 to 97% of the 16 to 44 year olds to 71% of those aged 75 years and over.

Figure 6.4.1, Table 6.4.1

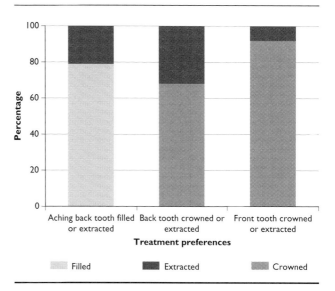

Fig 6.4.1 **Treatment preferences among dentate adults**

Table 6.4.2. shows the differences in treatment preferences for the countries of the United Kingdom and the different regions of England. Dentate adults in England were the least likely to say they would prefer extraction of an aching back tooth rather than having a filling (20%) in comparison with 24% of those in Northern Ireland, 26% of those in Scotland and 30% of those in Wales. Within England those in the South were the least likely to opt for extraction of an aching back tooth (17%) in comparison to the North (21%) or Midlands (26%). More people said they would opt for extraction if the alternative was to have a back tooth crowned; 31% of those in England and in Wales, 38% of those in Scotland and 39% of those in Northern Ireland would prefer to have a back tooth taken out rather than crowned. Again those in the South of England were the least likely to prefer an extraction to a crown for a back tooth (29% compared with 325 of those in the North and 35% in the Midlands.

Dentate adults in Wales were more likely than those in other parts of the UK to say they would have a front tooth extracted rather than crowned (14%), and those in England and in Northern Ireland were the least likely to express this as a preference (8%). Within England, dentate adults in the South (6%) were less likely than those in the North (9%) or Midlands (10%) to prefer extraction of a front tooth to having a crown fitted.

Table 6.4.2

There were no significant differences between the treatment preferences of men and women for any of the three options that were put to them. However, dentate adults from a non-manual working background were the least likely to prefer an extraction for a back or front tooth under each of the circumstances that were presented to them and those from an unskilled working background were the most likely to prefer an extraction. The differences between social classes in treatment preference were slightly greater when considering a back tooth rather than a front tooth. For an aching back tooth, adults from an unskilled manual working background were almost three times as likely to prefer an extraction (34%) as those from a non-manual household (13%). When comparing extraction or a crown for a back tooth, just under half (43%) of the adults from an unskilled manual background would prefer the tooth to be taken out rather than crowned, compared with a quarter (25%) of those from non-manual households.

A small proportion (5%) of adults from non-manual households would prefer extraction for a front tooth rather than a crown compared with 11% from skilled manual and 14% from unskilled manual backgrounds.

Figure 6.4.2, Table 6.4.3

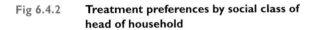

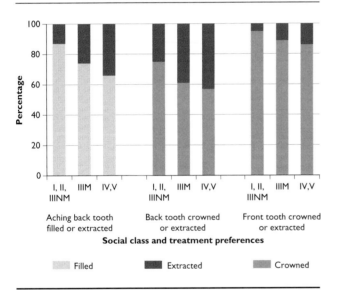

Fig 6.4.2   Treatment preferences by social class of head of household

Aching back tooth filled or extracted

Back tooth crowned or extracted

Front tooth crowned or extracted

Social class and treatment preferences

Filled   Extracted   Crowned

6.4.3 Treatment preferences and visiting the dentist

Adults' treatment preferences also varied with respect to their usual reason for dental attendance. When considering the choice of a filling or extraction for an aching back tooth, those adults who attended only with trouble were almost three times as likely to prefer an extraction (37%) as those who reported attending for regular check-ups (13%). There was less of a difference between different types of attender when the alternative to extraction was a crown; 44% of adults attending only with trouble would prefer a back tooth to be

extracted rather than crowned compared with just over a quarter of those who said they went for regular dental check-ups (26%).

Only a small proportion of dentate adults seeking regular check-ups would prefer a front tooth to be extracted rather than crowned (6%) compared with 14% of those who attend only with trouble.

Table 6.4.4

The proportion of adults preferring a back tooth to be taken out was lower among those who had recently been to the dentist. Of those adults who reported attending the dentist less than one year previously, only 15% said they would prefer to have an aching back tooth extracted rather than filled. This compared with around a third of adults last visiting the dentist between one and five years previously and between 5 and 10 years previously (30% and 35% respectively). Over half of dentate adults (55%) who attended more than 10 years previously preferred extraction to filling for an aching back tooth. When given the choice of a crown or extraction for a back tooth, the proportion preferring extraction was much higher (28%) among those who had been to the dentist within the previous year than it was when the choice involved an extraction or a filling for an aching back tooth (15%).

When comparing adults' treatment preference for the treatment of a front tooth, those whose last visit had been more than 10 years previously were around three times as likely to prefer an extraction rather to a crown (30%) than those who had been within the previous five years (6% to 13%).

Table 6.4.5

6.4.4 Treatment preferences by dental status and condition

Comparison of the treatment preferences of adults who have natural teeth only and those with natural teeth and dentures showed that adults with denture experience were more likely to prefer extraction. For an aching back tooth, twice as many adults with dentures in addition to their natural teeth would prefer an extraction to a filling for an aching tooth (36%) compared with those who have natural teeth only (18%). When considering the choice of a crown or extraction for a back tooth, there was a similar size of difference between those preferring an extraction to a crown; 50% of adults with natural teeth and some dentures would have a back tooth extracted rather than crowned compared with 28% of adults who had natural teeth only.

Almost all adults (96%) who had natural teeth only would prefer a front tooth to be restored with a crown rather than extracted, while less than three quarters (74%) of those adults who had a denture as well as natural teeth would prefer a crown to an extraction.

Table 6.4.6

Table 6.4.7 shows that for dentate adults, the more natural teeth they possess, the less likely they are to want a tooth extracted rather than restored. For example, only 15% of those with 27 or more teeth wanted an aching back tooth extracted rather than filled compared with just over half (51%) of those with between 1 and 8 teeth. Similarly, when given the option of an extraction rather than a crown for a front tooth, only 1% of those with 32 teeth preferred extraction compared with 39% of those with between 1 and 8 teeth.

Table 6.4.7

The relationship between current numbers of teeth and treatment preferences is probably due in part to a combination of past experience influencing current preferences and also preferences in the past having contributed to current dental status. This can also be seen in the final two tables in this chapter which look at the variation in treatment preference with respect to the numbers of teeth which were filled and the numbers which were crowned.

The likelihood of an adult preferring an extraction to a filling for an aching back tooth decreased considerably from 38% of those who had no existing fillings, to 15% of those who had between 6 and 11 filled teeth, and further, to 4% of those who had 18 or more filled teeth. Treatment preferences between extractions and crowns for back and front teeth followed the same pattern; 4% of dentate adults with 18 or more filled teeth would prefer a front tooth to be extracted rather than crowned, compared with 13% of those without any fillings. For a back tooth the differences were less marked, 31% of those with 18 or more filled teeth would prefer an extraction to a crown compared with 36% of those with no fillings.

Table 6.4.8

Adults with crowned teeth were more likely in all situations to prefer conservative treatment of a tooth rather than an extraction. For example, the proportion of those preferring a back tooth to be extracted rather than crowned decreased from 35% of those without crowns to about a quarter (24%) of those with between one and five crowns and 9% of those with 6 or more crowned teeth.

Table 6.4.9

Table 6.4.1   **Treatment preferences by age**

Dentate adults   *United Kingdom*

Treatment preferences	Age							All
	16-24	25-34	35-44	45-54	55-64	65-74	75 and over	
	%	%	%	%	%	%	%	%
Would prefer an aching back tooth to be:								
taken out	24	16	20	18	22	30	37	21
filled	76	84	80	82	78	70	63	79
Would prefer a front tooth to be:								
taken out	5	3	5	9	13	21	29	8
crowned	95	97	95	91	87	79	71	92
Would prefer a back tooth to be:								
taken out	29	29	28	29	37	44	55	32
crowned	71	71	72	71	63	56	45	68
Base	*701*	*1151*	*1099*	*1016*	*645*	*456*	*213*	*5821*

1988 Table 22.1

Table 6.4.2   **Treatment preferences by English region and country**

Dentate adults

Treatment preferences	English region			Country				
	North	Midlands	South	England	Wales	Scotland	Northern Ireland	United Kingdom
	%	%	%	%	%	%	%	%
Would prefer an aching back tooth to be:								
taken out	21	26	17	20	30	26	24	21
filled	79	74	83	80	70	74	76	79
Would prefer a front tooth to be:								
taken out	9	10	6	8	14	10	8	8
crowned	91	90	94	92	86	90	92	92
Would prefer a back tooth to be:								
taken out	32	35	29	31	31	38	39	32
crowned	68	65	71	69	69	62	61	68
Base	*823*	*731*	*1456*	*3010*	*682*	*953*	*636*	*5281*

1988 Table 22.2

Table 6.4.3   **Treatment preferences by gender and social class**

Dentate adults   *United Kingdom*

Treatment preferences	Gender		Social class of head of household			All
	Men	**Women**	**I, II, IIINM**	**IIIM**	**IV,V**	
	%	%	%	%	%	%
Would prefer an aching back tooth to be:						
taken out	22	20	13	26	34	21
filled	78	80	87	74	66	79
Would prefer a front tooth to be:						
taken out	9	7	5	11	14	8
crowned	91	93	95	89	86	92
Would prefer a back tooth to be:						
taken out	32	32	25	39	43	32
crowned	68	68	75	61	57	68
Base	*2406*	*2875*	*2483*	*1431*	*915*	*5281*

1988 Table 22.3 and 22.4 combined

Table 6.4.4   **Treatment preferences by usual reason for dental attendance**

Dentate adults   *United Kingdom*

Treatment preferences	Usual reason for dental attendance			All
	Regular check-up	**Occasional check-up**	**Only with trouble**	
	%	%	%	%
Would prefer an aching back tooth to be:				
taken out	13	19	37	21
filled	87	81	63	79
Would prefer a front tooth to be:				
taken out	6	5	14	8
crowned	94	95	86	92
Would prefer a back tooth to be:				
taken out	26	30	44	32
crowned	74	70	56	68
Base	*3121*	*575*	*1572*	*5281*

1988 Table 22.5

Table 6.4.5 Treatment preferences by time since last visit to the dentist

Dentate adults   *United Kingdom*

Treatment preferences	Time since last visit to dentist				All
	Up to 1 year	Over 1 year, up to 5	Over 5 years, up to 10	Over 10 years	
	%	%	%	%	%
Would prefer an aching back tooth to be:					
taken out	15	30	35	55	21
filled	85	70	65	45	79
Would prefer a front tooth to be:					
taken out	6	9	13	30	8
crowned	94	91	87	70	92
Would prefer a back tooth to be:					
taken out	28	39	42	59	32
crowned	72	61	58	41	68
Base	*3781*	*978*	*135*	*374*	*5281*

1988 Table 22.6

Table 6.4.6 Treatment preferences by dental status

Dentate adults   *United Kingdom*

Treatment preferences	Dental status		All
	Natural teeth only	Natural teeth and dentures	
	%	%	%
Would prefer an aching back tooth to be:			
taken out	18	36	21
filled	82	64	79
Would prefer a front tooth to be:			
taken out	4	26	8
crowned	96	74	92
Would prefer a back tooth to be:			
taken out	28	50	32
crowned	72	50	68
Base	*4188*	*1093*	*5281*

1988 Table 22.7

Table 6.4.7 Treatment preferences by the number of natural teeth

Dentate adults   *United Kingdom*

| Treatment preferences | Number of teeth | | | | | | All[†] |
	1 to 8	9 to 14	15 to 20	21 to 26	27 to 31	32	
	%	%	%	%	%	%	%
Would prefer an aching back tooth to be:							
taken out	51	43	35	17	15	15	20
filled	49	57	65	83	85	85	80
Would prefer a front tooth to be:							
taken out	39	37	21	5	3	1	8
crowned	61	63	79	95	97	99	92
Would prefer a back tooth to be:							
taken out	68	50	49	29	26	18	31
crowned	32	50	51	71	74	82	69
Base	*165*	*184*	*385*	*1218*	*1692*	*173*	*3817*

† Based on adults who were dentally examined. Percentages differ slightly from interview sample
1988 Table 22.8

Table 6.4.8 Treatment preferences by the number of filled (otherwise sound) teeth

Dentate adults   *United Kingdom*

| Treatment preferences | Number of filled (otherwise sound) teeth[†] | | | | | All |
	None	1 to 5	6 to 11	12 to 17	18 or more	
	%	%	%	%	%	%
Would prefer an aching back tooth to be:	39	26	16	8	3	20
taken out	61	74	84	92	97	80
filled						
Would prefer a front tooth to be:	13	12	6	2	4	8
taken out	87	88	94	98	96	92
crowned						
Would prefer a back tooth to be:	36	35	30	23	30	31
taken out	64	65	70	77	70	69
crowned						
Base	*308*	*1138*	*1571*	*700*	*100*	*3717*

† Based on 1998 criteria - see chapter 3.1. Does not include crowned teeth.
1988 Table 22.9

Table 6.4.9 **Treatment preferences by the number of crowned teeth**

Dentate adults   *United Kingdom*

Treatment preferences	Number of crowned teeth			All
	None	1 to 5	6 or more	
	%	%	%	%
Would prefer an aching back tooth to be:				
taken out	23	15	6	20
filled	77	85	94	80
Would prefer a front tooth to be:				
taken out	10	5	1	8
crowned	90	95	99	92
Would prefer a back tooth to be:				
taken out	36	24	9	31
crowned	64	76	91	69
Base	*2434*	*1190*	*193*	*3817*

1988 Table 22.11

6.5 The impact of dental disease

Summary

- Fifty-one percent of dentate adults reported having experienced one or more oral problems that had an impact on some aspect of their life occasionally or more often during the year preceding the survey.
- The most frequently reported problem was oral pain; 40% of dentate adults said they had experienced pain occasionally or more often in the previous 12 months. Two per cent of dentate adults experienced oral pain very often during the 12 month period.
- Problems with feeling tense or self conscious (classified as psychological discomfort) were reported by over a quarter of dentate adults (27%), while a fifth (19%) reported being embarrassed or finding it difficult to relax (classified as psychological disability) as a result of their oral condition occasionally or more often in the previous 12 months.
- The mean number of problems experienced by dentate adults, including those who experienced no problems, was 1.6. Dentate adults from unskilled manual backgrounds (1.9 problems), those with dentures and natural teeth (2.1 problems) and people who only attended the dentist with trouble (2.0 problems) reported the most problems on average.
- Dentate adults who reported having experienced one or more oral problems that had an impact on some aspect of their life had 1.8 fewer sound teeth and 0.6 more decayed teeth on average than those who did not report a problem.

6.5.1 Introduction

Previous chapters have reported the extent of dental disease which gives a clinical indication of the dental problems among the people of the United Kingdom. However, it may not directly reflect the problems people experience as a result of their dentition. This chapter considers people's perceptions of how they are affected as a whole by dental disease and conditions using the Oral Health Impact Profile (OHIP) developed by Slade & Spencer[1]. The aim of this index is to provide a comprehensive measure of self-reported dysfunction, discomfort and disability arising from oral conditions. Its choice for this survey was based on the topics it covered, which resembled the coverage of 13 items that were included in the 1988 survey questionnaire but were unreported in the survey report and the fact that it had a sound theoretical underpinning. This theoretical base was Locker's[2] adaptation of the World Health Organisation's classification of impairments, disabilities and handicaps[3]. In the WHO model, impacts are organised linearly to move from a biological to a behavioural to a social level of analysis. Slade and Spencer adapted this by proposing seven dimensions of impact of oral condition. These were functional limitation; physical pain; psychological discomfort; physical disability; psychological disability; social disability; and handicap. Each of the 7 dimensions in the original scale was assessed from seven questions on the type of problems experienced (a total of 49 questions). A shortened version (OHIP-14) was later developed[4] based on a subset of 2 questions for each of the 7 dimensions that are shown in Figure 6.5.1. The shorter version was the more practical to use in this survey. A weighting method is available for OHIP-14 to load dimensions in terms of their relative perceived impact. As these weights were not obtained from the UK population, unweighted scores have been presented. OHIP is intended to measure impacts of a person's general oral condition rather than the effects of specific disorders and is a measure of the burden of oral impairments that does not look at positive aspects of oral health. This is the first use of OHIP-14 in the national surveys of Adult Dental Health in the United Kingdom.

Figure 6.5.1

Fig 6.5.1 **Locker's conceptual model of oral health**

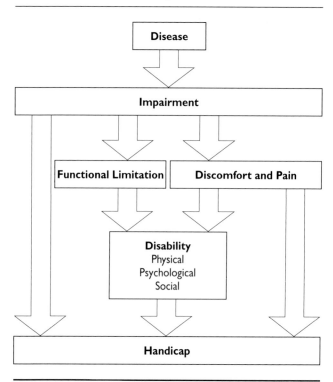

6.5.2 The impacts of oral health

Over a half of dentate adults (51%) experienced one or more of the problems included in OHIP-14 occasionally or more often in the previous 12 months; 35% occasionally, 9% fairly often and 7% very often. The most commonly reported problems were included in the categories of physical pain (40%), psychological discomfort (27%) and psychological disability (19%). Most of those reporting problems said that they had experienced them occasionally in the last 12 months, rather than fairly often or very often.

The type of problem designated as "handicap" was concerned with the impact of oral condition on overall quality of life. Seven per cent of dentate adults thought that their oral condition made their life less satisfying or led them to being unable to function occasionally or more often. In some cases oral condition was sufficient to cause severe problems; 76 people in the sample (1%) felt their problems with their oral condition had made them totally unable to function occasionally or more often in the preceding 12 months. Indicating that, to a few people, oral condition has a significant impact on overall quality of life.

Figure 6.5.2, Table 6.5.1

Fig 6.5.2   Frequency of reported problems related to oral conditions in the previous 12 months among dentate adults

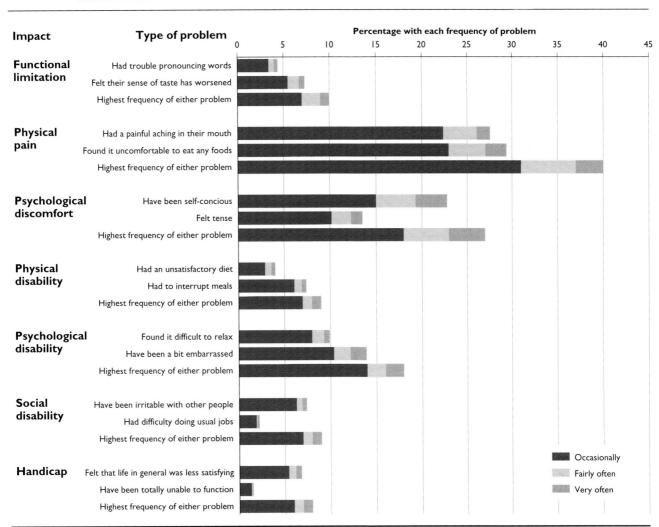

Dentate adults aged 55 years and over were less likely than those in the younger age groups to report having suffered problems within the categories of physical pain, psychological discomfort, psychological disability or social disability. For example, 21% of those aged 55 and over reported problems classified as psychological discomfort compared with 28% of 16 to 34 year olds and 30% of 35 to 54 year olds. The results for the older dentate adults will be affected by the loss of those who have been most affected by dental problems from this group, those who have lost all their natural teeth, who may well have been more affected by problems when they were younger. Nevertheless, it might still be expected that the dentate elderly would have more problems with their ageing dentition than many of those in the younger age groups. The finding that they do not may reflect a difference in expectations or values between the age groups.

Table 6.5.2

On the whole there were no large differences between dentate men and women in reporting OHIP-14 problems. There were small but statistically significant differences in reporting psychological discomfort (women 29%, men 25%) and psychological disability (women 21%, men 17%).

Table 6.5.3

Dentate adults from an unskilled working background were generally more likely than those from other working backgrounds to report having problems occasionally or more often with their oral condition.   Twice as many from an unskilled manual working background (14%) reported occasional or more frequent experience of functional limitation during the preceding 12 months compared with 7% of those from skilled non-manual working backgrounds. Over a quarter of those from an unskilled manual background reported a problem associated with psychological disability (26%) in comparison with 15% of those from a skilled non-manual working background. More people from an unskilled working background also reported experiencing problems associated with physical pain, psychological discomfort and physical disability than those from skilled non-manual households.

Table 6.5.4

There were marked differences in the reporting of nearly all types of oral problems in the previous 12 months between dentate adults who had dentures in combination with their natural teeth and those who were wholly reliant on natural teeth (the exception being in the reporting of problems associated with social disability). People with natural teeth and dentures were nearly three times as likely to report that they had some oral functional limitation (trouble pronouncing words, sense of taste) in the previous 12 months (21% compared with 7% of those with natural teeth only). People with dentures in combination with natural teeth were twice as likely to report problems classified as physical disability (16% compared with 8% of those with natural teeth only) and handicap (12% compared with 6%).

Table 6.5.5

People who, in general, only go to the dentist when they have trouble with their teeth were significantly more likely to report problems included in the OHIP-14 scale in the previous 12 months than those who sought check-ups on either a regular or occasional basis. Nevertheless, over a third of those who said they went for check-ups regularly (38%) or occasionally (37%) had been troubled by oral pain at some time in the previous 12 months. Ten per cent of those who only attend when they have some trouble with their teeth said their oral condition was sufficient to make their life less satisfying or made them unable to function (handicap) occasionally or more often in the last 12 months. Given that these attenders say they usually only attend when they have trouble with their teeth the next table examines the problems experienced in the last 12 months by them in comparison with whether they went to a dentist.

Dentate adults who reported usually only attending the dentist with trouble and who had been to the dentist in the last 12 months were more likely to have reported having experienced a dental problem in the same period than those who had not visited a dentist. Those who said they usually only attended with trouble and had attended in the last year were much more likely to report having experienced pain (62%), physical disability (24%) or some form of social disability (22%) caused by their oral condition in the last 12 months than those who did not attend (40%, 9% and 8% respectively). Nevertheless, substantial percentages of people who only seek dental care when they have some trouble with their teeth yet who said they had not visited the dentist in the previous 12

months reported experiencing problems (physical pain, 40%; psychological discomfort, 30%; psychological disability, 23%).

Tables 6.5.6–7

6.5.3 The number of problems experienced

The previous section concerned the experience of particular dimensions of oral problems defined by the OHIP-14 scale. This section examines the number of problems experienced. The experience of any problem defined in the OHIP-14 scale occasionally, fairly often or very often is counted as being a problem in this section.

Figure 6.5.3 shows the distribution of the number of OHIP-14 problems experienced occasionally or more often. This shows that 49% of dentate adults reported no oral problems in the previous 12 months. Most of those who experienced some problems reported two or more of them (33% of all dentate adults) and 11% reported experiencing more than 5 oral problems.

Figure 6.5.3

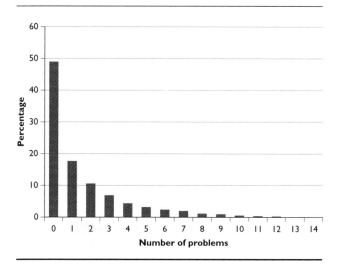

Fig 6.5.3   **The number of reported problems from OHIP - 14 experienced at least occasionally in the last 12 months among dentate adults in the United Kingdom**

The mean number of problems experienced by dentate adults, including those who experienced no problems, was 1.6. Dentate adults aged 65 to 74 were the age group least affected, according to their own reports, by the problems within OHIP-14. Dentate adults aged 35 to 54 were the most likely to be affected. A smaller proportion of dentate adults aged 65 to 74 than of those of other ages reported having any oral problems (39% compared with between 48% and 54% of the other age groups) and they reported fewer problems on average (1.1). Dentate adults aged 35 to 54 were more likely than others to report having experienced a problem (53% and 54%), and they reported more problems on average (1.7 and 1.8 respectively). There was no statistically significant difference between dentate men and women in the number of problems experienced.

Reported experience of oral problems differed between the countries of the United Kingdom. Dentate adults in Scotland were more likely to report experience of oral problems (55%) than others, and they reported more problems on average (1.7 compared with 1.6 in England, 1.4 in Wales and 1.4 in Northern Ireland).

There was an association between social class and reported experience of oral problems. Dentate adults from social classes IV and V were more likely than those from other social class backgrounds to report at least occasional experience of a problem in the OHIP-14 set (57%), and on average they had more problems (1.9).

Sixty per cent of dentate adults relying on dentures in combination with natural teeth reported experience of a problem in the previous 12 months, compared with 49% of dentate adults relying only on their natural teeth. They also had more problems on average; 2.1 compared with a mean of 1.5 problems reported by those who relied on their natural teeth.

Figure 6.5.4

Fig 6.5.4   **The impact of oral conditions in the last 12 months based on the frequency and type of reported problems by dental status**

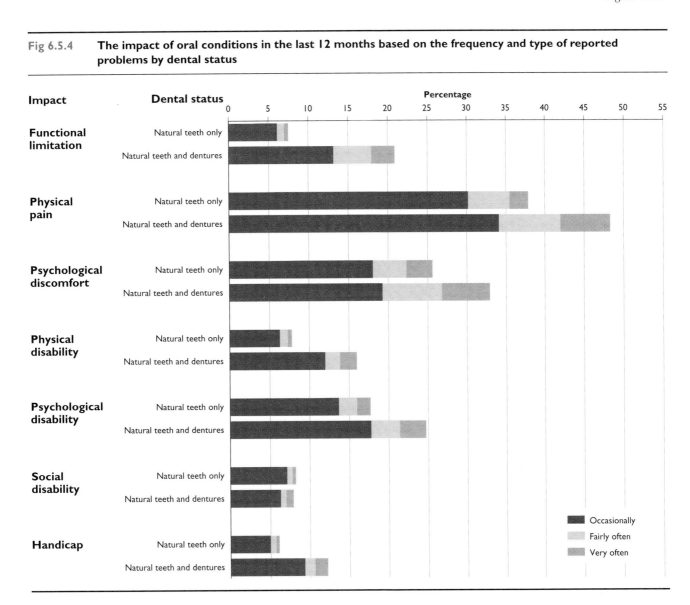

Dentate adults who said they only attend a dentist when they have some trouble with their teeth were more likely to report experience of a problem in the OHIP-14 set (58%) than those who said they sought dental check-ups occasionally or more often. They also reported more problems on average; 2.0 compared with 1.4 for those who said they attended for regular or occasional dental check ups. The people in the former group who had visited a dentist in the previous 12 months reported particularly high levels of oral problems: 72% had experienced one or more problems occasionally or more often in the previous 12 months and the mean number of problems reported was 3.0. Conversely, the fact that 28% of those who only attend when they have some dental trouble reported no problems in terms of the OHIP-14 statements but went to the dentist during the period suggests that the OHIP-14 scale may not be entirely comprehensive in terms of defining problems which motivate dental visits.

Table 6.5.9

Table 6.5.10 presents the reported frequency of oral problems by people with different attitudes and expectations in relation to oral health and care. Almost two-thirds (65%) of dentate adults who expected to need complete dentures at some time during their lifetime reported experiencing one or more problems in the OHIP-14 set occasionally or more often in the previous 12 months. The same group reported twice as many problems on average (2.6) as dentate adults who did not expect to require dentures (1.3). A similar level of reporting problems in the previous 12 months was found for dentate adults who thought they would need treatment if they went to the dentist tomorrow; 2.2 problems reported on average compared with 1.1 among those who did not think they would need treatment. Attitudes to the possibility of having to have full dentures did not affect reported experience of oral problems in the previous 12 months.

Table 6.5.10

Having asked dentate adults about their experience of dental problems in the preceding year it is informative to see how these relate to the clinical measures of dental condition that were assessed during the survey dental examination. In Table 6.5.11 the condition of teeth and supporting tissues of those who had no problems in the OHIP-14 set are compared with those who reported one or more problems occasionally or more often in the preceding year. Fewer of those who had one or more self-reported problem (80%) had 21 or more teeth in comparison to those who did not experience problems

(86%). Dentate adults who reported an OHIP-14 problem were also less likely to have 18 or more sound teeth (36%) and more likely to have some decayed teeth (60%) than those who did not report a problem of whom 45% had 18 or more sound teeth and 50% had some with decay. Of the clinical indicators selected, those which were most markedly related to the experience of an OHIP-14 problem were the number of sound and the number of decayed teeth. Those who reported an OHIP-14 problem had, on average, 1.8 decayed teeth and 14.5 sound teeth in comparison to an average of 1.2 decayed teeth and 16.3 sound teeth among those who did not report a problem.

Table 6.5.11

The final table in this chapter compares the OHIP-14 responses for the group who were satisfied with the appearance of their teeth and those who were not. More of those who were dissatisfied with their tooth appearance were also more likely to report a problem from each of the 7 dimensions of the OHIP-14 scale. In some instances these may be the same problem, for example, dissatisfaction with tooth appearance is likely to be related to psychological discomfort which concerns feeling self conscious or tense because of oral problems. However, significantly more of those who were dissatisfied with the appearance of their teeth also experienced some oral pain in the last year compared with those who were satisfied with their tooth appearance (53% compared with 35%). These results support the earlier findings shown in Figure 6.5.1 that many oral problems tend not to occur in isolation.

Table 6.5.12

Notes and References

1. Slade G (ed). *Measuring oral health and quality of life* Chapel Hill: University of North Carolina, Dental Ecology 1997.

2. Locker D. *Measuring oral health: a conceptual framework*. Community Dental Health 1988; 5: 5-13.

3. World Health Organisation. *International classification of impairments disabilities and handicaps: a manual of classification*. Geneva: World Health Organisation, 1980.

4. Slade GD. *Derivation and validation of a short-form oral health impact profile*. Community Dentistry Oral Epidemiology 1997; 25: 284-290.

Table 6.5.1   Frequency of reported problems related to oral conditions in the preceding 12 months

Dentate adults   *United Kingdom*

Type of problem[†]	Frequency of problem			Percentage experiencing either problem occasionally or more often
	Occasionally	Fairly often	Very often	
Functional limitation				
had trouble pronouncing words	% 3	1	0	} 10
felt their sense of taste has worsened	% 6	1	1	
Physical pain				
had a painful aching in their mouth	% 22	4	2	} 40
found it uncomfortable to eat any foods	% 23	4	2	
Psychological discomfort				
have been self-conscious	% 15	4	4	} 27
felt tense	% 10	2	1	
Physical disability				
had an unsatisfactory diet	% 3	1	0	} 9
had to interrupt meals	% 6	1	0	
Psychological disability				
found it difficult to relax	% 8	1	1	} 19
have been a bit embarrassed	% 10	2	2	
Social disability				
have been irritable with other people	% 6	1	0	} 8
had difficulty doing usual jobs	% 2	0	0	
Handicap				
felt that life in general was less satisfying	% 5	1	1	} 7
have been totally unable to function	% 1	0	0	
At least one problem	% 35	9	7	51
Base[††]	5281			

† The statements and their groupings are derived from the Oral Health Impact Profile (OHIP-14) (See section 6.5.1).
†† The same base is used for all percentages

Table 6.5.2 **The impact of oral conditions in the preceding 12 months based on the frequency and type of reported problems by age**

Dentate adults   *United Kingdom*

Impact and frequency of problems	Age			All
	16-34	35-54	55 and over	
			percentage	
Functional limitation				
Occasionally or more often	8	11	11	10
Occasionally	6	9	7	7
Fairly often	1	1	2	2
Very often	1	1	2	1
Physical pain				
Occasionally or more often	40	42	36	40
Occasionally	31	33	27	31
Fairly often	6	6	5	6
Very often	3	3	4	3
Psychological discomfort				
Occasionally or more often	28	30	21	27
Occasionally	19	20	14	18
Fairly often	4	6	4	5
Very often	4	4	3	4
Physical disability				
Occasionally or more often	9	10	9	9
Occasionally	8	7	7	7
Fairly often	1	1	2	1
Very often	1	1	1	1
Psychological disability				
Occasionally or more often	20	21	14	19
Occasionally	15	17	10	14
Fairly often	3	2	2	2
Very often	2	2	1	2
Social disability				
Occasionally or more often	9	10	5	8
Occasionally	8	8	4	7
Fairly often	1	1	0	1
Very often	1	0	1	0
Handicap				
Occasionally or more often	7	8	7	7
Occasionally	6	6	5	6
Fairly often	1	1	0	1
Very often	1	1	1	1
Base	*1852*	*2115*	*1314*	*5281*

Table 6.5.3 **The impact of oral conditions in the preceding 12 months based on the frequency and type of reported problems by gender**

Dentate adults   *United Kingdom*

Impact and frequency of problems	Gender		All
	Men	Women	
		percentage	
Functional limitation			
Occasionally or more often	10	9	10
Occasionally	8	7	7
Fairly often	2	1	2
Very often	1	1	1
Physical pain			
Occasionally or more often	39	41	40
Occasionally	31	30	31
Fairly often	5	7	6
Very often	3	4	3
Psychological discomfort			
Occasionally or more often	25	29	27
Occasionally	18	19	18
Fairly often	5	5	5
Very often	3	5	4
Physical disability			
Occasionally or more often	8	10	9
Occasionally	7	8	7
Fairly often	1	2	1
Very often	1	1	1
Psychological disability			
Occasionally or more often	17	21	19
Occasionally	14	15	14
Fairly often	2	3	2
Very often	1	3	2
Social disability			
Occasionally or more often	7	9	8
Occasionally	6	8	7
Fairly often	1	1	1
Very often	0	1	0
Handicap			
Occasionally or more often	8	7	7
Occasionally	6	6	6
Fairly often	1	1	1
Very often	0	1	1
Base	*2406*	*2875*	*5281*

Table 6.5.4 **The impact of oral conditions in the preceding 12 months based on the frequency and type of reported problem by social class of head of household**

Dentate adults   *United Kingdom*

Impact and frequency of problems	Social class of head of household			All
	I, II, IIINM	IIIM	IV,V	
		percentage		
Functional limitation				
Occasionally or more often	7	13	14	10
Occasionally	6	9	9	7
Fairly often	1	2	3	2
Very often	0	1	2	1
Physical pain				
Occasionally or more often	37	42	43	40
Occasionally	30	33	31	31
Fairly often	5	6	6	6
Very often	2	4	5	3
Psychological discomfort				
Occasionally or more often	25	28	32	27
Occasionally	18	18	20	18
Fairly often	4	5	6	5
Very often	3	5	6	4
Physical disability				
Occasionally or more often	7	11	12	9
Occasionally	6	9	10	7
Fairly often	1	1	1	1
Very often	0	1	1	1
Psychological disability				
Occasionally or more often	15	21	26	19
Occasionally	12	17	18	14
Fairly often	2	2	5	2
Very often	1	2	3	2
Social disability				
Occasionally or more often	7	10	8	8
Occasionally	6	8	6	7
Fairly often	0	1	1	1
Very often	0	1	1	0
Handicap				
Occasionally or more often	7	7	9	7
Occasionally	6	6	7	6
Fairly often	1	1	1	1
Very often	0	1	1	1
Base	*2483*	*1431*	*915*	*5281*

Table 6.5.5 **The impact of oral conditions in the preceding 12 months based on the frequency and type of problem by dental status**

Dentate adults   *United Kingdom*

Impact and frequency of problems	Dental status		All
	Natural teeth only	Natural teeth and dentures	
		percentage	
Functional limitation			
Occasionally or more often	7	21	10
Occasionally	6	13	7
Fairly often	1	5	2
Very often	0	3	1
Physical pain			
Occasionally or more often	38	48	40
Occasionally	30	34	31
Fairly often	5	8	6
Very often	2	6	3
Psychological discomfort			
Occasionally or more often	26	33	27
Occasionally	18	19	18
Fairly often	4	8	5
Very often	3	6	4
Physical disability			
Occasionally or more often	8	16	9
Occasionally	6	12	7
Fairly often	1	2	1
Very often	0	2	1
Psychological disability			
Occasionally or more often	18	25	19
Occasionally	14	18	14
Fairly often	2	4	2
Very often	2	3	2
Social disability			
Occasionally or more often	8	8	8
Occasionally	7	6	7
Fairly often	1	1	1
Very often	0	1	0
Handicap			
Occasionally or more often	6	12	7
Occasionally	5	9	6
Fairly often	1	1	1
Very often	0	2	1
Base	*4188*	*1093*	*5281*

Table 6.5.6 **The impact of oral conditions in the preceding 12 months based on the frequency and type of reported problems by usual reason for dental attendance**

Dentate adults   *United Kingdom*

Impact and frequency of problems	Usual reason for dental attendance			All
	Regular check-up	Occasional check-up	Only with trouble	
	percentage			
Functional limitation				
Occasionally or more often	8	7	14	10
Occasionally	6	5	10	7
Fairly often	1	2	2	2
Very often	1	0	2	1
Physical pain				
Occasionally or more often	38	37	45	40
Occasionally	30	29	33	31
Fairly often	5	6	7	6
Very often	2	2	5	3
Psychological discomfort				
Occasionally or more often	24	27	33	27
Occasionally	17	21	20	18
Fairly often	4	4	6	5
Very often	3	2	6	4
Physical disability				
Occasionally or more often	8	8	12	9
Occasionally	6	6	10	7
Fairly often	1	1	1	1
Very often	1	0	1	1
Psychological disability				
Occasionally or more often	16	16	26	19
Occasionally	13	13	18	14
Fairly often	2	2	4	2
Very often	1	1	4	2
Social disability				
Occasionally or more often	7	7	11	8
Occasionally	6	7	9	7
Fairly often	1	0	1	1
Very often	0	0	1	0
Handicap				
Occasionally or more often	6	5	10	7
Occasionally	5	5	7	6
Fairly often	1	1	2	1
Very often	1	0	1	1
Base	3121	575	1572	5281

Table 6.5.7 **The impact of oral conditions in the preceding 12 months among those who only attend with trouble with their teeth by whether they attended in the last year**

Dentate adults attending only with trouble   *United Kingdom*

Impact and frequency of problems	Whether had attended in the last year		All
	Visited dentist last year	Not visited in last year	
	percentage		
Functional limitation			
Occasionally or more often	19	12	14
Occasionally	12	9	10
Fairly often	3	2	2
Very often	4	1	2
Physical pain			
Occasionally or more often	62	40	45
Occasionally	41	31	33
Fairly often	11	6	7
Very often	10	3	5
Psychological discomfort			
Occasionally or more often	44	30	33
Occasionally	26	19	20
Fairly often	9	6	6
Very often	9	6	6
Physical disability			
Occasionally or more often	24	9	12
Occasionally	20	7	10
Fairly often	1	1	1
Very often	3	1	1
Psychological disability			
Occasionally or more often	34	23	26
Occasionally	24	16	18
Fairly often	5	4	4
Very often	5	3	4
Social disability			
Occasionally or more often	22	8	11
Occasionally	18	6	9
Fairly often	2	1	1
Very often	2	1	1
Handicap			
Occasionally or more often	17	8	10
Occasionally	12	6	7
Fairly often	4	1	2
Very often	2	1	1
Base	400	1171	1572

Table 6.5.8   **The number of reported problems based on OHIP-14 experienced at least occasionally in the preceding 12 months**

Dentate adults   *United Kingdom*

Number of problems reported	Percentage
None	49
At least one	51
At least two	33
At least five	11
Base	*5281*

Table 6.5.9   **Number and frequency of reported problems related to oral conditions by age, gender, social class of head of household, country, dental status, usual reason for dental attendance and most recent visit**

Dentate adults   *United Kingdom*

	Problems experienced occasionally or more often					Base
	Mean number of problems	Percentage mentioning at least one problem	Highest reported frequency of any problem			
			Occasionally	Fairly often	Very often	
All	1.6	51	35	9	7	*5281*
Age						
16-24	1.5	51	36	9	6	*701*
25-34	1.6	52	35	9	8	*1151*
35-44	1.7	53	37	10	5	*1099*
45-54	1.8	54	37	9	8	*1016*
55-64	1.4	48	35	9	4	*645*
65-74	1.1	39	28	6	5	*456*
75 and over	1.3	49	27	12	11	*213*
Gender						
Men	1.5	50	36	9	6	*2406*
Women	1.7	52	35	10	8	*2875*
Country						
England	1.6	51	35	9	7	*3010*
Wales	1.4	49	35	6	8	*682*
Scotland	1.7	55	35	12	8	*953*
Northern Ireland	1.4	47	33	9	5	*636*
Social class of head of household						
I,II,IIINM	1.4	47	35	8	4	*2483*
IIIM	1.8	52	37	10	8	*1431*
IV,V	1.9	57	35	11	12	*915*
Dental status						
Natural teeth only	1.5	49	35	8	6	*4188*
Natural teeth and dentures	2.1	60	34	14	12	*1093*
Usual reason for dental attendance						
Regular check-up	1.4	48	35	8	5	*3121*
Occasional check-up	1.4	50	36	10	4	*575*
Only with trouble	2.0	58	36	11	11	*1572*
Those who only attend with trouble						
Visited dentist in last year	3.0	72	37	16	18	*400*
Not visited in last year	1.7	54	36	9	9	*1171*

Table 6.5.10 **Number and frequency of reported problems related to oral health conditions by attitudes to dental care**

Dentate adults   *United Kingdom*

Attitudes to dental care	Problems experienced occasionally or more often					Base
	Mean number of problems	Percentage mentioning at least one problem	Highest reported frequency of any problem			
			Occasionally	Fairly often	Very often	
Expectation of tooth loss						
Expects to need full dentures	2.6	65	36	15	15	473
Expects to keep some natural teeth	1.3	46	35	7	4	3371
Whether would need treatment if went to dentist tomorrow						
Yes	2.2	64	40	13	10	2309
No	1.1	41	31	6	4	2773
Whether would find the need for full dentures very upsetting						
Yes	1.5	49	36	8	5	2601
No	1.5	49	34	9	6	1567

Table 6.5.11 Condition of teeth and supporting structures by whether reported any oral health problems

Dentate adults   *United Kingdom*

	Whether reported oral health problems		All
	Reported at least one problem	**Did not report any problems**	
Proportion with:		*percentage*	
21 or more teeth	80	86	83
18 or more sound and untreated teeth	36	45	40
Any decayed or unsound teeth	60	50	55
12 or more restored (otherwise sound) teeth	25	28	27
Exposed, worn, decayed or filled roots	67	66	66
Any periodontal loss of attachment 4mm or more[†]	45	41	43
Any periodontal pockets of 4mm or more[†]	57	50	54
Mean number of:			
Teeth	24.3	25.4	24.8
Sound and untreated teeth	14.5	16.3	15.3
Decayed or unsound teeth	1.8	1.2	1.5
Restored (otherwise sound) teeth	7.9	7.9	7.9
Exposed, worn, decayed or filled roots	6.3	6.5	6.4
Teeth with any periodontal loss of attachment 4mm or more[†]	2.5	2.3	2.4
Teeth with any periodontal pockets of 4mm or more[†]	3.3	2.8	3.1
		Base	
Condition of teeth and roots	2006	1809	3817
Periodontal conditions	1838	1667	3507

† 310 respondents were excluded from the periodontal part of the dental examination, therefore the base for those with any loss of attachment or pockets is different to the base for the other conditions of teeth and roots

Table 6.5.12 Frequency and type of problem by satisfaction with appearance of teeth

Dentate adults   *United Kingdom*

Impact and frequency of problem	Satisfied with appearance of teeth and/or dentures		
	Yes	**No**	**All**
		percentage	
Functional limitation			
Occasionally or more often	8	16	10
Occasionally	6	12	7
Fairly often	1	3	2
Very often	1	1	1
Physical pain			
Occasionally or more often	35	53	40
Occasionally	28	38	31
Fairly often	4	10	6
Very often	2	5	3
Psychological discomfort			
Occasionally or more often	17	54	27
Occasionally	14	31	18
Fairly often	2	12	5
Very often	1	11	4
Physical disability			
Occasionally or more often	7	16	9
Occasionally	6	12	7
Fairly often	1	2	1
Very often	0	2	1
Psychological disability			
Occasionally or more often	12	37	19
Occasionally	10	25	14
Fairly often	1	6	2
Very often	1	6	2
Social disability			
Occasionally or more often	6	14	8
Occasionally	5	12	7
Fairly often	0	2	1
Very often	0	1	0
Handicap			
Occasionally or more often	4	15	7
Occasionally	4	12	6
Fairly often	0	2	1
Very often	0	2	1
Base	3854	1416	5281

6.6 Attitudes, expectations and experiences in relation to dentures

Summary

- Among adults who had natural teeth only, nearly two thirds (61%) thought that the need for complete replacement of their teeth by dentures would be very upsetting, but a lower proportion (27%) found the idea of partial replacement very upsetting.
- Eighty-one per cent of adults who currently have natural teeth only, expect to retain some of these for their lifetime. There were no differences by gender but those aged 16 to 24 were less likely than those aged 35 or more to expect to retain their natural teeth (77% compared with between 83% and 85%).
- Among people who had an upper denture in combination with natural teeth, 91% said that they had worn their denture in the previous four weeks. A lower proportion (80%) of those with a lower denture in combination with natural teeth had worn this denture in the last 4 weeks.
- Among people with a combination of natural teeth and dentures a higher proportion wore their upper denture at night (56%), than wore their lower denture at night (45%).
- In total, 30% of people with a combination of natural teeth and dentures said they had worn their denture in the last four weeks, but reported problems of some sort with them.
- Upper dentures were more likely to be worn, and lower partial dentures were more likely to be associated with problems and more likely to be removed.
- Having complete dentures seem to be the greater leveller in terms of the experience of oral problems: there were no significant differences in the reporting of problems with eating, drinking, speaking or other problems by gender, social background, age group or past attendance behaviour (unlike the case among those with natural teeth).
- Overall 40% of edentate adults had some problem with their dentures. The most frequent problem was with difficulties with eating or drinking (26%).

6.6.1 Introduction

Although the proportion of people who are wholly reliant on dentures has been falling progressively since the first survey of adult dental health in 1968, over a quarter (28%) of people in the United Kingdom in 1998 are still completely or partly reliant on dentures. This section considers people's attitudes to the prospects of needing dentures in future among those who are wholly reliant on their natural teeth and the experiences with dentures among those who have them.

6.6.2 People with natural teeth only

Table 6.6.1 compares the anticipated reactions of those who currently have no dentures to the need for them in future. Among adults who had no experience of dentures the majority (61%) thought that the need for complete replacement of their teeth by dentures would be very upsetting, but a lower proportion (27%) found the idea of partial replacement similarly upsetting.

Older people, aged 45 and over, who were wholly reliant on their own teeth were more likely than young people, under the age of 45, to say they would be very upset by complete or partial replacement of their teeth. Men were less likely than women to think they would be very upset by the need for complete replacement or partial replacement of their natural teeth; 51% of men said they would be very upset by the need for complete dentures compared with 72% of women and only 18% of men thought partial dentures would be very upsetting in comparison with 35% of women. There were no significant differences in the proportion of people in the regions and countries of the United Kingdom who would be very upset by the need for complete or partial replacement of their natural teeth by dentures. Adults with natural teeth only from non-manual working backgrounds were more likely to be upset by the prospect of complete replacement of their natural teeth by dentures (66%) than those from a manual working background (54% to 55%) However, there was no statistically significant difference between these groups their anticipated reaction to partial dentures. Those who said they usually go for regular dental check-ups were also more likely to find the thought of the need for complete replacement to be very upsetting than those who usually only attend when they have some trouble with their teeth (66% compared with 59% of those reporting attending for occasional check-ups and 52% of those who reported attending only with trouble).

Figure 6.6.1, Table 6.6.1

A related issue to people's attitude to the potential need for dentures is their opinion of the likelihood of needing some

or all of their natural teeth replaced by dentures. About a tenth (9%) of people who currently had no dentures thought they would need partial replacement of their natural teeth by dentures in the next 5 years and a similar proportion (10%) thought they would need complete replacement at some time after this.

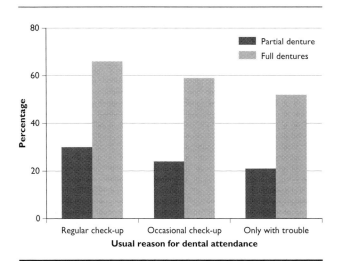

Fig 6.6.1 **Proportion of adults with natural teeth only who would be very upset by the thought of having dentures by usual reason for dental attendance**

A lower proportion of younger people anticipated requiring some dentures in the next 5 years; for example only 4% of the 16 to 24 age group expected to have a partial replacement, compared with 11% of those 65 and over. However, many more of the youngest age group expected to need all of their teeth replaced by dentures at some time after 5 years (12%) than among those aged 65 and over (2%).

There was no difference between the proportions of men and women with no current dentures in their anticipation of dentures. Nor were there any statistically significant differences between the regions and countries of the United Kingdom in the proportion of people anticipating some replacement of their teeth in the next 5 years or complete replacement at some time after this. More of those from a manual background expected to require partial dentures in the next five years (11% of those from skilled manual backgrounds said this and 12% from unskilled manual backgrounds compared with 7% of those from non-manual backgrounds) or complete dentures at some time after five years (12% for both manual groups compared with 8%). Those who said they usually attended a dentist when they had some trouble with their teeth were more likely than those who said they went for regular check-ups to think they would need complete replacement of their teeth at some time after 5 years; for example 12% of those who usually attend when having some trouble with their teeth anticipated requiring complete replacement of their teeth at some time after 5 years time compared with 8% of regular attenders.

Table 6.6.2

Thirteen per cent of the UK population at the time of this survey had none of their natural teeth left. It is anticipated that this will continue to decline. But how many people expect to keep some of their natural teeth? Table 6.6.3 shows that overall, 81% of adults who currently have natural teeth only, expect to retain some of these for their lifetime. The other 19% either expected to lose their teeth (11%) or did not know what they thought would happen (8% - figures not shown). So people in the UK are perhaps slightly more pessimistic about retaining some teeth over their lifetime than the trend figures suggest they need to be. There were no differences by gender but those aged 16 to 24 were less likely than those aged 35 or more to expect to retain their natural teeth (77% compared with between 83% and 85%).

Table 6.6.3

The expectation of requiring some dentures will partly reflect the way in which people would manage their own tooth loss. People were asked if they would have missing teeth at the back of their mouth replaced by a denture. More than half (58%) said they would not have a denture fitted. The proportion of adults with natural teeth only who would prefer to manage without a partial denture to replace any missing back teeth increased from 43% in the youngest age group to 78% in the oldest age group showing that those people in older age groups who have managed to survive without the need for dentures intend to try to do so for longer.

Men were more likely than women to say they would manage without a denture to replace missing back teeth, 60% compared with 55%. Across the United Kingdom the proportion of adults with only natural teeth who would prefer to manage without dentures if they had missing back teeth ranged from 57% in England to 63% in Scotland. Within England, those in the Midlands were more likely to prefer not to have dentures replace missing teeth (63%) than those in the North (56%) or South (54%). More of those from an unskilled manual working background (64%) would do without a replacement denture for missing back teeth in comparison to those from a non-manual working background (54%). The same was also the case for those who only attend a dentist when they have trouble with their teeth in comparison to those who seek regular dental check-ups (61% compared with 56%).

Table 6.6.4

6.6.3 People with natural teeth and dentures

Ownership of a denture does not necessarily imply that it is worn regularly. Different people use their dentures in different ways, and a proportion of dentures that are made are not worn on a regular basis. This section looks briefly at when and how people with natural teeth and dentures use their dentures, and to what degree this is influenced by problems with them.

People with dentures were asked if they had worn them in the past four weeks. The questions were asked for upper and lower dentures separately. Among people who had an upper denture in combination with natural teeth, 91% of people said that they had worn their denture in the previous four weeks. A lower proportion (80%) of those with a lower denture in combination with natural teeth had worn this denture in the last 4 weeks. When these data were analysed by denture type, 96% of people with a full upper denture had worn it in the last 4 weeks compared with 89% of those with a partial upper denture and only 79% of those with a partial lower. Younger people were a little less likely to wear their upper denture (87%) than older people (91%) but the difference was not statistically significant. A significantly smaller proportion of people who attend the dentist only when they have trouble said they had worn their dentures in the last 4 weeks (86%) compared with those who attend for check-ups (93%).

Table 6.6.5

People who wear dentures are generally advised to remove them at night for reasons of hygiene and tissue health. Among people who said they had worn their denture in the last 4 weeks, 63% with an upper complete denture and 53% with an upper partial denture said they had worn their denture at night. Among lower partial denture wearers the equivalent figure was 44%.

Table 6.6.6

Some people take their denture out at different stages throughout the day as well as, or instead of, taking it out at night. Overall, over half of upper denture wearers (56%) and 45% of the lower denture wearers who had worn their denture in the last four weeks, said that they kept their denture in at night. In total, 92% of upper dentures and 88% of lower dentures were worn throughout the day. Although the large majority of both upper (96%) and lower (92%) denture wearers kept their dentures in when eating, the rest removed them for this purpose. Not all denture wearers wore their dentures around the house (94% of upper wearers and 93% of lower wearers wore their denture in the house), although, almost all (99%) of upper dentures and 98% of lower dentures were worn to go out. These data suggest that, for a small proportion of people (probably around 5%), their dentures perform an almost exclusively social role, being removed when at home around the house or for eating.

Figure 6.6.2, Table 6.6.7

Fig 6.6.2   **The current use of upper and lower dentures among adults with natural teeth who have worn a denture in the last four weeks**

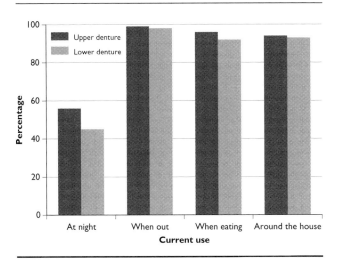

There are various reasons why a denture may not be worn at certain times. It may be that it causes pain or discomfort or that it drops during speech and causes social embarrassment. On the other hand the person may just feel that they do not like wearing it even if there are no specific problems. Among all people who said they had natural teeth and dentures, 11% had not worn their dentures at all in the last four weeks, representing a large number of unworn dentures. The remaining 89% had worn their dentures and 54% said that they had no specific problems and that they wore their dentures all day, while 21% said they wore their dentures all day despite having problems. A further 8% said that they had problems and did not wear their denture all day, while 5% had no particular problems but did not wear their dentures all day.

These proportions varied markedly according to the combination of dentures used. Proportionally more people with a complete upper denture and no lower partial denture said that they wore their denture all day despite having problems (29%). By contrast, people with a complete upper denture and a lower partial denture were the least likely to wear their dentures when they had problems (19%). A similar pattern was found for partial dentures, where most upper partial dentures alone caused no problems (61%) and of which a high proportion tended to be worn, despite problems (20%).

A small proportion (3%) caused problems and were removed for part of the day. This is markedly different from people who had partial upper and lower dentures. Less than half of these people (48%) were problem free and wore their dentures all day, and 17% had dentures that caused problems and were removed for part of the day. Once again it is likely that the difference in denture wearing habits is largely related to problems with lower dentures, which can often be removed if they cause problems without having a major impact on appearance.

Figure 6.6.3, Table 6.6.8

Fig 6.6.3   **Proportion of adults with natural teeth and dentures who wear their denture all day and have no problem by type of denture provision in the upper and lower jaws**

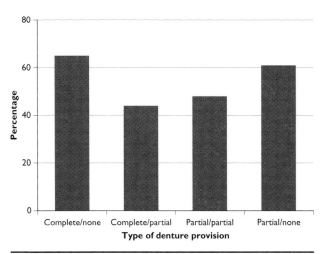

Looked at by attendance pattern there was some difference in the frequency of problems with denture wearing between those attending for check-ups on a regular or occasional basis and people who said they attended only when having trouble with their teeth. People who attended the dentist for check-ups were significantly more likely to report not having any problems with their dentures (58% to 59%) than people who attend only with trouble (47%). However, people who attend only with trouble (14%) were significantly more likely not to have worn their dentures at all than those going for regular check-ups (8%). Those who attended for occasional check-ups were similar to those seeking regular check-ups for some attributes (having problem free dentures) and similar to those who attend only with trouble for others (not wearing dentures which had been provided). The proportions for all three groups were very similar when it came to dentures not being worn because of problems (6% to 8%).

Table 6.6.9

Overall, three quarters (75%) of the dentate people who wore dentures in conjunction with natural teeth said they were satisfied with the appearance of their teeth. The youngest (those aged 16 to 34) were the least likely to be satisfied (56%), and the oldest (those aged 55 years and over) were the most likely to be satisfied (78%). Attendance pattern was related to satisfaction with appearance; of those who attended regularly for check-ups, 78% were satisfied compared with only 67% of those who attended only with trouble possibly reflecting the impact of older and less well maintained dentures.

Table 6.6.10

6.6.4 Edentate adults

Table 6.6.1 showed how people who have never experienced wearing dentures thought they would feel if they need their teeth fully replaced by dentures. Table 6.6.11 considers the reactions of the edentate to the loss of the last of their natural teeth. Overall, 27% of those who were edentate said that the loss of the last of their natural teeth had been very upsetting. More women were very upset at the loss of all their teeth than men; 31% were very upset compared with 20% of men. There were no significant differences between people from different social classes. The data suggest that more of those who lost the last of their teeth in middle age rather than later and more of those who had attended the dentist for check-ups found the loss of all their teeth upsetting, but small base sizes meant these differences were not significant.

Table 6.6.11

People who had lost all their natural teeth were also asked whether they were satisfied with the appearance of their denture teeth, whether their dentures caused any problems with speaking, eating or drinking, whether they had any other problems and whether they planned to go to a dentist with a current denture problem. Many people had some problem with their dentures (41%) but only 13% felt the problem required a visit to a dentist. The most frequently reported problem was with eating or drinking; over a quarter (26%) of people with no natural teeth said they experienced some problem when eating or drinking. Those who had become edentate within the last 10 years were more likely to report problems (53%) than those who had been edentate for a longer period (37% to 41%). This may represent a real difference in the experience of problems or it may reflect some variation in expectations related to reliance on dentures.

Table 6.6.12

Dentures for the upper and lower jaws need not necessarily have been made at the same time, so people were asked the age of the oldest of the dentures they were currently wearing.

The age of the oldest denture that was currently being worn showed little relationship to whether problems were experienced; there were no statistically significant differences between groups.

Table 6.6.13

Having complete dentures seem to be the greater leveller in terms of experiencing oral problems: there were no significant differences in the reporting of problems with eating, drinking, speaking or other problems by gender, social background, age group or past attendance behaviour (unlike the case among those with natural teeth). Notably, there was no difference between edentate men and women in their intention to visit a dentist; 14% of edentate men said they intended to visit the dentist because of problems with their dentures compared with 12% of women. This contrasts with the quite marked difference in attendance behaviour between men and women who have some teeth (see Table 6.1.3).

Table 6.6.14

The specific problems that edentate people have with their dentures are shown in Table 6.6.15. The most frequently reported problems were functional ones associated with eating, mainly having trouble with food lodging under the denture plate (12%) and finding the denture was too loose when eating (9%). Eight per cent of those with no natural teeth said the denture hurt their gums when eating and 3% said their denture made their gums sore generally, 3% also complained of mouth ulcers.

Table 6.6.15

Table 6.6.1   **Attitudes to the thought of having dentures by age, gender, country, English region, social class of head of household and usual reason for dental attendance**

Adults with natural teeth only   *United Kingdom*

	Very upset if partial replacement were needed	Very upset if full replacement were needed	Base†
	%	%	
All	27	61	*4013*
Age			
16-24	20	58	*691*
25-34	23	57	*1088*
35-44	27	61	*940*
45-54	33	67	*721*
55-64	37	69	*323*
65 and over	34	66	*250*
Gender			
Men	18	51	*1814*
Women	35	72	*2199*
Country			
England	27	61	*2362*
Wales	32	61	*500*
Scotland	23	62	*666*
Northern Ireland	23	57	*485*
English Region			
North	26	65	*602*
Midlands	27	60	*565*
South	27	60	*1195*
Social class of head of household			
I, II, IIINM	28	66	*1934*
IIIM	23	55	*1053*
IV,V	25	54	*654*
Usual reason for dental attendance			
Regular check-up	30	66	*2404*
Occasional check-up	24	59	*466*
Only with trouble	21	52	*1135*

†Excludes 175 dentate adults who had some time had a denture fitted they no longer had
1988 Table 17.22, 17.23, 17.24.

Table 6.6.2 **Expectation of needing dentures by age, gender, social class of head of household, country, English region and usual reason for dental attendance**

Adults with natural teeth only   *United Kingdom*

	Likely to need partial denture in next 5 years	Likely to need full dentures:		Base†
		in next 5 years	sometime (after 5 years)	
		percentage		
All	9	2	10	4013
Age				
16-24	4	1	12	691
25-34	7	1	13	1088
35-44	9	1	10	940
45-54	14	2	6	721
55-64	16	3	3	323
65 and over	11	4	2	250
Gender				
Men	9	1	10	1814
Women	9	2	10	2199
Country				
England	9	1	9	2362
Wales	9	2	9	500
Scotland	11	2	11	666
Northern Ireland	13	1	10	485
English Region				
North	8	1	9	602
Midlands	9	3	10	565
South	9	1	9	1195
Social class of head of household				
I, II, IIINM	7	1	8	1934
IIIM	11	2	12	1053
IV,V	12	3	12	654
Usual reason for attendance				
Regular check-up	8	1	8	2404
Occasional check-up	5	0	9	466
Only with trouble	13	4	12	1135

† Excludes 175 dentate adults who had at some time had a denture fitted which they no longer had
1988 Tables 17.2, 17.27

Table 6.6.3 **Expectation of retaining natural teeth by gender and age**

Adults with natural teeth only†   *United Kingdom*

Gender	Age					All
	16-24	25-34	35-44	45-54	55 and over	
	Percentage expecting to keep some natural teeth					
Men	78	80	85	82	88	82
Women	76	79	81	85	81	80
All	77	80	83	83	85	81
	Base					
Men	*325*	*446*	*441*	*330*	*272*	*1814*
Women	*366*	*642*	*499*	*391*	*301*	*2199*
All	*691*	*1088*	*940*	*721*	*573*	*4013*

† *Excludes 175 dentate adults who had at some time had a denture fitted which they no longer had*

Table 6.6.4 **Adults who would prefer to manage without a partial denture**

Adults with natural teeth only   *United Kingdom*

	Would prefer to manage without a partial denture	*Base*
All	58	*4188*
Age		
16-24	43	*696*
25-34	53	*1110*
35-44	58	*976*
45-54	63	*778*
55-64	76	*353*
65 and over	78	*275*
Gender		
Men	60	*1893*
Women	55	*2295*
Country		
England	57	*2463*
Wales	61	*530*
Scotland	63	*700*
Northern Ireland	61	*495*
English Region		
North	56	*641*
Midlands	63	*584*
South	54	*1238*
Social class of head of household		
I, II, IIINM	54	*2022*
IIIM	62	*1099*
IV,V	64	*684*
Usual reason for dental attendance		
Regular check-up	56	*2512*
Occasional check-up	54	*485*
Only with trouble	61	*1180*

1988 Tables 17.14, 17.16, 17.17, 17.18

Table 6.6.5 **Denture wearing in the last four weeks by type of denture, age and usual reason for dental attendance**

Adults with natural teeth and dentures   *United Kingdom*

	Has upper denture	Has lower denture	Base Has upper denture	Has lower denture
	Percentage who have worn upper denture in last 4 weeks	*Percentage who have worn lower denture in last 4 weeks*		
All	91	80	999	382
Type of denture				
Complete upper	96	..	299	..
Partial upper	89	..	700	..
Partial lower	..	79	..	358
Age				
16-34	87	-	43	7
35-54	91	78	326	98
55 and over	91	81	630	277
Usual reason for dental attendance				
Regular/occasional check-up	93	85	631	239
Only with trouble	86	73	366	143

1988 Table 16.24

Table 6.6.6 **Denture wearing at night by type of dentures worn in the last 4 weeks**

Adults with natural teeth who have worn a denture in the last 4 weeks   *United Kingdom*

Whether denture worn at night	Type of denture worn in the last 4 weeks[†]		
	Complete upper	Partial upper	Partial lower
		Percentage	
Upper worn at night	63	53	..
Lower worn at night	..	..	44
		Base	
Upper worn in last 4 weeks	*289*	*619*	*..*
Lower worn in last 4 weeks	*..*	*..*	*284*

† Only 22 people with natural teeth wore a full lower denture, so this group have been excluded.
1988 Table 16.25

Table 6.6.7 **The current use of upper and lower dentures**

Adults with natural teeth who have worn a denture in the last 4 weeks   *United Kingdom*

Current use	Worn denture in the last 4 weeks	
	Upper denture	Lower denture
	%	%
Wears denture at night	56	45
Wears denture from getting up to going to bed	92	88
Wears denture when goes out	99	98
Wears denture when eating	96	92
Wears denture around the house	94	93
Base	*908*	*308*

1988 Table 16.26

Table 6.6.8 **Denture wearing and whether problems experienced by type of denture provision**

Adults with natural teeth and dentures   *United Kingdom*

Denture wearing and whether has problems	Current type of denture provision					All†
	Complete - none	Complete - partial	Partial - partial	Partial - none	Other	
	%	%	%	%	%	%
Denture not worn at all in last four weeks	3	6	9	11	12	11
Denture not worn all day, has problems	1	21	17	3	9	8
Denture not worn all day, has no problems	1	10	6	5	5	5
Denture worn all day, has problems	29	19	20	20	27	21
Denture worn all day, has no problems	65	44	48	61	46	54
Base	178	118	176	508	88	1093

† Includes those for whom denture pattern not known.
1988 Table 16.27

Table 6.6.9 **Denture wearing and whether experienced problems by usual reason for dental attendance**

Adults with natural teeth and dentures   *United Kingdom*

Denture wearing and whether has problems	Usual reason for dental attendance			All
	Regular check-up	Occasional check-up	Only with trouble	
	%	%	%	%
Denture not worn at all in last four weeks	8	17	14	11
Denture not worn all day, has problems	8	6	8	8
Denture not worn all day, no problems	5	4	6	5
Denture worn all day, has problems	21	13	25	21
Denture worn all day, has no problems	58	59	47	54
Base	609	90	392	1093

1988 Table 16.28

Table 6.6.10 **Satisfaction with the appearance of teeth or dentures by age, type of denture provision and usual reason for dental attendance**

Adults with natural teeth and dentures   *United Kingdom*

		Satisfied with appearance of teeth	Not satisfied with appearance of teeth	Base
All	%	75	25	*1093*
Age				
16-34	%	56	44	*46*
35-54	%	70	30	*361*
55 and over	%	78	22	*686*
Pattern of denture provision				
Complete – none	%	75	25	*178*
Complete – partial	%	83	17	*118*
Partial – partial	%	76	24	*176*
Partial – none	%	73	27	*508*
Usual reason for dental attendance				
Regular check-up	%	78	22	*609*
Occasional check-up	%	84	16	*90*
Only with trouble	%	67	33	*392*

1988 Table 16.30

Table 6.6.11 **Edentate adults who found the loss of their teeth very upsetting**

Edentate adults   *United Kingdom*

	Loss of teeth was very upsetting	Base
All	27	*923*
Age when last teeth were lost		
Under 34	0	*370*
35 – 54	38	*381*
Over 55	25	*150*
Gender		
Men	20	*360*
Women	31	*563*
Social class of head of household		
I, II, IIINM	27	*249*
IIIM	25	*298*
IV,V	29	*306*
Usual reason for dental attendance		
Regular check-up	33	*165*
Occasional check-up	32	*111*
Only with trouble	24	*638*

Table 6.6.12 **Problems with dentures by the length of time since becoming edentate**

Edentate adults   *United Kingdom*

Problems with dentures	Time since became edentate				All
	Under 10 years	10 – 19 years	20 – 29 years	30 years or more	
	percentage with each problem				
Have any of the problems listed below	53	41	41	37	41
Not satisfied with appearance	16	11	14	11	12
Have trouble speaking clearly	15	6	10	4	7
Have trouble eating or drinking	40	31	22	23	26
Have other problems	15	10	17	13	13
	percentage planning to visit the dentist				
Planning to visit the dentist	28	12	12	8	13
Base	96	142	227	458	923

1988 Table 15.7

Table 6.6.13 **Problems with dentures by age of oldest denture currently worn**

Edentate adults   *United Kingdom*

Problems with dentures	Age of oldest denture currently worn				All
	0 – 5 years	5 – 10 years	10 – 19 years	20 years or more	
	percentage with each problem				
Have any of the problems listed below	44	44	38	39	41
Not satisfied with appearance	7	13	15	13	12
Have trouble speaking clearly	11	6	5	6	7
Have trouble eating or drinking	29	35	22	24	26
Have other problems	13	15	17	12	13
	percentage planning to visit the dentist				
Planning to visit the dentist	14	16	13	9	13
Base	198	142	207	361	923

1988 Tables 15.8, 15.9

Table 6.6.14 **Problems with dentures by age, gender, social class of head of household and usual reason for dental attendance**

Edentate adults   *United Kingdom*

	Problem with dentures						Base
	Have any of the following problems	Not satisfied with appearance	Have trouble speaking	Have trouble eating or drinking	Have other problems	Planning to visit dentist	
	percentage with each problem						
All	40	12	7	26	13	13	923
Age							
16-34	-	-	-	-	-	-	8
35-54	44	15	13	22	11	12	105
55 and over	39	12	6	26	14	12	810
Gender							
Men	40	11	9	29	11	14	360
Women	40	13	5	24	15	12	563
Social class of head of household							
I, II, IIINM	38	10	6	23	13	11	249
IIIM	39	13	8	25	13	13	298
IV, V	44	13	5	28	14	16	306
Usual reason for dental attendance							
Regular check-up	36	11	3	21	14	14	165
Occasional check-up	31	6	2	25	6	11	111
Only with trouble	43	14	9	27	15	15	638

1988 Tables 15.12, 15.13

Table 6.6.15 **Specific problems with dentures**

Edentate adults   *United Kingdom*

Problem	%
Problem with speaking	
Loose denture/slips when talking	4
Alters or slurs speech	3
Other speaking problems	2
No problems	93
Problem with eating	
Food sticks under denture plate	12
Loose denture/slips when eating	9
Hurts gums	8
Cannot chew or bite well	5
Other eating problems	3
No problems	74
Other denture problems	
Loose dentures	5
Gets ulcers	3
Sore gums/plate rubs gums	3
Denture worn down	1
Other denture problem	4
No problems	87
Base	*923*

6.7 Dental hygiene behaviour

Summary

- Seventy four percent of dentate adults claimed to clean their teeth at least twice a day, 22% once a day and 4% less than once a day.
- The proportion of adults cleaning their teeth at least twice a day decreased with age (from 79% of 16 to 24 year olds to 67% of those aged 65 and over) and was higher for women than men (83% and 64%) and for those from non-manual households (78% compared with 69% and 66% of those in skilled and unskilled manual households).
- Over half (52%) of dentate adults said they used a method of tooth cleaning in addition to tooth brushing.
- The two most frequently used methods were dental floss (28%) and mouthwash (24%).
- Nearly twice as many dentate adults from a non-manual background as those from an unskilled manual background used dental floss (33% compared with 17%) but the use of mouthwash showed no significant variation (25% compared with 22%).
- Over a third (38%) of the population could not recall having been given advice about gum care or having been shown how to brush their teeth by a dentist or a member of the practice staff.
- People who reported cleaning their teeth at least twice a day were less likely to have dental plaque than those who cleaned their teeth less frequently but even among those who cleaned at least twice a day 69% had visible plaque.

6.7.1 Introduction

All dentate adults were asked about their dental hygiene behaviour, for example how often they cleaned their teeth and what methods they used other than toothpaste and brush. The answers to these questions do not provide direct information about how effectively people clean their teeth, but taken together give some indication of motivation towards dental hygiene. Adults were also asked whether they had ever been shown how to clean their teeth or look after their gums, either by a dentist or a member of the dental practice staff.

6.7.2 Frequency and time of tooth cleaning

Seventy four percent of dentate adults claimed to clean their teeth at least twice a day, 22% said they clean their teeth once a day and 4% less than once a day. Frequency of tooth cleaning declined with age; 79% of those in the youngest age group (16 to 24) claimed to clean their teeth at least twice a day compared with 67% of dentate adults aged 65 and over. There was a corresponding rise with age in the proportion of adults who cleaned their teeth once a day.

A smaller proportion of men (64%) than women (83%) said they brushed their teeth twice a day or more. Differences between men and women who claimed to clean their teeth at least twice a day were found within each age group, and increased with age. Among 16 to 24 year olds 72% of men claimed to clean their teeth at least twice day, compared with 86% of women. In the oldest age group comparable figures for men and women were 51% and 78% respectively, a drop of 21% for men but of only 8% for women. (Table not shown.) There were some small but statistically significant differences between the countries of the United Kingdom; adults in England and in Wales were more likely to say they brushed their teeth twice a day or more (both 74%) than those living in Scotland or Northern Ireland (70% and 71% respectively). Some variation was also found between English regions with 76% of adults in the South saying that they cleaned their teeth at least twice a day, compared with 72% in the North and 72% in the Midlands. Within social class, dentate adults from a non-manual background were significantly more likely than those from a manual background to clean their teeth at least twice a day, and less likely to brush less than once a day.

People with natural teeth only were more likely to claim to brush at least twice daily (75%) than adults with natural teeth and a denture (69%). Tooth brushing frequency was also related to dental attendance; 80% of adults who said they went to the dentist for regular check-ups also claimed to clean their teeth at least twice a day compared with 61% of people who visit the dentist only when they have some trouble with their teeth. This last finding is consistent with the expectation that people who make regular visits to the dentist are more likely to take care of their teeth, but the association does not always hold when comparing different groups. Young men were more likely than older men to clean their teeth at least twice a day, but as was shown in Chapter 6.1 were the least likely to have regular dental check-ups.

Table 6.7.1

Table 6.7.2 shows the percentage of people who reported cleaning their teeth at particular times of day. The most popular time was last thing at night (74%) followed by after breakfast (46%) and before breakfast (40%).

The data relating to the time of day people cleaned their teeth also showed some variation when analysed by age, gender and social class, and by dental attendance pattern. In most respects the findings mirror those for the overall frequency of toothbrushing in Table 6.8.1. However, there were differences within two groups in the pattern of time of day when teeth were brushed: the proportion of dentate adults who brushed before breakfast increased with age from 38% of those aged 16 to 24 to 47% of those aged 75 and over. Dentate adults aged 65 and over were less likely than younger adults to brush last thing at night. Also a higher percentage of women than men said they cleaned their teeth last thing at night or after breakfast, whereas men were more likely than women to say they did so before breakfast.

In general, there was little difference between the preferred time for cleaning teeth of those living in different parts of the United Kingdom; although adults in Scotland and Northern Ireland were more likely to do so after breakfast (54% and 51% respectively) than their counterparts in England (46%) and Wales (45%), and were correspondingly less likely to do so before breakfast. Within England, the main regional variation was in the proportion of adults who said they cleaned their teeth last thing at night: 77% of those living in the South, compared with 72% in the North and 71% in the Midlands.

The time of day when people brush their teeth was related to overall frequency of brushing. A key difference between those brushing once a day and those brushing twice or more was whether brushing was undertaken last thing at night; 92% of those who brush twice a day or more, said they brushed their teeth last thing at night compared with 22% of those who brushed once a day.

Figure 6.7.1, Table 6.7.2

Fig 6.7.1   Time of day dentate adults clean their teeth by gender

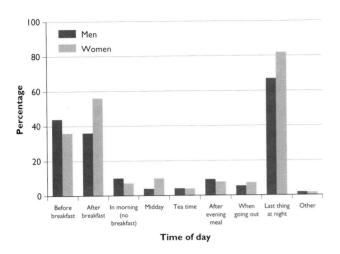

6.7.3 Use of methods other than ordinary toothpaste and brush

Dentate adults who cleaned their teeth were asked whether they used anything other than toothpaste and brush; 52% said that they did. The use of methods in addition to standard toothpaste and brush was most commonly reported by adults aged 25 to 54, around 57% of whom claimed to use something else. Young adults aged 16 to 24 (42%) and people in the two oldest age groups were markedly less likely to say they used additional tooth cleaning methods (40% and 31% for the two older age groups respectively).

Women (57%) were more likely than men (46%) to say they used an additional method to clean their teeth as were adults from a non-manual working background (56%) when compared with those from a manual working background (45%). Dentate adults living in Northern Ireland were less likely to use additional methods of cleaning than adults living in either England, Wales or Scotland. People living in the North of England or the Midlands were also less likely to use additional methods than those in the South.

The use of an additional method of tooth cleaning was associated with dental status and usual reason for visiting a dentist. Adults with natural teeth only (54%) were more likely to say they used an additional method for tooth cleaning than those who had some denture teeth (42%) and more of those adults who said they visited the dentist for regular check-ups (58%) said they used an additional tooth cleaning method than those who only go when they have some trouble with their teeth (39%).

Table 6.7.3

The methods reported to be used in addition to brushing are given in Table 6.7.4. These responses were unprompted and included some items that may have debatable significance as tooth cleaning methods. Dental floss and mouthwash were mentioned more often than other methods. In the United Kingdom as a whole, dental floss was mentioned by 28% of adults and mouthwash by 24%. No other method was mentioned by more than 5% of adults who brushed their teeth.

Table 6.7.4

Table 6.7.5 shows the pattern of use of the two most popular additional cleaning methods. Around one third of adults aged between 25 and 54 claimed to use dental floss (31% to 35%), but significantly fewer of those aged 16 to 24 (18%) or 55 and over (12% to 25%) said that they did. Mouthwash was more popular among 25 to 44 year olds (30% to 32%) than other age groups (6% to 23%), and was more likely than dental floss to be used by young adults aged between 16 and 24. Women (35%) were much more likely than men (20%) to say they used dental floss. The same was true, although to a lesser extent, for mouthwash which 26% of women said they used in comparison with 22% of men. Among women dental floss was considerably more popular than mouthwash (35% compared with 26%), but no similar pattern was found among men.

Adults in England who cleaned their teeth were more likely to say they used floss in addition to tooth brushing (29%) than were those in the other countries of the United Kingdom (24% to 25%), whereas people in Scotland were more likely to say they used mouthwash (29%) than those from other countries (23% to 24%). People living in the South of England were more likely to use dental floss (32%) than their counterparts in the North (23%) and Midlands (26%), and together with adults in the Midlands were more likely to use dental floss than mouthwash. Those in the Midlands were the least likely to say they used mouthwash in addition to brushing (18%).

Dental floss was also considerably more likely to be mentioned by adults from a non-manual working background (33%) than by those from a manual working background (17%) but differences between social classes in respect of the use of mouthwash were small. Comparing preference for either method, those from non-manual working backgrounds favoured dental floss (33%) to mouthwash (25%) whereas those from a manual working background tended to favour mouthwash (22%) to floss (17%).

Those who reported visiting the dentist for regular check-ups were much more likely to say they used floss (35%) than those who only attended with trouble (13%). As noted for some other groups there was a difference in preference for these two additional methods of tooth cleaning between those whose motives for visiting a dentist are dissimilar. People who only attended when having trouble with their teeth favoured the use of mouthwash (21%) to floss (13%) whereas those who sought regular dental check-ups were more likely to say they used floss (35%) to mouthwash (25%).

Figure 6.7.2

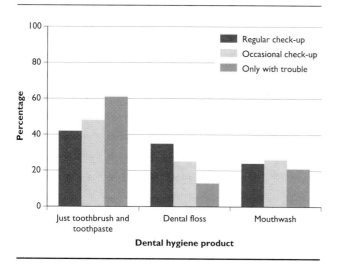

Fig 6.7.2   **The use of dental hygiene products among dentate adults by usual reason for dental attendance**

Those who have natural teeth only were much more likely (30%) than those who have dentures with natural teeth (15%) to use floss. The same was true for the use of mouthwash by these groups; 26% of those with natural teeth only said they used mouthwash compared with 14% of those with natural teeth and dentures.

Reported use of an additional method of tooth cleaning was associated with reported tooth brushing frequency. Almost a third of those who said they brushed their teeth twice a day or more also said they used dental floss (32%) in comparison with 14% of those brushing once a day and 10% of those brushing less than once a day. The preference for mouthwash over dental floss shown to exist in young adults aged 16 to 24, people in Scotland, those from a manual working background, and in people who reported attending only with trouble was also found among those who said they brush their teeth less than once a day.

Table 6.7.5

6.7.4 Advice and information about dental hygiene from dental staff

Dentists and other dental practice staff play an important role in instilling good dental hygiene behaviour among their patients. Adults who took part in the survey were asked whether they had ever been shown how to clean their teeth, and whether they had received advice on how to care for their gums, by a dentist or one of his or her staff.

Overall 62% of dentate adults in the United Kingdom could recall having been given some advice or information about tooth brushing or gum care; 54% said that they had been shown how to clean their teeth, and 45% said they had received advice about gum care. Older adults were the least likely to say they been given advice or information about brushing or gum care; 51% of those aged 55 and over said they had been shown how to clean their teeth or been given advice about gum care in comparison with 65% of those aged 16 to 34 years old. Fewer men recalled having been shown how to brush their teeth (52%) than women (56%). Men were also less likely to say they had been given advice about gum care (42%) than women (48%).

There was some variation between countries within the United Kingdom. Adults living in England were most likely both to have been given a teeth cleaning demonstration (55%) whilst those in Northern Ireland were the least likely (46%). Less variation was found for advice about for gum care; 46% of those in England could recall having been given advice on how to look after their gums compared with 42% of those in Northern Ireland. Some small but significant differences were also found between English regions; more of those in the South reported having been shown how to clean their teeth (58%) than those in either the Midlands (54%) or the North (47%). The same pattern was also found for information about gum care; 50% of adults in the South compared with 42% in the North and 41% in the Midlands.

Those from manual working backgrounds were also significantly less likely to recall having been given advice or information (52%) than those from non-manual backgrounds (68%). The same was true for those who only attend when having trouble with their teeth (50%) in comparison to those who seek regular dental check-ups (68%). The pattern was repeated for those with dentures in conjunction with natural teeth of whom less than half recalled having been given advice or information (49%), compared with 65% of those who had natural teeth only.

Table 6.7.6

6.7.5 The inter-relationship of oral health advice and information, reported dental hygiene behaviour, and oral cleanliness.

This section briefly examines whether tooth brushing and

gum care is associated with better cleaning behaviour and oral cleanliness. Table 6.7.7 shows the relationship between whether adults in the survey had received information about dental hygiene and three tooth cleaning behaviours; tooth brushing twice or more per day; the reported use of any additional tooth cleaning methods; and, more specifically, the use of dental floss.

For all three behaviours, there were significant differences between those who said they had been given information or advice and those who said they had not. Adults who had received some advice or information about gum care or tooth brushing from a dentist or a member of their staff were more likely to say they brushed their teeth twice a day or more than those who had not and were also more likely to use an additional method of tooth cleaning . In particular, almost twice as many who had been given gum advice (38%) or shown how to brush their teeth (35%) said they used dental floss compared with 18% of those who had no recall of having been given gum advice and 19% of those who said they had not been given a demonstration of how to brush their teeth.

Table 6.7.7

The survey dental examination recorded the presence of plaque, calculus (hardened plaque deposits) and dental decay, each of which can be used as indicators of the effectiveness with which teeth are cleaned. Dental plaque and calculus are discussed in more detail in Chapter 3.3 and dental decay in Chapter 3.1. Table 6.7.8 looks at the relationship of reported cleaning behaviour with each of these. Those who brushed their teeth less than once a day were more likely to have had some dental plaque detected in their mouths at the oral examination (87%) than those claiming to clean their teeth twice a day or more (69%). Those who said they brushed twice a day or more often were also less likely to have some calculus (71%) than those who brushed less than once a day (78%) and less likely to have some decay (35%) than those who brushed less than once a day (63%). However, the high level of plaque among those who claimed to clean their teeth twice a day suggests that plaque removal techniques even among those who say they brush at least twice a day are not particularly effective.

Reported use of dental floss was also associated with a lower likelihood of having some plaque or calculus but, again, the comparatively high levels of plaque and calculus found in the group who claimed to use floss suggests its use is often not wholly effective.

Table 6.7.8 also compares these three measures for those who had been given some advice about gum care or shown how to brush their teeth and those who had not. Although

statistically significant, the difference in the proportion having some plaque or calculus or decay in these groups was not particularly marked; for example 70% of those who could remember having been shown how to brush their teeth had some plaque in their mouths at the oral examination compared with 74% of those who could not recall having been shown how to brush.

Figure 6.7.3, Table 6.7.8

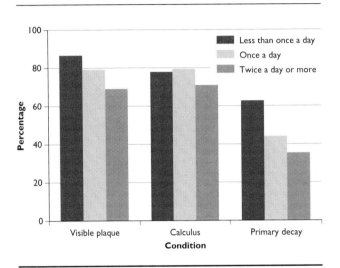

Fig 6.7.3   **The condition of teeth by reported frequency of teeth cleaning**

Table 6.7.1   **Reported frequency of tooth cleaning by characteristics of dentate adults**

Dentate adults   *United Kingdom*

		Reported frequency of tooth cleaning				Base
		Never	Less than once a day	Once a day	Twice a day or more often	
All	%	0	4	22	74	*5281*
Age						
16-24	%	0	4	18	79	*701*
25-34	%	0	3	20	77	*1151*
35-44	%	0	4	20	75	*1099*
45-54	%	0	4	24	71	*1016*
55-64	%	0	4	24	71	*645*
65-74	%	1	3	29	67	*456*
75 and over	%	2	5	26	67	*213*
Gender						
Men	%	1	6	28	64	*2406*
Women	%	0	1	15	83	*2875*
Country						
England	%	0	4	22	74	*3010*
Wales	%	0	4	21	74	*682*
Scotland	%	1	6	22	70	*953*
Northern Ireland	%	1	6	22	72	*636*
English region						
North	%	0	4	24	72	*823*
Midlands	%	1	3	24	72	*731*
South	%	0	4	20	76	*1456*
Social class **of head of household**						
I, II, IINM	%	0	2	20	78	*2483*
IIM	%	0	6	24	69	*1431*
IV,V	%	1	8	26	66	*915*
Usual reason for **dental attendance**						
Regular check-up	%	0	2	18	80	*3121*
Occasional check-up	%	0	2	20	78	*575*
Only with trouble	%	1	8	30	61	*1572*
Dental status						
Natural teeth only	%	0	4	21	75	*4188*
Natural teeth and dentures	%	0	5	26	69	*1093*

1988 Tables 26.1-26.5

Table 6.7.2   Time of day adults clean their teeth[†] by characteristics of dentate adults

Dentate adults[††]   *United Kingdom*

		Time of day adults clean their teeth									Base
		Before breakfast	After breakfast	In morning - no breakfast	Midday	Tea time	After evening meal	When going out	Last thing at night	Other	
All	%	40	46	9	7	4	9	6	74	2	5216
Age											
16-24	%	38	45	14	4	7	11	8	75	1	693
25-34	%	36	46	12	7	4	6	6	79	1	1142
35-44	%	41	47	9	7	6	9	7	75	2	1087
45-54	%	42	46	8	7	3	9	6	74	2	1001
55-64	%	42	48	3	10	2	7	4	74	2	641
65-74	%	45	47	2	10	2	10	5	66	3	445
75 and over	%	47	48	1	10	3	11	7	66	2	206
Gender											
Men	%	44	36	10	4	4	9	6	67	2	2355
Women	%	36	56	7	10	4	8	7	82	2	2860
Country											
England	%	41	46	9	7	4	8	6	74	2	2978
Wales	%	43	44	9	10	4	10	10	74	1	671
Scotland	%	31	54	8	9	5	11	7	75	2	942
Northern Ireland	%	36	51	4	9	4	10	10	73	1	624
English region											
North	%	42	44	8	8	4	10	7	72	2	818
Midlands	%	43	42	8	5	6	9	7	71	1	714
South	%	40	48	10	7	3	7	5	77	2	1446
Social class of head of household											
I, II, IIINM	%	40	52	7	8	4	8	6	78	1	2469
IIIM	%	42	40	10	6	4	11	7	70	2	1408
IV,V	%	41	40	10	5	5	9	6	66	2	889
Usual reason for dental attendance											
Regular check-up	%	38	52	7	8	4	9	6	80	2	3109
Occasional check-up	%	45	43	8	7	4	11	7	77	1	572
Only with trouble	%	42	35	13	4	4	8	7	63	2	1525
Reported frequency of tooth cleaning											
Less than once a day	%	25	16	8	1	2	6	6	34	12	197
Once a day	%	37	28	10	1	0	2	2	22	1	1128
Twice a day or more often	%	42	53	8	9	5	11	8	92	2	3890

Percentages may add to more than 100% as respondents could give more than one answer
† Respondents were asked at what times during the day they cleaned their teeth, irrespective of how often they did so
†† Excludes those who never cleaned their teeth
1988 Tables 26.9-26.12, 26.14

Table 6.7.3 Use of dental hygiene products other than an ordinary toothbrush and toothpaste by characteristics of dentate adults

Dentate adults†   *United Kingdom*

	Percentage using products other than ordinary toothbrush and toothpaste	Base
All	52	*5216*
Age		
16-24	42	*693*
25-34	56	*1142*
35-44	58	*1087*
45-54	57	*1001*
55-64	50	*641*
65-74	40	*446*
75 and over	31	*206*
Gender		
Men	46	*2355*
Women	57	*2861*
Country		
England	52	*2978*
Wales	52	*672*
Scotland	53	*942*
Northern Ireland	47	*624*
English region		
North	46	*818*
Midlands	47	*714*
South	56	*1446*
Social class		
of head of household		
I, II, IIINM	56	*2469*
IIIM	45	*1408*
IV,V	45	*890*
Usual reason for		
dental attendance		
Regular check-up	58	*3109*
Occasional check-up	52	*572*
Only with trouble	39	*1526*
Dental status		
Natural teeth only	54	*4139*
Natural teeth and dentures	42	*1077*

† Excludes those who never cleaned their teeth
1988 Tables 26.16-26.20

Table 6.7.4 The use of dental hygiene products†

Dentate adults††   *United Kingdom*

Dental hygiene products used	%
Just ordinary toothbrush & toothpaste	48
Other products	52
Dental floss	28
Mouthwash	24
Interdens toothpick/woodstick	5
Toothpaste for smokers or sensitive teeth	4
Interspace brush	3
Chewing gum	2
Other whitener or polish	2
Dental disclosing tablets	1
Denture cleaning product	1
Other	2
Base	*5216*

Percentages may add to more than 100% as respondents could give more than one answer

† Respondents were asked whether they used any dental hygiene products other than an ordinary toothbrush and toothpaste. 4% mentioned an electric toothbrush as another product. This figure may underestimate the use of electric toothbrushes as some respondents may consider them to be 'ordinary' toothbrushes

†† Excludes those who never cleaned their teeth
1988 Table 26.15

Table 6.7.5   **The use of dental hygiene products by characteristics of dentate adults**

Dentate adults†   *United Kingdom*

		Dental hygiene product used			Base
		Just toothbrush and toothpaste	Dental floss	Mouthwash	
All	%	48	28	24	5216
Age					
16-24	%	58	18	23	693
25-34	%	44	32	32	1142
35-44	%	42	35	30	1087
45-54	%	43	31	23	1001
55-64	%	50	25	15	641
65-74	%	60	16	12	446
75 and over	%	69	12	6	206
Gender					
Men	%	54	20	22	2355
Women	%	43	35	26	2861
Country					
England	%	48	28	23	2978
Wales	%	48	25	24	672
Scotland	%	47	24	29	942
Northern Ireland	%	53	24	23	624
English region					
North	%	54	23	25	818
Midlands	%	53	26	18	714
South	%	44	32	24	1446
Social class of head of household					
I, II, IIINM	%	44	33	25	2469
IIIM	%	55	22	22	1408
IV,V	%	55	17	22	890
Usual reason for dental attendance					
Regular check-up	%	42	35	24	3109
Occasional check-up	%	48	25	26	572
Only with trouble	%	61	13	21	1526
Dental status					
Natural teeth only	%	46	30	26	4139
Natural teeth and dentures	%	58	15	14	1077
Reported frequency of tooth cleaning					
Less than once a day	%	64	10	21	198
Once a day	%	62	14	16	1128
Twice a day or more often	%	44	32	26	3890

† Excludes those who never cleaned their teeth
Percentages may add to more than 100% as respondents could give more than one answer
1988 Tables 26.16-26.20

Table 6.7.6 **Advice and information about dental hygiene received from dental staff by characteristics of dentate adults**

Dentate adults   *United Kingdom*

	Type of advice or information:			Base
	Demonstration of brushing	Advice on gum care	Either	
	percentage who had received advice/information			
All	54	45	62	*5281*
Age				
16-24	56	43	65	*701*
24-34	58	48	66	*1151*
35-44	56	52	66	*1099*
45-54	57	53	66	*1016*
55-64	50	40	57	*645*
65-74	44	33	52	*456*
75 and over	29	14	31	*213*
Gender				
Men	52	42	60	*2406*
Women	56	48	64	*2875*
Country				
England	55	46	63	*3010*
Wales	52	45	59	*682*
Scotland	51	44	61	*953*
Northern Ireland	46	42	57	*636*
English region				
North	47	42	58	*823*
Midlands	54	41	59	*731*
South	58	50	67	*1456*
Social class of head of household				
I, II, IIINM	60	52	68	*2483*
IIIM	48	38	57	*1431*
IV,V	45	36	52	*915*
Usual reason for dental attendance				
Regular check-up	58	53	68	*3121*
Occasional check-up	57	49	67	*575*
Only with trouble	45	30	50	*1572*
Dental status				
Natural teeth only	57	48	65	*4188*
Natural teeth and dentures	41	32	49	*1093*

1988 Tables 26.23-26.25

Table 6.7.7 **Reported dental hygiene behaviour by receipt of advice/information on dental hygiene**

Dentate adults[†]   *United Kingdom*

Reported dental hygiene behaviour	Received type of advice or information:			
	Demonstration of brushing		Advice on gum care	
	Yes	No	Yes	No
	percentage reporting behaviour			
Brushes twice a day or more	77	71	79	70
Uses dental hygiene products other than toothbrush and toothpaste	58	44	63	42
Uses dental floss	35	19	38	18
Base	2752	2514	2364	2901

† Excludes those who never clean their teeth

Table 6.7.8   **The condition of teeth by dental hygiene behaviour and whether received advice/information on dental hygiene**

Dentate adults   *United Kingdom*

	Condition of teeth			Base	
	Visible plaque	Calculus[†]	Primary decay	*Main examination*	*Periodontal examination[†]*
	percentage with condition				
All	72	73	39	3817	*3507*
Reported frequency of tooth cleaning					
Less than once a day	87	78	63	181	*169*
Once a day	79	79	44	801	*741*
Twice a day or more	69	71	35	2826	*2588*
Used other dental hygiene products[††]					
Yes	69	72	36	2001	*1842*
No	75	74	41	1771	*1623*
Used dental floss[§]					
Yes	65	70	30	1078	*998*
No	74	74	42	2694	*2467*
Given demonstration on brushing					
Yes	70	72	37	2007	*1871*
No	74	75	40	1798	*1626*
Given advice on gum care					
Yes	69	71	36	1744	*1625*
No	75	75	41	2066	*1875*
Given either demonstration on brushing or advice on gum care					
Yes	70	72	37	2352	*2185*
No	75	75	41	1453	*1312*

† The proportion of dentate adults with some calculus present is based on the periodontal part of the examination; 310 people were excluded from this part of the examination (see section 3.3.1)

†† Other than an ordinary toothbrush and toothpaste. Excludes those who never clean their teeth

§ Excludes those who never clean their teeth

6.8 Other opinions about dental health and dentistry

Summary

- When asked if they would like to make any other comments about dental health and dentistry, 31% of people did so.
- The cost of dental treatment was the most frequently mentioned topic (10%). Generally these were that treatment cost too much or that treatment costs were not made clear.
- The next two most frequent comments referred directly to aspects of the NHS; 7% of people made general comments about a perceived drift away from NHS dentistry or that the service had changed or that it was difficult to find or keep an NHS dentist.
- People in England and in Wales, those in the South of England and those in the middle age ranges were more likely than other people to have made a comment about dental health or dentistry.
- Comments on cost or on aspects of NHS dentistry were made by an equal proportion of those who said they went for regular dental check-ups (mentioned by 10% and 9% respectively); while among those who attended for occasional check-ups or only with trouble, treatment costs were mentioned by 10% and 12% respectively compared with 4% to 6% who mentioned aspects of the NHS.

6.8.1 Introduction

During the interview all respondents were asked the following:

We have asked you a lot about dental health and dentistry. Is there anything you would like to say that we haven't asked you about?

Where possible, comments were assigned to one or more pre-defined categories by the interviewer. Comments that did not fit into these categories were recorded verbatim. At the analysis stage these comments were either placed in one of the existing categories or new categories were devised to accommodate them.

6.8.2 The types of topics mentioned

Additional comments were made by 1797 respondents, (31% of those interviewed). Topics mentioned by more than 1% of the respondents are presented in Table 6.8.1 which shows that no single topic was mentioned by more than 10% of people. The most frequent comment concerned the cost of dental treatment, which Chapter 6.2 also showed was a principal concern to patients. These referred both to the fact that respondents thought dental treatment was too expensive or that there were circumstances when they did not know what the cost would be. It was not possible to tell from most of the answers in this category whether they referred to NHS or private treatment. However, the next two most frequently made comments referred directly to the NHS; 7% of people made general comments concerned with a perceived drift away from NHS dentistry or the way the service had changed and 5% specifically said that they had found it difficult to find or keep an NHS dentist.

Comments on experiences of poor service or suggestions about how to improve the service were made by 4% of adults. A further 4% felt that dental treatment should be free from charges paid directly by the patient.

Some (4%) who provided additional comments said that they were satisfied with their dental situation. Others said they were frightened of the dentist (2%), that dentistry was better than in the past (2%), or made general comments about surgery hours and the distance to the dentist (1%).

Table 6.8.1

Table 6.8.2 describes the population who made comments. A higher proportion of people in England and Wales than in Scotland and Northern Ireland made additional comments (32% and 35% compared with 23% and 18% respectively). Across England those in the South were the most likely to

make comments, 38%, compared with 23% of those in the North and 28% of those in the Midlands.

Those from the youngest and oldest age groups were less likely to offer comments than those in the three middle age-groups; 38% of 45 to 54 year olds making additional comments compared with 15% of 16 to 24 year olds and 22% of those aged 75 and over. There was no difference between men and women in the proportion who made comments but more of those from a non-manual background made comments, 34%, compared with 28% of those from manual backgrounds.

Dentate adults were more likely to make comments than the edentate, 32% compared with 20%. It is possible that this difference reflects the older age profile of the latter group. There was little variation with respect to the usual reason for attendance in the proportion who offered a comment; 33% of those who said they went for regular check-ups made additional comments compared with 30% of those who attended only with trouble.

Table 6.8.2

Table 6.8.3 shows the variation in the three most frequently mentioned topics with socio-demographic group and according to usual reason for attendance. These were: the cost of dental treatment; adverse comments on the perceived drift away from NHS dentistry or the way the service had changed; and the difficulties of finding or keeping an NHS dentist. In general, the variations were similar to those seen for the overall level of comments made. People in England and Wales were more likely to have commented on any of the three topics than those in Scotland and Northern Ireland. For example, comments relating to the cost of dental treatment or problems with ascertaining the cost were mentioned by 10% of people in England and in Wales compared with 6% in Scotland and 5% in Northern Ireland. The regional variation was similar to that seen overall with people in the South being more likely than people elsewhere in England to mention any of these three topics. The one exception was cost where there was no significant difference between those in the South and those in the Midlands in the proportion offering this comment (13% and 10% respectively).

People in the youngest and oldest age groups were the least likely to mention the topic of cost; 5% and 6% respectively compared with between 10% and 12% of the other age groups. Adverse comments relating to a perceived drift away from NHS dentistry or the way the service had changed were also less likely to be made by these age groups and those aged 25 to 34. There was no variation with respect to gender.

Although, overall, those from non-manual backgrounds were most likely to make any comments, there was no significant difference between the proportions of people from each of the social class groups who mentioned cost (10%, 9% and 9%) or the difficulty of finding or keeping an NHS dentist (6%, 4%, and 5%). However, a significantly higher proportion of those from a non-manual background made adverse comments with respect to a perceived drift away from NHS dentistry or the way the service had changed; 8% of the non-manual group compared with 5% of those from skilled manual backgrounds and 4% from unskilled backgrounds.

Adults who had some natural teeth were more likely than those who had lost their natural teeth to offer a comment on each of the three topics which may again reflect the older age profile of the latter group. For example, 10% of dentate adults mentioned problems with cost compared with 6% of those who were edentate. A similar proportion of dentate people who reported attending the dentist for a regular check-up mentioned the topics of cost and changes in the NHS (10% and 9%); while, among those who attended for occasional check-ups or only with trouble, cost was mentioned by twice as many people as who commented on changes in the NHS. It is possible that this difference reflects to some extent the greater likelihood of regular attenders having personal experience of the effect of any changes. Difficulties in finding or keeping a dentist were mentioned by an equal proportion of those who attended for regular or for occasional check-ups or only when having trouble with teeth (6%, 5% and 6%).

Table 6.8.3

Table 6.8.1 **Types of issues mentioned, when asked for other comments about dental health or dentistry**

All adults *United Kingdom*

Topics mentioned	%
Costs too much/don't know what the cost will be	10
Dislike drift from the NHS/way NHS service has changed	7
Difficult to find/keep NHS dentist	5
Comments on poor experience of treatment or service and suggested improvements	4
Treatment/checks should be free or free for certain groups	4
Satisfied	4
Can't get appointment/Dentist over-loaded	2
Frightened of the dentist	2
Better than in the past	2
Comments on prevention and need for more advice	1
Comments on surgery hours/distance to dentist	1
Other	4
All comments	31
Base	6204

Percentages may add to more than 100% as respondents may have given more than one answer

Table 6.8.2 **Adults who made a comment by age, gender, country, English region, social class of head of household, dental status and usual reason for dental attendance**

All adults *United Kingdom*

	Percentage who made a comment	Base
All	31	6204
Age		
16-24	15	702
25-34	30	1158
35-44	35	1114
45-54	38	1106
55-64	36	833
65-74	32	758
75 and over	22	533
Gender		
Male	31	2766
Female	30	3438
Country		
England	32	3436
Wales	35	830
Scotland	23	1204
Northern Ireland	18	734
English region		
North	23	975
Midlands	28	856
South	38	1605
Social class of head of household		
I, II, IIINM	34	2732
IIIM	28	1729
IV,V	28	1221
Dental status		
Dentate	32	5281
Edentate	20	923
Usual reason for dental attendance†		
Regular check-up	33	3121
Occasional check-up	29	575
Only with trouble	30	1572

† Dentate adults only

Table 6.8.3   **Adults who made comments relating to costs of going to the dentist or about NHS dentistry**

All adults   *United Kingdom*

	Costs too much / don't know what cost will be	Dislike drift from **NHS** / **NHS** service has changed	Difficult to find / keep **NHS** dentist	Base
	percentage mentioning each issue			
All	10	7	5	6204
Age				
16-24	5	2	1	702
25-34	11	5	5	1158
35-44	12	9	7	1114
45-54	11	10	8	1106
55-64	10	9	6	833
65-74	11	8	6	758
75 and over	6	3	3	533
Gender				
Male	10	6	5	2766
Female	10	7	6	3438
Country				
England	10	7	6	3436
Wales	10	9	4	830
Scotland	6	3	1	1204
Northern Ireland	5	1	0	734
English Region				
North	6	3	1	975
Midlands	10	6	4	856
South	13	10	10	1605
Social class of head of household				
I, II, IIINM	10	8	6	2732
IIIM	9	5	4	1729
IV, V	9	4	5	1221
Dental status				
Dentate	10	7	6	5281
Edentate	6	3	3	923
Usual reason for dental attendance[†]				
Regular check-up	10	9	6	3121
Occasional check-up	10	4	5	575
Only with trouble	12	6	6	1572

† *Dentate adults only*

6.9 Trends in dental attitudes, experience and behaviour

Summary

- In 1998, a higher proportion of dentate adults in the United Kingdom reported going to the dentist for regular dental check-ups than in the past. In 1978, 43% of dentate adults reported attending for a regular check-up, by 1998, this had increased to 59%.

- There was little significant change since 1978 in the proportion of 16 to 24 year olds who said they went to the dentist for regular check-ups but the proportion of those aged 55 years and over who attend for regular check-ups has more than doubled during the same period, from 32% in 1978 to 66% in 1998.

- In both 1988 and 1998, 53% of adults in the United Kingdom said that they went to the dentist about as often as they did five years previously. The proportion saying that they attended more often rose slightly, from 17% in 1988 to 20% in 1998.

- There has been a marked increase in the proportion of respondents saying they would prefer a back tooth to be filled rather than extracted, from 65% in 1978 to 79% in 1998. It would appear, however, that the rate of change is slowing down: the increase between 1988 and 1998, from 76% to 79%, was much less than the change between 1978 and 1988.

- Reported use of private dental services increased from 6% of dentate adults in 1988 to 18% in 1998. Changes in the nature of treatments received and the types of attender using private dental services suggest that there has been a move away from the use of private dentistry to treat dental emergencies, as reported in earlier surveys in the series, towards its use for the provision of comprehensive dental care.

- In 1998, 85% of dentate adults said that they would use the same dental practice on their next visit, a figure little changed from the 82% who said this ten years previously.

- There was a decrease in the proportion of dentate 16 to 24 year olds who had their teeth filled at some time in their life from 91% in 1988 to 78% in 1998, reflecting the fall in incidence of caries among children seen in the 1980s. This was in marked contrast to the other age groups where there was either no significant change or a slight increase in experience of fillings. The largest increase in having teeth filled at some time was among those aged 75 and over from 80% in 1988 to 93% in 1998.

- The proportion of dentate adults who had no teeth filled or extracted at their most recent visit almost doubled between 1978 and 1998, from 31% to 60%.

- Twenty-seven per cent of dentate adults were dissatisfied with the appearance of their teeth. This proportion had not changed significantly since 1978. In 1998, as in 1988, the most common reason for dissatisfaction was the colour of the teeth: in 1998, 48% of those dissatisfied gave this as a reason, compared with 38% in 1988.

- The frequency of experience of specific problems with dentures among adults with no natural teeth was unchanged from 1988. In 1988 and in 1998, 41% of edentate adults experienced a problem with their dentures.

6.9.1 Introduction

Earlier chapters (in Part 5) have shown that there have been changes in various aspects of oral health since this series of surveys began with the first survey of England and Wales in 1968. These are likely to be accompanied by changes in people's attitudes towards their dental health and behaviour related to it. This chapter considers various aspects of people's attitudes and behaviour in 1998 and compares them with responses to similar questions asked in previous surveys.

6.9.2 Trends in usual reason for dental attendance

Since 1968 dentate adults interviewed in the national surveys of adult dental health have been asked whether, in general, they went to the dentist for a regular dental check-up, an occasional dental check-up or only when they were having trouble with their teeth. There has been a marked change in the reason given for attending a dentist in the thirty years since the first of these decennial surveys of adult dental health. In 1968, only 40% of dentate adults in England and Wales said that they usually went to the dentist for a regular dental check-up: by 1998, this had risen to 60%. The extent of the improvement in attendance is strongly related to people's age: there has been little or no change in the proportion of dentate 16 to 24 year olds who reported to attend for regular dental check-ups since 1968 but the proportion of those aged 55 years and over who attend for regular check-ups has more than doubled over the same period, from 27% in 1968 to 67% in 1998.

An element of the change in the usual reason for attendance at the dentist may reflect a 'cohort effect'; the introduction of the National Health Service in 1948 enabled many people their first opportunity to seek affordable dental check-ups. These people are now entering the older age groups and may account for some of the increase. However, the proportion of dentate adults who report going for regular check-ups has been rising above the level that would be predicted by the cohort effect alone. Following through the cohort aged 16 to 24 years in 1968 through to 1998 shows that 46% of 16 to 24 year olds reported going for regular dental check-ups; this rose to 50% among those aged 25 to 34 years in 1978, to 59% in those aged 35 to 44 years in 1988 and finally 64% of those aged 45 to 54 years in 1998. The process whereby those who are worse off in terms of dental health in each age group drop out through becoming edentate is unlikely to be a contributory factor in the increase in reported attendance in this cohort since levels of total tooth loss are so low for these age groups. These changes are likely to reflect improvements in oral health over the same period. Although the data are not available for 1968, those from 1978 onwards show an increase not only in the proportion of older adults with natural teeth but also an increase in the proportion with a functioning dentition (those with 21 or more natural teeth) and one which had been subject to restorative treatment. At the same time levels of dental disease fell among 16 to 24 year olds (see chapters 5.1, 5.2 and 5.3).

Table 6.9.1

The 1968 Adult Dental Health Survey took place in England and Wales followed by a survey of Scotland in 1972. From 1978 the surveys have covered the whole of the United Kingdom. Table 6.9.2 shows, for the United Kingdom, the proportion of dentate adults in 1978, 1988 and 1998 who said they usually went for regular dental check-ups. The general pattern of change in relation to age is similar to that of England and Wales since 1968. Both men and women showed similar increases in the proportion reporting attendance for regular check-ups. In all three surveys, women were much more likely than men to say that they went to the dentist for check-ups on a regular basis.

In 1978, dentate adults from non-manual backgrounds were twice as likely as were those from unskilled manual backgrounds to report attending the dentist for regular dental check-ups (56% and 28% respectively). In 1998, the difference, although still evident, was much less marked: the proportion saying they attended for regular check-ups had increased across all the social classes but was most marked among those from manual backgrounds; in 1998, 65% of those from non-manual backgrounds said they went to the dentist for regular check-ups compared with 49% of those from unskilled manual backgrounds.

The proportion of dentate adults who said they went to a dentist for regular check-ups increased between 1978 and 1998 in all countries and each of the three regions of England, but the increase was most marked in Wales and the North of England. For example, in 1978, 39% of dentate adults in Wales reported attending the dentist for a regular check-up compared with 48% in 1988 and 59% in 1998.

Figure 6.9.1, Table 6.9.2

Fig 6.9.1 **Dentate adults who reported attending the dentist for regular check-ups, 1978-98**

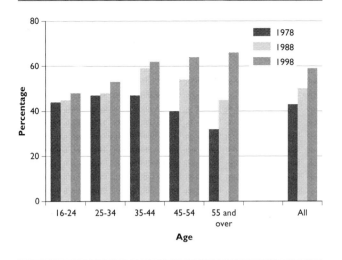

6.9.3 Trends in the change of frequency of dental attendance compared with five years previously

To gain some insight into the patterns of changes in dental attendance, dentate adults were asked whether they currently went to the dentist more often, less often or about the same as they did five years previously. This question was also asked in 1988, but not in earlier surveys.

There was little difference between responses in 1988 and 1998. In both years, 53% of dentate adults in the United Kingdom said that they went to the dentist about the same as five years previously. The proportion saying that they attended more often rose slightly, from 17% in 1988 to 20% in 1998 and there was a corresponding fall in the proportion saying that they went less often, from 30% to 27%. On the whole, the shift to more frequent attendance than five years ago was more marked among older than among younger respondents.

In both 1988 and 1998 dentate men were a little less likely than dentate women to say that there had been no change in their dental attendance pattern; in 1998, 51% of dentate men and 55% of dentate women said they attended as often as they did 5 years previously.

Just over half of dentate adults in all the different countries in the United Kingdom and English regions said that they currently attended the dentist about the same as they did five years ago: there were no significant differences between regions or countries in either survey or any significant changes in any country or region between 1988 and 1998.

There was, however, an association between change in frequency of dental attendance and social class of head of household. The change towards more frequent attendance than five years previously was more marked among those from non-manual backgrounds than among other respondents.

Table 6.9.3

6.9.4 Trends in opinions about dental visits

In 1998 dentate adults were shown a series of statements related to going to the dentist. For each statement respondents had to decide whether they 'definitely felt like that', 'felt like that to some extent' or 'did not feel like that'. In 1988, dentate adults were asked about their feelings towards going to the dentist. Ten of the statements shown to respondents in 1998 were the same as those shown to respondents during the 1988 survey. This has allowed limited comparisons to be made between the opinions held by dentate adults in 1988 and in 1998. The comparisons must be treated with caution, as there is only limited data available on people's opinions from 1988: there is no information on the proportions of adults who did not identify with the statements or who were unable to state an opinion. The statements related to different aspects of going to the dentist, and four factors were used to group the statements: fear, cost, long-term value and organisational aspects of going to the dentist.

Two of the statements related to the costs of dental treatment appear to be more of a concern to dentate adults in 1998 than was the case ten years previously. In 1998 half of dentate adults definitely felt that they would like to be given an estimate without commitment compared with just over a third (36%) of dentate adults in 1988. There was also a larger proportion of dentate adults who definitely felt that they would like to pay for treatment by instalments; 29% definitely felt this way in 1998 compared with 17% in 1988. However, the proportion of dentate adults who found NHS dentate treatment expensive in 1998 was not significantly different from the proportion in 1988; in both surveys just over 50% of dentate adults thought NHS dental treatment was expensive. As previously mentioned there is no information from 1988 on the number of people who did not identify with the cost statements or could not express an opinion; in 1998 27% of dentate adults did not feel NHS dental treatment was expensive and 22% could not give an opinion (not shown in table).

In 1988, 62% of dentate adults felt like, either definitely or to some extent, the statement 'the worst part of going to a dentist is the waiting'. A smaller proportion (50%) felt this same way in 1998. There could be a variety of reasons for this difference: aspects of dental surgeries or the nature of dental care may have changed; people may have less time to wait to see the dentist; or there may be other aspects of going to the dentist that people now think are worse than the waiting or there may have been a change in the proportion of adults who were not able to express an opinion.

There were no differences in the proportion of people who agreed with the remaining statements in 1988 and in 1998.

Table 6.9.4

6.9.5 Trends in reported dental hygiene behaviour

Dentate adults were asked how often they brushed their teeth. Chapter 6.7 examines frequency of teeth cleaning in more detail and discusses its association with motivation towards dental hygiene and the uptake of dental advice rather than as an indicator of the cleanliness of the teeth. Overall, in 1998, a larger proportion of dentate adults reported that they cleaned their teeth at least twice a day than in 1978, 74% compared with 64%.

There was little significant change between 1978 and 1998 in the frequency of teeth cleaning among dentate adults who go to the dentist for regular check-ups. For example, in 1978, 78% of regular attenders cleaned their teeth at least twice a day and 80% did so in 1998. There were, however, improvements in the reported teeth cleaning behaviour among dentate adults who only went to the dentist when they had trouble with their teeth. In 1978, 49% of those who only went to the dentist with trouble reported that they cleaned their teeth twice a day. This proportion had increased to 61% in 1998, which was still much lower than among those who attended for regular check-ups (80%).

Table 6.9.5

Dentate adults who cleaned their teeth were also asked if they used any products other than an ordinary toothbrush and toothpaste for dental hygiene purposes. In 1998, over half (52%) of dentate adults used such products compared with under a quarter (22%) in 1978. In 1998, dental floss and mouthwash were the two most frequently mentioned products respondents used (28% and 24% respectively). The proportion of dentate adults who attended the dentist for regular check-ups and who used dental floss increased from 13% in 1978 to 35% in 1998. Although the use of dental floss among those who attended the dentist only when they had trouble had also increased between 1978 and 1998, by 1998 the proportion of dentate adults who attended only with

trouble and who used dental floss had only increased to that observed among regular attenders in 1978 (13%). No data is available on the use of mouthwash in 1978, however the proportion of dentate adults who use mouthwash has increased from 11% in 1988 to 24% in 1998. There was little significant difference in the the use of mouthwash with respect to usual reason for dental attendance in either 1988 or 1998.

Figure 6.9.2, Table 6.9.6

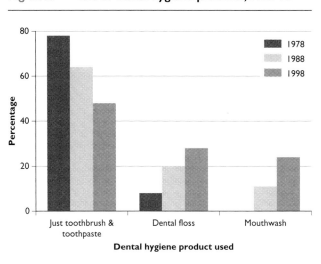

Fig 6.9.2 **Use of dental hygiene products, 1978-98**

6.9.6 Trends in treatment preferences of dentate adults

In all of the adult dental health surveys, dentate adults have been asked to indicate their treatment preferences given particular situations. One question included in every survey was "If you went to the dentist with an aching back tooth, would you prefer the dentist to take it out or fill it?" There has been a marked increase in the proportion of respondents saying they would prefer the tooth to be filled rather than taken out, from 52% in 1968 to 80% in 1998 in England and Wales. It would appear, however, that the rate of change is slowing down: the increase between 1988 and 1998, from 76% to 80%, was less than between 1978 and 1988. A similar trend was observed in the treatment preferences of dentate adults in the United Kingdom from 1978 to 1998.

Table 6.9.7

Two further questions about treatment preferences were asked of dentate adults in 1988 and 1998: "If the dentist said a back tooth would have to be taken out or crowned, which would you prefer?" and "If the dentist said a front tooth would have to be taken out or crowned, which would you prefer?" In both cases, the proportion preferring a crown rather than an extraction rose slightly between 1988 and 1998, from

65% to 68% for a back tooth, and from 89% to 92% for a front tooth.

The increases between 1988 and 1998 in the proportion of dentate adults preferring crowns were most marked among those aged 45 to 74 years: the oldest age group (aged 75 and over) showed virtually no change in their treatment preferences.

In 1998, as in 1988, women were a little more likely than men to say they would prefer a front tooth to be crowned. There was no clear pattern of change in relation to other socio-demographic characteristics, other than in relation to the respondent's usual reason for going to the dentist: a higher proportion of those who said they only went to the dentist when they had trouble with their teeth would opt for a crown rather than an extraction in 1998 (80%) compared with 1988 (86%).

Table 6.9.8

6.9.7 Trends in the type of dental service used

Since the 1988 Adult Dental Health Survey there have been changes in the payment system for dentists within the NHS and also in the charges to patients (see Chapter 6.3). As was the case at the last survey, people who are not exempt from some or all patient charges pay a proportion of their NHS dental treatment costs. This requirement may confuse some adults, as it is possible that some may consider this as "private" treatment. Similarly some paying adult patients may think they are getting NHS treatment when some or all of their course of treatment was private. Nevertheless, since 1978 all adults have been asked whether their most recent treatment was "under the NHS, was it private, or was it something else?". Responses were not prompted by the interviewer. In all three surveys, the majority of respondents said that the NHS had provided their most recent completed dental treatment; 83% in 1978, 90% in 1988 and 77% in 1998.

Table 6.9.9

Table 6.9.10 shows that the overall decrease in reported use of NHS dental services and an increase in the use of private dental services occurred for all types of treatment. However, the level of change was not similar. Between 1988 and 1998 there were small increases in the provision of extractions (10% in 1988 to 17% in 1998), treatment relating to abscesses (13% in 1988 to 18% in 1998) and treatments relating to dentures (12% in 1988 to 15% in 1998 for denture repairs) in privately funded dental visits. In contrast there were much larger increases between 1988 and 1998 in the provision of

treatments such as a scale and polish (6% to 20%) and teeth filled (5% to 19%) in privately funded visits. This suggests there has been a move away from the use of private dentistry to treat dental emergencies, as reported in earlier surveys in the series, towards its use for comprehensive dental care.

Table 6.9.10

The proportion of dentate adults who reported that their most recent course of treatment was wholly private rose from 6% in 1988 to 18% in 1998. Furthermore this occurs across men and women and people of all ages, social backgrounds and reasons for dental attendance and across all of the United Kingdom, although the rise in Northern Ireland was not statistically significant. The increase was most marked among those living in the South of England (from 6% in 1988 to 24% in 1998), those from non-manual backgrounds (from 5% in 1988 to 21% in 1998), and among those who said they attended the dentist for regular check-ups (from 4% in 1988 to 19% to 1998).

Figure 6.9.3, Table 6.9.11

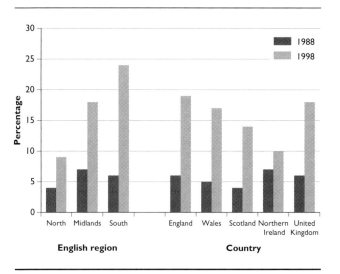

Fig 6.9.3 **Dentate adults who received private dental treatment by English region and country, 1988-98**

The increase in the reported use of private dental services differs among dentate adults who attend a dentist for different reasons. In 1988, those who attended for regular check-ups had been half as likely as those who attended with trouble to report having had private treatment at their most recent course of treatment (4% and 8% respectively). In 1998, there was no significant difference in the proportions of these two groups who reported having treatment privately (19% of those who reported regular dental check-ups compared with 16% of those who attended only with trouble). As with the case of the analysis of treatment received privately this further suggests that private dental care is increasingly being used to provide comprehensive dental care rather than emergency treatment.

Table 6.9.12

6.9.8 Trends in intentions to visit the same dental practice

In 1988 and 1998 dentate adults were asked whether or not they would use the same dental practice as they visited most recently, for their next visit to the dentist. In 1998, 85% of dentate adults said that they would use the same dental practice on their next visit, a similar figure as ten years previously (82%). Indeed, there were no significant differences between 1988 and 1998 in the proportion of dentate adults who said they would return to the same dentist next time according to any of the demographic characteristics shown, nor in relation to their usual reason for dental attendance. In 1998, as in 1988, people who went for regular check-ups, and those who had been to the dentist most recently, were more likely than other respondents to say that they would go back to the same dentist next time.

Table 6.9.13

Those who expected to change their dentist at the next visit were asked why this was so. In 1998, as in 1988, the reason given most often, by about two-fifths of these respondents, was that the dental practice they last went to was no longer convenient to visit. The next most common reason for intended change of dental practice was dissatisfaction with dental treatment received, this was mentioned by 20% of dentate adults who expected to change dental practice in 1998 and 17% in 1988. In 1998, 5% of those who expected to change said they were doing so because their previous dentist no longer undertook treatment under the NHS or were doing so for reasons of cost (4%); in 1988 these reasons had not been reported sufficiently often to be identified separately.

Table 6.9.14

6.9.9 Trends in lifetime dental treatment

In 1988 and in 1998, respondents were asked what types of dental treatment they had received during their life. In 1998, 89% of all adults had at some time had teeth filled, compared with 84% in 1988. The proportion of adults who had ever had dental x-rays taken increased markedly, from 64% in 1988 to 81% in 1998. The proportion of adults who had ever been treated for an abscess also increased from 29% to 36% over the same period. The most striking increase among the dentate population between 1988 and 1998 was in the proportion of people who had received treatment from a hygienist at some time - which rose from 20% in 1988 to 37% in 1998.

Table 6.9.15

Table 6.9.16 shows that there were also increases between 1978 and 1988 in the proportion of both dentate and edentate adults who had ever had a filling. Between 1988 and 1998, however, there was no further changes in the proportion of dentate adults who had ever had a filling, which remained at 93%, but there was an increase from 49% to 60% in the proportion of the edentate population who had had a tooth filled at sometime during their life. The increase in the overall proportion of adults who have ever had fillings between 1978 and 1998 is affected by the increase in proportion of edentate adults reporting they had fillings at some time during their life and also by the fact that a larger proportion of the population were dentate in 1998 than had been the case in 1978.

Table 6.9.16

The effect of the fall in the incidence of caries in children in the 1980s can now be seen in the experience of treatment among young adults. There was a fall from 91% in 1988 to 78% in 1998 in the proportion of 16 to 24 year olds who had their teeth filled at some time in their life. This was in marked contrast to the other age groups where there was either no significant change or a slight increase in experience of fillings. The largest increase in having teeth filled at some time was among those aged 75 and over from 80% in 1988 to 93% in 1998.

There was an overall decrease in the proportion of dentate adults who had extractions at some time from 86% in 1988 to 81% in 1998. Most of this change occurred among the younger age groups (those aged under 45 years). For example, 82% of those aged 25 to 34 years had some extraction in 1988 compared with 71% in 1998. Among those aged 45 years and over, the vast majority (over 90%) reported having had some extractions at some time during their life in 1988 and in 1998.

In 1998 a larger proportion of those aged 16 to 24 years in 1998 had worn a brace (an orthodontic appliance) at some time (40%) than those of a comparable age in 1988 (28%). On the other hand, the increase in experience of dental x-rays over a lifetime was greater among older dentate adults. For example, the proportion of dentate adults aged 75 years and over who had received a dental x-ray increased from 49% in 1988 to 69% in 1998 compared with a increase from 74% to 80% over the same period among those aged 16 to 24 years.

Figure 6.9.4, Table 6.9.17

Fig 6.9.4   **Denate adults who have ever had extractions by age, 1988-98**

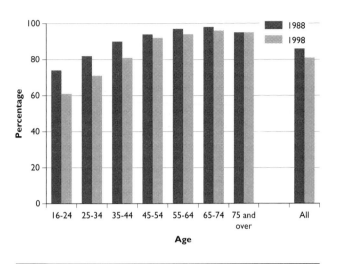

Table 6.9.18 shows that the increase in the proportion of dentate adults who had a dental x-ray was most marked among those who said they only went to the dentist when they had trouble with their teeth: 77% of this group in 1998 had a dental x-ray at some time compared with 39% in 1978. In 1998, the vast majority (93%) of dentate adults who attended for regular check-ups reported having had a dental x-ray at some time in 1998 compared with 74% of those in 1978.

Table 6.9.18

6.9.10 Trends in treatment received at most recent dental visit

As well as being asked if they had ever received various types of dental treatment, respondents were also asked about the treatment they received at their most recent visit to the dentist. The proportion of dentate adults who had no teeth filled or extracted at their most recent visit almost doubled between 1978 and 1998, from 31% to 60%. There were corresponding decreases in the proportions having teeth filled, extracted or both over the same period. The proportion of dentate adults who had extractions but no fillings at their most recent visit fell sharply, from 23% in 1978 to 10% in 1998, and the proportion that had both fillings and extractions fell from 9% to 4% over the same period. The proportion that had some teeth filled but none extracted fell from 37% in 1978 to 26% in 1998.

Changes in the types of treatment received at the most recent visit were observed in all age groups, and those aged 16 to 24 years continued to be the most likely age group to have had neither fillings nor extractions at the most recent visit: in 1998, 69% of those aged 16 to 24 years had no teeth filled

or taken out during this visit, compared with under 60% in each of the other age groups. The proportion who had extractions but no fillings fell most sharply among older respondents: in 1978, 37% of those aged 55 and over had extractions but no fillings, but by 1998 this had reduced to 15%.

Table 6.9.19

The increases in the proportion of dentate adults who had no fillings or extractions during their most recent visit were similar in all geographical areas (approximately doubling over the period 1978 to 1998). In 1978 Scotland had a much lower proportion of people who had neither fillings nor extractions than in England or Wales (no separate data are available for Northern Ireland) and this was still the case in 1998. In 1978, 25% of dentate adults in Scotland had neither fillings nor extractions at their most recent dental visit compared with 50% in 1998. In Northern Ireland the proportion of dentate adults who did not have any fillings or extractions in 1998 was similar to that observed in Scotland (49%).

Table 6.9.20

Treatment received at the most recent visit was also considered in relation to the usual reason for dental attendance. The proportion of dentate adults whose most recent visit had involved neither fillings nor extractions increased, not only among those who attended for regular check-ups (from 46% in 1978 to 71% in 1998), but also among those who attended for occasional check-ups (from 34% in 1978 to 61% in 1998) and among those who only went to the dentist when they had trouble with their teeth (from 14% in 1978 to 40% in 1998). In 1998, as in the previous surveys, this latter group were much more likely than other respondents to have extractions but no fillings during their most recent course of treatment. However, the proportion doing so fell markedly between 1978 and 1988, from 46% to 27%, but there was no further significant fall in 1998.

Figure 6.9.5, Table 6.9.21

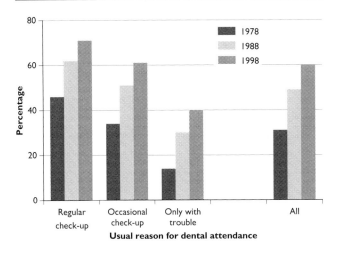

Fig 6.9.5   **Dentate adults who had no extractions or fillings at most recent visit to the dentist by usual reason for dental attendance, 1978-98**

not significantly different from that found in 1988, but the proportion who thought they would find partial dentures very upsetting increased in England and Wales but decreased slightly in Scotland and decreased markedly in Northern Ireland.

In both 1988 and 1998 older people who were wholly reliant on their own teeth were more likely than younger people to say they would be very upset by complete or partial replacement of their teeth. This may reflect that these older adults have already managed to fend off the need for dentures for some time and represent the "dental survivors" of their age group. In 1998, people of all ages were more likely to feel they would be very upset if they needed partial dentures than those in 1988, but the marked changes between surveys were among those in groups aged 45 and over.

Table 6.9.24

6.9.11 Trends in satisfaction with the appearance of the teeth

Dentate adults were asked whether or not they were satisfied with the appearance of their teeth (or their dentures if they had any). In both surveys, just over one quarter of dentate adults were dissatisfied with the appearance of their teeth (28% in 1988 and 27% in 1998). There were no significant changes between 1988 and 1998 with respect to age, gender, or usual reason for dental attendance. In 1998, as in 1988, those who went to the dentist only when they had trouble with their teeth were more likely to be dissatisfied with the appearance of their teeth than those who attended for check-ups.

Table 6.9.22

Those who were not satisfied with the general appearance of their teeth (or dentures) were asked why. In 1998, as in 1988, the most common reason for dissatisfaction was the colour of the teeth: in 1998, 48% of those dissatisfied gave this as a reason, compared with 38% in 1988. The proportions giving gaps in their teeth, and broken or chipped teeth as reasons for their dissatisfaction also increased between 1988 and 1998.

Table 6.9.23

6.9.12 Trends in attitudes to requiring dentures in future

Table 6.9.24 compares the anticipated reactions of those who had no dentures to the need for them in the future. In 1998 the proportion of adults wholly reliant on their own teeth who would find complete replacement very upsetting was

A related issue to people's reactions to dentures is how likely they think it is that they will need them in the future. There was no significant change between 1988 and 1998 in the proportion of adults who expect to need a partial or complete replacement of their teeth by dentures in the next 5 years. There were also no differences between countries or English regions in 1988 and 1998 in the proportion of people anticipating some replacement of their teeth in the 5 years following the survey. Similarly there were no marked differences in expectations with respect to gender, age, social class of head of household or usual reason for dental attendance in 1988 and in 1998.

Table 6.9.25

The overall level of expectation of retaining some natural teeth among dentate adults who were wholly reliant on their own teeth in 1998 (81%) was not significantly different from that in 1988 (82%) but represents a considerable improvement from the findings in 1978 when only 67% expected to retain some teeth for the rest of their lifetime. The apparent decrease in expectations among the youngest age group from 84% in 1988 to 77% in 1998 is partly explained by a higher proportion of those in 1998 saying they did not know if they would retain all their natural teeth (10% compared with 4% in 1988 - figures not shown).

Table 6.9.26

The expectation of requiring some dentures will in some respects reflect the way in which people would choose to manage their own tooth loss. People were therefore asked if they had some missing teeth at the back of their mouth, whether they would have these replaced by a denture or whether they would manage without. There was no significant difference between 1988 and 1998 in the overall proportion of people who said they would prefer not to have dentures to replace any missing back teeth. This was also the case for the different countries and English regions; among men and women; among those in different age groups or social classes; and those with different reasons for attending a dentist in 1988 and in 1998.

Table 6.9.27

6.9.13 Trends in problems with dentures experienced by edentate adults

The type of problems experienced by those who have no teeth have not changed in pattern or in frequency of occurrence between 1988 and 1998. Forty-one per cent of those with no natural teeth said they experienced some problem with their dentures in 1988 and in 1998. The most frequently reported problem, experienced by almost a quarter of edentate adults was trouble with eating or drinking (21% reported this problem in 1988 and 26% in 1998). Similar proportions of edentate adults intended to consult a dentist about the problems they were having with their denture problems in 1998 (13%) as in 1988 (11%).

Table 6.9.28

Table 6.9.1 **Attending the dentist for a regular check-up by age, 1968-98**

Dentate adults   *England and Wales*

	1968	1978	1988	1998
	percentage who attend for a regular check-up			
All	40	46	50	60
Age				
16-24	46	47	45	49
25-34	45	50	48	54
35-44	42	50	59	63
45-54	35	45	55	64
55 and over	27	36	45	67

For bases see Tables Appendix
1988 Table 20.11

Table 6.9.2 **Attending the dentist for a regular check-up by characteristics of dentate adults, 1978-98**

Dentate adults   *United Kingdom*

	1978	1988	1998
	percentage who attend for a regular check-up		
All	43	50	59
Age			
16-24	44	45	48
25-34	47	48	53
35-44	47	59	62
45-54	40	54	64
55 and over	32	45	66
Gender			
Men	36	42	52
Women	50	58	66
Country			
England	44	50	60
Wales	39	48	59
Scotland	38	50	55
Northern Ireland[†]	..	42	51
English region			
North	38	49	58
Midlands	46	49	63
South	47	52	60
Social class of head of household			
I, II, IIINM	56	59	65
IIIM	36	44	57
IV,V	28	34	49

† Data for 1978 not available.
For bases see Tables Appendix.
1988 Tables 20.10, 20.11, 20.14, 20.15

Table 6.9.3 **Reported change in frequency of dental attendance compared with five years ago, by characteristics of dentate adults, 1988-98**

Dentate adults   *United Kingdom*

Characteristic	Reported change in frequency of dental attendance compared with five years ago						Base	
	More often		About the same		Less often			
	1988	1998	1988	1998	1988	1998	1988	1998
				percentage				
All	17	20	53	53	30	27	3531	5281
Age								
16-24	15	12	37	40	48	48	819	701
25-34	22	30	49	43	28	27	797	1151
35-44	20	23	60	55	21	22	745	1099
45-54	17	20	61	60	21	21	505	1016
55-64	13	20	64	60	22	20	375	645
65-74	12	11	62	69	26	20	215	456
75 and over	7	11	61	69	32	20	73	213
Gender								
Men	17	21	51	51	32	29	1789	2406
Women	18	20	54	55	27	25	1740	2875
Country								
England	18	20	52	53	30	27	2955	3010
Wales	15	21	57	55	28	24	622	682
Scotland	16	21	54	52	31	28	1145	953
Northern Ireland	16	18	63	52	22	31	558	636
Region								
Northern	18	21	51	51	31	28	752	823
Midlands	20	17	49	57	30	27	692	731
Southern	16	21	54	52	29	26	1524	1456
Social class of head of houshold								
I, II, IIINM	15	20	57	56	28	24	1766	2483
IIIM	19	22	50	52	31	26	1079	1431
IV,V	19	20	47	49	34	32	571	915
Usual reason for dental attendance								
Regular check-up	23	26	68	66	10	9	1763	3121
Occasional check-up	18	21	30	28	52	51	512	575
Only with trouble	10	9	42	37	49	54	1249	1572

Bases for 1988 as presented in the 1988 report with the exception of bases for England, Wales, Scotland and Northern Ireland which are unweighted.
1988 Tables 20.16-20

Table 6.9.4 **Feelings towards going to the dentist, 1988-98**

Dentate adults   *United Kingdom*

Statement		1988		1998	
Organisation		%		%	
I'd like to know more about what the dentist is going to do and why	Definitely felt like that	40	} 67	43	} 71
	Felt like that to some extent	27		28	
I'd like to be able to drop in at the dentist without an appointment	Definitely felt like that	38	} 66	40	} 69
	Felt like that to some extent	28		29	
The worst part of going to the dentist is the waiting	Definitely felt like that	33	} 62	22	} 50
	Felt like that to some extent	29		28	
Costs					
I'd like to be given an estimate without commitment	Definitely felt like that	36	} 60	50	} 70
	Felt like that to some extent	24		20	
I find NHS dental treatment expensive	Definitely felt like that	25	} 53	28	} 51
	Felt like that to some extent	28		23	
I'd like to be able to pay for my dental treatment by instalments	Definitely felt like that	17	} 38	29	} 49
	Felt like that to some extent	21		20	
Fear					
I'm nervous of some kinds of dental treatment	Definitely felt like that	37	} 68	31	} 64
	Felt like that to some extent	31		32	
I always feel anxious about going to the dentist	Definitely felt like that	30	} 56	24	} 49
	Felt like that to some extent	26		24	
Long term value					
I don't want fancy (intricate) treatment by instalments	Definitely felt like that	24	} 48	24	} 47
	Felt like that to some extent	24		23	
I don't see any point in visiting the dentist unless I need to[†]	Definitely felt like that	..		19	} 36
	Felt like that to some extent	..		17	
It will cost me less in the long run if I only go when I'm having trouble	Definitely felt like that	19	} 37	19	} 35
	Felt like that to some extent	18		15	
Base		5280		5281	

† *Respondents were not asked about their feelings towards this statement in 1988.*
1988 Table A21.1

Table 6.9.5 Reported frequency of teeth cleaning by usual reason for dental attendance, 1978-98

Dentate adults   *United Kingdom*

Reported frequency of teeth cleaning	Usual reason for dental attendance									All		
	Regular check-up			Occasional check-up			Only with trouble					
	1978	1988	1998	1978[†]	1988	1998	1978	1988	1998	1978	1988	1998
	%	%	%	%	%	%	%	%	%	%	%	%
Twice a day or more	78	76	80	..	68	78	49	56	61	64	67	74
Once a day	20	22	18	..	29	20	34	32	30	27	26	22
Less than once a day	2	3	2	..	3	2	13	10	8	7	5	4
Never	0	0	0	..	0	0	4	2	1	2	1	0
Base	1404	1763	3121	..	513	575	1392	1258	1572	3262	3584	5281

† Data from 1978 not available.
Bases for 1978 and 1988 as presented in the 1988 report.
1988 Table 26.28

Table 6.9.6 Use of dental hygiene products by usual reason for dental attendance, 1978-98

Dentate adults[†]   *United Kingdom*

Dental hygiene products used	Usual reason for dental attendance									All		
	Regular check-up			Occasional check-up			Only with trouble					
	1978	1988	1998	1978	1988	1998	1978	1988	1998	1978	1988	1998
	%	%	%	%	%	%	%	%	%	%	%	%
Dental floss	13	29	35	8	21	25	3	8	13	8	20	28
Toothpicks/woodsticks	14	7	6	9	4	5	4	2	2	9	5	5
Mouthwash[††]	..	11	24	..	9	26	..	11	21	..	11	24
Just toothbrush & toothpaste	70	56	42	80	69	48	87	73	61	78	64	48
Base	1401	1762	3113	443	513	573	1341	1258	1544	3205	3583	5239

Percentages may add to more than 100% as respondents could give more than one answer.
† Excludes those who stated they never cleaned their teeth.
†† Data not collected in 1978 survey.
Bases for 1978 and 1988 as presented in 1988 report.
1988 Table 26.30.

Table 6.9.7 Preference for having an aching back tooth filled rather than taken out, 1968-98

Dentate adults

	1968	1978	1988	1998
	percentage who would prefer the tooth filled			
England and Wales	52	65	76	80
United Kingdom[†]	..	65	76	79

† 1968 survey covered England and Wales only.
For bases see Tables Appendix.
1988 Table 22.2

Table 6.9.8 Preference for having teeth crowned rather than taken out by characteristics of dentate adults, 1988-98

Dentate adults   *United Kingdom*

	Would prefer a back tooth to be crowned		Would prefer a front tooth to be crowned		Base	
	1988	*1998*	*1988*	*1998*	*1988*	*1998*
			percentage			
All	65	68	89	92	*3585*	*5281*
Age						
16-24	68	71	96	95	*831*	*701*
25-34	70	71	94	97	*811*	*1151*
35-44	70	72	93	95	*751*	*1099*
45-54	63	71	87	91	*508*	*1016*
55-64	53	63	76	87	*379*	*645*
65-74	43	56	68	79	*213*	*456*
75 and over	43	45	70	71	*70*	*213*
Gender						
Men	65	68	88	91	*1813*	*2406*
Women	64	68	91	93	*1748*	*2875*
Country						
England	65	69	89	92	*2955*	*3010*
Wales	67	69	90	86	*622*	*682*
Scotland	60	62	89	90	*1145*	*953*
Northern Ireland	62	61	91	92	*558*	*636*
English region						
Northern	62	68	88	91	*757*	*823*
Midlands	60	65	89	90	*706*	*731*
Southern	69	71	90	94	*1537*	*1456*
Social class of head of houshold						
I, II, III NM	74	75	92	95	*1779*	*2483*
III M	56	61	88	89	*1084*	*1431*
IV,V	52	57	82	86	*583*	*915*
Usual reason for dental attendance						
Regular check-up	74	74	95	94	*1757*	*3121*
Occasional check-up	69	70	92	95	*510*	*575*
Only with trouble	49	56	80	86	*1249*	*1572*

Bases for 1988 as presented in the 1988 report with the exception of bases for England, Wales, Scotland and Northern Ireland which are unweighted.
1988 Tables 22.1-22.5

Table 6.9.9   **Type of dental service by dental status, 1978-98**

Dentate adults†   *United Kingdom*

Type of dental service	Dental status						All		
	Natural teeth only			Natural teeth and dentures					
	1978	1988	1998	1978	1988	1998	1978	1988	1998
	%	%	%	%	%	%	%	%	%
NHS	83	91	76	82	88	80	83	90	77
Private	10	5	19	12	8	15	10	6	18
NHS and private	1	1	2	3	3	3	1	2	2
School/Community dental service	1	1	0	0	0	0	1	1	0
Armed Forces	1	1	1	0	0	0	1	1	1
Dental Hospital	2	1	0	1	1	0	2	1	0
Something else	2	0	1	2	0	0	2	0	1
Dentist at workplace††	..	..	0	..	..	0	..	..	0
Through insurance††	..	..	0	..	..	1	..	..	1
With a dental plan††	..	..	1	..	..	1	..	..	1
Base	2283	2525	3861	958	801	1024	3241	3326	4885

† Excludes those who have never been to the dentist or were in the middle of a course of treatment.
†† Not included as separate categories in 1978 and 1988.
Bases for 1978 and 1988 as presented in the 1988 report.
1988 Table 24.49

Table 6.9.10 **Type of dental service used by type of treatment, 1988-98**

Dentate adults†   *United Kingdom*

Treatment received			Type of dental service			Base
			NHS	**Private**	**Other**	
Scale and polish	1988	%	91	6	4	1773
	1998	%	75	20	6	2817
Teeth filled	1988	%	90	5	4	1282
	1998	%	77	18	5	1504
X-rays	1988	%	87	7	6	841
	1998	%	73	17	6	1233
Teeth extracted	1988	%	84	10	6	574
	1998	%	77	17	6	715
Crown fitted	1988	%	88	5	6	220
	1998	%	72	21	7	334
Abscess treated	1988	%	81	13	5	141
	1998	%	77	18	5	234
Denture fitted	1988	%	87	7	6	244
	1998	%	79	15	6	265
Denture repaired	1988	%	84	12	4	49
	1998	%	81	15	4	108
All	1988	%	90	6	4	3326
	1998	%	77	18	5	4885

† Excludes those who have never been to the dentist or were in the middle of a course of treatment
Bases for 1988 as presented in the 1988 report.

Table 6.9.11 **Private dental treatment by characteristics of dentate adults, 1988-98**

Dentate adults†   *United Kingdom*

Characteristic	Proportion who received wholly private treatment		Base	
	1988	1998	1988	1998
All	6	18	3326	4885
Age				
16-24	3	10	780	648
25-34	7	19	745	1060
35-44	5	19	684	1015
45-54	8	21	476	937
55-64	7	21	357	590
65 and over	8	18	209	635
Gender				
Men	7	19	1702	2232
Women	5	18	1623	2653
Country				
England	6	19	2797	2802
Wales	5	17	164	628
Scotland	4	14	277	874
Northern Ireland	7	10	89	581
English region				
North	4	9	708	758
Midlands	7	18	656	675
South	6	24	1433	1369
Social class of head of household				
I, II, III NM	5	21	1651	2294
III M	7	15	1020	1318
IV,V	5	15	550	849
Usual reason for dental attendance				
Regular check-up	4	19	1624	2856
Occasional check-up	7	20	490	535
Only with trouble	8	16	1205	1485

† Excludes those who have never been to the dentist or were in the middle of a course of treatment.
Bases for 1988 as presented in the 1988 report.
1988 Table 24.47

Table 6.9.12 **Type of dental service by usual reason for dental attendance, 1988-98**

Dentate adults†   *United Kingdom*

Type of service	Usual reason for dental attendance						All	
	Regular check-up		Occasional check-up		Only with trouble			
	1988	1998	1988	1998	1988	1998	1988	1998
	%	%	%	%	%	%	%	%
NHS	94	76	90	76	85	79	90	77
Private	4	19	7	20	8	16	6	18
NHS and private	1	2	2	2	2	1	2	2
Other	2	4	2	1	4	4	2	4
Base	1624	2856	490	535	1205	1485	3319	4885

† Excludes those who have never been to the dentist or were in the middle of a course of treatment.
Bases for 1988 as presented in the 1988 report.
1988 Table 24.48

Table 6.9.13 **Intention to visit the same dental practice next time as last time by characteristics of dentate adults, 1988-98**

Dentate adults†   *United Kingdom*

	Proportion intending to visit same dental practice		Base	
	1988	**1998**	*1988*	*1998*
All	82	85	*3533*	*5268*
Age				
16-24	80	81	*819*	*698*
25-34	79	80	*800*	*1149*
35-44	85	86	*745*	*1095*
45-54	88	87	*505*	*1015*
55-64	83	89	*374*	*645*
65-74	77	89	*215*	*454*
75 and over	75	84	*73*	*212*
Gender				
Men	80	82	*1793*	*2398*
Women	84	87	*1740*	*2870*
Social class of **head of houshold**				
I, II, III NM	84	85	*1767*	*2482*
III M	81	85	*1081*	*1425*
IV,V	81	82	*574*	*911*
Usual reason for dental **attendance**				
Regular check-up	94	95	*1763*	*3121*
Occasional check-up	85	86	*513*	*574*
Only with trouble	65	63	*1251*	*1564*
Time since last visit **to the dentist**				
Less than 6 months	94	96	*1802*	*3015*
Over 6 months up to 1 year	86	89	*582*	*766*
Over 1 year up to 2 years	77	78	*295*	*400*
Over 2 years up to 5 years	70	64	*475*	*578*
Over 5 years up to 10 years	45	36	*191*	*135*
Over 10 years	33	41	*190*	*374*

† Excludes those who have never been to the dentist.
Bases for 1988 as presented in 1988 report.
1988 Table 25.15

Table 6.9.14 **Reasons given for wanting to change dental practice, 1988-98**

Dentate adults intending to change dental practice   *United Kingdom*

Reasons for wanting to change dental practice	**1988**	**1998**
	%	%
No longer convenient	40	43
Dissatisfied with dental treatment	17	20
Dentist has moved, died or retired	12	13
Heard of better dentist	2	1
Never go regularly to dentist	1	8
Last visit not to usual dentist	3	5
No longer in system/on practice patient list†	9	11
Other reason	10	5
Present dentist no longer offers NHS treatment††	..	5
Cost††	..	4
Difficult to get appointment††	..	2
Base	*615*	*702*

† In 1988 'no longer in system' covered respondents who had left the CDS or who had changed their workplace. It is likely that in some cases in 1998 it also included people whose NHS status with their dentist had lapsed.
†† In 1988 interviewers recorded respondents' answers to this question and were subsequently coded to the set of codes shown in the upper part of the table. In 1998 interviewers recorded respondents' answers based on this set of codes. Answers which did not fit these codes were specified and recoded to produce three new codes shown in the lower part of the table.
1988 Table 25.16

Table 6.9.15 **Dental treatment received over lifetime by current dental status, 1988-98**

All adults   *United Kingdom*

| Dental treatment | Dental status | | | | All | |
| | Dentate | | Edentate | | | |
	1988	1998	1988	1998	1988	1998
	percentage who have ever had treatment					
Fillings	93	93	49	60	84	89
An x-ray	75	88	20	32	64	81
Extractions	86	81	100†	100†	89	83
Injection (gum) for extraction	74	69	68	74	73	70
Injection (gum) for filling	87	89	47	48	80	83
Gas	58	50	62	69	59	52
A scale and polish (dentist)	85	87	30	28	73	79
Treatment from a hygienist	24	42	3	4	20	37
A brace	15	20	2	2	12	17
An abscess	31	38	24	30	29	36
A crown	29	36	3	4	24	32
A bridge	5	7	2	2	5	6
Base	5280	5281	1545	923	6825	6204

† By definition.
1988 Table 18.1

Table 6.9.16 **Whether adults have ever had a filling by dental status, 1978-98**

All adults   *United Kingdom*

| Dental status | Proportion who have ever had a filling | | |
	1978	1988	1998
Dentate	90	93	93
Edentate	47	49	60
All	77	84	89
Base			
Dentate	3262	3585	5281
Edentate	1376	952	923
All	4638	4538	6204

Bases for 1978 and 1988 as presented in the 1988 report.
1998 Table 18.11

Table 6.9.17 **Dental treatment received over lifetime by age, 1988-98**

Dentate adults   *United Kingdom*

Type of dental treatment	Age														All	
	16-24		25-34		35-44		45-54		55-64		65-74		75 and over			
	1988	1998	1988	1998	1988	1998	1988	1998	1988	1998	1988	1998	1988	1998	1988	1998
	percentage who have ever had treatment															
Fillings	91	78	95	96	95	97	95	98	91	96	89	92	80	93	93	93
An x-ray	74	80	80	92	80	91	80	92	66	87	56	83	49	69	75	88
Extractions	74	61	82	71	90	81	94	92	97	94	98	96	95	95	86	81
Injection (gum) for extraction	55	50	66	56	78	66	89	82	90	89	93	89	89	86	74	69
Injection (gum) for filling	80	66	92	92	92	94	91	96	85	94	80	87	61	84	87	89
Gas	45	23	63	38	71	54	61	73	58	64	47	57	35	53	58	50
Scale and polish	75	68	87	88	92	92	92	92	86	93	80	90	71	80	85	87
Treatement from a hygienist	18	24	25	42	31	48	27	51	21	42	19	42	7	32	24	42
A brace	28	40	20	28	13	16	7	14	3	8	2	4	3	5	15	20
An abscess	18	15	28	28	37	42	42	53	37	50	29	45	33	36	29	38
A crown	15	11	32	29	37	40	37	51	30	48	28	42	22	36	29	36
A bridge	1	1	4	3	7	8	9	11	7	15	6	8	6	7	5	7
Base	*832*	*701*	*813*	*1151*	*754*	*1099*	*513*	*1016*	*380*	*645*	*217*	*456*	*74*	*213*	*3585*	*5281*

Bases for 1988 as presented in the 1988 report.
1988 Table 18.3.

Table 6.9.18 **Whether dentate adults have ever had an x-ray by usual reason for dental attendance, 1978-98**

Dentate adults   *United Kingdom*

Usual reason for dental attendance	Proportion who have ever had an x-ray		
	1978	1988	1998
Regular check-up	74	86	93
Occasional check-up	62	80	88
Only with trouble	39	59	77
All	57	76	88
Base			
Regular check-up	*1404*	*1764*	*3121*
Occasional check-up	*444*	*513*	*575*
Only with trouble	*1392*	*1257*	*1572*
All	*3262*	*3585*	*5281*

Bases for 1978 and 1988 as presented in the 1988 report.
1988 Table 18.14

Table 6.9.19 **Summary of most recent treatment by age, 1978-98**

Dentate adults†   *United Kingdom*

Age			Summary of treatment received				Base
			No extractions and:		Some extractions and:		
			No fillings	Some fillings	No fillings	Some fillings	
16-24	1978	%	35	43	13	9	672
	1988	%	60	29	6	4	772
	1998	%	69	23	4	4	648
25-34	1978	%	32	41	19	8	773
	1988	%	44	41	9	6	743
	1998	%	59	28	8	5	1060
35-44	1978	%	28	39	23	10	636
	1988	%	45	40	10	5	682
	1998	%	59	29	10	3	1015
45-54	1978	%	25	37	28	8	476
	1988	%	44	36	13	5	478
	1998	%	59	27	12	3	937
55 and over	1978	%	32	22	37	9	495
	1988	%	48	21	24	6	626
	1998	%	58	22	15	4	1225
All	1978	%	31	37	23	9	3056
	1988	%	49	34	12	5	3311
	1998	%	60	26	10	4	4885

† Excludes those who have never been to the dentist or were in the middle of a course of treatment.
Bases for 1978 and 1988 as presented in 1988 report.
1988 Table 24.43.

Table 6.9.20 **Summary of most recent treatment by country and English region, 1978-98**

Dentate adults†

Age			Summary of treatment received				Base
			No extractions and:		Some extractions and:		
			No fillings	Some fillings	No fillings	Some fillings	
United Kingdom	1978	%	31	37	23	9	3056
	1988	%	49	34	12	5	3311
	1998	%	60	26	10	4	4885
Country							
England	1978	%	32	38	22	8	2583
	1988	%	50	33	11	5	2787
	1998	%	61	25	10	4	2802
Wales	1978	%	29	35	27	9	341
	1988	%	50	32	14	4	161
	1998	%	62	27	12	3	628
Scotland	1978	%	25	37	27	11	817
	1988	%	41	36	15	8	275
	1998	%	50	31	15	5	874
Northern Ireland††	1978	%	..	..	..	..	..
	1988	%	33	40	19	8	89
	1998	%	49	37	9	5	581
English region							
North	1978	%	32	33	25	9	696
	1988	%	48	31	14	6	708
	1998	%	62	25	9	4	758
Midlands	1978	%	33	30	29	8	630
	1988	%	52	30	13	5	652
	1998	%	62	22	12	3	675
South	1978	%	31	44	17	8	1357
	1988	%	50	35	9	4	1421
	1998	%	61	26	9	4	1369

† Excludes those who have never been to the dentist or were in the middle of a course of treatment.
†† Data not available for 1978.
Bases for 1978 and 1988 as presented in the 1988 report.
1988 Table 24.42

Table 6.9.21 **Summary of most recent treatment by usual reason for dental attendance, 1978-98**

Dentate adults[†]   *United Kingdom*

Usual reason for dental attendance			Summary of treatment received				Base
			No extractions and:		**Some extractions and:**		
			No fillings	**Some fillings**	**No fillings**	**Some fillings**	
Regular check-up	1978	%	46	46	4	4	1287
	1988	%	62	33	3	2	1609
	1998	%	71	25	3	1	2856
Occasional check-up	1978	%	34	49	8	9	424
	1988	%	51	40	5	4	488
	1998	%	61	30	6	3	535
Only with trouble	1978	%	14	25	46	14	1342
	1988	%	30	32	27	10	1200
	1998	%	40	26	25	9	1485
All	1978	%	31	37	23	9	3056
	1988	%	49	34	12	5	3303
	1998	%	60	26	10	4	4885

† Excludes those who have never been to the dentist or were in the middle of a course of treatment.
Bases for 1978 and 1988 as presented in 1988 report.
1988 Table 24.44

Table 6.9.22 **Dissatisfaction with appearance of teeth (or dentures), by characteristics of dentate adults, 1988-98**

Dentate adults   *United Kingdom*

	Proportion dissatisfied with appearance		Base	
	1988	**1998**	*1988*	*1998*
All	28	27	3581	5281
Age				
16-24	28	26	832	701
25-34	32	28	812	1151
35-44	29	27	754	1099
45-54	30	30	512	1016
55-64	24	28	380	645
65-74	24	20	217	456
75 and over	23	20	74	213
Gender				
Men	26	25	1823	2406
Women	31	29	1758	2875
Usual reason for dental attendance				
Regular check-up	24	23	1763	3121
Occasional check-up	29	24	512	575
Only with trouble	34	36	1257	1572

Bases for 1988 as presented in the 1988 report.
1988 Tables 27.1-2

Table 6.9.23 **Reason for dissatisfaction with appearance of teeth, 1988-1998**

Dentate adults dissatisfied with appearance of teeth (or dentures)   *United Kingdom*

Reason for dissatisfaction	1988	1998
	%	%
Colour of teeth	38	48
Crooked/slanting/protruding/irregular teeth	36	34
Gaps/spaces in mouth	13	18
Fillings/colour of fillings	11	10
Decayed teeth	7	5
Size/shape of teeth	6	9
Broken/chipped teeth	5	11
Need filling	0	3
Other	10	8
Base	956	1416

Percentages may add to more than 100% as respondents could give more than one answer.
Base for 1988 as presented in the 1988 report.
1988 Table 27.10

Table 6.9.24 **Attitudes to the thought of having dentures, 1988-1998**

Adults with only natural teeth   *United Kingdom*

| | **Proportion who would be:** | | | |
| | **Very upset if partial replacement were needed** | | **Very upset if full replacement were needed** | |
	1988	1998 †	1988	1998 †
All	22	27	63	61
Age				
16–24	16	20	60	58
25–34	22	23	63	57
35–44	24	27	65	61
45–54	26	33	66	67
55–64	24	37	57	69
65 and over	27	34	69	66
Gender				
Men	14	18	55	51
Women	29	35	71	72
Country				
England	20	27	62	61
Wales	29	32	62	61
Scotland	26	23	66	62
Northern Ireland	32	23	70	57
English region				
North	18	26	58	65
Midlands	19	27	61	60
South	22	27	65	60
Social class of head of household				
I, II, IIINM	23	28	67	66
IIIM	20	23	57	55
IV,V	21	25	58	54
Usual reason for dental attendance				
Regular check-up	25	30	70	66
Occasional check-up	18	24	61	59
Only with trouble	17	21	53	52

† Excludes 175 dentate adults who had some time had a denture fitted they no longer had

For bases see Tables appendix

1988 Tables 17.21, 17.22, 17.23, 17.24.

Table 6.9.25 **Expectation of needing dentures by age, gender, social class of head of household, country, English region and usual reason for dental attendance, 1988-98**

Adults with only natural teeth   *United Kingdom*

	Adults with only natural teeth who think they are:					
	Likely to need partial denture in next 5 years		Likely to need full dentures:			
			In next 5 years		Sometime (after 5 years)	
	1988	1998[†]	1988	1998[†]	1988	1998[†]
			percentage			
All	9	9	2	2	13	10
Age						
16–24	4	4	1	1	13	12
25–34	9	7	2	1	17	13
35–44	10	9	1	1	12	10
45–54	15	14	3	2	10	6
55–64	18	16	6	3	8	3
65 and over	17	11	7	4	7	2
Gender						
Men	10	9	3	1	13	10
Women	9	9	1	2	13	10
Country						
England	9	9	2	1	13	9
Wales	8	9	1	2	12	9
Scotland	14	11	2	2	13	11
Northern Ireland	9	13	2	1	13	10
English region						
North	10	8	4	1	17	9
Midlands	9	9	2	3	13	10
South	9	9	1	1	11	9
Social class of head of household						
I, II, IIINM	8	7	1	1	7	8
IIIM	12	11	4	2	16	12
IV,V	12	12	2	3	18	12
Usual reason for dental attendance						
Regular check-up	7	8	1	1	10	8
Occasional check-up	6	5	1	0	14	9
Only with trouble	15	13	4	4	17	12

† Excludes 175 dentate adults who had at some time had a denture fitted which they no longer had
For bases see Tables appendix
1988 Tables 17.2, 17.27

Table 6.9.26 **Expectation of retaining natural teeth by age, 1978-98**

All adults with only natural teeth　　　　　　　*United Kingdom*

Age	Proportion who expect to keep some natural teeth all their lives:		
	1978	1988	1998†
		percentage	
16–24	70	84	77
25–34	67	77	80
35–44	63	84	83
45–54	63	83	83
55 and over	68	81	85
All	67	82††	81

† Excludes 175 dentate adults who had at some time had a denture fitted which they no longer had

†† Correction from 1988 report

For bases see Tables appendix

1988 Table 17.33

Table 6.9.27 **Adults with only natural teeth who would prefer to manage without a partial denture, 1988-98**

Adults with only natural teeth　　　　　　　*United Kingdom*

	Would prefer to manage without a partial denture	
	1988	1998
	percentage	
All	60	58
Age		
16–24	52	43
25–34	55	53
35–44	65	58
45–54	72	63
55–64	71	76
65 and over	79	78
Gender		
Men	64	60
Women	56	55
Country		
England	59	57
Wales	64	61
Scotland	65	63
Northern Ireland	68	61
English region		
North	62	56
Midlands	56	63
South	59	54
Social class of head of household		
I, II, IIINM	58	54
IIIM	63	62
IV,V	63	64
Usual reason for dental attendance		
Regular check-up	59	56
Occasional check-up	54	54
Only with trouble	65	61

For bases see Tables appendix.

1988 Tables 17.14–18

Table 6.9.28 **Problems with dentures by the length of time since becoming edentate, 1988-98**

Edentate adults   *United Kingdom*

Problems with dentures	1988	1998
	percentage with problem	
Have any of the problems listed below	41	41
Not satisfied with appearance	10	12
Have trouble speaking clearly	7	7
Have trouble eating or drinking	21	26
Have other problems	18	13
	percentage planning to visit the dentist	
Planning to visit the dentist	11	13
Base	*935*	*923*

1988 Table 15.7

Part 7

Reports by country

7.1 Dental Health in England

Summary

- The proportion of adults in England with no natural teeth (the edentate) reduced from 28% in 1978 to 20% in 1988 and to 12% in 1998.

- The number and condition of teeth of dentate adults in England has changed over the past two decades, although many of the changes between 1988 and 1998 were not significant. In 1998 older adults accounted for a higher proportion of the dentate population than in previous decades and have affected the measures of oral health related to age. For example, the average number of teeth increased from 23.3 in 1978 to 24.4 in 1988, but did not change significantly between 1988 and 1998.

- There were improvements in the number and condition of teeth of dentate adults within different age groups. For example, in 1978 dentate adults aged 35 to 44 years had, on average, 23.2 teeth of which 12.4 were sound and untreated. In 1988 they had, on average, 25.5 teeth and 13.3 sound and untreated teeth and by 1998, they had 26.9 teeth and 16.1 were sound and untreated.

- The average number of decayed or unsound teeth decreased from 1.9 in 1978 to 1.0 in 1988, but did not change between 1988 and 1998. There were also no significant changes in the average number of decayed or unsound teeth in any of the age groups between 1988 and 1998.

- In 1994 the Department of Health published 'An Oral Health Strategy for England' which set out a number of targets relating to the oral health of adults to be met by 1998. Data from the 1998 Adult Dental Health survey indicates that these targets have been met.

- In 1998 there were regional variations both in the proportion of edentate adults and in the condition of teeth among dentate adults. For example, almost two-thirds (65%) of dentate adults in the North had at least one decayed or unsound tooth compared with just over half of those living in the Midlands (52%) and the South (51%).

- Dentate adults from the most deprived areas (as measured by the Jarman underprivileged area score, see Section 7.1.1) had the lowest average number of restored (otherwise sound) teeth (6.5) whereas those from the least deprived areas had the highest (8.6).

- The proportion of dentate adults who went to the dentist for a regular check-up has increased from 44% in 1978 to 50% in 1988 and to 60% in 1998. Those groups least likely to go for a regular check-up are dentate adults aged 16 to 24 (49%) or aged 25 to 34 (54%); men (53%); those from unskilled manual backgrounds (50%); and those living in the most deprived areas (51% from Jarman Area 1 and 52% from Area 2).

- In 1998 nearly two-thirds (61%) of dentate adults had neither fillings nor extractions at their most recent visit to the dentist compared with a third (32%) in 1978 and half in 1988.

- Between 1988 and 1998 there was a three-fold increase in the proportion of dentate adults who reported having private treatment (from 6% in 1988 to 19% in 1998). In 1988 there was little regional variation in the proportion of dentate adults reporting to have treatment as a private patient. However, in 1998 dentate adults living in the South were most likely to report having been a private patient; 24% compared with 18% in the Midlands and 9% in the North.

- The costs of private dental treatment were higher than for NHS treatment: 4% of those who had their treatment under the NHS paid over £100 for their treatment compared with 16% who had private treatment.

- Dentate adults living in the most deprived areas (Jarman Area 1) were least likely to have paid for their most recent dental treatment (48%) and were most likely to have thought they were entitled to free NHS treatment (37%).

- The proportion of dentate adults who reported cleaning their teeth twice a day or more increased from 67% in 1988 to 74% in 1998. There was also a substantial increase in the proportion of dentate adults who reported using dental hygiene products other than toothpaste and toothbrush, such as dental floss. In 1978 under a quarter (22%) used such products compared with over a half (52%) in 1998.

7.1.1 Introduction

The United Kingdom comprises four separate countries, each with their own government health department setting individual health strategies including strategies relating to oral health. The four chapters in this part of the report focus on the dental health of adults within each of these four countries. Each chapter presents only selected information, giving an overview of the oral health and dental behaviour of adults and how it has changed over the period covered by the series of Adult Dental Health Surveys. The chapters also present information on specific issues of interest in the individual countries, such as targets set for oral health relating to that country.

England, together with Wales, has formed part of the series of Adult Dental Health surveys since their inception in 1968. Combined data for England and Wales only are available from the earliest survey in 1968. The first separate data for England are available from 1978 and changes in the oral health and dental behaviour for adults in England since then until 1998 are presented in this chapter.

England is the largest country within the United Kingdom, both in terms of geography and population, so the results for England tend to be very similar to those for the UK as a whole.

The larger sample size in England make some regional analyses possible. In the previous chapters of the report, England was divided into three large areas: the North; Midlands; and South, and are referred to as 'ADH Regions' in this chapter. In 1999, following the survey fieldwork, the National Health Service in England was divided into eight Regional Health Offices (RHOs). Limited analyses of the data for these RHOs was possible, but was severely restricted by the sample sizes with each of these areas, and are therefore subject to larger margins of error (see Appendix F). Wherever possible, data for individual RHOs as well as the ADH regions are presented in this chapter.

In 1994, the Department of Health published *An Oral Health Strategy for England*. This document set out a number of targets relating to adult dental health to be met by 1998. Some of these targets have already been referred to in the relevant sections of the report (for example, those relating to total tooth loss were given in Chapter 2.1). However, for completeness, section 7.1.8 shows estimates based on the 1998 survey data for each of the targets for England.

7.1.2 Tooth loss in England, 1978-98

In 1998, 12% of adults living in England had lost all their natural teeth. This showed a continued improvement in the retention of natural teeth since 1978; the proportion of adults with no natural teeth (the edentate) reduced from 28% in 1978 to 20% in 1988 and to 12% in 1998.

By 1998 there were very few adults (less than one percent) under the age of 45 who had lost all their natural teeth. In all other age groups (adults aged 45 years and over) there were considerable reductions in the proportion of adults who had lost their natural teeth between 1978 and 1998.

The proportion of edentate adults had fallen over the past twenty years for both men and women, and across all social classes and regions. It also appeared that the difference in total tooth loss between men and women had narrowed. In 1978 there was an eight percentage point difference in the proportion of edentate men and women. By 1998, the difference was only two percentage points.

Between 1978 and 1998 all social classes had shown improvements in the retention of natural teeth, however in 1998 there were still differences between the social classes. In 1998, 7% of adults from non-manual backgrounds had lost all their natural teeth compared with 14% of adults from skilled manual backgrounds, the same proportion as observed among those from non-manual backgrounds in 1988. In 1998, 21% of adults from unskilled manual backgrounds were edentate, a level similar to that seen among those from non-manual backgrounds in 1978.

Regional variations in the proportion of edentate adults were evident in 1978, 1988 and 1998, with lower proportions among those living in the South of England compared with those in the North or the Midlands. In 1998, 10% of adults in the South had lost all their natural teeth compared with 14% in the North and 15% in the Midlands. The results for the South of England appear to be consistently about 10 years ahead of the rest of England. In 1978, 23% of those from the South were edentate, a level not reached by people in the Midlands until 1988, and ahead of the point reached by the North of England in 1988 (27% edentate). In 1988, 16% of the population of the South of England had no natural teeth, very similar to the Midlands (15%) and the North of England (14%) ten years later in 1998.

Figure 7.1.1, Table 7.1.1

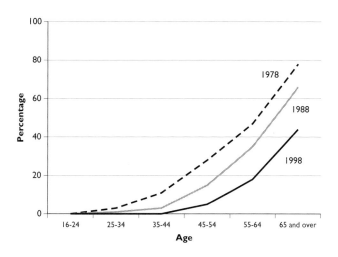

Fig 7.1.1 **Proportion of edentate adults in England by age, 1978-1998**

deprivation in the area where they lived. Sixteen per cent of adults living in the most deprived areas (Jarman Area 1) had lost all their natural teeth compared with 10% those from the least deprived areas (Jarman Areas 4 and 5).

Table 7.1.2

7.1.3 The number of natural teeth among dentate adults in England, 1978-98

The number of teeth retained in the mouth is an important aspect of oral health. People with a small number of teeth are at risk of becoming edentate in the future. Moreover, a minimum number of teeth are considered necessary to have an adequate oral function without the need for partial dentures (see Chapter 2.2). In 1998 dentate adults in England had, on average, 24.9 teeth. When the 1978, 1988 and 1998 data are compared, they show that the proportion of dentate adults with 21 or more teeth increased from 74% in 1978 to 81% 1988, and then did not change significantly between 1988 and 1998 (83% in 1998).

The age structure of the dentate population has changed since 1978. Adults are keeping their teeth for longer and the underlying population structure has changed due to longer life expectancy and lower birth rates. In 1998 older adults, therefore, accounted for a higher proportion of the dentate population than they did in previous decades. However, the number of natural teeth is, like many aspects of dental health, strongly associated with age, with older people having on average fewer teeth. The change in age structure of the dentate population has resulted in the therefore overall increase in the average number of teeth (from 23.3 teeth in 1978 to 24.9 teeth in 1998) being smaller than it would have been had the age structure remained the same. (Chapter 5.2 gives more details on the changes in age structure of the dentate population).

Although the dentate population of England had, on average, a similar number of teeth as in 1988 (24.9 and 24.4 respectively), there were changes among individual age groups over this period, particularly among those aged 35 years and over. Dentate adults aged 35 to 44 years had, on average, 25.5 teeth in 1988 and 26.9 teeth in 1998; an increase of 1.4. Those aged 55 years and over showed an increase of almost two in the average number of teeth over the same period, from 16.9 to 18.8.

Figure 7.1.2, Tables 7.1.3-4

The regional variations in the prevalence of total tooth loss are even more pronounced when the data are analysed by the individual Regional Health Offices (RHOs). There was a 13 percentage point difference between the Offices with the highest and lowest proportions of edentate adults; ranging from 9% adults from the London RHO to 21% of those in the Trent RHO.

There is general concern about health inequalities between people from different sections of society, including inequalities in oral health. Part 4 showed, in detail, the variation in several clinical measures of oral health with respect to the social characteristics of the respondents, such as social class. Another method of investigating possible inequalities is to classify respondents according to the relative living standards of the area they lived in (sometimes known as geographical deprivation), rather than their individual circumstances. There are various measures of geographical deprivation available, since none of them cover the United Kingdom as a whole they were not included in the analysis in Part 4.

In 1998 the data for England have been analysed using a measure of geographical deprivation known as the Jarman underprivileged area score (referred to as 'Jarman Area'). The respondents from the survey were grouped according to the relative living standards of the ward they lived in with Area 1 being the most deprived area and Area 5 the least deprived. Appendix F gives further details on the Jarman unprivileged area score.

The prevalence of total tooth loss was analysed by Jarman areas and showed that the proportion of adults who had lost all their natural teeth varied according to the level of

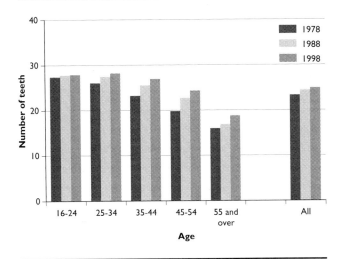

Fig 7.1.2   **Mean number of teeth among dentate adults in England by age, 1978-1998**

In 1998, dentate men and women had similar numbers of teeth (25.1 and 24.7 respectively). Dentate adults from non-manual backgrounds had, on average, more teeth than those from other backgrounds; an average of 25.5 teeth compared with 24.3 for those from skilled manual households and 23.8 for those from unskilled manual households. There was very little change in the average number of teeth between 1988 and 1998 with respect to gender or social class of head of household.

Tables 7.1.5-6

The average number of teeth among dentate adults who went to the dentist only when they had trouble increased from 22.7 in 1988 to 24.3 in 1998. In contrast, the average number of teeth among regular attenders did not change over this period (25.0 teeth). The difference in the average number of teeth between these groups which had narrowed slightly between 1978 and 1988, narrowed even further between 1988 and 1998.

Table 7.1.7

Table 7.1.8 shows that in 1998 there was little or no significant variation in the average number of teeth among dentate adults living in different parts of the country or with respect to the relative deprivation of the area (as measured by Jarman).

Table 7.1.8

7.1.4 Reliance on natural teeth and dentures in England, 1978-98

Not only are more adults keeping some of their natural teeth than previously but as people get older they are also keeping more of their teeth. Coinciding with this improvement in the

retention of teeth, the proportion of adults relying on dentures, either complete or partial, has decreased and the proportion of adults with natural teeth only has increased. In 1978 only half (51%) of adults had their natural teeth only and did not have dentures compared with almost three-quarters (73%) in 1998. In 1998, 15% had natural teeth and dentures and 12% had lost all their natural teeth and therefore required complete dentures.

Since 1978 in the North and Midlands, as the proportion of edentate adults reduced and the proportion of adults with only natural teeth increased, the proportion with natural teeth and dentures has remained at a fairly constant level (17% to 19%). Only the South of England has shown a net reduction in the proportion of adults with natural teeth and dentures from 23% in 1978 to 13% in 1998.

Table 7.1.9

The variations in the proportions of adults who were wholly reliant on their own teeth reflected the variations in tooth loss, in those who had lost all their natural teeth and those who were still dentate but had less than 21 teeth. Chapters 2.2 and 2.3 showed that, in general, the dentate adults who were least likely to have dentures were also those who were most likely to have 21 or more teeth. These patterns can also be seen among adults living in England. For example, the proportion of adults who were wholly reliant on their own teeth and did not have any dentures varied with age, from almost all adults aged 16 to 24 years (99%) to only a quarter (24%) of those aged 65 and over.

Table 7.1.2

7.1.5 The condition of the natural teeth based on the criteria used in 1998

For the 1998 Adult Dental Health survey, changes were made to the criteria used in the earlier surveys for assessing dental caries. In 1998, teeth with untreated visual caries were classified as decayed, whereas previously they were recorded as sound and untreated. Restored teeth with recurrent visual caries were also treated as decayed in the 1998 classification, but were defined as teeth with restored (otherwise sound) by pre-1998 criteria (see Chapter 3.1). Analyses using both the 1998 and the pre-1998 criteria are shown in this chapter. The 1998 criteria are used in this section to give as full a picture as possible of the level of decay among dentate adults in England in 1998. However, in order to compare accurately the condition of teeth in 1998 with those from 1978 and 1988, the 1998 data had to be re-classified according to the pre-1998 criteria and those trend data are discussed in sections 7.1.6 and 7.1.7.

In 1998 dentate adults in England had, on average, 15.6 sound and untreated teeth, 1.5 decayed or unsound teeth, and 7.8 restored (otherwise sound) teeth.

Table 7.1.10 shows how the distributions of these three different classifications of teeth varied by age. The youngest dentate adults, aged between 16 and 24 years, were the most likely to have 18 or more sound and untreated teeth (92%) compared with 68% of 25 to 34 year olds, for example. Thirty-three per cent of dentate 16 to 24 year olds had no restored (otherwise sound) teeth compared with only 4% those aged 25 to 34 years. Findings from the surveys of Children's Dental Health[1] show a similar cohort effect; in 1993, 48% of 15 years olds (who would be 20 years old in 1998) had no restored (otherwise sound) teeth, compared with 15% of 15 year olds in 1983 (who would be 30 years old in 1998). While these figures are for the United Kingdom as a whole (as there are no separate trend data available for England from the 1993 survey of Children's Dental Health survey[1]) it is likely, for reasons previously described, that the trends for England would be similar to that of the United Kingdom.

Among dentate adults in the middle and older age groups the average number of sound and untreated teeth reduced with age, from 15.9 among those aged 35 to 44 years to 8.6 among those aged 65 years and over. Dentate adults aged 45 to 54 years had the highest average number of restored (otherwise sound) teeth (10.9). The older age groups had fewer restored (otherwise sound) teeth because they tended to have fewer teeth overall. Also, dentate adults in the two oldest age groups were more likely to have no restored (otherwise sound) teeth. Two per cent of dentate adults aged 35 to 44 years and 45 to 54 years had no restored (otherwise sound) teeth compared with 8% of dentate adults aged 55 to 64 years and 12% of dentate adults aged 65 and over.

There was less variation with respect to age in the proportions of decayed or unsound teeth than with the number of sound and untreated teeth or restored (otherwise sound) teeth. Overall, 56% of dentate adults in England had at least one decayed or unsound tooth; in each age group at least half of the dentate adults had one or more decayed or unsound teeth and the 25 to 34 age group had the highest proportion, 61%.

Table 7.1.10

On average, dentate men had a higher number of decayed or unsound teeth (1.7) than dentate women (1.2). Dentate men also had, on average, fewer restored (otherwise sound) teeth; 7.3 compared with 8.3 for women. These differences may, in part, reflect gender differences in the attendance at the dentist.

Section 7.1.10 gives more details on variations in the usual reason for dental attendance in England.

Table 7.1.11

Tables 7.1.12 and 7.1.13 show the condition of natural teeth by ADH region and RHO. There were no significant regional differences in the number of sound and untreated teeth; however there were differences in the numbers of decayed or unsound teeth and restored (otherwise sound) teeth. Almost two-thirds (65%) of dentate adults in the North of England had at least one decayed or unsound tooth compared with just over half of dentate adults from the Midlands (52%) or South (51%). The North ADH region is, in the main, made up of two RHOs: Northern and Yorkshire, and North West. These two RHOs had the highest proportions of dentate adults with decayed or unsound teeth (65% and 64% respectively). In comparison, the South West had the lowest proportion of dentate adults with decayed or unsound teeth (41%). Dentate adults from the North of England also had, on average, a lower number of restored (otherwise sound) teeth (7.3 compared with 8.2 in the South). The average number of restored (otherwise sound) teeth also varied by RHO but the picture was more complex. Although the Northern and Yorkshire RHO had the lowest average (6.9), the North West had the same average as England overall (7.8), and although the South East had the highest average (8.8), London had a relatively low average (7.1).

Tables 7.1.12-13

There was little significant difference in the number of sound and untreated teeth with respect to social class of head of household; however there were differences in the numbers of decayed or unsound teeth and restored (otherwise sound) teeth. Dentate adults from non-manual backgrounds were less likely to have decayed or unsound teeth than those from partly skilled or unskilled backgrounds. For example, 50% of dentate adults from non-manual households had decayed or unsound teeth compared with 63% of those from unskilled manual households. Dentate adults from non-manual households were almost twice as likely as those from unskilled manual households to have 12 or more restored (otherwise sound) teeth (31% and 16% respectively).

Table 7.1.14

The condition of natural teeth of dentate adults also varied by Jarman Area. The proportion of dentate adults with 18 or more sound and untreated teeth varied from 52% in the most deprived wards (Area 1) to 36% of dentate adults from the least deprived wards (Area 5). The most deprived wards also have the highest prevalence of dentate adults with decayed or unsound teeth (69%) and the lowest average number of restored (otherwise sound) teeth (6.5). In comparison, the least deprived wards had the lowest proportion of dentate adults with decayed or unsound teeth (47%) and the highest average number of restored (otherwise sound) teeth (8.6).

Figure 7.1.3, Table 7.1.15

Fig 7.1.3

Fig 7.1.3 **The condition of teeth among dentate adults in England by Jarman area**

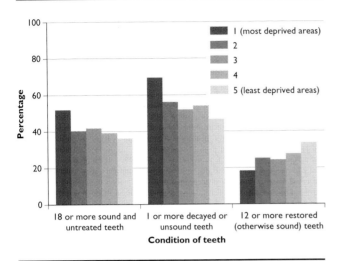

The condition of teeth of dentate adults varied according to the usual reason for their attendance at the dentist and may reflect the fact that those who go to the dentist regularly have more opportunities to have their teeth treated by a dentist than those who go less often. Regular dental attenders are likely to have had some restorative treatment for disease at an earlier stage than that identified by the survey examination (see Chapter 3.1). This leads to them having fewer teeth that can be classified as sound and untreated. Dentate adults who went to the dentist for a regular check-up had, on average, 9.0 restored (otherwise sound) teeth and 14.8 sound and untreated teeth. In comparison, those who went to the dentist only with trouble had, on average, 5.5 restored (otherwise sound) teeth and 16.4 sound and untreated teeth. Dentate adults who went to the dentist only with trouble also had a higher average number of decayed or unsound teeth (2.3) than those who went to the dentist for a check-up, either regularly (1.1) or occasionally (1.4).

Table 7.1.16

7.1.6 The effect of changing the criteria for measuring decay

As mentioned above, changes were made to the criteria used in the earlier surveys for assessing dental caries. In 1998, teeth with untreated visual caries were classified as decayed, whereas previously they were recorded as sound and untreated. Restored teeth with recurrent visual caries were

also treated as decayed in the 1998 classification, but were defined as restored (otherwise sound) by pre-1998 criteria (see Chapter 3.1). Table 7.1.17 shows how the new criteria affect estimates of the average number of teeth that were sound and untreated, decayed or unsound, and restored (otherwise sound).

The overall effect of using the new criteria is to:

- provide a more conservative estimate of the average number of sound and untreated teeth; 15.6 compared with 15.9 using pre-1998 criteria
- increase the estimate of the average number of decayed or unsound teeth from 1.0 to 1.5
- decrease the estimate of the average number of restored (otherwise sound) teeth from 8.0 to 7.8

However the size of the difference varied within specific groups of the population. For example, younger adults are more likely than older people to have visual caries only and the average number of teeth classified as decayed or unsound increased from 0.9 among 16 to 24 year olds using pre-1998 criteria to 1.6 based on the new criteria. In contrast, the new criteria had less effect on the average number of decayed or unsound teeth found in those aged 65 years and over; this figure increased from 1.0 to 1.3.

Table 7.1.17 also shows that dentate adults who went to the dentist only when they had trouble not only had more decayed or unsound teeth than those who went for a regular check-up, but also that they had more teeth with cavitated caries. Using the 1998 criteria dentate adults who went only with trouble had an average of 2.3 decayed or unsound teeth compared with 1.1 for those who went for a regular check-up. If the pre-1998 criteria are used, which included only cavitated caries, the average numbers of decayed or unsound teeth for those who went only with trouble and those who went regularly were 1.6 and 0.7 respectively.

Table 7.1.17

7.1.7 Trends in the condition of natural teeth

Tables 7.1.18 to 7.1.22 show the 1998 data on the condition of natural teeth classified according to the pre-1998 criteria together with the corresponding data for 1978 and 1988.

The average number of sound and untreated teeth increased between 1988 and 1998, although the change was not as large as had been observed between 1978 and 1988. In 1998, dentate adults in England had on average 15.9 sound and untreated teeth compared with 15.0 in 1988 and 13.2 in 1978.

The proportion of dentate adults who had 18 or more sound and untreated teeth also increased between 1988 and 1998 by six percentage points, from 37% to 43%. However, the rate of increase had slowed because between 1978 and 1988 there had been an increase of ten percentage points.

Both the average number of decayed or unsound teeth and the proportion of dentate adults who had decayed or unsound teeth decreased between 1978 and 1988 but did not change between 1988 and 1998. In 1978, three-fifths (60%) of dentate adults had at least one decayed or unsound tooth compared with just over two-fifths in 1988 (43%) and 1998 (43%).

The average number of restored (otherwise sound) teeth has remained at a similar level over the past twenty years (8.1 in 1978, 8.4 in 1988 and 8.0 in 1998). One slight change in the pattern of restorative treatment is that the proportion of dentate adults who had no restored (otherwise sound) teeth decreased from 14% in 1978 to 10% in 1988 and remained around this level in 1998 (9%). This was particularly due to an increase in the number of restored otherwise sound teeth among the older age groups (see below and Chapter 5.3).

Table 7.1.18

As the age structure of the dentate population has changed between 1978 and 1998 it is therefore important to look at the changes in condition of the natural teeth for each age group individually. If the data from 1978, 1988 and 1998 for the three youngest age groups (those aged 16 to 24, 25 to 34 and 35 to 44 years) are compared, they show that the prevalence of dental disease, and therefore the need for restorative treatment, has decreased over the past twenty years. In 1978, 53% of dentate adults aged 16 to 24 years had 18 or more sound and untreated teeth compared with 84% in 1988 and 94% in 1998. Among dentate adults aged 35 to 44 years, the proportion of dentate adults with 18 or more sound and untreated teeth almost doubled over the past decade, from 22% in 1988 to 42% in 1998. Also, the average number of restored (otherwise sound) teeth among dentate adults in these three age groups decreased over the past ten years. For example, in 1988 dentate adults aged 25 to 34 years had, on average, 9.9 restored (otherwise sound) teeth; this decreased to 7.1 in 1998.

The changes in the condition of teeth of older dentate adults show a different pattern. In 1998, dentate adults in these age groups were keeping more of their teeth than in previous decades, but were more likely to have them restored. Between 1978 and 1998, the average numbers of teeth and restored (otherwise sound) teeth among dentate adults aged 45 to 54 years and 55 years and over increased. However, the average numbers of sound and untreated teeth in these groups were

either similar to what they had been in 1978 (those aged 55 and over) or had shown a smaller increase. For example, in 1978 dentate adults aged 45 to 54 years had an average of 19.8 teeth of which 10.6 were sound and untreated and 7.2 were restored (otherwise sound). In 1998, dentate adults in this age group had 24.3 teeth; 12.1 were sound and untreated, and 11.2 were restored (otherwise sound).

As well as there being no overall change in the average number of decayed or unsound teeth between 1988 and 1998, there were also no significant changes in the average number of decayed or unsound teeth in any of the age groups during this time.

Figures 7.1.4-6, Table 7.1.19

Fig 7.1.4 **Proportion of dentate adults in England with 18 or more sound and untreated teeth by age, 1978-98**

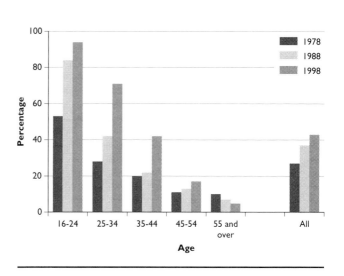

Fig 7.1.5 **Proportion of dentate adults in England with some decayed or unsound teeth by age, 1978-98**

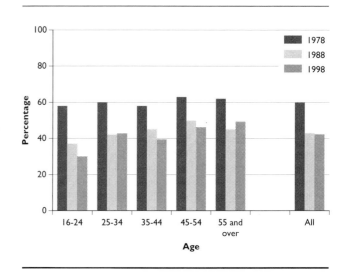

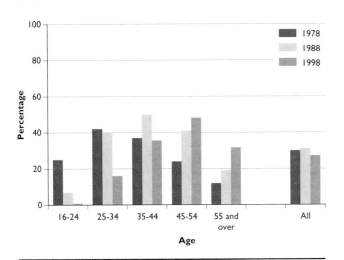

Fig 7.1.6   Proportion of dentate adults in England with 12 or more restored (otherwise sound) teeth by age, 1978-98

When the condition of teeth is analysed by gender and social class of head of household, there are differences in the number of sound and untreated teeth but there were few significant changes in the number of decayed or unsound teeth or restored (otherwise sound) teeth. Dentate men and women showed similar rates of change in the average numbers of sound and untreated teeth between 1988 and 1998. Over this period the average number of sound and untreated teeth increased by 0.9, from 15.5 to 16.4 for men and 14.5 to 15.4 for women. In 1978 and 1988 there were large differences in the proportion of dentate adults with 18 or more sound and untreated teeth with respect to social class of head of household but by 1998 these differences had disappeared. For example, in 1988, 35% of those from non-manual households had 18 or more sound and untreated teeth compared with 44% of those from unskilled households. In 1998 the proportion of those from non-manual backgrounds with 18 or more sound and untreated teeth increased to 42% whereas among those from unskilled households it remained at the 1988 level (44%).

Tables 7.1.20-21

The changes in the condition of teeth between 1978 and 1998 were different for dentate adults who went to the dentist for regular check-ups and those who went only with trouble. Dentate adults who went to the dentist for a regular check-up had, on average, a similar number of teeth in 1978 as in 1998 (24.6 in 1978 and 25.0 in 1998). Over this same period, the average number of restored (otherwise sound) teeth decreased (from 11.3 in 1978 to 9.2 in 1998) and the average number of sound and untreated teeth increased (from 12.2 in 1978 to 15.0 in 1998). In contrast the average number of teeth among dentate adults who went to the dentist only with trouble increased from 21.4 in 1978 to 24.3 in 1998. Between

1978 and 1988 the average number of restored (otherwise sound) teeth increased (from 4.4 to 5.9) but did not change significantly between 1988 and 1998 (5.7) and was still considerably lower than for regular attenders (9.2). Like regular attenders, those who went only with trouble showed an increase in the average number of sound and untreated teeth between 1978 and 1998 (from 14.1 to 16.9) and continued to have a higher average number of sound and untreated teeth than regular attenders. The average number of decayed or unsound teeth among both regular attenders and those who went only with trouble decreased between 1978 and 1998 but did not change significantly between 1988 and 1998. Over this twenty-year period, those who went to the dentist only with trouble had, on average, more decayed and unsound teeth than regular attenders.

Table 7.1.22

7.1.8 English oral health targets

In 1994 the Department of Health publication *An oral health strategy for England* set out a number of targets relating to the oral health of adults to be met by 1998:

● 33% of adults aged over 75 years old should have some natural teeth;

● 10% of adults aged over 75 years should have more than 20 teeth;

● 75% of 50 year olds should have more than 20 natural teeth;

● 50% of 30 year olds should have more than 20 natural teeth which are sound and unfilled

● The percentage of dentate adults aged over 45 years old with at least one deep periodontal pocket (greater than 6mm) should be reduced to 10%.

Table 7.1.23 shows data from this survey that can be used to measure the achievement or progress towards each of these targets. In 1998, among adults aged 75 years and over 44% had retained some of their teeth with 10% having more than 20 teeth, indicating that these targets have been met.

Two of the targets relate specifically to single-year age groups (those aged 30 and 50 years). Estimates for these targets based on data from the survey are subject to a large margin of error due to the small sample sizes of adults of these ages. The two ten-year age groups that encompass the adults of interest (25 to 34 years and 45 to 54 years) have much larger sample sizes and are therefore subject to a much smaller margin of error. Estimates for these age groups are therefore presented in Table 7.1.23. (See Appendix F for further details on the precision of estimates from the survey).

Seventy-nine per cent of all adults aged 45 to 54 in England had 21 or more teeth, confirming that the target appears to have been met.

Achievement towards the target relating to the proportion of 30-year-olds who had more than 20 teeth which were sound and unfilled is more difficult to assess because, as explained in Section 7.1.5, in the 1998 survey two different criteria for defining a tooth as sound and untreated were used. The target was set in 1994 and was based upon data from the 1988 Adult Dental Health survey which were consistent with the pre-1998 criteria used to analyse the 1998 data. Using the pre-1998 criteria, 48% of dentate adults aged 25 to 34 years have more than 20 teeth that are sound and untreated. If the 1998 criteria are used, which classify fewer teeth as sound and untreated, this proportion reduces to 44%. Although these figures are below the target of 50%, they are estimates based on a sample of individuals rather than the full population, and are therefore subject to error (see Appendix F). There is a 95% chance that in the population as a whole the proportion of dentate adults aged 25 to 34 years with more than 20 sound and untreated teeth lies between 44% and 53% using the pre-1998 criteria, and between 40% and 49% using the 1998 criteria.

In England in 1998, 10% of dentate adults aged 45 and over had some periodontal pockets measuring 6mm or more which equals the target set.

Table 7.1.23

7.1.9 Reported dental behaviour in England

The previous sections have shown that the teeth condition of dentate adults in England varied not only with the social characteristics of the individuals but also with one aspect of their dental behaviour, their usual reason for dental attendance. Part 4 of this report discusses the inter-relationships between a person's social characteristics, their dental related behaviour and their oral health. The rest of this chapter will describe the dental behaviours of dentate adults in England as reported during the interview, how they varied with social characteristics and how they have changed since 1978.

As many of the findings for England are similar to those for the United Kingdom, this chapter only presents selected information on the different types of dental behaviour. Also, the analyses on the changes of behaviour from 1978 to 1998 have been restricted by the information available from previous reports; therefore only trend data for England as a

whole and for the ADH regions are shown in some of the subsequent sections.

7.1.10 Usual reason for dental attendance

In 1998, six out of ten dentate adults in England stated they went to the dentist for a regular check-up, while three out of ten went to the dentist only when they were experiencing trouble with their teeth. The remaining dentate adults (10%) stated they went to the dentist for an occasional check-up.

Since 1978, for England as a whole, the proportion of dentate adults who went to the dentist for a regular check-up has increased from 44% in 1978 to 50% in 1988 and to 60% in 1998. Over the same period, the proportion of dentate adults attending only with trouble and the proportion who went for an occasional check-up decreased.

Figure 7.1.7, Table 7.1.24

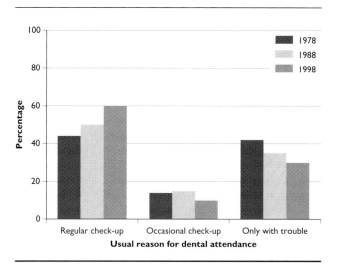

Fig 7.1.7 **Usual reason for dental attendance among dentate adults in England, 1978-98**

In 1998, there were variations by age, gender and social background in the proportion of dentate adults who went to the dentist for a regular check-up. The groups least likely to go for a regular check-up and most likely to only go with trouble were: dentate adults aged 16 to 24 years (49%) or aged 25 to 34 years (54%); men (53%); people from unskilled manual backgrounds (50%); and those living in the most deprived areas (51% in Jarman area 1 and 52% in Area 2). There was little significant variation in the usual reason for dental attendance in the different regions (either ADH regions or RHOs) of England.

Table 7.1.25

Some of the differences in the number and condition of the teeth among dentate adults with different characteristics may, to some extent, be affected by differences in the dental attendance of these groups. For example, dentate adults from unskilled manual households were more likely to have teeth that were decayed or unsound. This is consistent with the fact that they were also more likely to go to the dentist only when they have trouble with their teeth. In contrast, dentate adults from non-manual social classes were more likely to go to the dentist for a regular check-up and they had, on average, a higher number of restored (otherwise sound) teeth than other dentate adults.

Tables 7.1.14, 16, 25

7.1.11 Lifetime dental treatment

Respondents were asked whether they had received certain types of dental treatment during their life, including as a child. Table 7.1.26 shows the proportion of dentate adults in England who had experienced 12 of these dental treatments, and how these proportions have changed since 1988.

The most common treatments dentate adults had received over their lifetime were fillings (94%), an x-ray (89%), an injection for a filling (90%) and a scale and polish (88%). The proportion of dentate adults who had fillings, an injection for a filling and a scale and polish were the same or similar to those in 1988 (93%, 86% and 85% respectively). However, the proportion of dentate adults who had ever had an x-ray increased by 13 percentage points between 1988 and 1998.

In 1998 dentate adults were also more likely than in 1988 to report having had treatment from a dental hygienist at some time, the proportion increased from 25% in 1988 to 44% in 1998. Other treatments that were less common but had shown an increase since 1988 in the proportion of dentate adults who received them were: having had a brace (an increase from 16% to 19%), having an abscess treated (31% to 38%) and having a crown fitted (29% to 37%).

Extractions and having gas for extractions were the only treatments that showed a downward trend. In 1988, 86% of dentate adults reported having had an extraction at some time (for teeth other than wisdom teeth) compared with 81% reporting this in 1998. As the proportion of dentate adults having experienced extractions had decreased, the proportions of dentate adults having had an injection or gas for extractions also decreased. However, the fall in the proportion of dentate adults who had experienced gas for extractions, from 61% in 1988 to 51% in 1998, was larger than might be expected from the decrease in extractions alone. This decrease also reflects the fact there has been a general move away from using gas for extractions in recent years.

There were regional variations in the types of treatment dentate adults had received and also in the changes since 1988. For example, in 1998 the South had the highest proportion of dentate adults who had had an x-ray (92%). However, the North and the Midlands showed larger increases than the South in the proportion of dentate adults who had x-rays (from 68% in 1988 to 86% in 1998 in the North and from 66% in 1988 to 87% in 1998 in the Midlands) and were now at a level similar to that seen in the South in 1988.

Although there was a decrease in the proportion of dentate adults who had ever had extractions in all three ADH regions, there was no significant change in the proportion who had ever had gas for extractions in the North over this period (65% in 1988 and 64% in 1998). In contrast, the proportion of dentate adults who reported having had gas for extractions in the Midlands and South fell over this period. This suggests a difference in the dental practices of dentists in the North (as well as regional differences in the treatments needed by the dentate adults in both 1988 and 1998).

There were also regional variations in the proportion of dentate adults reporting having had treatment from a dental hygienist in 1988 and 1998, with those from the South most likely to have reported doing so in both 1988 (32%) and in 1998 (49%).

Table 7.1.26

7.1.12 The most recent dental visit

Adults were also asked about their most recent dental visit. They were first asked how long it had been since they last visited the dentist. In 1998, 71% of dentate adults said they had visited the dentist in the last year; which was similar to the proportion reporting doing so in 1988. In 1988, dentate adults in the South had been more likely than those in the North or Midlands to have visited the dentist in the last year whereas in 1998 there were no significant regional differences in the time since the most recent dental visit.

Table 7.1.27

Dentate adults were also asked about the course of treatment they had received at their most recent visit to the dentist, regardless of how long ago their last visit to the dentist had been. As 95% of dentate adults had attended the dentist in the last ten years, this provides a more direct picture of changes in treatment over the last ten and twenty years than the lifetime treatment comparisons discussed in the previous sections.

Most courses of treatment start with an examination and this part of the treatment is not identified or discussed separately. Some respondents reported that they had more than one type of treatment. The types of treatment received have therefore been summarised depending on whether a filling or an extraction or both took place. The majority of dentate adults, 61%, had neither fillings nor extractions at their last visit to the dentist. Dentate adults who had neither extractions nor fillings were mainly those who had an examination or a scale and polish, although a small proportion had more complex treatment such as an abscess treated, a partial denture made or a crown fitted. (The proportions of dentate adults who received specific treatments at their most recent visit are shown in table 7.1.29.)

The types of treatment dentate adults had received at the most recent visit have changed over the past twenty years. Overall, there has been a decrease in the proportion of dentate adults who have had their teeth filled or extracted. The proportion of dentate adults who had their teeth filled at their last visit, decreased from 46% in 1978 to 38% in 1988 and 29% in 1998. The proportion of dentate adults who had some teeth extracted almost halved between 1978 and 1988, from 30% to 17%, and there was a further reduction, although a much smaller one, to 13% between 1988 and 1998. There has also been an increase in the proportion of dentate adults who received a scale and polish, from 52% in 1978 and 54% in 1988 to 59% in 1998.

There were few regional differences in the treatment that dentate adults had received at their most recent visit. Between 1978 and 1998 there had been an increase in the proportion of dentate adults in the Midlands and the North who had an x-ray taken, but the proportion in the South had remained at a similar level over this period. In 1978 there were regional differences in the proportion of dentate adults who had some teeth extracted at their most recent visit, 47% in the North, 37% in the Midlands and 25% in the South. This proportion had decreased in all three regions so that by 1998 there was little difference between the regions (12% to 15%).

Tables 7.1.28-29

7.1.13 Type of dental services used in England

Since the previous Adult Dental Health Survey there have been changes in the payment system for dentists within the NHS and also in the charges to patients. As was the case in the last survey, people who are not exempt from some or all patient charges pay a proportion of their NHS dental treatment costs (see Chapter 6.3). This requirement may confuse some adults, as it is possible that some consider this as being the

same as having "private" treatment. Similarly some paying adult patients may think they are getting NHS treatment when some or all of their course of treatment was private. Notwithstanding this all adults were asked whether the treatment they received at their most recent visit was "under the NHS, was it private, or was it something else?".

In England in 1998, three-quarters (76%) of dentate adults reported that their most recent course of treatment was through the NHS and almost a fifth (19%) reported being treated as a private patient. The remainder (5%) reported that they received their treatment through some other type of dental service, including a combination of NHS and private care, the Community Dental Services, through insurance, or a dental plan.

Dentate adults aged 16 to 24 years were more likely than any other age group to have had their treatment through the NHS rather than privately; 85% of 16 to 24 year olds had their treatment through the NHS compared with between 73% and 76% of the other age groups. This reflects in part the fact that a high proportion of the youngest age group would have been exempt from NHS charges.

There were also marked regional differences in the proportion of dentate adults who had NHS and private treatment. Dentate adults living in the South were least likely to report having received their treatment under the NHS and most likely to have been a private patient; 24% of those living in the South reported they had their treatment done privately compared with 18% in the Midlands and 9% in the North. The same pattern was also observed in the individual RHOs. Dentate adults from the more northern RHOs (Northern and Yorkshire, North West and Trent) had the lowest proportions of dentate adults reporting they were treated as private patients (between 8% and 11%) compared with the South East and South West RHOs which had the highest reported proportions having private treatment (26% and 28% respectively).

The proportion of dentate adults who reported having had their treatment through the NHS or privately also varied by social class of head of household and by Jarman area. Dentate adults from non-manual backgrounds were less likely to report their most recent treatment as being under the NHS (73%) than those from manual backgrounds (80%). Dentate adults from the more deprived areas (Areas 1 to 3) were more likely to report they had NHS treatment than those from the least deprived areas (78%, 83% and 78% compared with 69% in Area 4 and 70% in Area 5). This may reflect that many people living in the more deprived areas would not be able to afford private treatment and also may be more likely to be exempt from some or all of the patient charges of NHS treatment.

Table 7.1.30

As mentioned earlier, there have been considerable changes to the NHS dental service since the last survey was carried out in 1988. When the data from the 1988 survey are compared with those from the most recent survey, there has been a three-fold increase in the proportion of dentate adults reporting that their treatment was carried out privately; from 6% in 1988 to 19% in 1998. In 1988, there was little regional variation in the proportion of dentate adults having private treatment; however, by 1998 those in the South were over twice as likely as those in the North to have reported that their most recent treatment was carried out privately.

Table 7.1.31

Reports from the previous Adult Dental Health surveys asserted that much of the private dentistry at that time was to treat 'one-off' emergencies or for more intricate dental treatments. Table 7.1.32 shows the types of treatments received by dentate adults according to whether they had the treatment under the NHS or privately. The data suggest that in 1998 this no longer seems to be the case and the private dentistry is being used to provide comprehensive dental care. The only significant difference was that dentate adults were more likely to have had a scale and polish if they had their treatment done privately than if they had their treatment under the NHS (65% and 57% respectively).

Table 7.1.32

7.1.14 Costs of dental treatment

All dentate adults who had been to the dentist were asked about the costs of their most recent course of treatment. It should also be noted that there is potential for some confusion over what was reported as being the cost of the treatment (see Chapter 6.3 for more details). The analysis of the costs of treatment is restricted to those whose most recent course of treatment was within the previous two years.

Table 7.1.33 shows whether, at their most recent visit, dentate adults had paid anything towards their treatment; 62% said they had paid something, 29% said they had not paid anything and the remaining 9% were unable to give an answer or did not know the costs. The proportion of dentate adults in England who had paid something towards the costs of their treatment had increased since 1988 (from 50%), probably reflecting both an increase in private dental care and changes in the charges of NHS treatment. For example, at the time of the 1988 survey check-ups under the NHS were free for patients, whereas now people generally have to pay for them.

A similar pattern of change between 1988 and 1998 was evident in all three ADH regions.

Table 7.1.33

As might be expected, the proportion of dentate adults who had paid something towards their treatment was higher among those who had their treatment carried out privately (85%) compared with those who were treated under the NHS (57%). Dentate adults living in the South were the most likely to have paid for their dental treatment; 65% reported they had done so compared with 56% of those from the North. These differences were also reflected in the proportion of dentate adults who paid for their treatment within the individual RHOs. Whether or not dentate adults paid anything for their treatment varied according to the social class of head of household and Jarman Area. The proportion of dentate adults who said they had paid something towards their treatment decreased from 71% among dentate adults living in the least deprived areas (Area 5) to 48% in the most deprived areas (Area 1). These figures may in part reflect the higher proportion of people having private dental care in the least deprived areas and the higher proportions of adults likely to be exempt from NHS charges in the more deprived areas.

Table 7.1.34

This survey could not assess respondents' eligibility for exemption from some or all of the charges for NHS dental treatment, however dentate adults were asked whether, if they visited a dentist in the next four weeks, they thought they would be entitled to free NHS treatment. Overall, 19% of dentate adults in England thought they would be entitled to free NHS treatment. This information was based on the respondent's own understanding of the eligibility criteria, and therefore some people may have been mistaken in determining their eligibility. However, the variations in the proportion of dentate adults who thought they would be entitled to free treatment reflected both the eligibility rules and the variations in the proportions of those who had not paid towards their treatment at their most recent visit. For example, dentate adults who lived in the most deprived areas were more likely to have said they were entitled to free treatment than dentate adults living in other areas; 37% of dentate adults living in the most deprived areas (Area 1) thought they were entitled to free treatment compared with between 13% and 20% in other areas.

Table 7.1.35

Table 7.1.36 provides details of the reported costs of treatment for dentate adults who had visited the dentist in the previous two years (excluding those who did not know or were unable to give an estimate of the cost of their treatment). There were there were large differences in the costs of NHS and private treatment; for example only 4% per cent those who reported having NHS treatment paid over £100 for their treatment compared with 16% who reported paying privately. Nearly a third (31%) of dentate adults who had their most recent treatment under the NHS did not pay anything. Eight per cent of those who reported that their most recent treatment was carried out privately also said this, although some of these people may not have been considering the money they had paid into a dental plan or health insurance. There were variations in costs with respect to other characteristics of the respondents. Some of the variations observed will have been due to the differences in the type of service used by these groups and also that the eligibility for free NHS treatment would be different among these groups.

Table 7.1.36

7.1.15 Reported dental hygiene

All dentate adults were asked about how often they clean their teeth and whether they used any dental hygiene products other than toothpaste and toothbrush. This does not provide direct information about how effectively people clean their teeth, but gives some indication of their motivation towards dental hygiene.

Overall, 74% of dentate adults in England stated they cleaned their teeth twice a day or more. A further 22% of dentate adults cleaned their teeth once a day and 4% less than once a day. Some dentate adults said they cleaned their teeth more frequently than others. The dentate adults most likely to state they clean their teeth twice a day or more were: women (84%); those from non-manual backgrounds (78%); those who went for check-ups either regularly (80%) or occasionally (79%); and those in the least deprived areas (78% in Jarman Area 4 and 77% in Jarman Area 5).

Table 7.1.37

Since 1988, the proportion of dentate adults who stated they cleaned their teeth twice a day or more has increased in the three ADH regions. In England this proportion increased from 67% in 1988 to 74% in 1998.

Table 7.1.38

Dentate adults who cleaned their teeth were also asked whether they used any products other than an ordinary toothpaste and toothbrush (for example, dental floss or mouthwash) for dental hygiene purposes. In 1998, over half (52%) of dentate adults used these sorts of products; 28% of dental adults reported using dental floss and 23% reported using mouthwash. Other individual products were mentioned by only 5% or less of dentate respondents, and are not shown separately (see Chapter 6.8 for details).

The groups who were most likely to report cleaning their teeth twice a day or more were also among those groups saying that they used other dental hygiene products. These groups were women, those from non-manual backgrounds and those who went to the dentist for check-ups, either regularly or occasionally. In addition, dentate adults aged between 25 and 54 years were more likely to report using other dental hygiene products than the youngest or oldest adults. For example, 36% of those aged 35 to 44 years said they used dental floss compared with 18% of those aged 16 to 24 years and 15% of those aged 65 and over. Those living in the South were also more likely than those from the North or the Midlands to report using other dental hygiene products (56% in the South compared with 46% in the North and 47% in the Midlands).

Table 7.1.39

Over the past twenty years, there has been a substantial increase in the use of other types of dental hygiene products. In 1978, under a quarter (22%) of dentate adults stated that they used products other than a toothbrush and toothpaste; by 1998 this increased to over half (52%). The proportion of dentate adults stating they used dental floss increased from 8% in 1978 to 21% in 1988, and then increased less sharply between 1988 to 1998 to 28%. No data on the use of mouthwash were collected in the 1978 survey. Since 1998, the reported use of mouthwash more than doubled from 10% in 1988 to 24% in 1998.

Figure 7.1.8, Table 7.1.40

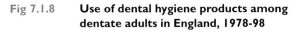

Fig 7.1.8   Use of dental hygiene products among dentate adults in England, 1978-98

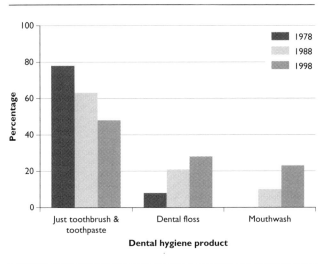

Notes and references

1.   O'Brien M, *Children's dental health in the United Kingdom 1993,* HMSO, London, 1994

Table 7.1.1   **Edentate adults by region, 1978-98**

All adults   *England*

| Dental status | ADH Region | | | | | | | | | All | | |
| | North | | | Midlands | | | South | | | | | |
	1978	1988	1998	1978	1988	1998	1978	1988	1998	1978	1988	1998
						percentage edentate						
All	34	27	14	31	23	15	23	16	10	28	20	12
Age												
16-24	0	0	0	0	0	0	0	0	0	0	0	0
25-34	5	2	1	4	1	1	2	0	0	3	1	0
35-44	16	4	0	10	6	1	9	1	1	11	3	1
45-54	32	22	6	33	23	11	22	8	3	28	15	5
55-64	63	53	25	47	40	26	37	22	11	47	35	18
65-74	} 92	} 80	} 50	} 83	} 69	} 45	} 68	} 57	} 41	} 78	} 66	34 } 44
75 and over												56 }
Gender												
Men	29	20	12	27	18	12	18	12	7	24	16	11
Women	38	33	15	35	27	17	26	19	12	32	25	13
Social class of head of household												
I, II, IIINM	24	17	9	23	17	10	17	11	5	20	14	7
IIIM	35	32	14	30	25	14	19	17	14	27	24	14
IV, V	38	38	27	38	28	24	36	28	17	37	31	21

1988 Table 3.6-3.9
For bases see Tables Appendix

Table 7.1.2   **Reliance on natural teeth and dentures by characteristics of adults**

All adults   *England*

		Dental status			Base
		Natural teeth only	Natural teeth with dentures	Edentate	
All	%	73	15	12	3436
Age					
16-24	%	99	1	0	352
25-34	%	98	2	0	704
35-44	%	91	9	1	637
45-54	%	78	16	5	606
55-64	%	48	34	18	450
65 and over	%	24	32	44	687
Gender					
Male	%	75	16	11	1562
Female	%	71	15	13	1874
ADH region					
North	%	69	17	14	975
Midlands	%	68	17	15	856
South	%	77	13	10	1605
Regional Health Office					
Northern & Yorkshire	%	65	19	16	527
Trent	%	62	17	21	326
Eastern	%	69	19	11	317
London	%	79	12	9	442
South East	%	79	13	18	669
South West	%	75	12	13	372
West Midlands	%	72	17	11	335
North West	%	74	14	12	448
Social class of head of household					
I, II, IIINM	%	79	14	7	1611
III M	%	69	16	14	948
IV,V	%	61	18	21	628
Jarman Area					
1 (most deprived)	%	71	13	16	497
2	%	71	16	13	655
3	%	73	16	12	989
4	%	77	13	10	663
5 (least deprived)	%	73	17	10	627

Table 7.1.3 **Distribution of number of teeth, 1978-98**

Dentate adults   *England*

Number of teeth	1978	1988	1998	
	%	%	%	
1-8	}13	}10	4	}8
9-14			4	
15-20	13	9	9	
21-26	34	32	29	
27-31	37	46	48	
32	3	4	6	
Proportion with 21 or more teeth	74	81	83	
Mean	23.3	24.4	24.9	
Base	2350	2444	2186	

1988 Table 5.17

Table 7.1.4 **Distribution of number of teeth by age, 1978-98**

Dentate adults   *England*

Number of teeth	16-24			25-34			35-44			45-54			55 and over		
	1978	1988	1998	1978	1988	1998	1978	1988	1998	1978	1988	1998	1978	1988	1998
	%	%	%	%	%	%	%	%	%	%	%	%	%	%	%
1-14	1	0	0	3	2	0	12	5	2	22	11	6	40	38	26
15-20	1	0	0	6	1	1	11	7	3	27	14	10	30	25	26
21-26	26	25	20	36	26	16	44	36	31	37	47	43	23	31	36
27-31	67	70	72	50	62	71	31	49	56	14	27	40	6	7	13
32	5	6	8	5	10	12	2	3	8	0	1	1	1	0	0
Mean	27.4	27.8	27.9	26.0	27.4	28.2	23.2	25.5	26.9	19.8	22.7	24.3	16.0	16.9	18.8
Base	537	587	246	602	567	517	496	518	447	363	359	423	351	469	553

1988 Table 5.17 (part)

Table 7.1.5   **Distribution of number of teeth by gender, 1978-98**

Dentate adults   *England*

Number of teeth	Gender						All		
	Men			Women					
	1978	1988	1998	1978	1988	1998	1978	1988	1998
	%	%	%	%	%	%	%	%	%
1-14	13	10	8	12	10	8	13	10	8
15-20	13	8	9	13	9	9	13	9	9
21-26	34	30	27	34	33	31	34	32	29
27-31	36	47	49	38	44	48	37	46	48
32	4	5	7	3	4	4	3	4	6
Mean	23.1	24.6	25.1	23.4	24.3	24.7	23.3	24.4	24.9
Base	1176	1298	1028	1174	1203	1158	2350	2500	2186

1988 Table 5.18 (part)

Table 7.1.6   **Distribution of number of teeth by social class of head of household, 1978-98**

Dentate adults   *England*

Number of teeth	Social class of head of household									All†		
	I, II, IIINM			IIIM			IV,V					
	1978	1988	1998	1978	1988	1998	1978	1988	1998	1978	1988	1998
	%	%	%	%	%	%	%	%	%	%	%	%
1-14	11	7	5	13	12	10	15	13	12	13	10	8
15-20	11	8	8	16	9	10	15	10	12	13	9	9
21-26	35	30	30	32	34	31	34	33	28	34	32	29
27-31	39	50	51	36	40	44	34	40	44	37	46	48
32	4	4	6	3	4	4	2	4	4	3	4	6
Mean	23.8	25.0	25.5	23.0	23.8	24.3	22.7	23.7	23.8	23.3	24.4	24.9
Base	964	1306	1155	805	731	586	411	384	342	2350	2500	2186

† Includes those for whom social class of head of household was not known and armed forces
1988 Table 5.19 (part)

Table 7.1.7   **Distribution of number of teeth by usual reason for dental attendance, 1978-98**

Dentate adults   *England*

Number of teeth	Usual reason for dental attendance						All†		
	Regular check-up			Only with trouble					
	1978	1988	1998	1978	1988	1998	1978	1988	1998
	%	%	%	%	%	%	%	%	%
1-14	7	6	6	20	18	12	13	10	8
15-20	10	9	9	17	10	10	13	9	9
21-26	37	34	33	30	29	23	34	32	29
27-31	43	48	47	30	39	49	37	46	48
32	3	3	5	3	4	7	3	4	6
Mean	24.6	25.0	25.0	21.4	22.7	24.3	23.3	24.4	24.9
Base	1091	1504	1413	927	999	554	2350	2500	2186

† Includes those who visit the dentist for an occasional check-up
1988 Table 5.20 (part)

Table 7.1.8 **Distribution of number of teeth by characteristics of dentate adults**

Dentate adults   *England*

		Number of teeth						Proportion with 21 or more teeth	Mean number of teeth	Base
		1-8	9-14	15-20	21-26	27-31	32			
All	%	4	4	9	29	48	6	83	24.9	2186
Age										
16-24	%	0	0	0	20	72	8	100	27.9	246
25-34	%	0	0	1	16	71	12	99	28.2	517
35-44	%	1	1	3	31	56	8	95	26.9	447
45-54	%	1	5	10	43	40	1	84	24.3	423
55-64	%	10	10	22	43	15	1	58	20.1	259
65 and over	%	16	16	29	29	11	0	40	17.6	294
Gender										
Male	%	4	4	9	27	49	7	83	25.1	1028
Female	%	4	4	9	31	48	4	83	24.7	1158
ADH region										
North	%	4	5	10	30	47	4	82	24.6	617
Midlands	%	4	4	11	27	46	7	80	24.6	495
South	%	3	4	8	30	50	6	85	25.3	1074
Regional Health Office										
Northern & Yorkshire	%	5	5	10	26	49	5	80	24.4	307
Trent	%	2	6	14	26	45	7	78	24.8	174
Eastern	%	4	4	10	31	45	6	82	24.5	229
London	%	4	3	6	24	57	7	87	25.6	280
South East	%	3	3	8	31	48	7	85	25.4	478
South West	%	2	5	6	34	48	4	86	25.0	230
West Midlands	%	7	4	12	23	48	6	78	24.1	178
North West	%	2	4	10	35	46	3	84	24.8	310
Social class of head of household										
I, II, IIINM	%	2	3	8	30	51	6	87	25.5	1155
IIIM	%	5	5	10	31	44	4	80	24.3	586
IV, V	%	5	7	12	28	44	4	76	23.8	342
Jarman Area										
1 (most deprived)	%	4	4	6	26	51	8	85	25.1	282
2	%	3	6	9	33	44	6	82	24.7	427
3	%	3	4	11	26	50	7	82	25.1	627
4	%	4	3	8	30	50	5	85	25.1	445
5 (least deprived)	%	4	4	9	32	48	3	83	24.7	402
Usual reason for dental attendance										
Regular check-up	%	3	4	9	33	47	5	85	25.0	1413
Occasional check-up	%	1	4	7	22	59	8	89	26.4	214
Only with trouble	%	6	6	10	23	49	7	78	24.3	554

Table 7.1.9   **Reliance on natural teeth and dentures by ADH region, 1978-98**

All adults   *England*

Dental status	ADH Region									All		
	North			Midlands			South					
	1978	1988	1998	1978	1988	1998	1978	1988	1998	1978	1988	1998
						percentage						
Natural teeth only	47	56	69	50	59	68	54	65	77	51	61	73
Natural teeth with dentures	19	17	17	19	18	17	23	19	13	21	18	15
Edentate	34	27	14	31	23	15	23	16	10	28	20	12
Base	1105	1042	975	965	917	856	1763	1790	1605	3833	3751	3436

1988 Table 16.12

Table 7.1.10 **The condition of natural teeth by age**

Dentate adults   *England*

The condition of natural teeth (1998 criteria)	Age						All
	16-24	25-34	35-44	45-54	55-64	65 and over	
	%	%	%	%	%	%	%
Sound and untreated							
0	0	0	0	0	2	5	1
1-5	1	1	5	9	12	22	7
6-11	1	8	17	39	52	45	24
12-17	7	23	38	36	30	23	26
18-23	32	44	30	15	4	4	24
24 or more	59	25	9	1	1	1	17
Proportion with 18 or more	92	68	40	16	4	5	41
Mean	23.7	19.5	15.9	11.9	9.7	8.6	15.6
	%	%	%	%	%	%	%
Decayed or unsound							
0	50	39	50	44	45	45	45
1-5	43	53	44	52	51	51	49
6 or more	6	8	6	5	4	4	6
Proportion with 1 or more	50	61	50	56	55	55	56
Mean	1.6	1.8	1.4	1.5	1.3	1.3	1.5
	%	%	%	%	%	%	%
Restored (otherwise sound)							
0	33	4	2	2	8	12	9
1-5	53	35	20	12	22	30	29
6-11	13	46	44	39	33	33	36
12 or more	1	14	34	47	37	26	26
Mean	2.4	6.9	9.6	10.9	9.0	7.7	7.8
Base	246	517	447	423	259	294	2186

Table 7.1.11 The condition of natural teeth by gender

Dentate adults *England*

The condition of natural teeth (1998 criteria)	Gender		All
	Male	Female	
	%	%	%
Sound and untreated			
0	I	I	I
1-5	7	8	7
6-11	22	26	24
12-17	27	26	26
18-23	25	24	24
24 or more	19	15	17
Proportion with 18 or more	44	39	41
Mean	16.0	15.1	15.6
	%	%	%
Decayed or unsound			
0	43	48	45
1-5	49	48	49
6 or more	8	4	6
Proportion with 1 or more	57	52	56
Mean	1.7	1.2	1.5
	%	%	%
Restored (otherwise sound)			
0	11	8	9
1-5	31	26	29
6-11	35	37	36
12 or more	23	29	26
Mean	7.3	8.3	7.8
Base	*1028*	*1158*	*2186*

Table 7.1.12 The condition of natural teeth by ADH region

Dentate adults *England*

The condition of natural teeth (1998 criteria)	ADH region			All
	North	Midlands	South	
	%	%	%	%
Sound and untreated				
0	I	I	I	I
1-5	8	8	6	7
6-11	26	20	25	24
12-17	25	28	27	26
18-23	26	24	24	24
24 or more	15	18	17	17
Proportion with 18 or more	41	42	41	41
Mean	15.4	15.6	15.7	15.6
	%	%	%	%
Decayed or unsound				
0	35	48	49	45
1-5	57	47	46	49
6 or more	8	5	5	6
Proportion with 1 or more	65	52	51	56
Mean	1.9	1.4	1.3	1.5
	%	%	%	%
Restored (otherwise sound)				
0	11	9	8	9
1-5	30	30	28	29
6-11	38	37	35	36
12 or more	22	24	29	26
Mean	7.3	7.5	8.2	7.8
Base	*617*	*495*	*1074*	*2186*

Table 7.1.13 The condition of natural teeth by Regional Health Office

Dentate adults *England*

The condition of natural teeth (1998 criteria)	Regional Health Office								All
	Northern and Yorkshire	Trent	Eastern	London	South East	South West	West Midlands	North West	
	%	%	%	%	%	%	%	%	%
Sound and untreated									
0	1	0	1	1	2	0	2	1	1
1-5	9	7	9	5	6	6	9	6	7
6-11	26	19	27	23	26	26	19	26	24
12-17	20	34	25	19	30	30	26	30	26
18-23	27	21	23	29	22	23	25	25	24
24 or more	17	20	16	23	16	15	19	13	17
Proportion with 18 or more	44	41	38	52	38	38	44	38	41
Mean	15.6	16.0	14.8	17.0	15.1	15.4	15.6	15.1	15.5
	%	%	%	%	%	%	%	%	%
Decayed or unsound									
0	35	40	55	43	48	59	47	36	45
1-5	58	54	40	52	46	38	47	56	49
6 or more	7	6	5	5	6	3	6	8	6
Proportion with 1 or more	65	60	45	57	52	41	53	64	55
Mean	1.8	1.6	1.2	1.5	1.4	1.0	1.5	1.9	1.5
	%	%	%	%	%	%	%	%	%
Restored (otherwise sound)									
0	13	7	8	11	9	6	11	9	9
1-5	33	34	27	33	24	24	35	28	29
6-11	35	39	34	37	34	39	33	40	36
12 or more	19	20	32	20	33	32	22	24	26
Mean	6.9	7.1	8.4	7.1	8.8	8.5	7.0	7.8	7.8
Base	307	174	229	280	478	230	178	310	2186

1988 Table 6.16

Table 7.1.14 **The condition of natural teeth by social class of head of household**

Dentate adults   *England*

The condition of natural teeth (1998 criteria)	Social class of head of household			All†
	I, II, IIINM	IIIM	IV,V	
	%	%	%	%
Sound and untreated				
0	I	I	I	I
1-5	7	8	6	7
6-11	24	24	24	24
12-17	27	28	25	27
18-23	23	26	25	24
24 or more	18	13	18	16
Proportion with 18 or more	41	39	43	41
Mean	15.7	14.9	15.7	15.6
	%	%	%	%
Decayed or unsound				
0	50	43	37	46
1-5	46	50	53	49
6 or more	4	7	10	6
Proportion with 1 or more	50	57	63	56
Mean	1.2	1.7	1.9	1.5
	%	%	%	%
Restored (otherwise sound)				
0	8	7	13	9
1-5	24	32	35	29
6-11	36	39	36	37
12 or more	31	22	16	26
Mean	8.6	7.7	6.2	7.8
Base	*1155*	*586*	*342*	*2186*

† Includes those for whom the social class of head of household was not known and Armed Forces

Table 7.1.15 **The condition of natural teeth by Jarman Area**

Dentate adults *England*

The condition of natural teeth (1998 criteria)	Jarman Area					All
	1 Most deprived	2	3	4	5 Least deprived	
	%	%	%	%	%	%
Sound and untreated						
0	1	1	1	2	2	1
1-5	7	7	8	6	8	7
6-11	24	24	21	26	27	24
12-17	17	27	29	28	28	26
18-23	34	26	23	22	21	24
24 or more	18	15	19	17	15	17
Proportion with 18 or more	52	40	42	39	36	41
Mean	16.6	15.4	15.8	15.4	14.7	15.6
	%	%	%	%	%	%
Decayed or unsound						
0	31	44	48	46	53	45
1-5	61	49	46	52	42	49
6 or more	9	7	7	3	4	6
Proportion with 1 or more	69	56	52	54	47	56
Mean	2.0	1.6	1.5	1.3	1.2	1.5
	%	%	%	%	%	%
Restored (otherwise sound)						
0	14	8	8	9	9	9
1-5	37	32	30	23	24	29
6-11	31	35	38	40	34	36
12 or more	18	25	24	28	34	26
Mean	6.5	7.6	7.8	8.3	8.6	7.8
Base	282	427	627	445	402	2186

Table 7.1.16 **The condition of natural teeth by usual reason for dental attendance**

Dentate adults   *England*

The condition of natural teeth (1998 criteria)	Usual reason for dental attendance			All
	Regular check-up	Occasional check-up	Only with trouble	
	%	%	%	%
Sound and untreated				
0	1	0	1	1
1-5	7	3	8	7
6-11	27	22	18	24
12-17	28	23	25	26
18-23	22	29	26	24
24 or more	14	24	21	17
Proportion with				
18 or more	36	53	47	41
Mean	14.8	17.7	16.4	15.6
	%	%	%	%
Decayed or unsound				
0	51	44	33	45
1-5	46	51	55	49
6 or more	3	6	12	6
Proportion with				
1 or more	49	56	67	56
Mean	1.1	1.4	2.3	1.5
	%	%	%	%
Restored (otherwise sound)				
0	5	11	17	9
1-5	23	29	40	29
6-11	38	37	31	36
12 or more	33	23	12	26
Mean	9.0	7.2	5.5	7.8
Base	*1413*	*214*	*554*	*2186*

Table 7.1.17 Difference in condition of teeth using 1998 and pre-1998 criteria

Dentate adults *England*

	Average number of teeth that were:									Base
	Sound and untreated			Decayed or unsound			Restored (otherwise sound)			
	1998 criteria	pre-1998 criteria	Difference	1998 criteria	pre-1998 criteria	Difference	1998 criteria	pre-1998 criteria	Difference	
All	15.6	15.9	0.4	1.5	1.0	0.5	7.8	8.0	0.2	2186
Age										
16-24	23.7	24.3	0.6	1.6	0.9	0.7	2.4	2.6	0.2	246
25-34	19.5	20.1	0.6	1.8	1.0	0.8	6.9	7.1	0.2	517
35-44	15.9	16.1	0.2	1.4	0.9	0.5	9.6	9.8	0.2	447
45-54	11.9	12.1	0.2	1.5	1.1	0.4	10.9	11.2	0.3	423
55-64	9.7	9.9	0.2	1.3	1.0	0.3	9.0	9.1	0.1	259
65 and over	8.6	8.7	0.1	1.3	1.0	0.3	7.7	7.8	0.1	294
Gender										
Men	16.0	16.4	0.4	1.7	1.2	0.5	7.3	7.5	0.2	1028
Women	15.1	15.4	0.3	1.2	0.8	0.4	8.3	8.5	0.2	1158
ADH Region										
North	15.4	15.9	0.5	1.9	1.0	0.9	7.3	7.6	0.3	617
Midlands	15.6	16.0	0.4	1.4	0.9	0.5	7.5	7.7	0.2	495
South	15.7	15.8	0.1	1.3	1.0	0.3	8.2	8.3	0.1	1074
Regional Health Office										
Northern and Yorkshire	15.6	16.1	0.5	1.8	1.0	0.8	6.9	7.2	0.3	307
Trent	16.0	16.3	0.3	1.6	1.1	0.5	7.1	7.3	0.2	174
Eastern	14.8	15.0	0.2	1.2	0.8	0.4	8.4	8.5	0.1	229
London	17.0	17.3	0.3	1.5	1.0	0.5	7.1	7.3	0.2	280
South East	15.1	15.3	0.2	1.4	1.1	0.3	8.8	8.9	0.1	478
South West	15.4	15.5	0.1	1.0	0.8	0.2	8.5	8.6	0.1	230
West Midlands	15.6	16.2	0.6	1.5	0.7	0.8	7.0	7.2	0.2	178
North West	15.1	15.7	0.6	1.9	1.0	0.9	7.8	8.1	0.3	310
Social class of head of household										
I, II, IIINM	15.7	16.0	0.3	1.2	0.7	0.5	8.6	8.7	0.1	1155
IIIM	14.9	15.3	0.4	1.7	1.2	0.5	7.7	7.8	0.1	586
IV, V	15.7	16.1	0.4	1.9	1.2	0.7	6.2	6.4	0.2	342
Jarman Area										
1 (most deprived)	16.6	17.1	0.5	2.0	1.3	0.7	6.5	6.7	0.2	282
2	15.4	15.8	0.4	1.6	1.0	0.6	7.6	7.8	0.2	427
3	15.8	16.1	0.3	1.5	1.0	0.5	7.8	7.9	0.1	627
4	15.4	15.7	0.3	1.3	0.9	0.4	8.3	8.5	0.2	445
5 (least deprived)	14.7	14.9	0.2	1.2	0.9	0.3	8.6	8.8	0.2	402
Usual reason for dental attendance										
Regular check-up	14.8	15.0	0.2	1.1	0.7	0.4	9.0	9.2	0.2	1413
Occasion check-up	17.7	18.1	0.4	1.4	0.8	0.6	7.2	7.4	0.2	214
Only with trouble	16.4	16.9	0.5	2.3	1.6	0.7	5.5	5.7	0.2	554

Table 7.1.18 The condition of natural teeth, 1978-98

Dentate adults			England
The condition of natural teeth (1988 criteria)	1978	1988	1998
	%	%	%
Sound and untreated			
0	1	0	1
1-5	9	7	7
6-11	31	26	23
12-17	32	29	26
18 or more	27	37	43
Mean	13.2	15.0	15.9
	%	%	%
Decayed or unsound			
0	40	57	58
1-5	51	40	40
6 or more	9	3	3
Mean	1.9	1.0	1.0
	%	%	%
Restored (otherwise sound)			
0	14	10	9
1-5	23	25	28
6-11	33	34	36
12 or more	30	31	27
Mean	8.1	8.4	8.0
Mean number of teeth	23.3	24.4	24.9
Base	2350	2444	2186

1988 Table 6.20, 7.21,8.26

Table 7.1.19 **The condition of natural teeth by age, 1978-98**

Dentate adults   *England*

The condition of natural teeth (1988 criteria)	Age														
	16-24			25-34			35-44			45-54			55 and over		
	1978	1988	1998	1978	1988	1998	1978	1988	1998	1978	1988	1998	1978	1988	1998
	%	%	%	%	%	%	%	%	%	%	%	%	%	%	%
Sound and untreated															
0	0	0	0	1	1	0	0	0	0	2	1	0	4	1	3
1-5	2	0	0	6	2	0	9	6	4	12	10	9	25	21	18
6-11	14	2	2	26	18	6	40	36	16	48	40	38	39	46	48
12-17	31	14	4	39	37	23	31	36	38	27	35	36	22	24	27
18 or more	53	84	94	28	42	71	20	22	42	11	13	17	10	7	5
Mean	17.5	21.7	24.3	14.1	16.4	20.1	12.4	13.3	16.1	10.6	11.7	12.1	9.2	9.5	9.3
	%	%	%	%	%	%	%	%	%	%	%	%	%	%	%
Decayed or unsound															
0	42	63	70	40	58	57	42	55	60	37	50	54	38	55	51
1-5	48	34	26	51	38	40	50	42	37	55	48	43	52	41	47
6 or more	10	2	4	9	4	2	8	3	3	8	3	3	10	4	3
Mean	2.0	0.8	0.9	1.9	1.1	1.0	1.8	1.0	0.9	1.9	1.2	1.1	2.0	1.1	1.0
	%	%	%	%	%	%	%	%	%	%	%	%	%	%	%
Restored (otherwise sound)															
0	9	14	32	10	5	4	15	5	2	13	7	2	25	17	9
1-5	26	40	52	15	18	34	17	13	19	28	19	11	36	33	26
6-11	40	39	14	33	37	46	31	31	43	35	33	39	27	30	33
12 or more	25	7	1	42	40	16	37	50	36	24	41	48	12	19	32
Mean	7.8	5.2	2.6	10.0	9.9	7.1	9.0	11.2	9.8	7.2	9.8	11.2	4.9	6.3	8.4
Mean number of teeth	27.4	27.8	27.9	26.0	27.4	28.2	23.2	25.5	26.9	19.8	22.7	24.3	16.0	16.9	18.8
Base	537	587	246	602	567	517	496	518	447	363	359	423	351	469	553

1988 Table 6.20, 7.21, 8.26

Table 7.1.20 **The condition of natural teeth by gender, 1978-98**

Dentate adults   *England*

The condition of natural teeth (1988 criteria)	Gender						All		
	Men			Women					
	1978	1988	1998	1978	1988	1998	1978	1988	1998
	%	%	%	%	%	%	%	%	%
Sound and untreated									
0	2	0	0	1	0	1	1	0	1
1-5	10	8	6	9	7	7	9	7	7
6-11	28	25	22	35	28	25	31	26	23
12-17	30	27	26	32	31	26	32	29	26
18 or more	30	41	45	23	34	40	27	37	43
Mean	13.6	15.5	16.4	12.9	14.5	15.4	13.2	15.0	15.9
	%	%	%	%	%	%	%	%	%
Decayed or unsound									
0	36	53	55	44	61	61	40	57	58
1-5	53	42	41	49	37	38	51	40	40
6 or more	11	4	4	7	2	2	9	3	3
Mean	2.2	1.2	1.2	1.6	0.8	0.8	1.9	1.0	1.0
	%	%	%	%	%	%	%	%	%
Restored (otherwise sound)									
0	16	12	10	11	8	7	14	10	9
1-5	26	27	30	20	23	25	23	25	28
6-11	34	33	35	34	36	37	33	34	36
12 or more	24	28	24	35	31	30	30	31	27
Mean	7.3	7.9	7.5	8.9	8.9	8.5	8.1	8.4	8.0
Mean number of teeth	23.1	24.6	25.1	23.4	24.3	24.7	23.3	24.4	24.9
Base	*1176*	*1298*	*1028*	*1174*	*1203*	*1158*	*2350*	*2444*	*2186*

1988 Table 6.21, 7.22, 8.27

Table 7.1.21 The condition of natural teeth by social class of head of household, 1978-98

Dentate adults   *England*

The condition of natural teeth (1988 criteria)	Social class of head of household									All†		
	I, II, IIINM			IIIM			IV,V					
	1978	1988	1998	1978	1988	1998	1978	1988	1998	1978	1988	1998
	%	%	%	%	%	%	%	%	%	%	%	%
Sound and untreated												
0	1	1	1	1	0	1	2	0	1	1	0	1
1-5	10	8	6	8	8	7	7	4	7	9	7	7
6-11	35	27	23	31	26	24	27	28	24	31	26	23
12-17	34	30	27	31	29	28	32	24	24	32	29	27
18 or more	20	35	42	20	37	40	32	44	44	27	37	42
Mean	12.3	14.7	16.0	12.3	14.8	15.3	14.3	15.8	16.1	13.2	15.0	15.9
	%	%	%	%	%	%	%	%	%	%	%	%
Decayed or unsound												
0	43	62	63	40	51	54	34	47	52	40	57	58
1-5	51	36	36	50	45	42	55	46	43	51	40	39
6 or more	6	2	1	10	4	4	11	6	5	9	3	3
Mean	1.6	0.8	0.7	2.1	1.2	1.2	2.3	1.5	1.2	1.9	1.0	1.0
	%	%	%	%	%	%	%	%	%	%	%	%
Restored (otherwise sound)												
0	6	7	8	17	10	7	24	18	12	14	10	8
1-5	18	21	24	26	29	31	27	30	33	23	25	28
6-11	36	34	35	33	36	38	29	35	39	33	34	36
12 or more	40	38	33	24	26	24	20	18	16	30	31	27
Mean	9.9	9.4	8.7	7.2	7.7	7.8	6.1	6.4	6.4	8.1	8.4	8.0
Mean number of teeth	23.8	25.0	25.5	23.0	23.8	24.3	22.7	23.7	23.8	23.3	24.4	24.9
Base	964	1306	1155	805	731	586	411	384	342	2350	2444	2186

† Includes those for whom social class of head of household was not known and Armed Forces
1988 Table 6.22, 7.23, 8.28

Table 7.1.22   **The condition of natural teeth by usual reason for dental attendance, 1978-98**

Dentate adults   *England*

The condition of natural teeth (1988 criteria)	Usual reason for dental attendance						All †		
	Regular check-up			Only with trouble					
	1978	1988	1998	1978	1988	1998	1978	1988	1998
	%	%	%	%	%	%	%	%	%
Sound and untreated									
0	1	0	1	1	1	1	1	0	1
1-5	9	6	7	11	9	8	9	7	7
6-11	37	31	27	26	24	17	31	26	23
12-17	35	32	28	28	26	24	32	29	26
18 or more	18	30	37	34	40	50	27	37	43
Mean	12.2	14.2	15.0	14.1	15.1	16.9	13.2	15.0	15.9
	%	%	%	%	%	%	%	%	%
Decayed or unsound									
0	51	66	63	28	42	44	40	57	58
1-5	46	34	35	56	51	50	51	40	40
6 or more	3	1	1	16	7	6	9	3	3
Mean	1.1	0.6	0.7	2.9	1.7	1.6	1.9	1.0	1.0
	%	%	%	%	%	%	%	%	%
Restored (otherwise sound)									
0	1	4	5	30	18	16	14	10	9
1-5	12	18	22	36	35	40	23	25	28
6-11	39	35	39	23	32	31	33	34	36
12 or more	48	42	34	11	16	14	30	31	27
Mean	11.3	10.3	9.2	4.4	5.9	5.7	8.1	8.4	8.1
Mean number of teeth	24.6	25.0	25.0	21.4	22.7	24.3	23.3	24.4	24.9
Base	*1091*	*1504*	*1413*	*927*	*999*	*554*	*2350*	*2444*	*2186*

† Includes those who only visit the dentist for an occasional check-up
1988 Table 6.23, 7.24, 8.29

Table 7.1.23 **English oral health targets**

England

Oral health targets and ADH estimates[†]	Percentage	Confidence Intervals		Base 95%
33% of adults aged over 75 should have some natural teeth				
Adults aged 75 and over with natural teeth	44	38-50		*271*
10% of adults aged over 75 should have more than 20 teeth				
Adults aged 75 and over with more than 20 teeth[††]	10	..	..	..
75% of 50 year olds should have more than 20 teeth				
Adults aged 45-54 with more than 20 teeth[††]	79	..	..	..
50% of 30-year-olds should have more than 20 teeth which are sound and unfilled				
Dentate adults aged 25-34 with more than 20 teeth which are sound and untreated[§]				
pre-1998 criteria	48	44-53		*637*
1998 criteria	44	40-49		*637*
The proportion of adults aged over 45 with deep periodontal pockets (greater than 6mm) should be reduced to 10%				
Dentate adults aged 45 and over with deep periodontal pockets (6mm or more)[§§]	10	5-13		*846*

† The Oral Health Strategy for England used slightly different age groups than those used in the ADH series

†† The percentage of adults with more than 20 teeth was calculated by multiplying the percentage of adults who are dentate (taken from the interview sample) by the percentage of dentate adults with 21 or more teeth (taken from the examination sample). As these figures come from different samples the base and confidence intervals for the resulting percentage are not shown

§ Details of the two sets of criteria are given in Appendix A

§§ The Oral Health Strategy for England target related to deep periodontal pockets of greater than 6mm. The ADH series recorded periodontal pockets of 6mm or more

Table 7.1.24 **Usual reason for dental attendance by ADH region, 1978-98**

Dentate adults

England

Usual reason for dental attendance	ADH Region									All		
	North			Midlands			South					
	1978	1988	1998	1978	1988	1998	1978	1988	1998	1978	1988	1998
	%	%	%	%	%	%	%	%	%	%	%	%
Regular check-up	38	49	58	46	49	63	47	52	60	44	50	60
Occasional check-up	13	13	12	10	14	7	15	16	11	14	15	10
Only with trouble	48	38	31	44	37	30	38	32	29	42	35	30
Base	736	756	823	666	692	731	1359	1524	1456	2761	2972	3010

1988 Table 20.10 (part)

Table 7.1.25 **Usual reason for dental attendance by characteristics of dentate adults**

Dentate adults   *England*

		Usual reason for dental attendance			Base
		Regular check-up	Occasional check-up	Only with trouble	
All	%	60	10	30	3010
Age					
16-24	%	49	16	35	352
25-34	%	54	12	35	701
35-44	%	63	11	27	633
45-54	%	64	11	25	572
55-64	%	68	6	26	367
65 and over	%	67	5	28	385
Gender					
Male	%	53	10	37	1395
Female	%	67	11	22	1615
ADH region					
North	%	58	12	31	823
Midlands	%	63	7	30	731
South	%	60	11	29	1456
Regional Health Office					
Northern & Yorkshire	%	59	12	29	432
Trent	%	60	7	33	255
Eastern	%	69	9	22	284
London	%	50	16	34	402
South East	%	62	11	27	620
South West	%	66	6	28	327
West Midlands	%	59	7	34	299
North West	%	56	11	33	391
Social class of head of household					
I, II, IIINM	%	65	12	24	1488
IIIM	%	58	8	34	810
IV,V	%	50	9	42	489
Jarman Area					
1 (most deprived)	%	51	12	37	412
2	%	52	12	36	568
3	%	60	10	31	868
4	%	65	10	25	593
5 (least deprived)	%	69	10	21	564

Table 7.1.26 **Treatment history by ADH region, 1988-98**

Dentate adults   *England*

Have had at some time:	ADH Region						All	
	North		Midlands		South			
	1988	1998	1988	1998	1988	1998	1988	1998
	%	%	%	%	%	%	%	%
Fillings	93	94	88	93	94	94	93	94
An xray	68	86	66	87	85	92	76	89
Extractions††	91	87	85	82	84	77	86	81
Injection (gum) for extraction	68	73	72	70	77	67	73	69
Injection (gum) for filling	86	90	83	88	88	90	86	90
Gas	65	64	58	51	54	44	61	51
A scale and polish	86	88	81	89	87	87	85	88
Treatment from a hygienist	15	39	21	39	32	49	25	44
A brace	12	20	13	16	20	21	16	19
An abcess	27	40	28	39	34	37	31	38
A crown	25	37	26	36	32	38	29	37
A bridge	5	9	4	5	6	8	6	7
Base	764	823	707	731	1546	1456	3017	3010

Percentages may add to more than 100% as respondents could give more than one answer
† Other than exclusively wisdom teeth
1988 Table 18.4 (part)

Table 7.1.27 **Time since last visit to the dentist by ADH region, 1988-98**

Dentate adults†   *England*

Time since last visit to the dentist	ADH Region						All	
	North		Midlands		South			
	1988	1998	1988	1998	1988	1998	1988	1998
	%	%	%	%	%	%	%	%
Up to 1 year	65	71	66	71	69	71	67	71
Over 1 year, up to 2 years	7	7	8	6	9	8	8 95	7 95
Over 2 years, up to 5 years	14	12	15	13	14	10	13	12
Over 5 years, up to 10 years	7	5	5	5	5	6	6	6
Over 10 years, up to 20 years	4	3	4	3	3	3	3	3
Over 20 years	2	2	2	2	2	1	2	2
Base	760	758	701	675	1532	1369	2993	2802

† Excludes those who have never been to the dentist
1988 Table 24.3 (part)

Table 7.1.28 **Treatment received at most recent dental visit by ADH region, 1978-98**

Dentate adults†   *England*

Treatment received	ADH Region									All		
	North			Midlands			South					
	1978	1988	1998	1978	1988	1998	1978	1988	1998	1978	1988	1998
	%	%	%	%	%	%	%	%	%	%	%	%
No fillings/no extractions	32	48	62	33	52	62	31	50	61	32	50	61
Some fillings/no extractions	33	31	25	30	30	22	44	35	26	38	33	25
Some fillings/ some extractions	9	6	4	8	5	3	8	4	4	8	5	4
No fillings/ some extractions	25	14	9	29	13	12	17	9	9	22	11	10
Base	696	708	758	630	652	675	1357	1421	1369	2583	2787	2802

† Excludes those who have never been to the dentist or were in the middle of a course of treatment
1988 Table 24.42 (part)

Table 7.1.29 **Treatment received at most recent dental visit by ADH region, 1978-98**

Dentate adults†   *England*

Treatment received	ADH Region									All		
	North			Midlands			South					
	1978	1988	1998	1978	1988	1998	1978	1988	1998	1978	1988	1998
	%	%	%	%	%	%	%	%	%	%	%	%
Scale and polish	46	54	57	48	50	59	60	56	60	52	54	59
Teeth filled	42	37	29	38	35	26	52	40	30	46	38	29
X-rays	12	19	25	11	16	24	31	36	32	21	27	28
Teeth extracted	47	21	14	37	18	15	25	14	12	30	17	13
Crown fitted††	-	5	6	-	6	6	-	8	7	-	7	7
Abcess treated††	-	5	5	-	5	4	-	4	4	-	4	4
Denture fitted††	-	8	6	-	8	6	-	6	4	-	7	5
Denture repaired††	-	2	2	-	2	2	-	2	1	-	2	2
Base	696	708	758	630	652	675	1357	1421	1369	2583	2787	2802

Percentages may add to more than 100% as respondents could give more than one answer
† Excludes those who have never been to the dentist or were in the middle of a course of treatment
†† Data from 1978 not available
1988 Table 24.32 (part), 24.41 (part)

Table 7.1.30 Type of dental service by characteristics of dentate adults

Dentate adults†   *England*

		Type of dental service			Base
		NHS	**Private**	**Other**	
All	%	76	19	5	2802
Age					
16-24	%	85	12	4	331
25-34	%	75	20	5	650
35-44	%	73	20	6	588
45-54	%	73	22	6	533
55-64	%	73	22	6	335
65 and over	%	76	18	6	365
Gender					
Male	%	74	19	6	1297
Female	%	77	18	4	1505
ADH region					
North	%	87	9	4	758
Midlands	%	78	18	4	675
South	%	70	24	6	1369
Regional Health Office					
Northern & Yorkshire	%	85	10	5	398
Trent	%	84	11	5	240
Eastern	%	77	18	5	255
London	%	74	22	4	370
South East	%	67	26	7	590
South West	%	62	28	9	310
West Midlands	%	79	18	3	279
North West	%	89	8	3	360
Social class of head of household					
I, II, IIINM	%	73	22	5	1378
IIIM	%	80	15	5	749
IV,V	%	80	16	5	466
Jarman area					
1 (most deprived)	%	78	18	5	376
2	%	83	13	5	533
3	%	78	16	6	822
4	%	69	26	5	533
5 (least deprived)	%	70	23	6	533
Usual reason for dental attendance					
Regular check-up	%	75	20	6	1688
Occasional check-up	%	75	22	4	295
Only with trouble	%	78	16	5	813

† Excludes those who have never been to the dentist or were in the middle of a course of treatment

Table 7.1.31 Last treatment was wholly private by ADH region, 1988-98

Dentate adults[†]   *England*

ADH region	Percentage whose last treatment was wholly private	
	1988	**1998**
North	4	9
Midlands	7	18
South	6	24
All	6	19
Base	*2797*	*2802*

† Excludes those who have never been to the dentist or were in the middle of a course of treatment
1998 Table 24.47 (part)

Table 7.1.32 Treatment at last visit by type of dental service

Dentate adults[†]   *England*

Treatment received at last visit	Type of service		All[††]
	NHS	**Private**	
	%	%	%
Check-up	91	88	90
Scale and polish	57	65	59
Teeth filled	28	30	29
X-ray	27	32	28
Teeth extracted	13	12	13
Crown fitted	6	8	7
Abcess treated	4	4	4
Denture fitted	5	4	5
Denture repaired	2	1	2
Base	*2120*	*519*	*2802*

† Excludes dentate adults who have never been to the dentist or are in the middle of a course of treatment
†† Includes other types of dental service

Table 7.1.33 Whether or not respondents paid for dental treatment by ADH region, 1978-98

Dentate adults[†]   *England*

Whether or not respondent paid	ADH Region						All	
	North		**Midlands**		**South**			
	1988	**1998**	**1988**	**1998**	**1988**	**1998**	**1988**	**1998**
	%	%	%	%	%	%	%	%
Paid something	42	56	46	60	55	65	50	62
Paid nothing	47	33	41	30	34	27	39	29
Don't know	10	11	12	10	11	8	11	9
Base	*712*	*758*	*659*	*675*	*1434*	*1369*	*2805*	*2802*

† Excludes those who have never been to the dentist or were in the middle of a course of treatment
1988 Table 24.51 (part)

Table 7.1.34 **Whether or not respondents paid for dental treatment by characteristics of dentate adults**

Dentate adults†   *England*

		Whether or not respondents paid for dental treatment			Base
		Paid something	Paid nothing	Don't know	
All	%	62	29	9	2802
Type of service					
NHS	%	57	33	10	2120
Private	%	85	8	7	519
Age					
16-24	%	27	66	6	331
25-34	%	58	32	10	650
35-44	%	68	22	9	588
45-54	%	71	18	11	533
55-64	%	73	17	10	335
65 and over	%	75	15	10	365
Gender					
Male	%	63	27	10	1297
Female	%	60	31	9	1505
ADH region					
North	%	56	33	11	758
Midlands	%	60	30	10	675
South	%	65	27	8	1369
Regional Health Office					
Northern & Yorkshire	%	55	36	9	398
Trent	%	57	31	12	240
Eastern	%	67	26	7	255
London	%	61	33	6	370
South East	%	67	23	10	590
South West	%	65	27	8	310
West Midlands	%	60	33	8	279
North West	%	56	29	14	360
Social class of head of household					
I, II, IIINM	%	69	22	9	1378
IIIM	%	58	32	10	749
IV,V	%	51	40	8	466
Jarman Area					
1 (most deprived)	%	48	46	7	376
2	%	53	35	12	533
3	%	61	28	11	822
4	%	70	23	7	533
5 (least deprived)	%	71	20	9	533

† Excludes those who have never been to the dentist or were in the middle of a course of treatment

Table 7.1.35 **Whether respondents thought they were entitled to free NHS treatment by characteristics of dentate adults**

Dentate adults   *England*

		Respondents thought:			Base
		Treatment would be free	They would have to pay	Don't know	
All	%	19	76	5	3010
Age					
16-24	%	39	56	5	352
25-34	%	22	73	5	701
35-44	%	15	82	3	633
45-54	%	12	85	3	572
55-64	%	11	82	6	367
65 and over	%	15	76	9	385
Gender					
Male	%	15	79	5	1395
Female	%	23	73	4	1615
ADH region					
North	%	21	75	4	823
Midlands	%	22	74	5	731
South	%	17	78	5	1456
Regional Health Office					
Northern & Yorkshire	%	23	72	5	432
Trent	%	21	73	7	255
Eastern	%	19	76	5	284
London	%	25	70	5	402
South East	%	14	79	7	620
South West	%	14	84	2	327
West Midlands	%	24	73	3	299
North West	%	18	78	4	391
Social class of head of household					
I, II, IIINM	%	13	82	4	1488
IIIM	%	22	74	4	810
IV,V	%	30	64	6	489
Jarman Area					
1 (most deprived)	%	37	59	4	412
2	%	20	75	5	568
3	%	18	76	5	868
4	%	13	82	5	593
5 (least deprived)	%	14	81	5	564

Table 7.1.36 Cost of dental treatment by characteristics of dentate adults

Dentate adults who had been to the dentist in the last two years†

England

		Cost of dental treatment							Base
		Nothing	Under £10	£10 less than £20	£20 less than £30	£30 less than £50	£50 less than £100	£100 or more	
All	%	28	14	24	10	9	8	7	2049
Type of service									
NHS	%	31	17	27	8	6	6	4	1531
Private	%	8	4	14	19	20	19	16	410
Age									
16-24	%	66	8	14	4	3	4	2	231
25-34	%	30	15	22	10	10	8	6	458
35-44	%	22	14	27	10	10	8	8	441
45-54	%	19	12	26	13	10	11	9	402
55-64	%	17	15	24	13	13	9	9	250
65 and over	%	11	20	29	13	10	9	8	267
Gender									
Male	%	24	14	24	11	9	10	7	874
Female	%	31	14	23	10	9	7	7	1175
ADH region									
North	%	31	18	26	8	6	6	5	547
Midlands	%	28	13	27	10	9	8	5	481
South	%	26	12	21	11	11	10	9	1021
Regional Health Office									
Northern & Yorkshire	%	33	13	30	8	4	5	6	295
Trent	%	31	14	26	9	9	7	5	163
Eastern	%	24	13	22	11	10	12	8	200
London	%	31	8	19	10	13	8	11	267
South East	%	22	15	21	13	9	11	9	444
South West	%	28	14	24	10	12	8	5	237
West Midlands	%	27	10	32	11	9	6	4	191
North West	%	29	24	22	8	7	6	4	252
Social class of head of household									
I, II, IIINM	%	21	14	26	12	9	10	8	1066
IIIM	%	30	15	25	8	9	7	7	520
IV,V	%	40	14	17	10	9	6	4	313
Jarman Area									
1 (most deprived)	%	45	12	16	7	6	7	6	272
2	%	31	14	24	10	7	7	7	359
3	%	27	12	24	10	10	8	7	576
4	%	22	15	21	13	13	10	6	411
5 (least deprived)	%	20	15	30	10	9	9	8	427

† Excludes those who were in the middle of a course of treatment and those who did not know the cost of their dental treatment.

Table 7.1.37 **Reported frequency of teeth cleaning by characteristics of dentate adults**

Dentate adults   *England*

		Reported frequency of teeth cleaning				Base
		Twice a day or more	Once a day	Less than once a day	Never	
All	%	74	22	4	0	3010
Age						
16-24	%	80	16	4	0	352
25-34	%	78	20	3	0	701
35-44	%	75	20	4	0	633
45-54	%	71	25	4	0	572
55-64	%	72	24	4	0	367
65 and over	%	67	28	4	2	385
Gender						
Male	%	65	29	6	1	1395
Female	%	84	15	1	0	1615
ADH region						
North	%	72	24	4	0	823
Midlands	%	72	24	3	1	731
South	%	76	20	4	0	1456
Regional Health Office						
Northern & Yorkshire	%	73	24	3	0	432
Trent	%	68	27	2	2	255
Eastern	%	75	21	4	0	284
London	%	82	16	2	0	402
South East	%	74	23	3	0	620
South West	%	74	20	6	0	327
West Midlands	%	72	23	3	1	299
North West	%	72	23	5	0	391
Social class of head of household						
I, II, IIINM	%	78	20	1	0	1488
IIIM	%	70	24	6	0	810
IV, V	%	66	26	7	1	489
Jarman area						
1 (most deprived)	%	71	24	5	1	412
2	%	72	23	5	0	568
3	%	74	22	4	1	868
4	%	78	20	2	0	593
5 (least deprived)	%	77	21	2	0	564
Usual reason for dental attendance						
Regular check-up	%	80	18	2	0	1832
Occasional check-up	%	79	20	1	0	310
Only with trouble	%	62	29	8	1	859

Table 7.1.38 **Frequency of teeth cleaning by ADH region, 1988-98**

Dentate adults   *England*

Frequency of teeth cleaning	ADH Region						All	
	North		Midlands		South			
	1988	1998	1988	1998	1988	1998	1988	1998
	%	%	%	%	%	%	%	%
Twice a day or more	65	72	63	72	70	76	67	74
Once a day	27	24	30	24	25	20	27	22
Less than once a day	7	4	6	3	3	4	5	4
Never	1	0	1	1	1	0	1	0
Base	764	823	707	731	1545	1456	3016	3010

1998 Table 26.1 (part)

Table 7.1.39 **Use of dental hygiene products by characteristics of dentate adults**

Dentate adults†   *England*

		Types of dental hygiene products				Base
		Used just toothpaste & toothbrush	Used other dental hygiene products			
			All	Dental floss	Mouthwash	
All	%	48	52	28	23	2978
Age						
16-24	%	58	42	18	22	349
25-34	%	44	56	33	31	699
35-44	%	42	58	36	29	623
45-54	%	43	57	32	23	564
55-64	%	50	50	26	14	366
65 and over	%	63	37	15	9	377
Gender						
Male	%	54	46	20	21	1368
Female	%	43	57	36	25	1610
ADH region						
North	%	54	46	23	25	818
Midlands	%	53	47	26	18	714
South	%	44	56	32	24	1446
Regional Health Office						
Northern & Yorkshire	%	52	48	23	25	430
Trent	%	54	46	24	21	248
Eastern	%	53	47	28	14	279
London	%	42	58	33	31	399
South East	%	43	57	34	21	619
South West	%	45	55	26	25	322
West Midlands	%	51	49	25	18	293
North West	%	55	45	23	25	388
Social class of head of household						
I, II, IIINM	%	44	56	33	24	1481
IIIM	%	55	45	23	21	798
IV,V	%	54	46	18	21	478
Jarman area						
1 (most deprived)	%	45	55	25	32	405
2	%	52	48	26	20	559
3	%	50	50	28	23	857
4	%	46	54	32	24	588
5 (least deprived)	%	46	54	30	20	564
Usual reason for dental attendance						
Regular check-up	%	42	58	35	24	1827
Occasional check-up	%	46	54	27	26	309
Only with trouble	%	62	38	14	21	836

Percentages may add to more than 100% as respondents could give more than one answer
† Excludes those who stated they never cleaned their teeth

Table 7.1.40 **Use of dental hygiene products by ADH region, 1978-98**

Dentate adults†   *England*

Respondent used:	ADH Region									All		
	North			Midlands			South					
	1978	1988	1998	1978	1988	1998	1978	1988	1998	1978	1988	1998
	%	%	%	%	%	%	%	%	%	%	%	%
Dental floss	5	17	23	4	16	26	13	26	32	8	21	28
Toothpicks/woodsticks	7	4	4	6	3	3	14	6	6	10	5	5
Mouthwash††	-	11	25	-	8	18	-	11	24	-	10	24
Just toothbrush & toothpaste	84	67	54	84	72	53	71	58	44	78	63	48
Base	721	764	818	652	706	714	1344	1545	1446	2717	3015	2978

Percentages may add to more than 100% as respondents could give more than one answer
† Excludes those who stated they never cleaned their teeth
†† Data not collected in 1978 survey
1988 Table 26.16, 26.29

7.2 Dental health in Scotland

Summary

- In 1998, the level of total tooth loss among the Scottish population was 18% which represents a considerable improvement from 26% in 1988 and 44% in 1972.

- Progress is being made towards the oral health target that by 2010, 95% of 45 to 54 year olds will be dentate.

- The move away from the need for dentures is clearly shown by the finding that in 1972, 64% of the Scottish population were reliant on dentures to some extent and 36% had natural teeth only. By 1998, the opposite was true with 63% of the population wholly reliant on their own teeth.

- The average number of teeth in 1998 among dentate adults in Scotland was 23.8. This was a significant increase since 1978 when the average number was 21.6.

- The general improvement in dental health among dentate adults was indicated by an increase in the average number of sound and untreated teeth from 11.3 in 1978 to 13.0 in 1988 and to 14.1 in 1998.

- There had been little change in the average number of decayed teeth between 1988 and 1998 but there was a decline since 1978 (from 2.1 to 0.9).

- The percentage of dentate adults in Scotland with 12 or more restored (otherwise sound) teeth remained at a similar level from 1978 to 1998 at 32% and 34% respectively. The variation with age was similar to that seen overall in the United Kingdom with a decrease in the experience of restored teeth among the youngest dentate adults and an increase among older adults.

- Based on the 1998 criteria which included visual caries in the estimate of decay, 29% of the dentate adult population of Scotland had 18 or more sound and untreated teeth, 58% had one or more decayed or unsound teeth and 32% had 12 or more restored (otherwise sound) teeth.

- Adults aged 16 to 24 were more likely than those in other age groups to have 18 or more sound and untreated teeth.

- Forty-one percent of dentate adults in Scotland reported having had some dental pain in the previous 12 months and almost a third said they had either felt self-conscious or tense because of problems with their teeth, mouth or denture in the last 12 months. Almost 1 in 12 of those in Scotland reported that their oral condition had an occasional impact on their overall quality of life during the preceding 12 months. In just under 2% of cases this amounted to feeling totally unable to function on occasion.

- Dentate adults in Scotland were more likely to report attending the dentist for regular check-ups in 1998 than in 1988 (55% compared with 50%).

- Fourteen per cent of dentate adults in Scotland in 1998 said that the last course of dental treatment they received was private compared with 18% of adults in the United Kingdom as a whole.

- There has been no change in teeth cleaning frequency in the past ten years; 70% of dentate adults in Scotland reported cleaning their teeth at least twice a day. Of various additional dental hygiene products used, dental floss and mouthwash showed the largest increases in use from 1988 to 1998 to 24% and 29% respectively.

- The impact of oral health on people was distributed unequally with those in the poorest areas experiencing more unhealthy tooth conditions and reporting a greater burden on their life in general stemming from dental disease than those from more affluent areas.

7.2.1 Introduction

This chapter examines the clinical, behavioural and attitudinal information from a Scottish perspective in relation to the dental surveys of Scotland which have been undertaken since 1972 and in relation to the United Kingdom as a whole. Scotland has tended to be a country in which the level of dental health and uptake of good dental health behaviour has been poorer than that found in other areas of the United Kingdom. The previous surveys have shown improvements in dental health in Scotland but despite this, in 1988, when the previous survey was carried out, Scotland was effectively 10 years behind England in terms of the proportion of its population who had some teeth and the average number of missing teeth in dentate adults. This chapter looks at how Scotland is fairing in 1998.

The most appropriate point at which to begin an assessment of a nation's dental health is to look at the proportion of people who have lost all their teeth, which is a key indicator of oral health in a population, in order to examine whether past improvements are continuing.

7.2.2 Total tooth loss in Scotland 1972-98

Total tooth loss among the United Kingdom population as a whole has been falling since the first national surveys in 1968 (England and Wales) and 1972 (Scotland). In 1972 the level of total tooth loss among the Scottish population was 44%. By the time of the 1988 survey this proportion had dropped to 26%. The 1998 survey shows further improvement; 18% of the Scottish population had lost all of their teeth. This improvement can be attributed to a continuing low level of tooth loss in the younger age groups in past years. In 1972, only among those aged 16 to 24 were there fewer than 10% with no teeth. By 1998 this had increased to all those up to the age of 44. The cohort effect is equally impressive; among the cohort who were 16 to 24 year olds in 1978 the level of tooth loss has only risen from 2% to 4% from 1978 to 1998. The level of tooth loss in this cohort would suggest that by 2008 there is a very good chance that fewer than 10% of people up to the age of 55 will have lost all their teeth. It is perhaps noteworthy that those aged 16 to 24 in 1972 are the only cohort to comprise solely those who have had the opportunity for dental care within the National Health Service over their entire lifetime.

However, while total tooth loss is decreasing, it must be noted that the level of total tooth loss for those in Scotland in 1998 was significantly higher than in the United Kingdom as a whole (13%; see Chapter 2.1). As in previous surveys of Scotland a smaller proportion of men than women had lost

all their teeth; 14% compared with 21%. This is a finding common to other parts of the United Kingdom and is particularly noticeable in older adults. It is likely that this is partly the result of early tooth loss among women in the past and partly due to women living longer.

Figure 7.2.1, Table 7.2.1

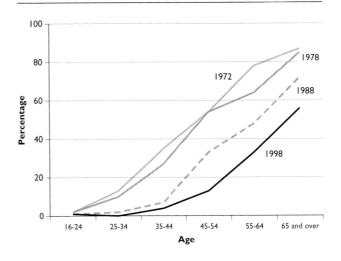

Fig 7.2.1 **Proportion of edentate adults in Scotland by age, 1972-98**

Table 7.2.2 shows that the decline in total tooth loss is not being supported by a growth in reliance on dentures in conjunction with natural teeth. The proportion of people in Scotland who rely on dentures in conjunction with their own teeth has more or less stayed constant, at around 20%, since 1972. The move away from the need for such dental prostheses is more clearly shown than by the finding that in 1972, 64% of the Scottish population were reliant on dentures to some extent and 36% had natural teeth only whereas by 1998, the opposite was true with 63% of the population wholly reliant on their own teeth.

Table 7.2.2

7.2.3 The oral health target for Scotland

In 1991, Scotland set a target for the year 2000 that less than 10% of 45 to 54 year olds should have lost all their natural teeth. The figure for 1998 was 13% of this age group with no natural teeth remaining which suggests that Scotland will be close to achieving this target. In 1999, Scotland set a new target that by 2010 more than 95% of this same age group (45 to 54 year olds) should have some natural teeth (or less than 5% should have lost them all). Among those aged 35 to 44 (who will be 45 to 54 in 2008) 96% had some natural teeth. Thus if the target is to be met no more than 1% of this age group can lose the last of their natural teeth over the next ten years. Looking at the exact cohort who will be 45 to 54 in 2010 (those aged 33 to 42 in 1998) 97% still had some natural teeth (data not shown). Based on this figure, no more than 2% of this age group can lose the last of their natural teeth over the next ten years if the target is to be met. Thus, it seems possible that Scotland will be close to achieving this oral health target in 2010.

7.2.4 The number of natural teeth among dentate adults in Scotland 1972-98

Dental health among dentate adults has traditionally been measured by the condition of teeth that are retained (sound, restored or decayed) or the number of teeth which are missing. As dental health has improved and more teeth are retained the focus has changed from the number of teeth that are missing to the number retained.

The variation in the numbers of teeth in the dentate population gives some indication of the rate at which people are losing teeth and an indirect measure of how many people may be likely to require dentures in addition to their own teeth in ensuing years. The average number of teeth among dentate adults has been increasing in Scotland since 1972; dentate adults then had an average of 21.2 teeth, ranging from 25.3 teeth among those aged 16 to 24 to 14.2 among those aged 55 and over. In 1998 the overall figure was 23.8 teeth, ranging from 27.8 for the youngest age group to 17.4 for the oldest. The proportion with 21 or more teeth had increased from 64% in 1972 to 78% in 1998. There had been little change in this proportion since 1988, although there had been significant increases within age groups above the age of 25. This apparent anomaly is the result of the changing age distribution of the dentate population as a result of people keeping their teeth for longer (see Section 5.2.2). Although there have been improvements in the numbers of teeth retained the figures are still slightly less than the level in England in 1988 (24.4 teeth, see Chapter 7.1) and that found in the United Kingdom as a whole in 1998 (24.7 teeth, see Chapter 2.2).

Figure 7.2.2, Tables 7.2.3–4

There was little difference between the average numbers of teeth among men and women and nor was there any significant difference between the genders in the proportion with 21 or more teeth (75% and 80%). There were no significant differences between the average numbers of teeth among dentate adults grouped by social class of the head of household in 1998 and similar increases since 1978 were seen for each group.

Tables 7.2.5–6

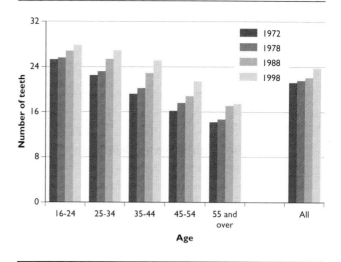

Fig 7.2.2 **Mean number of teeth among dentate adults in Scotland by age, 1972-98**

In 1998, those who attended the dentist only when they had trouble with their teeth had an average of 22.0 teeth in comparison with 24.4 among those who reported attending for regular check-ups. The overall increase in the numbers of teeth since 1978 was significant only among those who said they attended when they had some trouble with their teeth and not among those who claimed to seek regular dental check-ups.

Table 7.2.7

7.2.5 Change in the condition of natural teeth in Scotland 1972-98

The figures relating to the condition of the natural teeth presented in Tables 7.2.8 to 7.2.12 are based on the criteria used in the previous surveys of dental health (see Section 7.2.6).

The interpretation of the number and distribution of teeth which are sound and untreated or restored (otherwise sound) in the dentate population is not wholly straightforward. Sound and untreated teeth are those which have no decay and no fillings. Such teeth are often as much an indication of an absence of dental treatment as a sign of dental health. Thus, high numbers of sound and untreated teeth can be associated with those who avoid going to a dentist on a regular basis but can also be a sign of a well looked after mouth. Restored (otherwise sound) teeth are those which have a restoration (a filling or an artificial crown) and are not decayed nor have any major defect. Restored teeth indicate past experience of disease but they also reflect having received some dental care. The problem this creates is in the now routinely found result that fillings are more common in those who attend for regular dental check-ups than in those who only attend when they have some trouble with their teeth. This would seem to suggest that attending for check-ups is associated with a poor dental outcome or more disease experience, yet surveys, such as this one, repeatedly show that regular dental check-ups are associated with the positive dental outcome of the retention of natural teeth.

Overall, dentate adults in Scotland had on average 1.1 more sound and untreated teeth in 1998 than in 1988 and 2.8 more than 1978. Dentate adults aged 16 to 24 had almost 4 more sound and untreated teeth on average than their counterparts in 1988. The differences between 1988 and 1998 were successively lower in each of the remaining age groups and were only significantly different for those aged less than 45. It should be noted that older people bear the evidence of past disease levels and restorative treatment and that in the past many such people would have lost all their teeth. The proportion with 18 or more sound and untreated teeth had increased from 23% in 1972 to 34% in 1998. Those in the youngest age group showed the largest increase in this percentage - from 41% in 1972 to 86% in 1998.

Since 1972 the level of decayed or unsound teeth had fallen from an average of 2.4 teeth affected to 0.9 teeth in 1998. However, there was no significant decrease between 1988 and 1998, a finding which was also true for English dentate adults. The reductions since 1972 have been fairly consistent across all age groups. Among those aged 16 to 24 the average number of teeth with decay had decreased from 2.7 in 1972 to 1.2 in 1998 and among adults aged 55 and over it had dropped from 2.0 teeth in 1972 to 0.8 in 1998. The proportion with any decay also showed a steady decrease during this period from 67% in 1972 to 41% in 1998.

There has been little variation in the distribution of restored (otherwise sound) teeth among dentate adults in Scotland since 1978 (32% had 12 or more such teeth in 1978 compared with 34% in 1998). The variation with age was similar to that seen in the other countries of the United Kingdom with a marked decrease in the restorative experience of the youngest dentate adults (from 21% with 12 or more restored (otherwise sound) teeth in 1972 to 6% in 1998) and an increase among the older adults. For example, 13% of those aged 55 and over had 12 or more restored (otherwise sound) teeth in 1972 compared with 31% in 1998.

Figures 7.2.3–5, Tables 7.2.8–9

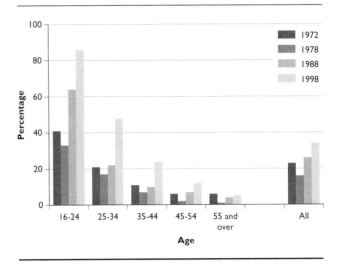

Fig 7.2.3   **Proportion of dentate adults in Scotland with 18 or more sound and untreated teeth by age, 1972-98**

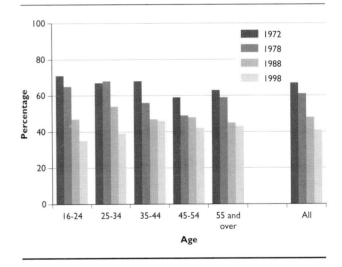

Fig 7.2.4   **Proportion of dentate adults in Scotland with some decayed or unsound teeth by age, 1972-98**

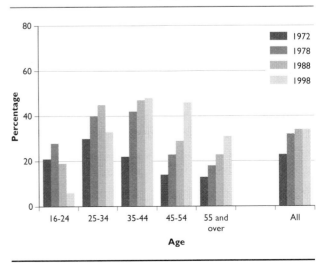

Fig 7.2.5   **Proportion of dentate adults in Scotland with 12 or more restored (otherwise sound) teeth by age, 1972-98**

In 1972, dentate men were more likely than dentate women to have 18 or more sound and untreated teeth, and less likely to have 12 or more restored (otherwise sound) teeth. By 1998, these differences had mostly disappeared. Among dentate men there had been no significant increase in the proportion with 18 or more sound and untreated teeth (29% in 1972 and 36% in 1998) while among dentate women this proportion had increased from 15% to 31%. Conversely dentate men showed a much larger increase in experience of restorative treatment. In 1998, dentate men had 1.2 decayed or unsound teeth compared with 2.8 such teeth in 1972. Similarly among dentate women the number of decayed or unsound teeth dropped from 2.1 in 1972 to 0.7 in 1998.

Table 7.2.10

There were no marked differences between 1988 and 1998 in the distribution of sound and untreated teeth for dentate adults within any social class. The only marked difference between the 1998 findings and those for 1978 was an increase in the number of sound and untreated teeth among those from non-manual households from 10.4 in 1978 to 13.6 in 1998 (data for 1972 were not available). For restored (otherwise sound) teeth none of the differences were significant, however, the data indicated increases in restorative treatment among those from manual backgrounds and a decrease in such treatment among those from non-manual backgrounds. Those from skilled manual backgrounds showed the largest reduction in the numbers of decayed or unsound teeth, an average of 2.6 teeth in 1978 to 0.8 teeth in 1998. However, in 1998, there were still differences between the groups; those from a non-manual background had 0.8 decayed or unsound teeth compared with 1.6 such teeth among those from an unskilled manual background, a figure which was similar to the level among the non-manual group twenty years previously.

Table 7.2.11

The average number of sound and untreated teeth showed a greater increase between 1978 and 1998 among those who attend for regular check-ups than among those who attend the dentist only when they have some trouble with their teeth (10.2 teeth in 1978 to 13.3 teeth in 1998 compared with 11.7 to 14.0). The 1998 average values for those who reported attending for regular check-ups and those who only attended with trouble were both lower than the overall average for Scotland. This is because there is a third group, those who attend for occasional check-ups, who were omitted from the table as no comparable data from previous surveys exists for them. The average number of sound and untreated teeth among this group was 17.5. The average level of decayed or unsound teeth among those who reported attending only with trouble dropped significantly from 3.3 teeth in 1978 to 1.6

teeth in 1998. There was no significant reduction among those who attend for regular check-ups during the same period.

It has already been noted that having restored teeth is associated with the usual reason for attendance. However, the gap between the two types of attenders appears to be closing. In 1998, those who attended for regular check-ups had fewer restored (otherwise) sound teeth than in 1978 (10.4 compared with 12.1) while the number among those who attend only with trouble had increased from 4.5 teeth in 1978 to 6.3 teeth in 1998. The reduction in the number of restored teeth in regular attenders may reflect less disease experience and a change in approach to the management of caries. The increase in restored (otherwise sound) teeth among those who only attend when they have some trouble may reflect a move away from extractions; this would be consistent with the reduction in the number of missing teeth (shown in Table 7.2.7) in this type of attender since 1978.

Table 7.2.12

7.2.6 The condition of the natural teeth in 1998 based on the new criteria

One of the most significant changes with respect to assessment of dental caries in this survey has been in the collection of dental caries data. In the previous surveys in this series, a tooth was recorded as decayed only if cavitated caries was present. At the same time an assessment was made of the extent of the decay, either cavitated decay or decay that was so extensive that the tooth was broken down, perhaps with pulpal involvement and unrestorable as a consequence. Both these categories were used in the 1998 survey but for the first time, an assessment of visual caries (that is caries that has caused demineralisation of the tooth but without cavitation) was also included in the assessment of decay. The change in criteria for assessing dental caries has implications for each of the definitions of tooth condition. In 1998, teeth with untreated visual caries were classified as decayed but in the 1988 survey and earlier such teeth were recorded as sound and untreated. Filled teeth with recurrent visual caries were assessed as decayed in the 1998 classification whereas they were defined as teeth with sound restorations when applying the pre-1998 criteria

Table 7.2.13 shows the effect of these changes on the Scottish results. Twice as many teeth were classified as decayed or unsound under the 1998 criteria in Scotland than would have been the case in 1988. The average number of teeth classified as restored (otherwise sound) and the average number classified as sound and untreated were reduced by 0.4 and 0.5 respectively. The inclusion of visual caries had a greater effect on the findings for younger age groups. Among those aged 16 to 24 the number of teeth classified as decayed or unsound almost doubled from 1.2 to 2.2 using the new criteria. However among those aged 65 or over the difference was less marked rising from 0.8 to 1.3 teeth.

Table 7.2.13

In brief, the condition of teeth in Scotland based on the 1998 criteria was as follows:

- Twenty-nine percent of the dentate adult population had 18 or more sound and untreated teeth - 79% of dentate adults aged 16 to 24, 38% of those aged 25 to 34 and 5% or less of people over 55 years of age. Overall, 58% of dentate adults had one or more decayed or unsound teeth, with little variation with age. One third (32%) of dentate adults had 12 or more restored (otherwise sound) teeth. This proportion was highest among those aged 35 to 54 (43% and 46%) and was lowest among those aged 16 to 24 (5%).

- Dentate men and women were equally likely to have 18 or more sound and untreated teeth, but 63% of men compared with 53% of women had one or more decayed or unsound teeth. There was no significant difference between dentate men and women in the proportion with 12 or more restored (otherwise sound) teeth.

- There were no statistically significant differences between the social classes in terms of the proportion with 18 or more sound and untreated teeth, but the proportion of dentate adults in non- manual classes with one or more decayed or unsound teeth was significantly lower than that for those from unskilled manual backgrounds (53% compared 66%). The converse was true for restorative experience with those from non-manual backgrounds being more likely to have 12 or more restored (otherwise sound) teeth (39% compared with 22%).

- There was no significant difference between regular attenders and those who attended only with trouble in the proportion of dentate adults with 18 or more sound and untreated teeth, but those who attended for occasional check-ups were more likely to have sound and untreated teeth. Half (51%) of attenders for regular check-ups had one or more decayed or unsound teeth, compared with just under three-quarters (71%) of people who reported visiting the dentist only with trouble. Twice as many attenders for regular check-ups as those who attended only with trouble had 12 or more restored (otherwise sound) teeth (40% compared with 19%).

Tables 7.2.14–17

7.2.7 The periodontal condition of dentate adults in Scotland

A factor in tooth retention other than dental decay is gum disease. Gum disease affects teeth by destroying the underlying bone and can lead to teeth becoming loose and can be assessed by measuring loss of attachment and depth of pocketing. In 1998, 41% of those in Scotland who had lost all their teeth said that gum disease was the main reason for the last of their teeth being extracted (table not shown).

Two-thirds (62%) of dentate adults in Scotland had some calculus and 60% had some visible plaque (table not shown). Table 7.2 18 shows that 47% had some pocketing of 4mm or more and 5% had pocketing of 6mm or more. More than half of dentate adults (56%) in Scotland had no loss of attachment greater than 4mm but 8% had loss of attachment of 6mm or more.

Table 7.2.18

7.2.8 The impact of dental disease on adults in Scotland

Knowledge of the extent of dental disease gives a clinical indication of the experience of dental problems but it does not necessarily reflect the problems that people experience as a result of the condition of their dentition. There are differences between clinicians' and the public's evaluation of oral health. For example, dentists often appear to be most concerned about the integrity of previous dental work whereas people often seem more concerned with the appearance of their teeth. It is becoming increasingly appreciated that the way a disease affects people's lives is just as important as epidemiological measures of its prevalence or incidence. This section considers the self-perceived impact on people of dental disease and conditions as a whole and the resultant effect on their quality of life.

The measure used to assess this was the 14-question version of Slade and Spencer's Oral Health Impact Profile (OHIP-14). A full description of this measure can be found in Chapter 6.5.

The three most important impacts of oral condition were pain, psychological discomfort and psychological disability. Forty-one per cent of Scottish dentate adults reported having had some dental pain in the previous 12 months. In most cases this was occasional but 2% said they had a painful aching in the mouth very often and 2% found it uncomfortable to eat any foods very often in the last 12 months. Almost a third (30%) of the dentate population in Scotland said they had either felt self-conscious or tense because of problems with their teeth, mouth or denture in the previous 12 months. Just over a fifth (22%) found their oral condition made it difficult to relax or felt a bit embarrassed by it. Fewer reported being affected functionally by their oral condition; 11% said they had had some trouble pronouncing words or that their sense of taste had changed and 10% felt that their diet had been unsatisfactory or that they had had to interrupt meals because of their oral condition.

Although some impacts were not reported very frequently they are nevertheless important as they concern people whose dental condition is such that it affects their life. Almost 1 in 12 dentate adults in Scotland reported that their oral condition had an occasional impact on their quality of life as a whole over the preceding 12 months. For 8% of people this was reflected in making their life in general less satisfying on occasion but in the case of a small proportion of people it amounted to feeling totally unable to function on occasion as a result of their oral condition.

Table 7.2.19

7.2.9 Usual reason for dental attendance in Scotland 1972-1998

Since the first survey of adult dental health in 1972 participants have been asked "In general do you go to a dentist for a regular check-up, an occasional check-up or only when you're having trouble with your teeth?" It has been made clear in previous reports that seeking regular dental check-ups is not intended to be interpreted as attending at a particular frequency. Follow-up work of the 1978 Adult Dental Health survey in Scotland has shown that observed frequency of attendance is not directly related to a persons stated motive for attending.[1] Nevertheless, this is a key variable in relation to people's attitudes towards dental care and one which has consistently shown relationships with dental disease and dental treatment received.

In 1972 only a third of dentate adults in Scotland reported to visit a dentist for regular check-ups. This proportion has increased steadily to over a half (55%) in 1998. It is notable that attending for check-ups is lowest among the youngest age group. In some respects this gives a misleading impression as seeking check-ups is strongly related to retaining teeth and those who have poorer dental health in each age group gradually drop out through becoming edentate leaving a progressively higher proportion of those who are more dentally motivated.

Table 7.2.20

7.2.10 Visiting the dentist among dentate adults in Scotland

Respondents were asked if their last course of dental treatment had been NHS, private or something else. The answers to this question have to be treated with some caution. Most adults are required to pay a contribution to the cost of their NHS dental care and this may cause confusion about whether their dental treatment was within the NHS. Overall 14% of all adults in Scotland said that the most recent course of dental treatment they received was private. This proportion showed little variation with the time since the most recent visit took place. The use of private dental services did not vary significantly with gender, social class or usual reason for dental attendance. Age, however, was a significant factor. Those aged 16 to 24 were more likely than other age groups to say their last course of treatment was provided by the NHS but many of these courses would have been when they were still eligible for free NHS care (see Chapter 6.3). Nevertheless there was also an indication that a greater proportion of dentate adults aged 45 or more had private dental care during their last course of treatment although the differences were not significant.

Tables 7.2.21–22

7.2.11 Opinions about dental visits among dentate adults in Scotland

The reasons underlying people's attendance behaviour vary; concerns can be due to cost, anxiety about treatment or a lack of belief in the benefits of seeking check-ups. Around a third of those who reported only attending with trouble definitely felt that it would cost less if they only visited when they had trouble (29%) and that NHS dental treatment was expensive (34%). A higher proportion of those who only attend when they have some trouble with their teeth definitely felt anxious about attending (41% compared with 14%) or nervous of some kinds of dental treatments (41% compared with 21%) than those who sought regular check-ups. A key factor for almost a half (48%) of those who said they only attend when they have trouble was that there is no point in visiting a dentist unless they feel they have a need for treatment. This was a view shared only by 5% of those who attend for check-ups regularly.

Table 7.2.23

7.2.12 Reported dental hygiene behaviour among dentate adults in Scotland

Effective teeth cleaning is important in the prevention and control of dental decay and gum disease. It is difficult to determine how effective cleaning is by self-report but the frequency of brushing and use of additional methods can be used as proxy measures.

There was no difference in reported frequency of tooth cleaning between 1988 and 1998. Most people in Scotland (93%) reported brushing at least daily with 70% reporting brushing at least twice each day in 1998. The use of dental floss among people in Scotland had increased from 17% to 24% in the 10 years between 1988 and 1998. Mouthwash was also being used more in Scotland; in 1988, 13% said they used mouthwash which had more than doubled to 29% by 1998.

A greater proportion of the dentate adults in Scotland in 1998 recalled having been given a demonstration of teeth cleaning than in 1988. In 1988 only 38% said they had been shown how to clean their teeth whereas by 1998, 51% said they had been given a teeth cleaning demonstration. The proportion who had been given advice on gum care had also risen from 29% in 1988 to 44% in 1998. Overall 61% of the Scottish population could recall having received either of these types of advice in 1998 which is similar to the findings in the United Kingdom as a whole.

Table 7.2.24

7.2.13 Deprivation and dental health, attitudes and behaviour in Scotland

The data collected for this survey enables a variety of competing analyses to be undertaken in relation to material deprivation. Information was gathered on car ownership, household income and social class. However, there is increasing use in Scotland of the Carstairs and Morris index of deprivation[2] as a means of classifying people on the basis of the postcode sector where they live. The index consists of 7 categories which are referred to as DEPCAT area 1 (least deprived) to DEPCAT area 7 (most deprived). (Appendix F gives details on the Carstairs and Morris index of deprivation.)

In many respects, however, this is likely to be a less discriminating measure of deprivation than those based on individual circumstances; the index can inappropriately classify individuals whose living standards may be different from that generally present in an area. Furthermore the importance or impact of some aspects of the Carstairs and Morris classification such as car ownership will differ from rural to urban areas. Nevertheless, household postcode is often the most easily available personal data and often has fewer problems relating to privacy for individuals than other measures.

The 1998 survey shows that many conditions in dental health were related to deprivation. Fewer of those in the more deprived areas (DEPCAT areas 5, 6 and 7) retained some of their own teeth in comparison to those in less deprived areas. Dentate adults in the most deprived areas also had fewer teeth on average than those in other areas, which suggests that more of them may lose all their teeth earlier than those who are least deprived.

People in the most deprived areas were more likely than those in the least deprived areas to have some teeth with cavitated decay and less likely to have 12 or more restored (otherwise sound) teeth. For example, 37% of those in DEPCAT areas 6 and 7 had one or more teeth with cavitated decay compared with 20% of those in DEPCAT areas 1 and 2.

Figure 7.2.6, Table 7.2.25

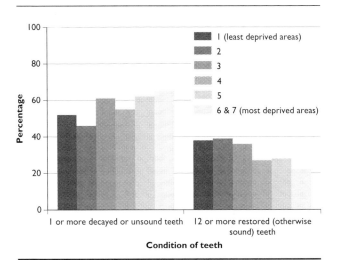

Fig 7.2.6 **The condition of teeth among dentate adults in Scotland by DEPCAT area**

Legend:
- 1 (least deprived areas)
- 2
- 3
- 4
- 5
- 6 & 7 (most deprived areas)

Condition of teeth

There was no clear relationship between the measures associated with gum (periodontal) health and deprivation category, although there was a significant difference in calculus levels which were lower in the more affluent areas; 61% of those in DEPCAT area 1 had calculus present compared with 79% of those in DEPCAT areas 6 and 7.

Table 7.2.26

People from the more deprived areas were also more likely to be affected by their oral condition than those from less deprived areas. Those from the most deprived areas reported experiencing an average of 2.2 different problems on the OHIP-14 scale during the last 12 months compared with 1.7 and 1.1 respectively among those living in DEPCAT areas 1 and 2. Overall, 61% of those in DEPCAT areas 6 and 7 said they were affected by their dental condition compared with 48% of those in DEPCAT area 1.

Table 7.2.27

Deprivation was also related to satisfaction with tooth appearance and concerns about dentistry. People living in the more deprived areas were more likely to be dissatisfied with the appearance of their teeth than those from less deprived areas. They were also more likely to have definitely felt that NHS treatment was expensive; to feel that it would cost less in the long run if they only go to the dentist when they had trouble and to feel that they would like to pay for their dental treatment by instalments.

The relationship between deprivation category and concerns about cost may reasonably be expected. Less obvious however are the findings concerning some other aspects of dental visits. A larger proportion of those who lived in the more deprived areas felt they were anxious about going to a dentist and nervous about some kinds of treatment. They were also more likely to feel that there is no point in visiting unless they are aware of some specific need.

Table 7.2.28

Given the concerns among those who live in the more deprived areas it is not surprising to see these concerns reflected in reported attendance behaviour. Over 70% of those in the least deprived areas said they attended for regular dental check-ups compared with 28% in the two most deprived categories. There was, however, little variation in the use of private dental care between the different DEPCAT groups.

Those in the more deprived areas were less likely to have been shown how to clean their teeth or have been given advice on how to look after their gums. Between 69% and 72% of those in the two least deprived areas had been given some oral care information compared with 47% of those in the most deprived areas. Dentate adults living in these areas were also less likely to brush their teeth twice a day and considerably less likely to use dental floss (9% compared with 41% and 39% of those in the two least deprived areas). However, an intriguing difference was that use of mouthwash was reported by significantly more of those in the more deprived areas (between 35% and 32%) compared with those in less deprived areas (19% to 24%).

Figure 7.2.7, Table 7.2.29

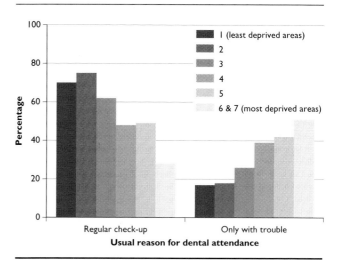

Fig 7.2.7 **Usual reason for dental attendance of dentate adults in Scotland by DEPCAT area**

This is the first analysis of national data for adult dental health by deprivation category. It shows that oral health like many other aspects of health are related to deprivation category. It also shows that the impact of oral health on people is distributed unequally with those in the poorest areas reporting a greater burden on their life in general stemming from dental disease.

Notes and references

1. Eddie S. *Frequency of attendance in the General Dental Service in Scotland a comparison with claimed attendance.* Br Dent J 1984; 157: 267-270.
2. Carstairs V, Morris R. *Deprivation and health in Scotland.* Aberdeen: Aberdeen University Press, 1991.

Table 7.2.1   Edentate adults, 1972-98

All adults   *Scotland*

	1972	1978	1988	1998
	percentage edentate			
All	44	39	26	18
Age				
16-24	2	2	1	1
25-34	13	10	2	0
35-44	35	27	7	4
45-54	54	54	33	13
55-64	78	64	48	33
65 and over	87	85	72	56
Gender				
Men	39	35	22	14
Women	48	42	31	21
Social class of head of household				
I, II, IIINM	33	32	18	12
IIIM	41	38	29	20
IV,V	53	45	38	28

For bases see Tables Appendix
1988 Table 3.3

Table 7.2.2   Natural teeth and dentures, 1972-98

All adults   *Scotland*

	1972	1978	1988	1998
	%	%	%	%
Natural teeth only	36	40	52	63
Natural teeth and dentures	20	21	21	20
No natural teeth	44	39	26	18
Base	2290	1420	1582	1204

1988 Table 16.12

Table 7.2.3   Distribution of number of teeth, 1972-98

Dentate adults   *Scotland*

Number of teeth	1972	1978	1988	1998
	%	%	%	%
1-14	18	18	13	11
15-21	18	16	12	12
21-26	39	39	35	32
27-31	24	26	39	43
32	1	1	2	3
21 or more	64	66	76	78
Mean	21.2	21.6	23.1	23.8
Base	1170	748	926	668

1988 Table 5.9

Table 7.2.4 **Distribution of number of teeth by age, 1972-98**

Dentate adults   *Scotland*

Number of teeth	Age																			
	16-24				25-34				35-44				45-54				55 and over			
	1972	1978	1988	1998	1972	1978	1988	1998	1972	1978	1988	1998	1972	1978	1988	1998	1972	1978	1988	1998
	%	%	%	%	%	%	%	%	%	%	%	%	%	%	%	%	%	%	%	%
1-14	2	1	0	0	12	11	6	1	24	22	12	5	40	31	30	18	48	51	34	32
15-20	7	5	2	2	15	15	7	3	26	22	14	9	30	29	21	16	31	25	26	28
21-26	50	50	33	26	45	41	35	30	34	36	41	34	25	35	36	39	19	18	28	28
27-31	40	43	62	65	27	32	49	60	15	19	32	51	4	5	13	26	2	6	13	11
32	1	1	2	7	1	1	3	5	1	1	1	1	1	0	0	1	0	0	0	1
21 or more	91	94	97	98	73	74	87	95	50	56	74	86	30	40	49	66	21	24	41	40
Mean	25.3	25.6	26.8	27.8	22.5	23.2	25.4	26.9	19.2	20.2	22.9	25.1	16.2	17.6	18.8	21.4	14.2	14.7	17.1	17.4
Base	382	211	252	80	297	212	244	157	234	144	216	142	152	86	116	132	104	95	156	157

1988 Table 5.9

Table 7.2.5 **Distribution of number of teeth by gender, 1972-98**

Dentate adults   *Scotland*

Number of teeth	Gender								All			
	Men				Women							
	1972	1978	1988	1998	1972	1978	1988	1998	1972	1978	1988	1998
	%	%	%	%	%	%	%	%	%	%	%	%
1-14	15	18	16	11	21	17	11	11	18	18	13	11
15-20	20	16	12	14	15	16	12	9	18	16	12	12
21-26	39	40	33	29	40	39	36	34	39	39	35	32
27-31	25	25	38	44	23	27	40	43	24	26	39	43
32	1	1	2	3	1	1	1	3	1	1	2	3
21 or more	65	66	73	75	64	67	77	80	64	66	76	78
Mean	21.6	21.4	22.8	23.7	20.8	21.7	23.4	23.9	21.2	21.6	23.1	23.8
Base	601	370	504	298	569	378	480	370	1170	748	984	668

1988 Table 5.10

Table 7.2.6   **Distribution of number of teeth by social class of head of household, 1978-98**

Dentate adults   *Scotland*

Number of teeth	Social class of head of household									All†		
	I, II, IIINM			IIIM			IV,V					
	1978	1988	1998	1978	1988	1998	1978	1988	1998	1978	1988	1998
	%	%	%	%	%	%	%	%	%	%	%	%
1-14	14	9	8	19	18	13	24	18	17	18	13	11
15-20	17	10	10	15	11	14	20	18	13	16	12	12
21-26	36	34	32	45	34	33	33	34	32	39	35	32
27-31	32	44	45	21	35	38	23	30	35	26	39	43
32	1	3	5	0	1	1	0	0	3	1	2	3
21 or more	69	81	82	66	70	72	56	64	70	66	76	78
Mean	22.4	24.2	24.5	21.1	22.2	22.9	20	21.5	22.4	21.6	23.1	23.8
Base	308	480	330	258	316	164	140	176	119	748	984	668

† Includes those for whom social class of head of household was not known and armed forces
1988 Table 5.11

Table 7.2.7   **Distribution of number of teeth by usual reason for dental attendance, 1978-98**

Dentate adults   *Scotland*

Number of teeth	Usual reason for dental attendance						All†		
	Regular check-up			Only with trouble					
	1978	1988	1998	1978	1988	1998	1978	1988	1998
	%	%	%	%	%	%	%	%	%
1-14	10	6	7	27	26	21	18	13	11
15-20	13	10	13	18	14	11	16	12	12
21-26	41	38	36	36	30	26	39	35	32
27-31	35	45	43	18	28	38	26	39	43
32	1	2	2	1	1	3	1	2	3
21 or more	77	85	68	55	59	68	66	76	78
Mean	23.2	24.7	24.4	19.5	20.6	22.0	21.6	23.1	23.8
Base	296	500	404	311	356	189	748	984	668

† Includes those who visit the dentist for an occasional check-up
1988 Table 5.12

Table 7.2.8   **The condition of natural teeth, 1972-98**

Dentate adults   *Scotland*

The condition of natural teeth (1988 criteria)	1972	1978	1988	1998
	%	%	%	%
Sound and untreated				
0	1	2	1	1
1-5	11	15	10	10
6-11	35	38	33	27
12-17	30	29	30	28
18 or more	23	16	26	34
Mean	12.3	11.3	13.0	14.1
	%	%	%	%
Decayed or unsound				
0	33	39	52	59
1-5	53	51	44	38
6 or more	14	10	4	3
Mean	2.4	2.1	1.2	0.9
	%	%	%	%
Restored (otherwise sound)				
0	23	16	8	9
1-5	27	23	24	24
6-11	27	29	34	33
12 or more	23	32	34	34
Mean	6.5	8.1	8.9	8.8
Mean number of teeth	21.2	21.6	23.1	23.8
Base	*1170*	*748*	*984*	*668*

1988 Table 6.12, 7.13, 8.18

Table 7.2.9   **The condition of natural teeth by age, 1972-98**

Dentate adults   *Scotland*

The condition of natural teeth (1988 criteria)	Age																			
	16-24				25-34				35-44				45-54				55 and over			
	1972	1978	1988	1998	1972	1978	1988	1998	1972	1978	1988	1998	1972	1978	1988	1998	1972	1978	1988	1998
	%	%	%	%	%	%	%	%	%	%	%	%	%	%	%	%	%	%	%	%
Sound and untreated																				
0	0	1	0	0	1	0	0	0	1	2	0	2	2	1	2	2	6	4	6	2
1-5	3	4	1	0	9	15	7	4	15	19	13	8	17	20	18	11	32	27	18	26
6-11	21	20	9	4	36	37	34	14	42	50	40	30	49	53	45	47	40	52	50	41
12-17	35	42	26	10	33	31	37	35	31	22	36	36	26	24	27	27	16	16	23	26
18 or more	41	33	64	86	21	17	22	48	11	7	10	24	6	2	7	12	6	1	4	5
Mean	15.8	15.0	18.6	22.5	12.4	11.8	13.3	16.5	10.6	9.4	11.2	13.2	9.2	8.8	9.4	10.4	7.7	7.5	8.7	8.6
	%	%	%	%	%	%	%	%	%	%	%	%	%	%	%	%	%	%	%	%
Decayed or unsound																				
0	29	35	53	65	33	32	46	61	32	44	53	54	41	51	52	58	37	41	55	57
1-5	54	54	41	30	49	57	49	34	58	47	44	45	53	40	46	40	53	50	41	40
6 or more	17	11	6	5	18	11	5	5	10	9	3	2	6	9	2	2	10	9	4	3
Mean	2.7	2.2	1.3	1.2	2.6	2.2	1.3	0.9	2.2	2.1	1.0	0.9	1.7	1.8	1.0	0.8	2.0	2.1	1.2	0.8
	%	%	%	%	%	%	%	%	%	%	%	%	%	%	%	%	%	%	%	%
Restored (otherwise sound)																				
0	20	7	7	25	20	12	5	1	25	16	6	2	26	23	13	7	39	34	15	15
1-5	27	25	37	46	24	20	16	22	20	23	15	14	30	23	21	17	27	27	28	26
6-11	32	40	38	23	26	28	33	44	23	19	32	35	30	31	37	30	21	21	33	28
12 or more	21	28	19	6	30	40	45	33	22	42	47	48	14	23	29	46	13	18	23	31
Mean	6.8	8.4	6.9	4.0	7.5	9.3	10.8	9.4	6.4	8.7	10.7	11.0	5.3	7.0	8.4	10.1	4.5	5.1	7.2	8.0
Mean number of teeth	25.3	25.6	26.8	27.8	22.5	23.2	25.4	26.9	19.2	20.2	22.9	25.1	16.2	17.6	18.8	21.4	14.2	14.7	17.1	17.4
Base	382	211	252	80	297	212	244	157	234	144	216	142	152	86	116	132	104	95	156	157

1988 Table 6.12, 7.13, 8.18

Table 7.2.10 **The condition of natural teeth by gender, 1972-98**

Dentate adults   *Scotland*

The condition of natural teeth (1988 criteria)	Gender								All			
	Men				Women							
	1972	1978	1988	1998	1972	1978	1988	1998	1972	1978	1988	1998
	%	%	%	%	%	%	%	%	%	%	%	%
Sound and untreated												
0	2	1	2	1	1	2	1	1	1	2	1	1
1-5	8	15	9	11	14	15	10	9	11	15	10	10
6-11	32	34	32	25	37	42	33	30	35	38	33	27
12-17	29	33	30	28	33	26	30	29	30	29	30	28
18 or more	29	17	27	36	15	15	25	31	23	16	26	34
Mean	13.2	11.8	13.3	14.3	11.3	10.9	12.8	13.8	12.3	11.3	13.0	14.1
	%	%	%	%	%	%	%	%	%	%	%	%
Decayed or unsound												
0	27	33	47	53	40	44	56	64	33	39	52	59
1-5	57	54	47	42	49	48	40	34	53	51	44	36
6 or more	16	13	6	5	11	8	3	2	14	10	4	3
Mean	2.8	2.5	1.4	1.2	2.1	1.8	1.0	0.7	2.4	2.1	1.2	0.9
	%	%	%	%	%	%	%	%	%	%	%	%
Restored (otherwise sound)												
0	28	18	10	11	19	13	6	6	23	16	8	9
1-5	29	27	27	24	25	19	20	23	27	23	24	24
6-11	26	28	34	33	28	30	35	34	27	29	34	33
12 or more	17	27	30	32	28	38	38	37	23	32	34	34
Mean	5.6	7.2	8.2	8.2	7.4	9.1	9.7	9.3	6.5	8.1	8.9	8.8
Mean number of teeth	21.6	21.4	22.8	23.7	20.8	21.7	23.4	23.9	21.2	21.6	23.1	23.8
Base	*601*	*370*	*504*	*298*	*569*	*378*	*480*	*370*	*1170*	*748*	*984*	*668*

1988 Table 6.13, 7.14, 8.19

Table 7.2.11 The condition of natural teeth by social class of head of household, 1978-98

Dentate adults   *Scotland*

The condition of natural teeth (1988 criteria)	Social class of head of household									All†		
	I, II, IIINM			IIIM			IV,V					
	1978	1988	1998	1978	1988	1998	1978	1988	1998	1978	1988	1998
	%	%	%	%	%	%	%	%	%	%	%	%
Sound and untreated												
0	2	2	1	1	0	2	1	3	2	2	1	1
1-5	14	9	12	16	11	6	14	10	9	15	10	10
6-11	46	35	26	33	33	36	33	27	28	38	33	27
12-17	27	32	30	31	29	24	33	30	30	29	30	28
18 or more	11	22	31	19	28	31	19	29	32 ·	16	26	34
Mean	10.4	12.8	13.6	11.8	12.9	13.5	11.9	13.1	13.5	11.3	13.0	14.1
	%	%	%	%	%	%	%	%	%	%	%	%
Decayed or unsound												
0	48	57	60	31	47	60	32	43	47	39	52	59
1-5	46	41	39	56	47	37	53	47	44	51	44	38
6 or more	6	2	2	13	6	3	15	10	9	10	4	3
Mean	1.5	0.8	0.8	2.6	1.4	0.8	2.7	1.8	1.6	2.1	1.2	0.9
	%	%	%	%	%	%	%	%	%	%	%	%
Restored (otherwise sound)												
0	8	4	7	20	11	7	25	16	11	16	8	9
1-5	16	19	15	27	26	28	31	31	33	23	24	24
6-11	29	33	37	28	36	33	30	36	31	29	34	33
12 or more	47	45	41	25	27	32	14	17	26	32	34	34
Mean	10.5	10.6	10.0	6.8	7.8	8.5	5.5	6.6	7.3	8.1	8.9	8.8
Mean number of teeth	22.4	24.2	24.5	21.1	22.2	22.9	20.0	21.5	22.4	21.6	23.1	23.8
Base	308	480	330	258	316	164	140	176	119	748	984	668

† Includes those for whom social class of head of household was not known and armed forces
1988 Tables 6.14, 7.15, 8.20

Table 7.2.12 **The condition of natural teeth by usual reason for dental attendance, 1978-98**

Dentate adults   *Scotland*

The condition of natural teeth (1998 criteria)	Usual reason for dental attendance						All[†]		
	Regular check-up			Only with trouble					
	1978	1988	1998	1978	1988	1998	1978	1988	1998
	%	%	%	%	%	%	%	%	%
Sound and untreated									
0	2	0	1	1	3	1	2	1	1
1-5	16	10	10	17	10	12	15	10	10
6-11	46	35	31	32	32	24	38	33	27
12-17	27	34	31	30	26	26	29	30	28
18 or more	9	21	27	20	28	37	16	26	34
Mean	10.2	12.8	13.4	11.7	12.8	14.0	11.3	13.0	14.1
	%	%	%	%	%	%	%	%	%
Decayed or unsound									
0	56	65	66	22	33	43	39	52	59
1-5	42	34	32	58	57	50	51	44	38
6 or more	2	1	2	20	11	7	10	4	3
Mean	1.0	0.6	0.6	3.3	2.1	1.6	2.1	1.2	0.9
	%	%	%	%	%	%	%	%	%
Restored (otherwise sound)									
0	1	1	4	31	18	16	16	8	9
1-5	12	17	17	33	35	35	23	24	24
6-11	29	34	37	25	32	30	29	34	33
12 or more	58	48	43	11	15	20	32	34	34
Mean	12.1	11.3	10.4	4.5	5.7	6.3	8.1	8.9	8.8
Mean number of teeth	23.2	24.7	24.4	19.5	20.6	22.0	21.6	23.1	23.8
Base	*296*	*500*	*409*	*311*	*356*	*189*	*748*	*984*	*668*

† Includes those who only visit the dentist for an occasional check-up
1988 Tables 6.15, 7.16, 8.21

Table 7.2.13 **Difference in the condition of teeth by age using 1998 and pre-1998 criteria**

Dentate adults   *Scotland*

| Age | Percentage with one or more decayed teeth | | | Average number of teeth that were: | | | | | | | | | Base |
| | | | | Sound and untreated | | | Decayed or unsound | | | Restored (otherwise sound) | | | |
	1998 criteria	pre-1998 criteria	Difference	1998 criteria	pre-1998 criteria	Difference	1998 criteria	pre-1998 criteria	Difference	1998 criteria	pre-1998 criteria	Difference	
All	58	41	17	13.6	14.1	0.5	1.8	0.9	0.9	8.4	8.8	0.4	668
16-24	56	35	21	21.7	22.5	0.8	2.2	1.2	1.0	3.8	4.0	0.2	80
25-34	58	39	19	15.7	16.5	0.8	2.2	0.9	1.3	8.9	9.4	0.5	157
35-44	61	46	15	12.9	13.2	0.3	1.6	0.9	0.7	10.5	11.0	0.5	142
45-54	61	42	19	10.1	10.4	0.3	1.6	0.8	0.8	9.7	10.1	0.4	132
55-64	53	44	9	9.1	9.3	0.2	1.5	0.9	0.6	8.0	8.4	0.4	75
65 and over	56	42	14	7.6	7.8	0.2	1.3	0.8	0.5	7.2	7.6	0.4	82

Table 7.2.14 **The condition of natural teeth by age**

Dentate adults   *Scotland*

| The condition of natural teeth (1998 criteria) | Age | | | | | | All |
	16-24	25-34	35-44	45-54	55-64	65 and over	
	%	%	%	%	%	%	%
Sound and untreated							
0	0	0	2	2	1	2	1
1-5	0	5	8	12	24	33	11
6-11	4	15	32	51	37	44	29
12-17	17	41	36	24	32	17	30
18-23	40	31	19	9	5	4	21
24 or more	38	7	3	2	0	0	8
Proportion with 18 or more	79	38	22	10	5	4	29
Mean	21.7	15.7	12.9	10.1	9.1	7.6	13.6
Decayed or unsound							
0	44	42	39	39	47	44	42
1-5	46	44	54	54	44	52	49
6 or more	10	14	7	7	9	4	9
Proportion with 1 or more	56	58	61	61	53	56	58
Mean	2.2	2.2	1.6	1.6	1.5	1.3	1.8
Restored (otherwise sound)							
0	27	2	3	8	8	22	9
1-5	46	22	16	18	38	19	25
6-11	22	45	38	29	22	32	34
12 or more	5	31	43	46	32	27	32
Mean	3.8	8.9	10.5	9.7	8.0	7.2	8.4
Base	80	157	142	132	75	82	668

Table 7.2.15 **The condition of natural teeth by gender**

Dentate adults *Scotland*

The condition of natural teeth (1998 criteria)	Gender		All
	Male	Female	
	%	%	%
Sound and untreated			
0	1	1	1
1-5	12	10	11
6-11	26	31	29
12-17	31	30	30
18-23	21	20	21
24 or more	9	8	8
Proportion with 18 or more	30	28	29
Mean	13.7	13.4	13.6
Decayed or unsound			
0	37	47	42
1-5	52	46	49
6 or more	11	7	9
Proportion with 1 or more	63	53	58
Mean	2.1	1.6	1.8
Restored (otherwise sound)			
0	12	7	9
1-5	25	25	25
6-11	34	33	34
12 or more	29	35	32
Mean	7.9	8.9	8.4
Base	298	370	668

Table 7.2.16 **Condition of natural teeth by social class of head of household**

Dentate adults *Scotland*

The condition of natural teeth (1998 criteria)	Social class of head of household			All[†]
	I, II, IIINM	IIIM	IV, V	
	%	%	%	%
Sound and untreated				
0	1	2	2	1
1-5	12	7	13	11
6-11	27	36	30	29
12-17	33	30	29	30
18-23	20	21	20	21
24 or more	7	5	6	8
Proportion with 18 or more	27	26	26	29
Mean	13.3	13.0	12.7	13.6
Decayed or unsound				
0	47	37	34	42
1-5	47	54	45	49
6 or more	6	8	21	9
Proportion with 1 or more	53	63	66	59
Mean	1.5	1.8	3.0	1.8
Restored (otherwise sound)				
0	7	8	11	9
1-5	16	29	36	25
6-11	38	33	31	34
12 or more	39	30	22	32
Mean	9.7	8.1	6.7	8.4
Base	330	164	119	668

† Includes those for whom the social class of the head of household was not known and Armed Forces

Table 7.2.17 **The condition of natural teeth by usual reason for dental attendance**

Dentate adults   *Scotland*

The condition of natural teeth (1998 criteria)	Usual reason for dental attendance			All
	Regular check-up	Occasional check-up	Only with trouble	
	%	%	%	%
Sound and untreated				
0	I	0	I	I
I-5	II	4	I4	II
6-11	32	18	26	29
12-17	31	28	29	30
18-23	17	28	24	21
24 or more	7	22	6	8
Proportion with				
18 or more	24	49	30	29
Mean	13.0	16.9	13.3	13.6
Decayed or unsound				
0	49	41	29	42
1-5	45	54	55	49
6 or more	6	5	16	9
Proportion with				
1 or more	51	59	71	58
Mean	1.4	1.5	2.7	1.8
Restored (otherwise sound)				
0	4	17	17	9
1-5	19	25	35	25
6-11	37	31	29	34
12 or more	40	27	19	32
Mean	9.9	7.3	6.0	8.4
Base	*404*	*75*	*189*	*668*

Table 7.2.18 **Peridontal pocketing and loss of attachment**

Dentate adults[†]   *Scotland*

Maximum Depth	Pocketing	Loss of attachment
	%	%
Up to 4mm	53	56
4mm - less than 6mm	42	36
6mm - less than 9mm	4	7
9mm and over	I	I
Base	*614*	*614*

† *Excludes 54 dentate adults who did not take part in the periodontal examination*

Table 7.2.19 Frequency of reported problems related to oral conditions in the last 12 months

Dentate adults   *Scotland*

Type of problem†		Frequency of problem			Percentage experiencing either problem at least occasionally
		Occasionally	Fairly often	Very often	
Functional limitation					
had trouble pronouncing words	%	4	0	0	} 11
felt their sense of taste has worsened	%	7	1	1	
Physical pain					
had a painful aching in their mouth	%	23	5	2	} 41
found it uncomfortable to eat any foods	%	22	5	2	
Psychological discomfort					
have been self-conscious	%	16	6	4	} 30
felt tense	%	10	2	1	
Physical disability					
had an unsatisfactory diet	%	3	0	0	} 10
had to interrupt meals	%	7	1	0	
Psychological disability					
found it difficult to relax	%	9	2	1	} 22
have been a bit embarrassed	%	12	2	2	
Social disability					
have been irritable with other people	%	6	1	1	} 9
had difficulty doing usual jobs	%	3	0	0	
Handicap					
felt that life in general was less satisfying	%	7	1	1	} 8
have been totally unable to function	%	1	0	0	
Base††		953			

† The statements and their groupings are derived from the Oral Health Impact Profile (OHIP-14) See section 6.5.1
†† The same base is used for all percentages

Table 7.2.20 Dentate adults who attended for a regular check-up by age, 1972-98

Dentate adults   *Scotland*

Age	1972	1978	1988	1998
	Percentage attending for a regular check-up			
16-24	33	40	51	46
25-34	34	35	48	52
35-44	36	42	55	55
45-54	33	34	44	62
55 and over	29	35	47	60
All	33	38	50	55

For bases see Tables Appendix
1988 Table 20.13

Table 7.2.21 **Type of dental service by characteristics of dentate adults**

Dentate adults†   Scotland

	Type of dental service			
	NHS	Private	Other	Base
	%	%	%	
All	81	14	5	875
Age				
16-24	90	6	4	125
25-34	83	14	4	181
35-44	82	11	7	190
45-54	75	19	6	157
55-64	77	19	4	106
65 and over	77	17	6	116
Gender				
Male	81	14	6	406
Female	82	14	5	469
Social class of head of household				
I, II, IIINM	77	17	6	405
IIIM	82	12	6	217
IV,V	86	11	4	163
Usual reason for dental attendance				
Regular check-up	81	13	6	483
Occasional check-up	86	13	1	101
Only with trouble	80	14	6	290

† Excludes those who have never been to the dentist or were in the middle of a course of treatment

Table 7.2.22 **Time since last visit to the dentist by type of dental service**

Dentate adults†   Scotland

Type of dental service	Time since last visit to the dentist				All
	1 year ago or less	Over 1 year up to 5 years ago	Over 5 years up to 10 years ago	Over 10 years ago	
	%	%	%	%	%
NHS	82	80	80	82	81
Private	13	16	16	7	14
Other	5	4	4	12	5
Base	605	187	39	43	875

† Excludes those who were in the middle of a course of treatment or had never been to the dentist

Table 7.2.23 **Concerns about dental visits by usual reason for attendance**

Dentate adults   *Scotland*

Concerns about dental visits	Usual reason for attendance		All
	Regular check-up	Only with trouble	
		percentage	
Fear			
Definitely feels that…			
I am nervous about some kind of treatment	21	41	29
I always feel anxious about going	14	41	25
Costs			
Definitely feels that…			
I would like to be given an estimate of cost	44	56	48
I find NHS dental treatment expensive	19	34	25
I would like to be able to pay by instalments	24	46	33
Long term value			
Definitely feels that…			
I do not want fancy (intricate) dental treatment	15	33	21
I do not see any point in visiting unless I need to	5	48	20
It would cost less if I only visit when having trouble	11	29	18
Organisation			
Definitely feels that…			
I would like to know what the dentist is doing	40	55	45
I would like to drop in for dental treatment without an appointment	34	46	40
The worst part about going is the waiting	20	24	22
Base	534	307	953

Table 7.2.24 **Reported dental hygiene behaviour, 1988-98**

Dentate adults   *Scotland*

Reported dental hygiene behaviour	1988	1998
Frequency of teeth cleaning	%	%
Twice a day or more	69	70
Once a day	24	22
Less than once a day	6	7
Use of other dental hygiene products[†]		
Dental floss	17	24
Toothpicks/woodsticks	4	3
Mouthwash	13	29
Advice given		
Demonstration of tooth cleaning	38	51
Advice on gum care	29	44
Either of above	-	61
Base	1145	923

† *Excludes those who never clean their teeth*
1988 Tables 26.1, 26.16, 26.23

Table 7.2.25 **Dental status and condition of the teeth by DEPCAT area**

Scotland

Dental status and condition of teeth (1998 criteria)	DEPCAT area					
	1 Least deprived	2	3	4	5	6&7 Most deprived
	%	%	%	%	%	%
All adults						
Natural teeth only	70	67	65	73	58	48
Natural teeth and dentures	16	20	21	16	19	27
No natural teeth	14	12	15	11	23	25
			mean number			
Dentate adults						
Mean number of:						
Teeth	24.7	25.2	23.9	24.9	22.8	21.3
Missing teeth	7.3	6.8	8.1	7.1	9.2	10.7
Sound and untreated teeth	14.2	15.3	12.7	14.4	12.7	13.7
Decayed or unsound teeth	1.1	1.2	2.2	2.0	2.0	1.9
Teeth with cavitated caries	0.3	0.3	0.4	0.8	0.4	0.9
Restored (otherwise sound) teeth	9.3	8.6	9.1	8.4	8.1	5.6
			percentage			
Proportion with:						
21 or more natural teeth	79	86	76	81	75	67
18 or more sound and untreated teeth	34	39	25	31	23	33
1 or more decayed or unsound teeth	52	46	61	55	62	65
1 or more teeth with cavitated caries	20	20	16	31	23	37
12 or more restored (otherwise sound) teeth	38	39	36	27	28	22
Base: All adults	75	170	281	183	333	129
Dentate adults	55	93	177	110	166	49

Table 7.2.26 **Peridontal health by DEPCAT area**

Dentate adults
Scotland

Perodontal health	DEPCAT area					
	1 Least deprived	2	3	4	5	6&7 Most deprived
			percentage			
Proportion with:						
Visible plaque	49	57	78	65	48	58
Calculus	61	45	64	54	71	79
Pocketing greater than 6mm	3	7	6	1	8	4
Base	51	83	163	100	153	48

Table 7.2.27 Impact of oral health by DEPCAT area

Dentate adults   *Scotland*

Impact of oral health	DEPCAT area					
	1 Least deprived	2	3	4	5	6&7 Most deprived
Mean OHIP-14[†] Total Score	1.7	1.1	1.5	2.0	2.0	2.2
Percentage mentioning 1 or more problems[††]	48	46	48	59	63	61
Base	62	144	232	158	243	91

† OHIP problems listed in chapter 6.5
†† At least occasionally

Table 7.2.28 Concerns about tooth appearance and dental visits by DEPCAT area

Dentate adults   *Scotland*

Statement	DEPCAT area					
	1 Least deprived	2	3	4	5	6&7 Most deprived
	%	%	%	%	%	%
Appearance						
Dissatisfied with tooth appearance	26	16	30	33	35	37
Fear						
Definitely felt like that…						
I'm nervous of some kinds of dental treatment	23	25	25	32	32	34
I always feel anxious about going to the dentist	19	21	18	25	33	29
Costs						
Definitely felt like that…						
I would like to be given an estimate without commitment	29	54	45	57	47	51
I find NHS treatment expensive	7	17	25	31	29	30
I would like to be able to pay for my dental treatment by instalments	11	30	25	42	36	46
Long term value						
Definitely felt like that…						
I don't want fancy (intricate) treatment	12	17	20	23	23	26
I don't see any point in visiting the dentist unless I have to	12	17	15	25	22	31
It will cost me less in the long run if I only go when I'm having trouble	18	17	11	19	22	25
Organisation						
Definitely felt like that…						
I'd like to know what the dentist is doing and why	28	46	42	40	50	54
I'd like to be able to drop in at the dentist without an appointment	23	35	41	32	48	42
The worst part of going to the dentist is the waiting	19	18	23	25	24	20
Base	62	144	232	158	243	91

Table 7.2.29 **Usual reason for dental attendance and dental hygiene by DEPCAT area**

Dentate adults   *Scotland*

Usual reason for dental attendance and dental hygiene	DEPCAT area					
	1 Least deprived	2	3	4	5	6&7 Most deprived
	%	%	%	%	%	%
Dental attendance						
Regular check-ups	70	75	62	48	49	28
Only with trouble	17	18	26	39	42	51
Type of dental treatment						
NHS	83	72	83	78	85	89
Private	14	18	12	19	11	6
Dental hygiene advice						
Given toothbrush demonstration	61	62	48	52	49	40
Given gum advice	54	60	45	44	41	22
Given either of above	69	72	59	61	58	47
Dental hygiene						
Brushes teeth twice or more per day	75	80	75	68	65	64
Uses additional method of teeth cleaning†	60	55	51	54	51	49
Uses floss†	41	39	24	23	19	9
Uses mouthwash†	19	23	24	35	34	32
Base	62	144	232	158	243	91

† *Excludes people who never cleaned their teeth*

7.3 Dental health in Wales

Summary

- Between 1978 and 1998, the proportion of adults in Wales with no natural teeth decreased from 37% to 17%, a change affecting all except the youngest age groups in which levels have remained very low. Differences between men and women have reduced over the twenty-year period.
- Although there has been significant improvement within each social class since 1978, the gap between them has remained. In non-manual households the proportion of edentate adults decreased from 29% in 1978 to 10% in 1998, and in unskilled manual households from 40% to 24% during this time.
- Between 1978 and 1998, the proportion of dentate adults in Wales with 21 or more natural teeth increased from 71% to 82%, most of which occurred between 1988 and 1998. The average number of teeth increased from 22.9 in 1988 to 24.2 in 1998.
- In general, the changes in the condition of the natural teeth reflected those which were seen overall in the United Kingdom. There has been a marked increase since 1978 in the proportion of dentate adults with 18 or more sound and untreated teeth: from 27% to 39%. The biggest improvements have occurred in the two youngest age groups: 16 to 24 years and 25 to 34 years.
- The average number of decayed and unsound teeth found in dentate adults decreased from 1.7 in 1978 to 0.7 in 1998 although there has been no significant change since 1988.
- The proportion of dentate adults in Wales with 12 or more restored (otherwise sound) teeth has remained at a similar level over the past two decades; 24% in 1978 and 27% in 1998. This trend conceals variations in the changes found within separate age groups. Among people over 35, there was an increase in the proportion with 12 or more restored (otherwise sound) teeth, most of which occurred between 1978 and 1988.
- Based on the 1998 criteria which included visual dental caries in the categorisation of decayed teeth, dentate adults in Wales had on average 15.1 teeth which were sound and untreated, 1.2 teeth which were decayed or unsound and 7.9 restored (otherwise sound) teeth.
- Fifty-nine percent of dentate adults in Wales in 1998 said that they visit the dentist regularly for check-ups, compared with 48% in 1988 and 39% in 1978.
- There were marked improvements in reported dental hygiene behaviour in Wales between 1988 and 1998. In the most recent survey, 74% of dentate adults reported cleaning their teeth at least twice a day, compared with 64% in 1988.
- In 1998, a quarter (25%) of dentate adults in Wales said they used dental floss and a similar proportion reported using mouthwash (24%), compared with 14% and 9% respectively in 1988.
- In 1998, 59% of dentate adults in Wales reported attending the dentist regularly for check-ups, compared with 48% in 1988 and only 39% in 1978. Over the same period there was a small decrease in the proportion of people who claimed to visit the dentist occasionally for check-ups, but a marked decline in the proportion who visit the dentist only when having trouble with their teeth (43% in 1978 compared with 29% in 1998).

7.3.1 Introduction

Wales has participated in all of the Adult Dental Health Surveys since 1968. In the first adult dental health survey[1], carried out in 1968, data for Wales were based on too small a sample to permit separate analysis and were combined with those for the South West of England. However, in 1978, 1988 and 1998 over-sampling in Wales has enabled separate analysis of the dental health of the principality to be carried out. This chapter draws together the data for Wales and comments on trends between 1978 and 1998 and other selected aspects of the dental health and behaviour of adults in Wales.

7.3.2 Total tooth loss in Wales, 1978-98

In 1978, 37% of adults living in Wales had no natural teeth (were edentate). By 1988, this proportion had fallen to 22% with a further reduction to 17% by 1998. Except among those aged 16 to 24 years and 25 to 34 years for whom low or non-existent levels were recorded in each of the survey years, the same pattern of decline was discernible within every age group. For example, in 1978 one in five people aged 35 to 44 years had no teeth compared with only one in a hundred in 1998. Over the same period the proportion of edentate adults aged 55 to 64 years fell from nearly two-thirds (64%) to one quarter (25%).

In 1978, there was a significant difference between men and women in respect of the proportion with no teeth (40% compared with 34%). By 1998 this difference was no longer so evident; 18% of women compared with 15% of men were found to be edentate. Although there has been a significant improvement within each social class since 1978, the gap between them has remained. In non-manual households the proportion of edentate adults decreased from 29% in 1978 to 10% in 1998, and in unskilled manual households from 40% to 24% during this time.

Figure 7.3.1, Table 7.3.1

7.3.3 The number of teeth among dentate adults in Wales, 1978-98

Between 1978 and 1998, the proportion of dentate adults in Wales with 21 or more natural teeth increased from 71% to 81%, the biggest change occurring between 1988 and 1998. There was a corresponding increase in the average number of teeth between 1988 and 1998, from 22.9 to 24.2.

Table 7.3.2

Care should be taken in the interpretation of the figures by age since small base numbers mean that apparent differences are not significant. However, in general, over time the change

within age groups was broadly similar to that seen in the United Kingdom (see Chapter 5.2) with the data indicating that the largest increases in the mean number of teeth found among dentate adults aged 45 to 54 years, from 17.3 in 1978 to 24.1 in 1998.

Figure 7.3.2, Table 7.3.3

The increase in the numbers of teeth did not vary by gender and was broadly similar for those from different social class backgrounds although very few of the differences were significant.

Table 7.3.4–5

The improvements in the average number of teeth and the proportion with 21 or more teeth were seen mainly among those who reported attending the dentist with trouble; this proportion increased from 58% in 1978 to 72% in 1998.

Tables 7.3 6

Fig 7.3.1 Proportion of edentate adults in Wales by age, 1978-98

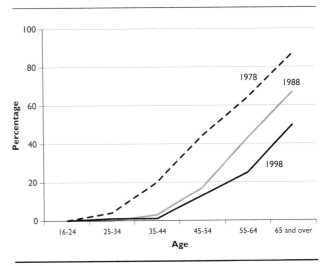

Fig 7.3.2 Mean number of teeth among dentate adults in Wales by age, 1978-98

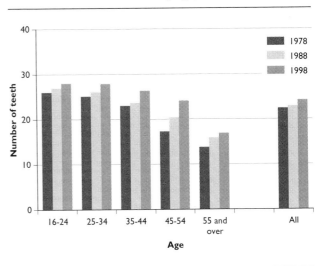

7.3.4 Reliance on natural teeth and dentures in Wales, 1978-98

There was little change from 1978 to 1998 in the proportion of adults in Wales wearing a denture in conjunction with natural teeth (19% in 1978 compared with 17% in 1998). This results from a combination of the increase in the dentate population and the decrease in the numbers of missing teeth among those dentate adults. Neither has there been any significant change from 1988 to 1998 in the different combinations of partial or complete dentures among dentate adults with natural teeth and dentures.

Tables 7.3.7–8

7.3.5 The condition of natural teeth in Wales, 1978-98

The figures relating to the condition of the natural teeth presented in Tables 7.3.9 to 7.3.13 are based on the criteria used in the previous surveys of dental health (see section 7.3.6).

Since 1978 the proportion of dentate adults classified as having 18 or more sound and untreated teeth has increased from 27% in 1978 to 39% in 1998. The most marked increases have occurred among younger people. In 1998, 88% of dentate adults aged 16 to 24 years had 18 or more sound and untreated teeth compared with 51% in 1978; and 74% of those aged 25 to 34 years were classified in this way in 1998 compared with only 31% in 1978.

The data also indicate other improvements in dental health in Wales. The average number of decayed and unsound teeth found in dentate adults living in Wales has decreased since 1978: from 1.7 to 0.7 in 1998 although there was no significant decrease in this from 1988 to 1998, these trends were evident across all age groups.

The proportion of dentate adults in Wales with 12 or more restored (otherwise sound) teeth has remained similar between 1978 and 1998; 24% in 1978 and 27% in 1998. There was variation with age as was seen in the United Kingdom (see Chapter 5.3) with a decrease in the experience of restored teeth among younger dentate adults and an increase among older adults. For example, the proportion of dentate adults aged 16 to 24 years with 12 or more restored (otherwise sound) teeth fell steadily from 19% in 1978 to 6% in 1998 while among those aged 45 to 54 years this proportion rose from 21% to 45% during this time.

Figures 7.3.3–5, Tables 7.3.9–10

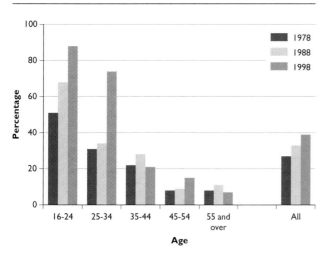

Fig 7.3.3 **Proportion of dentate adults in Wales with 18 or more sound and untreated teeth by age, 1978-98**

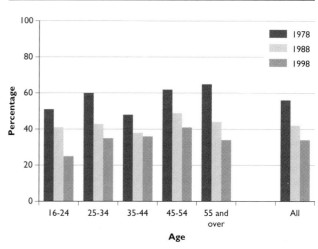

Fig 7.3.4 **Proportion of dentate adults in Wales with some decayed or unsound teeth by age, 1978-98**

Fig 7.3.5   **Proportion of dentate adults in Wales with 12 or more restored (otherwise sound) teeth by age, 1978-98**

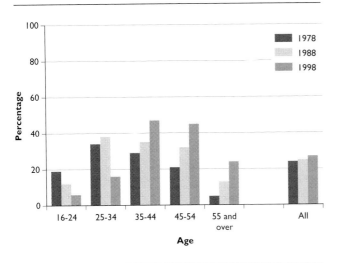

Analysis by gender, social class and usual reason for dental attendance showed similar changes to those seen overall in the United Kingdom, although small base numbers meant that many of the apparent differences were not significant. The increase in the proportion of dentate adults with 18 or more sound and untreated teeth showed little variation with respect to any of these factors with all groups showing a similar level of improvement. With the numbers of decayed and unsound teeth there were indications that there had been some improvement over the last decade among those from unskilled manual backgrounds and those who reported attending the dentist with trouble, but small base numbers meant these differences were not statistically significant.

Tables 7.3.11–13

7.3.6   The condition of the natural teeth based on the criteria used in 1998

For this survey, changes were made to the criteria used in the earlier surveys for assessing dental caries. In 1998, teeth with untreated visual caries were classified as decayed, whereas previously they were recorded as sound and untreated. Restored teeth with recurrent visual caries were also treated as decayed in the 1998 classification, but were defined as teeth with sound restorations by pre-1998 criteria (see Chapter 3.1). Table 7.3.14 shows how the new criteria affect estimates of the average number of teeth that were sound and untreated, decayed or unsound, and restored (otherwise sound).

The overall effect of using the new criteria is to:

● provide a more conservative estimate of the average number of sound and untreated teeth; 15.1 compared with 15.4 using pre-1998 criteria

● increase the estimate of the average number of decayed or unsound teeth from 0.7 to 1.2

● decrease the estimate of the average number of restored (otherwise sound) teeth from 8.1 to 7.9.

However the size of the difference varied within specific groups of the population. For example, younger adults are more likely than older people to have visual caries only and the average number of teeth classified as decayed or unsound increased from 0.4 among 16 to 24 year olds using pre-1998 criteria to 1.1 based on the new criteria. In contrast, the new criteria had less effect on the average number of decayed and unsound teeth found those aged 65 years and over; this figure increased from 0.6 to 0.8.

Table 7.3.14

Based on the 1998 criteria, dentate adults in Wales had, on average, 15.1 teeth which were sound and untreated, 1.2 teeth which were decayed or unsound and 7.9 restored (otherwise sound) teeth. The proportion of dentate adults with 18 or more sound teeth decreased with age from 85% of the youngest age group to 7% of those aged 65 and over, while the proportion with 12 or more restored (otherwise sound) teeth was highest among those in the middle age groups. The proportion with no decayed or unsound teeth showed little variation with age. As was the case for variation with respect to age, the variation with respect to gender, social class and usual reason for dental attendance was broadly similar to that seen overall for the United Kingdom (see Chapters 3.1 and 3.2).

Tables 7.3.15–18

Finally, 14% of dentate adults living in Wales in 1998 had three or more artificial crowns compared with 8% in 1988, and there was a significant drop in the proportion of dentate adults with no artificial crowns – from 74% in 1988 to 64% in 1998.

Table 7.3.19

7.3.7   The periodontal condition of dentate adults in Wales

In 1998, dental plaque was detected through oral examination in 52% of Welsh adults participating in this stage of the survey, and calculus was present on the teeth of 39% of these adults (table not shown).

In 1998, periodontal pocketing of between 4mm or more was found in 47% of dentate adults and pocketing of 6mm or more in 4%. Loss of attachment of teeth to the same depths was found in 34% and 7% of dentate adults respectively.

Table 7.3.20

7.3.8 Reported usual reason for dental attendance and visiting the dentist in Wales, 1988-98

In 1998, 59% of dentate adults in Wales reported attending the dentist regularly for check-ups, compared with 48% in 1988 and 39% in 1978. Over the same period there was a small decrease in the proportion of people who reported to visit the dentist occasionally for check-ups, but a marked decline in the proportion who visit the dentist only when having trouble with their teeth (43% in 1978 compared with 29% in 1998).

Figure 7.3.6, Table 7.3.21

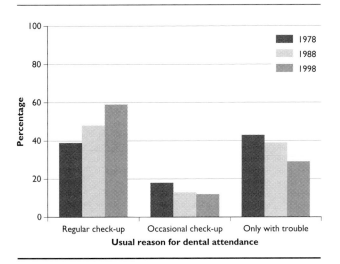

Fig 7.3.6 **Usual reason for dental attendance among dentate adults in Wales, 1978-98**

Nearly two-thirds (63%) of dentate adults who took part in the survey in 1998 said that they had attended the same dental practice for more than 5 years. The most frequently mentioned reasons given by those who said they were planning to change practice before their next visit to the dentist were that the practice was no longer convenient (38%), and that the dentist had retired, moved or died (20%). Seventeen percent of adults were planning to change practice because they were not satisfied with the treatment they were given. (These results are based on small numbers and should be treated with caution.)

Tables 7.3.22–23

In respect of their last visit to the dentist, 54% percent of adults in Wales said that they had paid something towards their treatment compared with 42% in 1988. Nearly a third

(32%) of those who worked had taken time off work to visit the dentist. More than two-thirds (68%) of dentate adults who were working or studying attended a dental practice that was closer to home than to work or college. Just under half (45%) of all dentate adults travelled more than two miles to their dentist.

Tables 7.3.24–27

7.3.9 Treatment preferences and satisfaction with appearance of teeth in Wales, 1988-98

Between 1988 and 1998, the proportion of dentate adults in Wales who said they would prefer to have an aching back tooth taken out rather than filled showed no significant change, 26% in 1988 and 30% in 1998. Neither were there significant changes in the proportions that said they would prefer to have a front tooth or a back tooth extracted rather than crowned during this time.

Table 7.3.28

In 1998, 76% of adults in Wales said that they were satisfied with the appearance of their teeth. There was no statistically significant change in this respect between 1988 and 1998.

Table 7.3.29

7.3.10 Reported dental hygiene behaviour in Wales, 1988-98

In 1998, 74% of dentate adults in Wales said that they cleaned their teeth at least twice a day, compared with 64% in 1988. There were also significant increases in reported use of dental floss and of mouthwash, the two dental hygiene products most commonly used other than toothpaste and a toothbrush. In 1998, a quarter (25%) of dentate adults in Wales said they used dental floss and a similar proportion reported using mouthwash (24%), compared with 14% and 9% respectively in 1988.

Figure 7.3.7, Tables 7.3.30–31

A larger proportion of dentate adults in 1998 reported having been shown how to clean their teeth by a dental professional than ten years previously (52% compared with 37%). There was a similar increase in the proportion who reported to have been given advice about gum care during this time (44% in 1998 compared 25% in 1988).

Table 7.3.32

Fig 7.3.7   **Use of dental hygiene products among dentate adults in Wales, 1978-98**

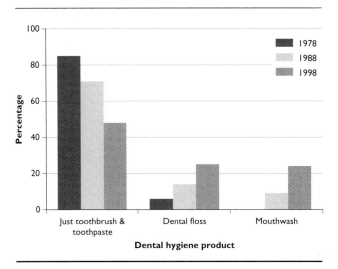

Notes and references

1.   See Chapter 1.1

Table 7.3.1   Edentate adults, 1978-98

All adults   *Wales*

	1978	1988	1998
	Percentage edentate		
All	37	22	17
Age			
16-24	0	0	0
25-34	4	0	1
35-44	20	3	1
45-54	44	17	13
55-64	64	43	25
65-74	} 87	56	39
75 and over		82	64
Gender			
Men	34	18	15
Women	40	26	18
Social class of head of household			
I, II, IIINM	29	14	10
IIIM	38	21	20
IV,V	40	35	24
Base	*580*	*807*	*830*

1988 Table 3.10

Table 7.3.2   Distribution of number of teeth, 1978-98

Dentate adults   *Wales*

	1978	1988	1998
	%	%	%
1-14	14	13	11
15-20	15	14	8
21-26	36	86	34
27-31	34	34	43
32	1	2	4
Proportion with 21 or more teeth	71	72	81
Mean	22.4	22.9	24.2
Base	*326*	*521*	*502*

1988 Table 5.21

Table 7.3.3   Distribution of number of teeth by age, 1978-98

Dentate adults   *Wales*

Number of teeth	Age														
	16-24			25-34			35-44			45-54			55 and over		
	1978	1988	1998	1978	1988	1998	1978	1988	1998	1978	1988	1998	1978	1988	1998
	%	%	%	%	%	%	%	%	%	%	%	%	%	%	%
1-14	1	1	0	4	2	0	8	8	0	31	18	7	54	40	38
15-20	3	0	0	11	6	0	18	16	4	32	30	8	24	29	24
21-26	42	36	18	34	39	28	45	47	47	29	36	46	22	23	33
27-31	54	59	74	50	50	63	28	28	46	8	15	38	0	8	5
32	0	4	7	1	3	9	1	2	3	0	2	2	0	0	1
Mean	26.0	26.9	28.0	25.1	26.0	27.9	23.0	23.6	26.3	17.3	20.4	24.1	13.8	15.9	16.9
Base	*72*	*140*	*69*	*94*	*128*	*90*	*77*	*128*	*96*	*48*	*80*	*109*	*36*	*120*	*138*

1988 Table 5.21

Table 7.3.4 **Distribution of number of teeth by gender, 1978-98**

Dentate adults *Wales*

Number of teeth	Gender						All		
	Men			Women					
	1978	1988	1998	1978	1988	1998	1978	1988	1998
	%	%	%	%	%	%	%	%	%
1-14	14	12	11	14	13	10	14	13	11
15-20	15	16	8	16	13	9	15	14	8
21-26	35	34	36	37	39	33	36	36	34
27-31	35	35	41	32	33	44	34	34	43
32	1	3	4	1	2	4	1	2	4
Mean	22.4	23.0	24.2	22.3	22.8	24.1	22.4	22.9	24.2
Base	163	308	236	163	284	266	326	521	502

1988 Table 5.22

Table 7.3.5 **Distribution of number of teeth by social class of head of household, 1978-98**

Dentate adults *Wales*

Number of teeth	Social class of head of household									All†		
	I, II, IIINM			IIIM			IV,V					
	1978	1988	1998	1978	1988	1998	1978	1988	1998	1978	1988	1998
	%	%	%	%	%	%	%	%	%	%	%	%
1-14	11	14	10	10	14	11	22	13	14	14	13	11
15-20	16	15	6	15	16	12	13	10	10	15	14	8
21-26	30	34	36	46	34	32	36	33	40	36	36	34
27-31	41	35	43	29	33	42	29	40	34	34	34	43
32	2	2	5	0	3	3	0	4	2	1	2	4
Mean	23.1	22.6	24.6	22.8	22.6	23.9	21.4	23.7	22.6	22.4	22.9	24.2
Base	129	272	251	116	220	146	63	92	81	326	521	502

† Includes those for whom social class of head of household was not known and armed forces
1988 Table 5.23

Table 7.3.6 Distribution of number of teeth by usual reason for dental attendance, 1978-98

Dentate adults   *Wales*

Number of teeth	Usual reason for dental attendance						All†		
	Regular check-up			Only with trouble					
	1978	1988	1998	1978	1988	1998	1978	1988	1998
	%	%	%	%	%	%	%	%	%
1-14	7	7	7	24	20	20	14	13	11
15-20	13	15	8	18	12	7	15	14	8
21-26	40	40	38	31	34	27	36	36	34
27-31	39	37	42	26	31	42	34	34	43
32	1	1	4	1	3	3	1	2	4
Mean	23.7	23.7	24.6	20.6	21.6	22.7	22.4	22.9	24.2
Base	140	284	326	158	228	123	326	521	502

† Includes those who visit the dentist for an occasional check-up
1988 Table 5.24

Table 7.3.7 Reliance on natural teeth and dentures, 1978-98

All adults   *Wales*

Age	1978	1988	1998
	%	%	%
Natural teeth only	44	57	66
Natural teeth and dentures	19	21	17
Edentate	37	22	17
Base	580	807	830

1988 Table 16.12

Table 7.3.8 The pattern of dentures for use with natural teeth, 1988-98

Adults with natural teeth and dentures   *Wales*

	1988	1998
	%	%
Full - none	18	14
Full - partial	12	11
Partial - partial	14	18
Partial - none	50	48
Other	5	9
Base	44	152

1988 Table 16.8

Table 7.3.9 The condition of natural teeth, 1978-98

Dentate adults   *Wales*

The condition of natural teeth (1988 criteria)	1978	1988	1998
	%	%	%
Sound and untreated			
0	1	1	0
1-5	7	9	7
6-11	34	28	24
12-17	31	30	29
18 or more	27	33	39
Mean	13.3	14.3	15.4
Decayed or unsound			
0	44	58	66
1-5	48	40	32
6 or more	8	3	2
Mean	1.7	0.9	0.7
Restored (otherwise sound)			
0	19	12	7
1-5	23	25	30
6-11	34	37	36
12 or more	24	25	27
Mean	7.3	7.7	8.1
Mean number of teeth	22.4	22.9	24.2
Base	326	521	502

1988 Tables 6.25, 7.25, 8.30

Table 7.3.10 **The condition of natural teeth by age, 1978-98**

Dentate adults   *Wales*

The condition of natural teeth (1988 criteria)	Age														
	16-24			25-34			35-44			45-54			55 and over		
	1978	1988	1998	1978	1988	1998	1978	1988	1998	1978	1988	1998	1978	1988	1998
	%	%	%	%	%	%	%	%	%	%	%	%	%	%	%
Sound and untreated															
0	0	0	0	0	0	0	1	2	0	2	1	0	5	3	1
1-5	1	0	0	3	1	0	7	8	4	10	11	5	27	26	22
6-11	13	5	3	32	27	4	37	33	25	55	48	33	44	38	47
12-17	35	27	8	34	38	22	33	30	50	25	32	48	16	22	22
18 or more	51	68	88	31	34	74	22	28	21	8	9	15	8	11	7
Mean	17.4	19.8	23.2	14.4	15.4	20.4	12.9	13.8	13.9	10.0	10.9	12.4	8.0	9.5	9.1
	%	%	%	%	%	%	%	%	%	%	%	%	%	%	%
Decayed or unsound															
0	49	59	75	40	57	65	52	62	64	38	51	59	35	56	66
1-5	45	36	25	50	40	34	38	37	31	58	45	41	57	42	31
6 or more	6	5	0	10	3	1	10	1	5	4	4	0	8	2	2
Mean	1.4	0.9	0.4	1.8	1.1	0.7	1.8	0.7	0.8	1.6	1.1	0.7	2.2	1.0	0.7
	%	%	%	%	%	%	%	%	%	%	%	%	%	%	%
Restored (otherwise sound)															
0	9	11	12	12	6	7	15	7	2	31	14	2	49	26	11
1-5	30	31	61	20	18	34	18	20	8	25	20	12	22	34	34
6-11	42	46	22	34	37	43	38	38	43	23	35	41	24	27	30
12 or more	19	12	6	34	38	16	29	35	47	21	32	45	5	13	24
Mean	7.2	6.2	4.3	8.9	9.6	6.8	8.3	9.1	11.5	5.6	8.4	11.0	3.6	5.3	7.1
Mean number of teeth	26.0	26.9	28.0	25.1	26.0	27.9	23.0	23.6	26.3	17.3	20.4	24.1	13.8	15.9	16.9
Base	*72*	*140*	*69*	*94*	*128*	*90*	*77*	*128*	*96*	*48*	*80*	*109*	*36*	*120*	*138*

1988 Table 6.24, 7.25, 8.30

Table 7.3.11 **The condition of natural teeth by gender, 1978-98**

Dentate adults   *Wales*

The condition of natural teeth (1988 criteria)	Gender						All		
	Men			Women					
	1978	1988	1998	1978	1988	1998	1978	1988	1998
	%	%	%	%	%	%	%	%	%
Sound and untreated									
0	I	I	I	I	I	0	I	I	0
1-5	8	6	7	7	7	8	7	9	7
6-11	28	26	22	39	28	26	34	28	24
12-17	32	31	29	29	31	30	31	30	29
18 or more	31	35	41	24	34	37	27	33	39
Mean	14.0	15.0	15.7	12.7	14.6	15.1	13.3	14.3	15.4
	%	%	%	%	%	%	%	%	%
Decayed or unsound									
0	42	51	59	47	65	72	44	58	66
1-5	48	46	38	48	33	27	48	40	32
6 or more	10	4	3	5	2	I	8	3	2
Mean	2.0	1.1	0.9	1.5	0.8	0.5	1.7	0.9	0.7
	%	%	%	%	%	%	%	%	%
Restored (otherwise sound)									
0	24	14	9	13	10	5	19	12	7
1-5	23	24	32	23	26	28	23	25	30
6-11	33	38	36	35	36	36	34	37	36
12 or more	20	24	24	29	29	31	24	25	27
Mean	6.5	7.3	7.6	8.1	8.1	8.5	7.3	7.7	8.1
Mean number of teeth	22.4	23.0	24.2	22.3	22.8	24.1	22.4	22.9	24.2
Base	*163*	*308*	*236*	*163*	*284*	*266*	*326*	*521*	*502*

1988 Table 6.25, 7.26, 8.31

Table 7.3.12 **The condition of natural teeth by social class of head of household, 1978-98**

Dentate adults   *Wales*

The condition of natural teeth (1988 criteria)	Social class of head of household									All†		
	I, II, IIINM			IIIM			IV, V					
	1978	1988	1998	1978	1988	1998	1978	1988	1998	1978	1988	1998
	%	%	%	%	%	%	%	%	%	%	%	%
Sound and untreated												
0	0	2	0	1	0	0	5	0	1	1	1	0
1-5	10	12	8	5	5	8	3	10	5	7	9	7
6-11	37	31	27	28	25	22	33	24	24	34	28	24
12-17	33	31	28	32	29	29	27	28	40	31	30	29
18 or more	20	24	37	34	41	41	32	39	29	27	33	39
Mean	12.4	13.0	14.9	14.4	15.6	15.6	13.8	15.1	14.6	13.3	14.3	15.4
	%	%	%	%	%	%	%	%	%	%	%	%
Decayed or unsound												
0	48	60	69	42	56	60	43	53	59	44	58	66
1-5	47	37	30	52	43	35	43	40	39	48	40	32
6 or more	5	3	0	4	1	4	14	6	2	8	3	2
Mean	1.4	0.9	0.5	1.5	0.9	1.0	4	1.3	0.8	1.7	0.9	0.7
	%	%	%	%	%	%	%	%	%	%	%	%
Restored (otherwise sound)												
0	12	10	5	17	14	9	27	16	6	19	12	7
1-5	15	22	22	26	26	36	35	32	32	23	25	30
6-11	38	36	38	34	39	32	27	34	40	34	37	36
12 or more	35	32	34	23	21	24	11	17	22	24	25	27
Mean	9.3	8.7	9.2	6.9	7.0	7.2	5.2	6.2	7.2	7.3	7.7	8.1
Mean number of teeth	23.1	22.6	24.6	22.8	22.6	23.9	21.4	23.7	22.6	22.4	22.9	24.2
Base	129	272	251	116	220	146	63	92	81	326	521	502

† Includes those for whom social class of head of household was not known and Armed Forces
1988 Table 6.26, 7.27, 8.32

Table 7.3.13 **The condition of natural teeth by usual reason for dental attendance, 1978-98**

Dentate adults   *Wales*

The condition of natural teeth (1988 criteria)	Usual reason for dental attendance						All†		
	Regular check-up			Only with trouble					
	1978	1988	1998	1978	1988	1998	1978	1988	1998
	%	%	%	%	%	%	%	%	%
Sound and untreated									
0	0	1	0	3	0	0	1	1	0
1-5	7	8	8	10	10	9	7	9	7
6-11	42	34	24	27	23	24	34	28	24
12-17	30	31	34	25	30	20	31	30	29
18 or more	21	26	33	35	37	47	27	33	39
Mean	12.1	13.3	14.7	14.0	15.2	15.6	13.3	14.3	15.4
	%	%	%	%	%	%	%	%	%
Decayed or unsound									
0	59	64	74	26	48	46	44	58	66
1-5	39	36	26	58	47	49	48	40	32
6 or more	2	0	0	16	5	5	8	3	2
Mean	0.8	0.6	0.4	2.9	1.3	1.3	1.7	0.9	0.7
	%	%	%	%	%	%	%	%	%
Restored (otherwise sound)									
0	1	4	3	41	23	12	19	12	7
1-5	13	18	22	32	34	46	23	25	30
6-11	45	40	38	17	31	31	34	37	36
12 or more	41	38	36	10	13	12	24	25	27
Mean	10.8	9.8	9.5	3.8	5.1	5.7	7.3	7.7	8.1
Mean number of teeth	23.7	23.7	24.6	30.6	31.6	22.7	22.4	22.9	24.2
Base	*140*	*284*	*326*	*158*	*228*	*123*	*326*	*521*	*502*

† *Includes those who only visit the dentist for an occasional check-up*
1988 Table 6.27, 7.28, 8.33

Table 7.3.14 Difference in condition of teeth using 1998 and pre-1998 criteria

Dentate adults *Wales*

	Mean number of teeth that were:									Base
	Sound and untreated			Decayed or unsound			Restored (otherwise sound)			
	1998 criteria	pre-1998 criteria	Difference	1998 criteria	pre-1998 criteria	Difference	1998 criteria	pre-1998 criteria	Difference	
All	15.1	15.4	0.3	1.2	0.7	0.5	7.9	8.1	0.2	502
Age										
16-24	22.6	23.2	0.6	1.1	0.4	0.7	4.2	4.3	0.1	69
25-34	19.8	20.4	0.6	1.4	0.7	0.7	6.7	6.8	0.1	90
35-44	13.6	13.9	0.3	1.4	0.8	0.6	11.3	11.5	0.2	96
45-54	12.3	12.4	0.1	1.0	0.7	0.3	10.8	11.0	0.2	109
55-64	9.6	9.7	0.1	1.1	0.7	0.4	7.9	8.2	0.3	73
65 and over	8.3	8.4	0.1	0.8	0.6	0.2	5.6	5.8	0.2	65
Gender										
Men	15.3	15.7	0.4	1.5	0.9	0.6	7.4	7.6	0.2	236
Women	14.8	15.1	0.3	0.9	0.5	0.4	8.4	8.5	0.1	266
Social class of head of household										
I, II, IIINM	14.7	14.9	0.2	0.8	0.5	0.3	9.0	9.2	0.2	251
IIIM	15.1	15.6	0.5	1.7	1.0	0.7	7.0	7.2	0.2	146
IV,V	14.3	14.6	0.3	1.2	0.8	0.4	7.1	7.2	0.1	81
Usual reason for dental attendance										
Regular check-up	14.5	14.7	0.2	0.7	0.4	0.3	9.4	9.5	0.1	326
Occasion check-up	17.6	18.1	0.5	1.3	0.5	0.8	6.5	6.7	0.2	52
Only with trouble	15.1	15.6	0.5	2.1	1.3	0.8	5.5	5.7	0.2	123

Table 7.3.15 **The condition of natural teeth by age**

Dentate adults   *Wales*

The condition of natural teeth (1998 criteria)	Age						All
	16-24	25-34	35-44	45-54	55-64	65 and over	
	%	%	%	%	%	%	%
Sound and untreated							
0	0	0	0	0	1	2	0
1-5	0	1	4	5	16	30	8
6-11	5	6	29	34	49	45	25
12-17	10	24	47	48	26	17	30
18-23	30	45	17	14	6	7	22
24 or more	55	24	2	0	1	0	15
Proportion with 18 or more	85	69	19	14	8	7	37
Mean	22.6	19.8	13.6	12.3	9.6	8.3	15.1
	%	%	%	%	%	%	%
Decayed or unsound							
0	54	53	49	50	48	61	52
1-5	42	41	45	47	48	36	43
6 or more	5	6	6	2	4	3	5
Proportion with 1 or more	46	47	51	50	52	39	48
Mean	1.1	1.4	1.4	1.0	1.1	0.8	1.2
	%	%	%	%	%	%	%
Restored (otherwise sound)							
0	12	8	2	2	8	15	7
1-5	62	34	8	14	27	43	31
6-11	22	43	46	41	36	26	36
12 or more	4	15	44	43	29	16	26
Mean	4.2	6.7	11.3	10.8	7.9	5.6	7.9
Base	*69*	*90*	*96*	*109*	*73*	*65*	*502*

Table 7.3.16 **The condition of natural teeth by gender**

Dentate adults *Wales*

The condition of natural teeth (1998 criteria)	Gender		All
	Male	Female	
	%	%	%
Sound and untreated			
0	1	0	0
1-5	7	8	8
6-11	24	27	25
12-17	29	31	30
18-23	23	21	22
24 or more	16	14	15
Proportion with 18 or more	39	35	37
Mean	15.3	14.8	15.1
	%	%	%
Decayed or unsound			
0	42	62	52
1-5	51	35	43
6 or more	6	3	5
Proportion with 1 or more	58	38	48
Mean	1.5	0.9	1.2
	%	%	%
Restored (otherwise sound)			
0	9	6	7
1-5	32	29	31
6-11	37	36	36
12 or more	22	30	26
Mean	7.4	8.4	7.9
Base	236	266	502

Table 7.3.17 **The condition of natural teeth by social class of head of household**

Dentate adults *Wales*

The condition of natural teeth (1998 criteria)	Social class of head of household			All[†]
	I, II, IIINM	IIIM	IV,V	
	%	%	%	%
Sound and untreated				
0	0	0	1	0
1-5	8	9	5	8
6-11	27	25	26	25
12-17	28	29	40	30
18-23	19	26	20	22
24 or more	17	12	8	15
Proportion with 18 or more	36	38	28	37
Mean	14.7	15.1	14.3	15.1
	%	%	%	%
Decayed or unsound				
0	55	44	50	52
1-5	45	45	47	43
6 or more	0	11	3	5
Proportion with 1 or more	45	56	50	48
Mean	0.8	1.7	1.2	1.2
	%	%	%	%
Restored (otherwise sound)				
0	5	9	6	7
1-5	23	37	33	31
6-11	40	32	40	36
12 or more	32	22	20	26
Mean	9.0	7.0	7.1	7.9
Base	251	146	81	502

† Includes those for whom the social class of the head of household was not known and Armed Forces

Table 7.3.18 The condition of natural teeth by usual reason for dental attendance

Dentate adults   *Wales*

The condition of natural teeth (1998 criteria)	Usual reason for dental attendance			All
	Regular check-up	Occasional check-up	Only with trouble	
	%	%	%	%
Sound and untreated				
0	0	2	0	0
1-5	8	0	10	8
6-11	25	27	26	25
12-17	34	22	22	30
18-23	19	22	27	22
24 or more	12	28	15	15
Proportion with				
18 or more	32	50	42	37
Mean	14.5	17.6	15.1	15.1
	%	%	%	%
Decayed or unsound				
0	60	53	35	52
1-5	39	44	51	43
6 or more	1	3	14	5
Proportion with				
1 or more	40	47	65	48
Mean	0.7	1.3	2.1	1.2
	%	%	%	%
Restored (otherwise sound)				
0	3	13	13	7
1-5	23	32	46	31
6-11	39	34	32	36
12 or more	34	21	9	26
Mean	9.4	6.5	5.5	7.9
Base	*326*	*52*	*123*	*502*

Table 7.3.19 Number of artificial crowns, 1988-98

Dentate adults   *Wales*

Number of crowns	1988	1998
	%	%
None	74	64
One	13	14
Two	8	8
Three or more	8	14
Base	*521*	*502*

1988 Table 9.2 (part)

Table 7.3.20 Peridontal loss of attachment and pocket depth

Dentate adults   *Wales*

Depth	Loss of attachment	Pocket depth
	percentage with condition	
4mm or more	34	47
6mm or more	7	4
Base	*459*	*459*

Table 7.3.21 Usual reason for dental attendance, 1978-98

Dentate adults   *Wales*

Usual reason for dental attendance	1978	1988	1998
	%	%	%
Regular check-up	39	48	59
Occasional check-up	18	13	12
Only with trouble	43	39	29
Base	*363*	*622*	*682*

1988 Table 20.10 (part)

Table 7.3.22 How long respondent had been visiting the same dentist

Dentate adults†   *Wales*

How long respondent had been visiting the same dentist	%
First time attender	10
Attended for less than 5 years	20
Attended for more than 5 years	64
Can't remember	7
Base	*680*

† Excluding those who have never been to the dentist

Table 7.3.23 Reasons for wanting to change to a different dental practice

Dentate adults expecting to change to a different dental practice — *Wales*

Reasons for wanting to change to a different dental practice	%
No longer convenient	38
Dentist retired, moved, died	20
Informant not satisfied with treatment	17
Difficult to get an appointment	1
Informant never goes regularly	5
Last visit was extraordinary	8
Heard of a better dentist	3
No longer in the system	7
No NHS dentist	13
Cost	3
Other	7
Base	80

Percentages may add up to more than 100% as respondents could give more than one answer

Table 7.3.24 Whether or not respondents paid for dental treatment, 1988-98

Dentate adults† — *Wales*

Whether or not respondent paid	1988	1998
	%	%
Paid something	42	54
Paid nothing	42	33
Don't know	16	13
Base	593	628

† *Excluding those who have never been to dentist or were in the middle of a course of treatment*
1988 Table 24.51 (part)

Table 7.3.25 Time taken off work to visit the dentist

Dentate adults currently working† — *Wales*

Time taken off work to visit the dentist	%
None	68
Under 1 hour	14
1 hour but less than 2	12
2 or more hours	6
Base	395

† *Excluding those who have never been to dentist*

Table 7.3.26 Location of dental practice

Dentate adults currently working or studying† — *Wales*

Location of dental practice	%
Nearer to home	68
Nearer to work/college	15
The same	17
Base	434

† *Excluding those who have never been to dentist*

Table 7.3.27 Estimated distance of dental practice from home/work/college

Dentate adults† — *Wales*

Estimated distance of dental practice from home/work/college	%
Up to half a mile	23
More than half up to 1 mile	15
More than 1 up to 2 miles	17
More than 2 up to 5 miles	26
More than 5 up to 10 miles	11
More than 10 miles up to 20 miles	4
More than 20 miles up to 30 miles	0
More than 30 miles	2
Other	2
Base	680

† *Excluding those never been to dentist*

Table 7.3.28 Treatment preferences, 1988-98

Dentate adults — *Wales*

Treatment preferences	1988	1998
Would prefer an aching back tooth to be:		
Taken out	26	30
Filled	74	70
Would prefer a front tooth to be:		
Extracted	10	14
Crowned	90	86
Would prefer a back tooth to be:		
Extracted	33	31
Crowned	67	69
Base	622	682

1988 Table 22.2

Table 7.3.29 **Satisfaction with appearance of teeth (and dentures), 1978-98**

Dentate adults *Wales*

Satisfied with appearance of teeth (and dentures)	1978	1988	1998
Yes	72	73	76
No	28	27	24
Base	363	622	682

1988 Table 27.9 (part)

Table 7.3.30 **Frequency of teeth cleaning, 1988-98**

Dentate adults *Wales*

Frequency of teeth cleaning	1988	1998
	%	%
Twice a day or more	64	74
Once a day	27	21
Less than once a day	8	4
Never	1	0
Base	622	682

1998 Table 26.1 (part)

Table 7.3.31 **Use of dental hygiene products, 1978-98**

Dentate adults† *Wales*

Respondent used:	1978	1988	1998
	%	%	%
Dental floss	6	14	25
Toothpicks/woodsticks	6	3	5
Mouthwash††	-	9	24
Just toothbrush & toothpaste	85	71	48
Base	353	622	682

Percentages may add to more than 100% as respondents could give more than one answer
† Excludes those who stated they never cleaned their teeth
†† Data not available for 1978
1988 Table 26.16, 26.29

Table 7.3.32 **Whether given a demonstration on toothbrush or advice on caring for gums, 1988-98**

Dentate adults *Wales*

	1988	1998
	%	%
Given demonstration on toothbrushing	37	52
Given advice on gum care	25	45
Given either or both †	-	59
Base	622	682

† Data not available for 1988
1988 Table 26.23

7.4 Dental health in Northern Ireland

Summary

- The proportion of adults in Northern Ireland with no natural teeth was 12% in 1998, close to the target of 10% by 2008 set in 1995 by the Department of Health and Social Services.

- A decrease in the proportion of edentate people since 1979 is evident in every age group, among both men and women and in every social class. Change has occurred more rapidly among younger adults, men and people from non-manual or unskilled manual backgrounds.

- In 1998, the average number of teeth among dentate adults in Northern Ireland was 24.5. This figure has increased with every survey since 1979. Eighty one percent of dentate adults had 21 or more teeth in 1998 compared with 74% in 1988 and 68% in 1979.

- Significant improvements in dental health are evident from 1979 to 1998 both in the proportion of dentate adults with 18 or more sound and untreated teeth – an increase from 13% in 1979 to 34% in 1998 – and the proportion with one or more decayed or unsound teeth – a decrease from 57% in 1979 to 41% in 1998.

- By 1998, both the proportion of dentate adults with 18 or more sound and untreated teeth (34%) and the mean number of sound and untreated teeth (14.3) were very near the targets of 35% and 15 teeth set for 2008.

- Improvements in the proportion of adults with 18 or more sound and untreated teeth have occurred mainly among younger people under 35 years of age. There is little evidence of a reduction in the difference between men and women of the proportion with one or more decayed or unsound teeth, but such differences between the social classes in this respect have decreased.

- The proportion of dentate adults in Northern Ireland with 12 or more restored (otherwise sound) teeth has remained similar between 1979 and 1998; 32% in 1979 and 35% in 1998. The variation with age was similar to that seen overall in the United Kingdom, with a decrease in the experience of restored teeth among the youngest dentate adults and an increase among older adults.

- Based on the 1998 criteria which included visual caries in the estimate of decay, 32% of the dentate adult population of Northern Ireland had 18 or more sound and untreated teeth, 60% had one or more decayed or unsound teeth and 33% had 12 or more restored (otherwise sound) teeth.

- Adults aged 16 to 24 years were more likely than those in other age groups to have 18 or more sound and untreated teeth.

- The numbers of decayed and unsound teeth were higher among men, those from manual backgrounds and those who reported attending the dentist only with trouble.

- In 1998, 48% of dentate adults in Northern Ireland had periodontal pocketing of 4mm or more and 4% had some pocketing of 6mm or more.

- Dentate adults in Northern Ireland were more likely to report attending the dentist for regular check-ups in 1998 than in 1988 (51% compared with 42%). There was also an increase in the proportion of dentate adults who visited the dentist in the previous two years – from 71% in 1988 to 81% in 1998.

- Nearly three quarters (72%) of dentate adults in Northern Ireland reported cleaning their teeth at least twice a day, 22% once a day. Just under half (47%) reported to use dental hygiene products other than toothpaste and brush compared with less than a quarter (23%) in 1988. Of various other products used, dental floss and mouthwash were the most popular.

7.4.1 Introduction

Dental disease in Northern Ireland presents a significant problem, especially in comparison with the rest of the United Kingdom. For example, in 1988 only 25% of dentate adults had 18 or more sound untreated teeth compared with 35% of dentate adults in the United Kingdom as a whole.

In 1995, the Department of Health and Social Services published an oral health strategy for Northern Ireland. This set targets for adult oral health for 2008, based on traditional benchmarks of adult oral health and were:

● to raise the proportion of dentate adults with 18 or more sound and untreated teeth from 25% to 35%
● to increase the average number of sound and untreated teeth from 12.6 to 15
● to reduce the number of adults with no remaining teeth from 18% to 10%.

This chapter focuses on key trends in adult dental health in Northern Ireland over the two decades prior to 1998, including progress towards these targets. Although Northern Ireland was included in the 1978 Adult Dental Health Survey, the sample was too small to allow separate analysis of the dental health of the country. The main starting point for the examination of patterns of change, therefore, is data collected in a separate Northern Ireland survey of adult dental health carried out in 1979[1].

7.4.2 Total tooth loss in Northern Ireland, 1979-98

In 1998, around one in eight (12%) of the adult population of Northern Ireland had no natural teeth (were edentate) compared with nearly one in five (18%) in 1988 and one in three (33%) in 1979. This represents considerable progress towards the target of 10% by 2008 set out in the 1995 DHSS strategy for Northern Ireland.

Among those aged 35 years or over the proportion of people with no natural teeth has declined markedly since 1979. The improvements were greatest among those in the middle age groups; for example, in 1979, 35% of adults aged 45 to 54 years were edentate compared with 16% in 1988 and 5% in 1998. Thus, by 1998 total tooth loss was mostly confined to adults aged 55 and over. Among adults aged less than 35 years, total tooth loss has remained relatively uncommon from 1979 to 1998.

The results of the 1998 survey show little difference between men and women with no natural teeth. This represents a narrowing of the gender gap that was evident in both 1979

and 1988, when women were more likely than men to have no teeth. In 1979, 37% of women and 30% of men had no natural teeth compared with 13% and 11% respectively in 1998.

Information about the relationship between total tooth loss and social class was not available in 1979. The proportion of adults with no natural teeth decreased in all social classes between 1988 and 1998. The data suggest that there were larger decreases among adults from non-manual households than in the other social classes. However, the sample sizes are small and some of the differences were not significant.

Figure 7.4.1, Table 7.4.1

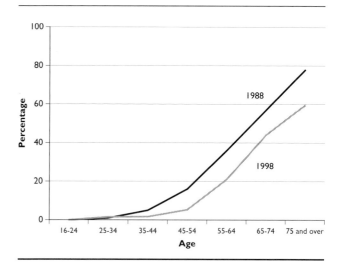

Fig 7.4.1 **Proportion of edentate adults in Northern Ireland by age, 1988-98**

7.4.3 The number of natural teeth in dentate adults in Northern Ireland, 1979-98

The number of teeth retained in the mouth provides an indication of the proportion of dentate adults at risk of becoming edentate in the future. A more specific indicator of oral health is the proportion of dentate adults with 21 or more natural teeth – which can be considered as the minimum number of teeth required for adequate oral function without the aid of dentures (see Chapter 2.2).

In 1998, the average number of teeth among dentate adults was 24.5. This figure has increased with each survey since 1979, despite the fact that older people formed a larger proportion of the adult dentate population in 1998 than they did in 1979. People are keeping more of their teeth for longer, and in 1998, 81% of dentate adults in Northern Ireland had 21 or more teeth, compared with 74% in 1988 and 68% in 1979.

Table 7.4.2

Table 7.4.3 shows the distribution of teeth by age. Numbers within age groups are small and figures should therefore be interpreted with some caution. In broad terms, the 1998 survey data indicate a clear divide between dentate adults under 45 years of age and those aged 45 years and over, for the average number of teeth per person. Since 1979, there have been improvements in the mean number of teeth within every age group, but most markedly among adults aged 25 to 34 years and 35 to 44 years. In 1998, over 90% of dentate adults under 45 years old had 21 or more teeth. In 1988, this was true only for adults under 35, and in 1979 only for adults aged under 25. In 1998, there was little difference between men and women both in respect of the average number of teeth per person, and in terms of the proportion with 21 or more teeth. This pattern has been stable since 1979.

Figure 7.4.2, Tables 7.4.3–4

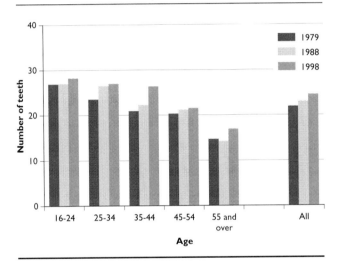

Fig 7.4.2 **Mean number of teeth among dentate adults in Northern Ireland by age, 1979-98**

Table 7.4.5 shows the distribution of teeth by social class of head of household in 1988 and 1998. Bases are small, and in 1998 there were no statistically significant differences between social classes in the proportion of dentate adults with 21 or more teeth. Nor were there any significant differences in this proportion in any social class between 1988 and 1998.

Table 7.4.5

Dentate adults who reported attending the dentist for regular check-ups had on average 24.9 teeth per person in 1998, compared with 23.0 among those who attend only when having trouble with their teeth; but the mean number of teeth in the latter group has risen since 1988 – from 20.4 to 23.0. This improvement was not seen among those who attend for regular check-ups. The difference in the number of teeth among dentate adults in these two groups has narrowed

considerably since 1988. In 1988, 58% of people who visit the dentist only when having trouble had 21 or more teeth compared with 84% of those who went for regular check-ups, a difference of 25 percentage points. In 1998, the comparative figures were 73% and 83% respectively, a difference of only 10 percentage points.

Table 7.4.6

7.4.4 Reliance on natural teeth and dentures in Northern Ireland

In 1998, 18% of all adults had dentures in conjunction with natural teeth. In general, this proportion increased with age reaching a peak of 44% in adults aged 55 to 64 years. People aged 65 years and over were more likely to have no natural teeth at all, and therefore the proportion with dentures and natural teeth was lower (33%) than in the 55 to 64 year age group.

There was no difference between men and women in terms of the proportion with both natural teeth and dentures, and neither were there significant differences between the social classes in this respect. For example, 21% of adults from non-manual households and 18% of those from manual households had dentures in conjunction with natural teeth.

Table 7.4.7

7.4.5 The condition of natural teeth in Northern Ireland, 1979-98

The data relating to the condition of the natural teeth presented in Tables 7.4.8 to 7.4.12 are based on the criteria used in the previous surveys of dental health (see section 7.4.6).

Since 1979 there have been continuing improvements in adult dental health in Northern Ireland, with an increasing proportion of dentate adults classified as having 18 or more sound and untreated teeth; from 13% in 1979 to 34% in 1998 - a figure approaching the target of 35% set for 2008. At 14.3, the mean number of sound and untreated teeth in 1998 was also nearing the target of 15 set for 2008. Improvements among younger people have been the most dramatic; in 1998, 74% of dentate adults aged 16 to 24 years had 18 or more sound and untreated teeth compared with 28% in 1979. The proportion of people aged 45 years and over with 18 or more sound and untreated teeth has not changed significantly over two decades and remains relatively low. It should be noted however that these older people bear the evidence of past disease levels and experience of restorative treatment and that previously, by this age, many would have lost all their teeth.

Other indicators also attest to general improvements in dental health in Northern Ireland. The average number of decayed and unsound teeth found in dentate adults in Northern Ireland has decreased from 1.7 in 1979 to 1.5 in 1988 to 0.8 in 1998. In fact Northern Ireland was the only country in the United Kingdom to show a decrease in the mean number of decayed or unsound teeth between 1988 and 1998.

The proportion of dentate adults in Northern Ireland with 12 or more restored (otherwise sound) teeth has remained at a similar level from 1979 to 1998; 32% in 1979, 33% in 1988 and 35% in 1998. The variation with age was similar to that seen overall in the United Kingdom (see Chapter 5.3) with a decrease in the experience of restored teeth among the youngest dentate adults and an increase among older adults. For example, the proportion of adults aged 16 to 24 years with 12 or more restored (otherwise sound) teeth decreased from 39% in 1979 to 10% in 1998 while among those aged 45 to 54 years this proportion rose from 32% to 49% during this time.

Figures 7.4.3–5, Tables 7.4.8–9

Fig 7.4.3   Proportion of dentate adults in Northern Ireland with 18 or more sound and untreated teeth by age, 1979-98

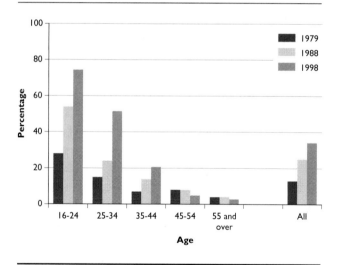

Fig 7.4.4   Proportion of dentate adults in Northern Ireland with some decayed or unsound teeth by age, 1979-98

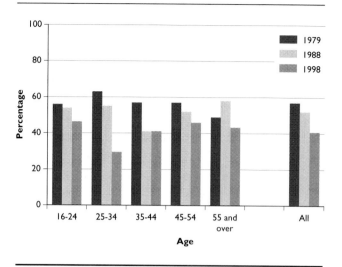

Fig 7.4.5   Proportion of dentate adults in Northern Ireland with 12 or more restored (otherwise sound) teeth by age, 1979-98

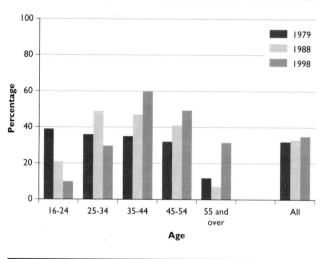

Both men and women showed similar improvements in the proportion with 18 or more sound and untreated teeth. In terms of decayed or unsound teeth and restored (otherwise sound) teeth the difference between men and women had largely disappeared by 1998. For example, in 1979, 24% of men and 39% of women had 12 or more restored (otherwise sound) teeth compared with 34% and 35% respectively in 1998.

Table 7.4.10

Analysis by social class and usual reason for dental attendance showed similar changes to those seen overall in the United Kingdom (see Chapters 5.2 and 5.3), although the small base numbers of the Northern Ireland samples meant that some of the differences were not significant. The increase in the proportion of dentate adults with 18 or more sound and untreated teeth showed little variation with respect to either of these factors, with all groups showing similar improvement. There were indications of decreased numbers of decayed and unsound teeth over the last decade in all social class groups and among those who reported attending the dentist with trouble but again small base numbers meant these differences were not significant. Although differences between social classes of the proportion of dentate adults with decayed or unsound teeth have persisted since 1988, the gap between non manual and unskilled manual classes closed considerably over the ten years to 1998. There was little change in the proportion of dentate adults with 12 or more restored (otherwise sound) teeth in any social class. The data suggested an increase in the number of restored (otherwise sound) teeth among those who attended the dentist only with trouble but the differences were not significant.

Tables 7.4.11–12

7.4.6 The condition of the natural teeth based on the criteria used in 1998

For this survey, changes were made to the criteria used in the earlier surveys for assessing dental caries. In 1998, teeth with untreated visual caries were classified as decayed, whereas previously they were recorded as sound and untreated. Restored teeth with recurrent visual caries were also treated as decayed in the 1998 classification, but were defined as teeth with sound restorations by pre-1998 criteria (see Chapter 3.1). Table 7.4.13 shows how the new criteria affect estimates of the average number of teeth that were sound and untreated, decayed or unsound, and restored (otherwise sound).

The overall effect of using the new criteria is to:

- provide a more conservative estimate of the average number of sound and untreated teeth; 13.9 compared with 14.3 using pre-1998 criteria
- increase the estimate of the average number of decayed or unsound teeth from 0.8 to 1.5
- decrease the estimate of the average number of restored (otherwise sound) teeth from 9.3 to 9.0.

However the size of the difference varied within specific groups of the population. For example, younger adults are more likely than older people to have visual caries only and the average number of teeth classified as decayed or unsound increased from 0.8 among 16 to 24 year olds using pre-1998 criteria to 2.0 based on the new criteria. In contrast, the new

criteria had less effect on the average number of decayed and unsound teeth found in those aged 65 years and over; this figure increased up from 0.6 to 0.9.

Table 7.4.13

In brief, the condition of teeth in Northern Ireland based on the 1998 criteria was as follows:

- Thirty-two percent of the dentate adult population had 18 or more sound and untreated teeth. This ranged from 71% of adults aged 16 to 24 years, to 45% of those aged 25 to 34 years and to 5% or less of people aged 45 years or over.
- Sixty percent of dentate adults had one or more decayed or unsound teeth. Around three-quarters (72%) of the youngest adults (aged 16 to 24 years) had one or more decayed or unsound teeth, but this was not statistically significantly more than in any other age group.
- One third (33%) of dentate adults had 12 or more restored (otherwise sound) teeth in 1998. This proportion increased with age up to those aged 54 years from 9% of 16 to 24 year olds to 48% of those aged 45 to 54 years.
- Dentate men and women were almost equally likely to have 18 or more sound and untreated teeth, but 63% of men compared with 56% of women had one or more decayed or unsound teeth. There was no significant difference between the numbers of restored (otherwise sound) teeth.
- There were no significant differences between the social classes in the proportion with 18 or more sound and untreated teeth. Half (51%) of dentate adults from non-manual households had one or more decayed or unsound teeth compared with two-thirds of the those from manual households (skilled manual households, 68%; partly-skilled and unskilled households, 67%). The proportion of dentate adults from non-manual households with 12 or more restored (otherwise sound) teeth was significantly higher than among those from an unskilled manual background (41% compared with 22%).
- There were no significant differences in the proportion of dentate adults with 18 or more sound and untreated teeth with respect to the usual reason for dental attendance. However, under half (48%) of dentate adults who attended for regular check-ups had one or more decayed or unsound teeth, compared with around three-quarters (76%) of those who visit the dentist only when having trouble with their teeth. Dentate adults who attended for regular check-ups were far more likely to have 12 or more restored teeth than those who attended with trouble; 42% compared with 18%.

Tables 7.4.14–17

7.4.7 The periodontal condition of adults in Northern Ireland

In the 1998, 66% of dentate adults in Northern Ireland were found to have visible plaque and 67% had some calculus (table not shown).

The depth of any periodontal pocketing and the extent of any loss of attachment were measured to assess the condition of the structures which support natural teeth (gums). Table 7.4.18 shows that 39% of dentate adults had loss of attachment of 4mm or more and 6% had extensive loss of attachment of 6mm or more. Forty-eight per cent of those examined had pocketing of 4mm or more and 4% had deep pockets of 6mm or more.

Table 7.4.18

7.4.8 Usual reason for dental attendance, dental treatment and dental hygiene

In 1998, half of dentate adults in Northern Ireland (51%) said that they visit the dentist for regular check ups, this is a significant increase from 42% in 1988. There were corresponding decreases in the proportion of adults who said that they visit the dentist only for an occasional check up or when having trouble with their teeth.

Figure 7.4.6, Table 7.4.19

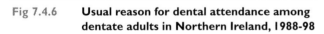

Fig 7.4.6 **Usual reason for dental attendance among dentate adults in Northern Ireland, 1988-98**

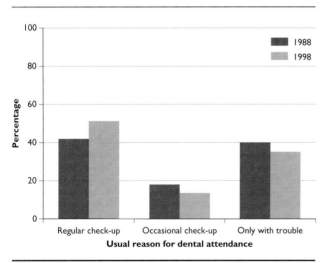

Between the surveys in 1988 and 1998, there was a significant increase in the proportion of dentate adults who said that they had visited the dentist within the previous two years; 71% in 1988 compared with 81% in 1998. In 1998, half (49%) of those interviewed said that they had no fillings or

extractions at their most recent visit to the dentist compared with one third (33%) in 1988. Sixty two percent said that their teeth had been scaled and polished, compared with 50% in 1988 and 41% said that they had had teeth filled, compared with 48% in the previous survey. In 1998, the proportion of adults who said that they had had teeth extracted was half that in 1988 (14% compared with 28%). Respondents were also asked about the costs, if any, of the treatment they had at their last visit to the dentist. In 1998, 56% of dentate adults said they had paid something towards their treatment compared with 43% in 1988.

Tables 7.4.20–23

Dentate adults were asked about the frequency with which they clean their teeth. Nearly three quarters (72%) reported to clean their teeth at least twice a day and 22% once a day. There has been little change in this pattern since 1988. Respondents who cleaned their teeth (1% reported never to clean their teeth) were asked whether they used anything other than toothpaste and brush for dental hygiene purpose. In 1998 just under half (47%) said that they did, which is double the proportion who said this in 1988 (23%). In the ten years between 1988 and 1998, dental floss has become increasingly popular and in the most recent survey was as likely to be mentioned as mouthwash.

Figure 7.4.7, Tables 7.4.24–25

Fig 7.4.7 **Use of dental hygiene products among dentate adults in Northern Ireland, 1988-98**

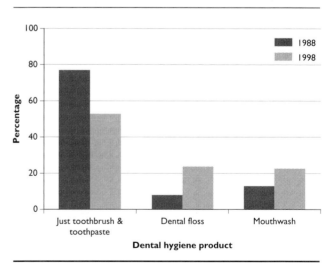

Notes and references
1. See Chapter 1.1

Table 7.4.1 Edentate adults, 1979-98

All adults   *Northern Ireland*

	1979	1988	1998
		percentage edentate	
All	33	18	12
Age			
16-24	} 4	0	0
25-34		1	2
35-44	17	5	2
45-54	35	16	5
55-64	64	36	21
65-74	67	57	44
75 and over	74	78	60
Gender			
Male	30	15	11
Female	37	22	13
Social class of head of household			
I, II, IIINM	-	16	6
IIIM	-	15	12
IV,V	-	26	20

For bases see Tables Appendix and 1979 report.
1988 Table 3.11, 2.14

Table 7.4.2 Distribution of number of teeth, 1979-98

Dentate adults   *Northern Ireland*

Number of teeth	1979	1988	1998
	%	%	%
1-14	16	13	9
15-20	16	13	10
21-26	40	35	32
27-31	26	38	42
32	2	1	6
Proportion with 21 or more teeth	68	74	81
Mean	21.9	23.0	24.5
Base	592	450	461

1988 Table 5.25 (part)

Table 7.4.3 Distribution of number of teeth by age, 1979-98

Dentate adults   *Northern Ireland*

Number of teeth	16-24			25-34			35-44			45-54			55 and over		
	1979	1988	1998	1979	1988	1998	1979	1988	1998	1979	1988	1998	1979	1988	1998
	%	%	%	%	%	%	%	%	%	%	%	%	%	%	%
1-14	0	0	0	5	1	3	20	10	0	15	12	15	50	50	36
15-20	2	2	0	18	5	3	15	19	8	33	22	20	24	21	26
21-26	40	31	23	49	33	27	44	50	37	37	48	47	22	21	34
27-31	54	64	63	26	59	62	20	21	51	15	18	15	3	7	5
32	4	2	14	2	1	6	1	0	5	0	0	2	1	0	0
Mean	26.9	27.0	28.2	23.5	26.5	27.0	20.9	22.2	26.4	20.3	21.2	21.5	14.7	14.2	16.9
Base	144	120	96	143	108	90	123	84	96	86	54	82	96	84	97

1988 Table 5.25

Table 7.4.4   **Distribution of number of teeth by gender, 1979-98**

Dentate adults   *Northern Ireland*

Number of teeth	Gender						All		
	Men			**Women**					
	1979	**1988**	**1998**	**1979**	**1988**	**1998**	**1979**	**1988**	**1998**
	%	%	%	%	%	%	%	%	%
1-14	14	14	10	17	14	9	16	13	9
15-20	20	14	11	13	12	9	16	13	10
21-26	39	34	33	41	36	32	40	35	32
27-31	24	38	38	28	38	47	26	38	42
32	3	0	9	1	1	3	2	1	6
Mean	21.9	23.0	24.5	21.9	23.0	24.5	21.9	23.0	24.5
Base	274	234	183	318	216	278	592	450	461

1988 Table 5.26

Table 7.4.5   **Distribution of number of teeth by social class of head of household, 1988-98**

Dentate adults   *Northern Ireland*

Number of teeth	Social class of head of household						All†	
	I, II, IIINM		**IIIM**		**IV,V**			
	1988	**1998**	**1988**	**1998**	**1988**	**1998**	**1988**	**1998**
	%	%	%	%	%	%	%	%
1-14	14	9	14	8	10	15	13	9
15-20	10	10	13	12	18	11	13	10
21-26	37	30	32	35	37	28	35	32
27-31	37	45	40	40	34	40	38	42
32	1	6	0	5	0	5	1	6
Mean	23.0	24.7	22.8	24.5	22.8	23.2	23.0	24.5
Base	180	190	144	129	108	83	450	461

† Includes those for whom social class of head of household was not known and armed forces
1988 Table A3.15

Table 7.4.6   **Distribution of number of teeth by usual reason for dental attendance, 1988-98**

Dentate adults   *Northern Ireland*

Number of teeth	Usual reason for dental attendance				All†	
	Regular check-up		Only with trouble			
	1988	1998	1988	1998	1988	1998
	%	%	%	%	%	%
1-14	6	5	25	17	13	9
15-20	10	12	17	10	13	10
21-26	38	35	28	32	35	32
27-31	46	43	31	32	38	42
32	0	4	0	8	1	6
Mean	24.6	24.9	20.4	23.0	23.0	24.5
Base	192	257	174	137	450	461

† Includes those who visit the dentist for an occasional check-up
1988 Table A3.16

Table 7.4.7   **Reliance on natural teeth and dentures**

All adults   *Northern Ireland*

		Dental status			
		Natural teeth only	Natural teeth with dentures	Edentate	Base
All	%	70	18	12	734
Age					
16-24	%	99	1	0	120
25-34	%	96	3	2	126
35-44	%	89	9	2	129
45-54	%	64	31	5	127
55-64	%	35	44	21	98
65 and over	%	17	33	50	134
Gender					
Male	%	70	18	11	298
Female	%	69	18	13	436
Social class of head of household					
I, II, IIINM	%	73	21	6	261
IIIM	%	70	18	12	208
IV,V	%	62	18	20	157

Table 7.4.8 The condition of natural teeth, 1979-98

Dentate adults   *Northern Ireland*

The condition of natural teeth (1988 criteria)	1979	1988	1998
	%	%	%
Sound and untreated			
0	1	2	2
1-5	13	11	7
6-11	39	32	31
12-17	34	30	26
18 or more	13	25	34
Mean	11.4	12.6	14.3
	%	%	%
Decayed or unsound			
0	43	48	59
1-5	50	47	39
6 or more	7	5	2
Mean	1.7	1.5	0.8
	%	%	%
Restored (otherwise sound)			
0	13	9	5
1-5	20	22	23
6-11	35	36	37
12 or more	32	33	35
Mean	8.4	8.9	9.3
Mean number of teeth	21.9	23.0	24.5
Base	*592*	*450*	*461*

1988 Table 6.28, 7.29, 8.34

Table 7.4.9   **The condition of natural teeth by age, 1979-98**

Dentate adults   *Northern Ireland*

The condition of natural teeth (1988 criteria)	Age														
	16-24			25-34			35-44			45-54			55 and over		
	1979	1988	1998	1979	1988	1998	1979	1988	1998	1979	1988	1998	1979	1988	1998
	%	%	%	%	%	%	%	%	%	%	%	%	%	%	%
Sound and untreated															
0	0	0	0	1	2	0	1	4	1	2	5	4	2	2	4
1-5	1	2	0	8	6	2	17	9	6	12	8	12	30	36	20
6-11	22	6	1	39	33	23	49	51	37	45	40	54	48	44	53
12-17	49	37	25	37	36	23	26	23	36	33	39	26	16	13	20
18 or more	28	54	74	15	24	52	7	14	21	8	8	5	4	4	3
Mean	15.0	17.8	21.4	11.8	13.1	16.7	9.9	10.7	12.9	10.5	10.8	9.6	8.1	7.6	8.4
	%	%	%	%	%	%	%	%	%	%	%	%	%	%	%
Decayed or unsound															
0	44	46	54	37	45	70	43	59	59	43	48	54	51	42	57
1-5	46	48	46	55	47	26	54	40	39	55	48	44	39	53	43
6 or more	10	6	0	8	8	4	3	2	2	2	4	1	10	6	0
Mean	2.0	1.6	0.8	1.8	1.7	0.8	1.4	0.9	0.8	1.4	1.2	1.0	1.9	1.8	0.6
	%	%	%	%	%	%	%	%	%	%	%	%	%	%	%
Restored (otherwise sound)															
0	6	7	6	7	2	5	11	4	2	15	8	3	35	27	7
1-5	17	25	44	16	11	20	18	17	8	17	24	14	33	37	29
6-11	38	47	41	41	38	45	36	33	30	36	27	34	20	28	32
12 or more	39	21	10	36	49	30	35	47	60	32	41	49	12	7	32
Mean	9.5	7.6	5.9	9.5	11.6	9.5	9.1	10.6	12.6	8.3	9.3	11.0	4.2	4.9	7.9
Mean number of teeth	26.9	27.0	28.2	23.5	26.5	27.0	20.9	22.2	26.4	20.3	21.2	21.5	14.7	14.2	16.9
Base	*144*	*120*	*96*	*143*	*108*	*90*	*123*	*84*	*96*	*86*	*54*	*82*	*96*	*84*	*97*

1988 Table 6.28, 7.29, 8.34

Table 7.4.10 **The condition of natural teeth by gender, 1979-98**

Dentate adults   *Northern Ireland*

The condition of natural teeth (1988 criteria)	Gender						All		
	Men			Women					
	1979	1988	1998	1979	1988	1998	1979	1988	1998
	%	%	%	%	%	%	%	%	%
Sound and untreated									
0	1	2	2	2	2	1	1	2	2
1-5	10	11	6	15	12	8	13	11	7
6-11	37	28	31	41	37	31	39	32	31
12-17	36	34	27	31	26	25	33	30	26
18 or more	16	25	33	11	23	35	14	25	34
Mean	12.2	13.0	14.5	10.7	12.1	14.2	11.4	12.6	14.3
	%	%	%	%	%	%	%	%	%
Decayed or unsound									
0	38	43	55	48	52	63	43	48	59
1-5	51	50	44	49	44	36	50	47	39
6 or more	11	7	2	3	4	1	7	5	2
Mean	2.2	1.7	0.9	1.4	1.2	0.7	1.7	1.5	0.8
	%	%	%	%	%	%	%	%	%
Restored (otherwise sound)									
0	19	11	4	8	7	5	13	9	5
1-5	34	25	27	32	19	20	20	22	23
6-11	23	38	34	21	35	39	35	36	37
12 or more	24	27	34	39	39	35	32	33	35
Mean	7.2	8.2	9.1	9.4	9.6	9.5	8.4	8.9	9.3
Mean number of teeth	21.9	23.0	24.5	21.9	23.0	24.5	21.9	23.0	24.5
Base	*274*	*234*	*183*	*318*	*216*	*278*	*592*	*450*	*461*

1988 Table 6.29, 7.30, 8.35

Table 7.4.11 The condition of natural teeth by social class of head of household, 1988-98

Dentate adults   *Northern Ireland*

The condition of natural teeth (1988 criteria)	Social class of head of household						All†	
	I, II, IIINM		IIIM		IV,V			
	1988	1998	1988	1998	1988	1998	1988	1998
	%	%	%	%	%	%	%	%
Sound and untreated								
0	4	2	1	2	1	2	2	2
1-5	12	8	12	5	9	8	12	7
6-11	34	36	33	31	30	29	32	31
12-17	28	24	30	27	38	27	30	26
18 or more	23	30	24	35	21	34	24	34
Mean	12.0	13.6	12.9	14.6	12.8	14.1	12.5	14.3
	%	%	%	%	%	%	%	%
Decayed or unsound								
0	58	64	42	55	37	55	47	59
1-5	37	36	51	42	60	42	48	39
6 or more	5	0	8	3	4	3	6	2
Mean	1.2	0.6	1.8	1.0	1.7	1.0	1.5	0.8
	%	%	%	%	%	%	%	%
Restored (otherwise sound)								
0	8	4	12	3	7	8	9	5
1-5	20	16	27	30	19	24	22	23
6-11	32	37	32	33	48	44	35	37
12 or more	40	43	30	33	27	24	34	35
Mean	9.8	10.5	8.1	8.9	8.3	8.0	8.9	9.3
Mean number of teeth	23.0	24.7	22.8	24.5	22.8	23.2	23.0	24.5
Base	180	190	144	129	108	83	450	461

† Includes those for whom social class of head of household was not known and Armed Forces
1988 Table A4.15, A5.15, A6.15

Table 7.4.12 **The condition of natural teeth by usual reason for dental attendance, 1988-98**

Dentate adults   *Northern Ireland*

| The condition of natural teeth (1988 criteria) | Usual reason for dental attendance | | | | All† | |
| | Regular check-up | | Only with trouble | | | |
	1988	1998	1988	1998	1988	1998
	%	%	%	%	%	%
Sound and untreated						
0	3	2	2	2	2	2
1-5	12	7	14	7	12	7
6-11	30	35	32	28	32	31
12-17	35	27	27	27	30	26
18 or more	20	30	24	35	24	34
Mean	12.1	13.7	12.4	14.6	12.6	14.3
	%	%	%	%	%	%
Decayed or unsound						
0	62	72	31	40	48	59
1-5	35	28	61	56	47	39
6 or more	3	0	8	4	5	2
Mean	0.9	0.4	2.2	1.4	1.5	0.8
	%	%	%	%	%	%
Restored (otherwise sound)						
0	1	1	19	11	8	5
1-5	14	15	34	34	22	23
6-11	34	41	36	34	36	37
12 or more	51	43	12	22	33	35
Mean	11.5	10.7	5.8	7.0	8.9	9.3
Mean number of teeth	24.6	24.9	20.4	23.0	23.0	24.5
Base	*192*	*257*	*174*	*137*	*450*	*461*

† Includes those who only visit the dentist for an occasional check-up
1988 Table A4.16, A5.16, A6.16

Table 7.4.13 **Difference in condition of teeth using 1998 and pre-1998 criteria**

Dentate adults   *Northern Ireland*

| | Average number of teeth that were: | | | | | | | | | Base |
| | Sound and untreated | | | Decayed or unsound | | | Restored (otherwise sound) | | | |
	1998 criteria	pre-1998 criteria	Difference	1998 criteria	pre-1998 criteria	Difference	1998 criteria	pre-1998 criteria	Difference	
All	13.9	14.3	0.4	1.5	0.8	0.7	9.0	9.3	0.3	461
Age										
16-24	20.5	21.4	0.9	2.0	0.8	1.2	5.6	5.9	0.3	96
25-34	16.1	16.7	0.6	1.7	0.8	0.9	9.2	9.5	0.3	90
35-44	12.6	12.9	0.3	1.5	0.8	0.7	12.3	12.6	0.3	96
45-54	9.4	9.6	0.2	1.3	1.0	0.3	10.8	11.0	0.2	82
55-64	9.3	9.4	0.1	0.9	0.6	0.3	8.3	8.5	0.2	54
65 and over	6.9	7.1	0.2	0.9	0.6	0.3	7.1	7.3	0.2	43
Gender										
Men	14.0	14.5	0.5	1.7	0.9	0.8	8.8	9.1	0.3	183
Women	13.8	14.2	0.4	1.3	0.7	0.6	9.2	9.5	0.3	278
Social class of head of household										
I, II, IIINM	13.3	13.6	0.3	1.1	0.6	0.5	10.3	10.5	0.2	190
IIIM	14.0	14.6	0.6	2.0	1.0	1.0	8.6	8.9	0.3	129
IV,V	13.6	14.1	0.5	1.8	0.9	0.9	7.7	8.0	0.3	83
Usual reason for dental attendance										
Regular check-up	13.5	13.7	0.2	0.9	0.4	0.5	10.4	10.7	0.3	257
Occasion check-up	15.2	15.7	0.5	1.5	0.7	0.8	9.3	9.6	0.3	67
Only with trouble	13.9	14.6	0.7	2.4	1.4	1.0	6.7	7.0	0.3	137

Table 7.4.14 **The condition of natural teeth by age**

Dentate adults   *Northern Ireland*

The condition of natural teeth (1998 criteria)	Age						All
	16-24	25-34	35-44	45-54	55-64	65 and over	
	%	%	%	%	%	%	%
Sound and untreated							
0	0	0	1	4	6	2	2
1-5	0	5	7	12	10	33	8
6-11	1	24	40	54	56	49	32
12-17	28	26	32	26	26	14	27
18-23	48	37	14	5	2	2	23
24 or more	23	9	7	0	0	0	9
Proportion with 18 or more	71	45	21	5	2	2	32
Mean	20.5	16.1	12.6	9.4	9.3	6.9	13.9
	%	%	%	%	%	%	%
Decayed or unsound							
0	28	43	43	42	43	58	40
1-5	65	48	52	55	57	42	54
6 or more	8	9	5	4	0	0	5
Proportion with 1 or more	72	57	57	58	57	42	60
Mean	2.0	1.7	1.5	1.3	0.9	0.9	1.5
	%	%	%	%	%	%	%
Restored (otherwise sound)							
0	6	5	2	4	8	12	5
1-5	49	20	10	16	23	37	25
6-11	36	46	32	32	38	25	36
12 or more	9	28	55	48	32	27	33
Mean	5.6	9.2	12.3	10.8	8.3	7.1	9.0
Base	96	90	96	82	54	43	461

Table 7.4.15 The condition of natural teeth by gender

Dentate adults   *Northern Ireland*

The condition of natural teeth (1998 criteria)	Gender		All
	Male	Female	
	%	%	%
Sound and untreated			
0	2	1	2
1-5	8	8	8
6-11	32	32	32
12-17	28	26	27
18-23	21	25	23
24 or more	10	7	9
Proportion with 18 or more	31	32	32
Mean	14.0	13.8	13.9
	%	%	%
Decayed or unsound			
0	37	44	40
1-5	57	52	54
6 or more	6	5	5
Proportion with 1 or more	63	56	60
Mean	1.7	1.3	1.5
	%	%	%
Restored (otherwise sound)			
0	4	6	5
1-5	30	21	25
6-11	34	38	36
12 or more	32	35	33
Mean	8.8	9.2	9.0
Base	*183*	*278*	*461*

Table 7.4.16 Condition of natural teeth by social class of head of household

Dentate adults   *Northern Ireland*

The condition of natural teeth (1998 criteria)	Social class of head of household			All[†]
	I, II, IIINM	IIIM	IV,V	
	%	%	%	%
Sound and untreated				
0	2	2	2	2
1-5	8	6	10	8
6-11	38	31	30	32
12-17	23	30	25	27
18-23	22	22	26	23
24 or more	7	9	6	9
Proportion with 18 or more	29	32	32	32
Mean	13.3	14.0	13.6	13.9
	%	%	%	%
Decayed or unsound				
0	49	32	33	40
1-5	49	58	61	54
6 or more	2	10	6	5
Proportion with 1 or more	51	68	67	60
Mean	1.1	2.0	1.8	1.5
	%	%	%	%
Restored (otherwise sound)				
0	4	4	8	5
1-5	16	32	29	25
6-11	39	33	41	36
12 or more	41	32	22	33
Mean	10.3	8.6	7.7	9.0
Base	*190*	*129*	*83*	*461*

† Includes those for whom the social class of the head of household was not known and Armed Forces

Table 7.4.17 The condition of natural teeth by usual reason for dental attendance

Dentate adults   *Northern Ireland*

The condition of natural teeth (1998 criteria)	Usual reason for dental attendance			All
	Regular check-up	Occasional check-up	Only with trouble	
	%	%	%	%
Sound and untreated				
0	2	0	2	2
1-5	7	10	8	8
6-11	35	25	31	32
12-17	28	22	27	27
18-23	22	35	20	23
24 or more	6	8	12	9
Proportion with				
18 or more	28	43	32	32
Mean	13.5	15.2	13.9	13.9
	%	%	%	%
Decayed or unsound				
0	52	39	24	40
1-5	48	57	63	54
6 or more	1	4	13	5
Proportion with				
1 or more	48	61	76	60
Mean	0.9	1.5	2.4	1.5
	%	%	%	%
Restored (otherwise sound)				
0	1	3	12	5
1-5	18	30	35	25
6-11	40	30	34	36
12 or more	42	37	18	33
Mean	10.4	9.3	6.7	9.0
Base	257	67	137	461

Table 7.4.18 The condition of supporting structures

Dentate adults[†]   *Northern Ireland*

Condition	Percentage
Any loss of attachment of 4mm or more	39
Any loss of attachment of 6mm or more	6
Any pockets of 4mm or more	48
Any pockets of 6mm or more	4
Base	427

† 34 people were excluded from the periodontal examination

Table 7.4.19 Usual reason for attending the dentist, 1988-98

Dentate adults   *Northern Ireland*

Usual reason for attending the dentist	1988	1998
	%	%
Regular check-up	42	51
Occasional check-up	18	14
Only with trouble	40	35
Base	558	636

1988 Table 20.1 (part)

Table 7.4.20 Time since last visit to the dentist, 1988-98

Dentate adults[†]   *Northern Ireland*

Time since last visit to the dentist	1988	1998
	%	%
Up to 1 year	61 ⎫	71 ⎫
Over 1 year, up to 2 years	10 ⎬ 94	10 ⎬ 97
Over 2 years, up to 5 years	18	10
Over 5 years, up to 10 years	5 ⎭	6 ⎭
Over 10 years, up to 20 years	3	2
Over 20 years	2	2
Base	558	636

† Excludes those who have never been to the dentist
1988 Table 24.3 (part)

Table 7.4.21 Treatment received at last dental visit 1988-98

Dentate adults† — *Northern Ireland*

Treatment received	1988	1998
	%	%
No fillings/no extractions	33	49
Some fillings/no extractions	40	37
Some fillings/ some extractions	8	4
No fillings/ some extractions	19	10
Base	534	581

† Excludes those who have never been to the dentist or were in the middle of a course of treatment
1988 Table 24.34 (part)

Table 7.4.22 Treatment received at last dental visit 1988-98

Dentate adults† — *Northern Ireland*

Treatment received	1988	1998
	%	%
Scale and polish	50	62
Teeth filled	48	41
X-rays	22	25
Teeth extracted	28	14
Crown fitted	9	8
Abcess treated	8	5
Denture fitted	7	6
Denture repaired	1	4
Base	534	581

Percentages may add to more than 100% as respondent may had given more than one answer
† Excludes those who have never been to the dentist or were in the middle of a course of treatment
1988 Table 24.32 (part)

Table 7.4.23 Whether or not respondents paid for dental treatment, 1988-98

Dentate adults† — *Northern Ireland*

Whether or not respondent paid	1988	1998
	%	%
Paid something	43	56
Paid nothing	47	33
Don't know	10	11
Base	534	581

† Excludes those who have never been to the dentist or were in the middle of a course of treatment
1988 Table 24.51 (part)

Table 7.4.24 Frequency of teeth cleaning, 1988-98

Dentate adults — *Northern Ireland*

Frequency of teeth cleaning	1988	1998
	%	%
Twice a day or more	69	72
Once a day	22	22
Less than once a day	7	6
Never	1	1
Base	558	636

1998 Table 26.1 (part)

Table 7.4.25 Use of dental hygiene products, 1988-98

Dentate adults† — *Northern Ireland*

Respondent used:	1988	1998
	%	%
Dental floss	8	24
Toothpicks/woodsticks	4	5
Mouthwash	13	23
Just toothbrush & toothpaste	77	53
Base	558	624

Percentages may add to more than 100% as respondents could give more than one answer
† Excludes those who stated they never cleaned their teeth
1988 Table 26.16

Appendices

Appendix A   Glossary of terms

Artificial (full coverage) crown

A tooth restoration which is cemented to the tooth and covers all the natural coronal surfaces. It is usually made of metal, porcelain or a combination of both materials.

Bridge

A prosthesis used to replace a tooth or teeth which is cemented on to a natural tooth or teeth nearby and which is not intended for removal by the patient. Two types of bridges were identified; conventional (which usually relies on adjacent teeth being cut down for crowns to which the replacement tooth is attached) and adhesive (which is attached to the adjacent teeth using adhesive techniques and which does not require the tooth to be extensively prepared).

Caries

See *dental caries*.

Carstairs & Morris Index

See DEPCAT.

Cavitated caries/decay

Decay present which has caused the lesion to cavitate.

Complete denture

A prosthesis which replaces all of the natural teeth in one jaw. In some cases there may be a few natural roots remaining, but the denture will cover these, so that all of the visible teeth are on the denture.

Coronal surfaces

The surfaces of the crown of the tooth.

Crown

The crown is the part of the tooth which, on a natural sound tooth, is covered in dental enamel. See *coronal surfaces* and *artificial crown*.

Decayed teeth

1998 criteria
Teeth with visual caries or cavitated caries or teeth that were so broken down, possibly with pulpal involvement, that they were unrestorable. It includes teeth that had restorations with recurrent caries but does *not* include teeth that had restorations which were lost, broken or damaged but where there was no recurrent caries.

1988 criteria
Teeth with cavitated caries or teeth that were so broken down, possibly with pulpal involvement, that they were unrestorable. It includes teeth that had restorations with recurrent cavitated caries but does *not* include teeth that had restorations which were lost, broken or damaged but where there was no recurrent caries.

Decayed and unsound teeth

1998 criteria
Teeth with visual caries, cavitated caries and teeth that were so broken down, possibly with pulpal involvement, that they were unrestorable. It includes teeth that had restorations with recurrent caries and restorations which were lost, broken or damaged.

1988 criteria
Teeth with cavitated caries and teeth that were so broken down, possibly with pulpal involvement, that they were unrestorable. It includes teeth that had restorations with recurrent cavitated caries and restorations which were lost, broken or damaged.

Dental caries/dental decay

A disease process that results in the demineralisation of the hard tissues of the tooth by microbial activity. The terms caries, dental caries, decay and dental decay are used interchangeably in this report.

Dentate

Having one or more natural teeth. (Compare with *edentate.*)

Dentine

The hard, calcified tissue which forms the major part of the tooth. It encloses the dental pulp, but is covered by enamel on the coronal surfaces.

DEPCAT

A system used in Scotland known more formally as the Carstairs & Morris index as a measure of material deprivation within a postcode sector. It is based on 4 factors in the postcode sector: 1) the level of overcrowding in households; 2) male unemployment in the area; 3) proportion of social classes IV or V in the area and 4) the proportion of persons in private households with no car. The index consists of 7 deprivation categories which can be referred to as DEPCAT1 (least deprived) to DEPCAT7 (most deprived). See Appendix F for more details.

Edentate

Having no natural teeth. (Compare with *dentate.*)

Enamel

The hard mineralised outer layer covering the *coronal surfaces* of the natural tooth.

Filled and decayed teeth

Teeth with a filling and some active decay.

Filled, but unsound teeth

Teeth with a filling which is damaged but not decayed.

Filled (otherwise sound) teeth

Teeth in which a filling has been placed but which are now sound with no active decay and no damage to the filling. (Compare with *restored (otherwise sound) teeth.*)

Fissure sealants

A material, usually a resin, which has been placed in the pits and fissures of teeth to protect against the development of caries. Sealants are also used in conjunction with filling materials.

Functional dentition

Oral health was defined in terms of function in the 1994 Department of Health publication *An oral health strategy for England*. The definition of oral health referred to the ability to "eat, speak and socialise without active disease, discomfort or embarrassment". Such attributes as eating comfortably and socialising without embarrassment can be related directly to the number and distribution of natural teeth, described as a functional dentition. From the point of view of analysis in this report a functional dentition was defined as having 21 or more standing teeth, although at an individual level the above attributes could be achieved with fewer.

Incisors

The four front teeth in each jaw.

Jarman

The Jarman under-privileged area score provides a measure of relative social deprivation for each Electoral ward in England, although it was originally devised to assess General Practitioner workload in England and Wales. The scores are derived from data from 1991 census relating to eight social factors including unemployment, overcrowding and social class. There are five categories with Area 1 containing the most deprived and Area 5 containing the least deprived wards. See Appendix F for more details.

Loss of attachment

See *periodontal attachment.*

Manual social class/manual groups

See *social class.*

Missing teeth

Teeth which were not present or visible in the mouth at the time of examination. Missing teeth includes those which had been extracted and those which were unerupted.

Molar

Large, grinding tooth situated at the back of the mouth. There are up to six molars in each jaw.

Non-manual social classes

See *social class.*

OHIP/OHIP14

Oral Health Impact Profile. A standardised measure of the overall impact of oral problems on an individual based on a conceptual model of oral health proposed by Locker. The form used in this survey was the short form consisting of 14 questions (see Slade GD. *Derivation and validation of a short-form oral health impact profile.* Community Dentistry Oral Epidemiology 1997; 25: 284-290).

Partial dentures (removable partial dentures)

A prosthesis which replaces some of the natural teeth in one jaw, and which can be removed by the patient. (Compare with *bridge* and *complete denture.*)

Periodontal attachment

The fibrous connection between the tooth root and the supporting bone and gum. Where gum (periodontal) disease has occurred some of this attachment between the tooth and supporting bone is lost. *This loss of attachment* begins around the neck of the tooth where the tooth projects into the mouth. Loss of attachment below the level of the gum margin results in a *periodontal pocket.* The loss of attachment which has taken place and the depth of the periodontal pocket can be gauged by using a graduated blunt probe held against the root of the tooth and gently placed under the gum as far as the base of the pocket.

Periodontal disease(s)

The group of diseases of the tissues which invest and support the teeth (gum disease).

Periodontal pocket

See *periodontal attachment.*

Plaque

The soft, sticky white bacterial material which collects around the teeth and which is implicated in causing dental caries and the periodontal diseases.

Premolar

A permanent tooth situated between the permanent canine and molar teeth.

Primary (coronal/root) caries

A tooth is described as having primary decay if it has any caries on a surface which has not been treated previously (for the purpose of this definition, sealants alone are not included as treatment). Other surfaces of the tooth may or may not have restorative treatment or recurrent decay. The terms *primary caries* and *primary decay* are used interchangeably in this report. (Compare with *recurrent caries*.)

Pulp (dental)

The vascular soft tissue which fills the pulp chamber and the root canals of a tooth. It is the innermost part of the tooth and includes connective tissue, blood vessels and nerves.

Recurrent caries/decay

A tooth is described as having recurrent decay if it has any caries on a surface which has been treated previously (for the purpose of this definition, sealants alone are not included as treatment). Other surfaces of the tooth may or may not have decay or restorations.

Restoration

The material end result of operative procedures that restore the form, function and appearance of a tooth. In this survey it was defined as a filling or artificial crown.

Restored (otherwise sound) teeth

1998 criteria
Teeth which include a restoration but are now sound with no visual or cavitated decay and no damage to the restoration.

1988 criteria
Teeth which include a restoration but are now sound with no cavitated decay and no damage to the restoration. Previous surveys in this series have described this category as 'filled (otherwise sound)'.

Root

The part of the tooth not covered by enamel and which is usually below the level of the gum. It may become exposed due to the recession of the gums associated with the loss of periodontal attachment, particularly with increasing age.

Root caries/root decay

Decay occurring on the roots of the teeth where there has been loss of periodontal attachment. (See also *dental caries*.)

Sealants

See *fissure sealants*

Social class

Based on the Registrar General's Standard Occupational Classification, Volume 3 HMSO (1991). Social class was ascribed on the basis of the occupation of the head of household. The classification used in the tables is as follows:

I,II,IIINM referred to in text as 'non-manual'. Professional and intermediate (Classes I and II). Skilled occupations, non manual (Class III non-manual).

IIIM referred to in text as 'skilled manual'. Skilled occupations, manual (Class III manual).

IV,V referred to in text as 'partly skilled and unskilled' or 'unskilled', these terms are used interchangeably. Partly skilled and unskilled occupations (Classes IV and V).

Social class was not determined for households where the head had never worked, was a full-time student, was in the Armed Forces or whose occupation was inadequately described.

Sound and untreated teeth

1998 criteria

Teeth with no evidence of visual or cavitated caries, nor any restorative treatment. It includes teeth with sealants which were sound or fractured but with no evidence of caries.

1988 criteria

Teeth with no evidence of cavitated caries, nor any restorative treatment. It includes teeth with visual caries. It includes teeth with sealants which were sound, fractured or with visual caries.

Tissue supported partial denture

A partial denture which rests only on the soft tissues of the mouth, so the natural teeth share none of the load of the partial dentures during function.

Tooth borne partial denture

A partial denture which rests on, and is supported by, some of the remaining natural teeth. A partial denture may be completely supported by the teeth or supported by both the teeth and the soft tissues. See *tissue supported partial denture.*

Unsound restoration

A restoration which has been lost, broken or damaged but is not decayed. Fillings and artificial crowns are both included here as forms of restoration.

Usual reason for dental attendance

Previous reports in this series have referred to this variable as 'dental attendance pattern'. A full description can be found in Chapter 6.1.

Visual caries/decay

Visible decay is present but it is not obviously cavitated.

Wear/tooth wear

Loss of tooth substance due to a non-bacterial cause. This may take the form of attrition (where the teeth in opposing arches have worn away each other), abrasion (where the teeth have been worn away mechanically by a foreign body, such as a toothbrush) or erosion (where there has been damage to the teeth from acids, usually dietary or gastric, not produced by bacteria).

Appendix B   Training and calibration of the dental examiners

B.1   Introduction

Undertaking nearly four thousand dental examinations on a random sample of the United Kingdom adult population is a substantial task, and in order to complete the work in a reasonable time several examiners are required. When more than one examiner is involved in a study there is immediately an issue of variation in the way in which they interpret the criteria they are asked to use, in the face of what they see during the examination. This is always a problem in large dental studies of this sort, but is additionally complicated in this survey because of the range of diseases and conditions which have to be measured in a population covering such a broad age span. To give a complete picture of the oral health of United Kingdom adults, the examination needed to record data on the number and distribution of teeth and dentures, the presence and condition of the natural teeth, including, caries, fillings, crowns and wear and the condition of the roots and the gums. Issues surrounding training and calibration are of most relevance for the data presented in Parts 3 and 5 of this report.

With a wide-ranging examination and a large sample, various strategies can be employed to minimise the biases caused by individual variation. These include:

- *the selection of an appropriate number of appropriately skilled examiners*
- *ensuring simple, clear and familiar criteria and codes for the examination*
- *putting in place a comprehensive and detailed training procedure*
- *calibration testing*

Each of these strategies is described below, in addition to the results of the calibration exercise.

B.2   The number of examiners and their selection

For an examination of this sort a small number of examiners could be highly trained and undertake many examinations, perhaps several hundred, each. This has advantages in terms of training, but there is a risk of introducing important bias to the results. One individual who scores high or low for any given condition can have a measurable and serious effect on the final results. If individual examiners undertake most of the examinations in one specific geographical area this can even result in inaccurate conclusions being drawn about regional variations in oral health.

The alternative strategy is to use a large number of examiners, each taking responsibility for a much smaller number of examinations. Although the spread of individual scores may be greater than for a more intensively trained small group, the risk of very serious biases in the data is much lower and with so many examiners, the impact of individual outliers is much less. The risk of aberrant regional values should also be less for the same reasons. However, it makes training and measuring calibration for the group a more complicated business and inevitably there is less time for rigorous one to one contact between trainer and examiner. Nevertheless, this strategy also has the advantage that the data are collected in a relatively short space of time, that the examiners themselves are only required to be away from their normal duties for a short period, and that the examinations can be completed soon after training without the need for periodic retraining. This strategy is the one employed in this survey, in which 70 examiners were used.

The source and skills of the different examiners was also of relevance. The availability of examiners experienced in the use of the most important criteria should help training and improve calibration. The examiners used were drawn from the Community Dental Services and almost all were very experienced in epidemiological work, and were familiar with many of the criteria used (see below).

B.3   The criteria and coding for the examination

One of the primary requirements of this survey was to maintain comparability with previous surveys. We were able to design the criteria for coronal condition to allow such comparability while using rules which were familiar to examiners with experience of other surveys, most notably the national surveys of children conducted by the British Association for the Study of Community Dentistry. In many cases we were able to use the same codes for coronal conditions. The ability to use familiar rules and codes was intended to reduce the amount of new material with which examiners were to be faced and thus reduce the likelihood of error.

Criteria and coding systems for other conditions used instruments and codes which would have been familiar to the examiners. Periodontal measurements used the CPI probe which is widely used in survey and clinical work. The codes for root decay were matched with those for coronal decay, although the detail of criteria was slightly different.

B.4   The training procedure

The examiners had to be trained in "field" conditions and a ready supply of volunteers was also required. With seventy examiners, a large number of subjects were required to make sure that no individual volunteer examinees were subject to an excessive number of examinations. The ONS offices at Newport in Wales were chosen because of the large and willing workforce while the availability of volunteers from the adjacent Patent Office, who had a slightly older profile, allowed a broad age range to be examined. Through the efforts of a number of committed individuals we were able to obtain more than enough material for training purposes. This was helped by our ability to offer a small fee for taking part.

The training and calibration took place in two waves over consecutive weeks in September 1998, each wave accommodating about 35 examiners and lasting 3 days. Initial papers and criteria lists were sent out in advance to the examiners to allow prior preparation, and then after a long briefing session on the first evening, two full days were available with subjects for training with the final day for calibration.

On the first day the examination was split into its component parts so examiners could spend time getting used to the different coding schemes before piecing them together into a full examination. At this stage the trainers offered help with the procedure of the examination and answered queries about coding, but did not compare scores. The trainers included the consortium members who had been involved in developing the criteria and the dentists who had acted as examiners during the pilot study (a small study conducted before the main fieldwork to test the procedures to be used in the field), who were themselves experienced in dental epidemiology field work. Where decision making was particularly awkward or informative, trainers were able to share the solution with many examiners. These problems and

other specific points were collected together and fed back to all examiners in a plenary session after each one and a half hour session. Once the examiners moved on to whole mouth examinations the paper data sheets were collected and collated. With each volunteer being examined by three or four examiners it was possible to identify areas of inconsistency and resolve disagreements between examiners whilst still feeding back to the whole group. Once again direct feedback and plenary sessions were used to sort out problems as they arose. By the end of the second full day all examiners had undertaken several full and many partial examinations, they had worked in pairs with colleagues in the first instance and then with the one of the social survey interviewers, with whom they would be working, acting as recorders and using the same computer recording system that was to be used in the field. All examiners also had their scores compared with those of other examiners for the same subject and inconsistencies addressed while the volunteer was still present.

B.5   Calibration testing

Calibrating 70 dentists is a difficult task. It involves as many dentists as possible all seeing the same volunteers. There are physical limits to the number of examinations any individual can have, in particular for the periodontal examination, which can be rather uncomfortable when repeated on many occasions. Calibration took place on the final morning of the two training courses. The examiners were divided into four groups of nine, each group examining nine subjects. The subjects were examined in strict sequence by the examiners, having the same quadrant of their mouth examined. Where possible, mixing of examiners from different geographical groups was undertaken. Thus, the strategy was to make available calibration results for eight groups (called A to H) of nine examiners who each examined nine subjects. In the event, some problems with recording (see below) and with the timing of the exercise led to some variability in the total amount of data available.

B.6   The results of the calibration exercise

Calibration data were calculated for coronal and root condition, tooth wear and contacts, but in view of the ethical issues of repeating an uncomfortable procedure, not for periodontal probing depths, and because of methodological issues, not for plaque levels.

Calibration data from the first two groups were collected using laptop computers but problems with the program lead to the subsequent groups using paper recording. As a result of the problems with the computers some information was lost for the first two groups. The paper records were entered onto a flat ASCII data file. A certain amount of data cleaning was necessary, including resolution of discrepancies relating to tooth identification. In the context of the analysis and the report, disagreement about, for example, whether a single premolar was a first or a second premolar was not relevant. However disagreements over whether that premolar was decayed or not, or whether or not it contained a sound or unsound filling were of considerable relevance. The protocol contained advice about assigning difficult-to-identify teeth, but in many cases correctly assigning a single tooth was impossible and unimportant, so in order to ensure that measures of agreement reflected the important reporting issues, any disagreements about the identification of individual teeth were identified and eliminated at the raw data stage. There were no disagreements about the number or type of teeth present.

Levels of agreement and disagreement are reported here as mean kappa[1] scores for each group. Kappa scores give an indication of the level of agreement between examiners, the closer the score is to one, the higher the agreement. For intermediate values it has been suggested that kappa values greater than 0.81 indicate excellent agreement and between 0.61 and 0.80 substantial agreement[1]. Kappa scores were calculated for each possible pair of examiners within a group, and then an overall mean kappa calculated for the group of 9 examiners.

The condition of the coronal surfaces (whether they were decayed, restored or sound) was one of the key elements of the analysis in Chapters 3.1 and 3.2, and also significant elements of Parts 4 and 5 and the chapters relating to different countries. High levels of agreement between examiners were of some importance. The kappa scores, when considering whole teeth, are given in Table B.1. The mean scores for the groups ranged from 0.88 to 0.96, representing very high levels of agreement. The lowest individual pair scored 0.64 which is still an acceptably high level of calibration. In order to calculate scores for whole teeth a certain amount of data processing is necessary. During this procedure some of the detail of the levels of agreement and disagreement may be lost because the five surface scores are amalgamated and one selected according to a strict hierarchy. In order to test the calibration at an even more detailed level, kappa scores were calculated for surfaces, giving five times the number of codes to compare. The scores for all of the groups of examiners remained high when surface scores were analysed, ranging from 0.85 to 0.94.

There are short sections on the condition of the roots in Chapters 3.1 and 3.2. The examination of root surfaces was

expanded on that for 1988 and the criteria clarified. Agreement on the scores for roots may be expected to be a little lower as the criteria would be less familiar to the examiners. Mean kappa scores for the groups ranged from 0.57 to 0.89, which are mostly within the range of substantial to excellent agreement. The lowest score for an individual pair was 0.28, and the highest was 1.00, but, as stated earlier, the large number of examiners used in the survey served to reduce any effect of individual outliers.

The examination of coronal wear was another new part of the examination. The mean kappa scores were not generally as high as those for the condition of the coronal surfaces, ranging from 0.44 to 0.96 and there was quite substantial individual variation within the groups. One group had insufficient cases to allow the calculation of mean kappa scores because of the recording problems discussed previously. These findings suggest that the wear data were likely to be more variable than those for the coronal surfaces. Thus detailed analyses of these data were not conducted. However, this is not to say that presentation of the prevalence levels are not useful and such data are reported in Chapter 3.1.

The measurement of tooth contacts was also a new part of the examination, included to give an indication of a functional dentition, and contacts are reported in Chapter 2.2. The calibration exercise measured the agreement for each contact zone, Kappa scores for this detailed measure ranged from a disappointing 0.29 up to a very good 0.83. There were only two mean scores of less than 0.6, and these were recorded for the two groups who had experienced technical difficulties with direct entry onto computers during the calibration exercise. This undoubtedly contributed to the lower levels of agreement recorded for these two groups. In fact the report presents the number of contacts, which is a much cruder measure than used in the calibration and therefore liable to less variation in measurement.

Table B1

Notes and references

1. When two examiners make decisions about the same sample to the same criteria, a statistic called Cohen's kappa $(K)^2$ can be used to measure their agreement. The kappa statistic measures the agreement between the two sets of results allowing for chance agreements. The observed level of agreement is compared to that generated by assuming that the distribution of results for each classification was the same as that for the overall distribution.

    For example, two dentists examine 50 teeth and classify each as decayed, filled or sound:

Dentist A	Dentist B			
	Decayed	Filled	Sound	Total
Decayed	8	2	1	11
Filled	1	11	1	13
Sound	1		25	26
Total	10	13	27	50

This data can be expressed as proportions of the total:

Dentist A	Dentist B			
	Decayed	Filled	Sound	Total
Decayed	.16	.04	.02	.22
Filled	.02	.22	.02	.26
Sound	.02		.50	.52
Total	.20	.26	.54	1

If all agreement was due to chance alone, the expected frequencies for each table could be calculated by multiplying row proportion by column proportion:

Dentist A	Dentist B			
	Decayed	Filled	Sound	Total
Decayed	.0.44	.0572	.1188	.22
Filled	.052	.0676	.1404	.26
Sound	.104	.1352	.2808	.52
Total	.20	.26	.54	1

The observed frequency of agreement is

$$p_0 = 0.16 + 0.22 + .50 = 0.88$$

while the frequency expected by chance is

$$p_e = .044 + .0676 + .2808 = 0.3924$$

and Cohen's kappa is calculated as:

$$K = \frac{p_o - p_e}{1 - p_e} = 0.80$$

A kappa value of one indicates complete agreement; a value of zero indicates the agreement expected by chance; a value of less than zero indicates a level of agreement less than that expected by chance. For intermediate values it has been suggested that kappa values greater than 0.81 indicate excellent agreement and between 0.61 and 0.80 substantial agreement[3]. See *Statistics in Dentistry*[4] for more discussion of this statistic.

2. Cohen J. *A coefficient for agreement of nominal scales* Educ Psychol Measurement 20, 37-46 1960
3. Landis JR, Koch GG. *The measurement of observer agreement for categorical data* Biometrics 33, 159-174 1977.
4. Bulman JS, Osborn JF. *Statistics in Dentistry* British Dental Association, London 1989

Table B.1   **Kappa scores for calibration results**

	Examiner group							
	A	**B**	**C**	**D**	**E**	**F**	**G[†]**	**H[†]**
Condition of teeth (coronal surfaces)								
Mean	0.88	0.92	0.88	0.90	0.92	0.90	0.89	0.96
Standard deviation	0.07	0.04	0.04	0.06	0.04	0.06	0.10	0.08
Coefficient of variation	0.08	0.05	0.05	0.07	0.05	0.07	0.11	0.08
Minimum	0.70	0.81	0.80	0.77	0.84	0.77	0.64	0.77
Maximum	0.98	1.00	0.97	1.00	1.00	1.00	1.00	1.00
Condition of teeth (individual coronal surfaces)								
Mean	0.90	0.87	0.92	0.89	0.89	0.94	0.85	0.93
Standard deviation	0.04	0.02	0.03	0.03	0.05	0.03	0.07	0.06
Coefficient of variation	0.04	0.02	0.03	0.04	0.05	0.03	0.08	0.06
Minimum	0.81	0.82	0.85	0.79	0.79	0.88	0.71	0.83
Maximum	0.98	0.91	0.96	0.95	0.97	0.99	0.95	1.00
Condition of roots								
Mean	0.57	0.66	0.69	0.57	0.62	0.69	0.76	0.89
Standard deviation	0.10	0.11	0.09	0.15	0.14	0.12	0.18	0.12
Coefficient of variation	0.17	0.17	0.13	0.26	0.22	0.17	0.24	0.14
Minimum	0.40	0.38	0.52	0.28	0.33	0.44	0.48	0.73
Maximum	0.79	0.89	0.87	0.91	0.96	0.91	1.00	1.00
Tooth wear								
Mean	0.61	0.66	0.44	0.90	0.62	0.69	0.96	-
Standard deviation	0.35	0.18	0.37	0.11	0.21	0.32	0.04	-
Coefficient of variation	0.58	0.27	0.83	0.13	0.34	0.46	0.04	-
Minimum	0.32	0.36	-0.16	0.67	0.17	-0.18	0.92	-
Maximum	1.00	1.00	1.00	1.00	1.00	1.00	1.00	-
The position of natural tooth contacts								
Mean	0.75	0.83	0.67	0.65	0.70	0.78	0.39	0.29
Standard deviation	0.19	0.09	0.19	0.19	0.13	0.19	0.24	0.27
Coefficient of variation	0.26	0.10	0.29	0.29	0.18	0.24	0.61	0.96
Minimum	0.31	0.64	0.28	0.26	0.40	0.27	0.09	-0.14
Maximum	0.97	1.00	0.96	1.00	0.91	1.00	1.00	0.69

†   Calibration data from the first two groups (labelled G and H) were collected using laptop computers but problems with the program meant that some information was lost for these groups. Subsequent groups used paper recording

Appendix C.1 The conduct of the examination and clinical criteria used for the assessments

Introduction

These criteria are written for the use of the dental examiners prior to and during training and for consultation purposes during the fieldwork.

The aim in setting these criteria has been to maintain comparability with the 1978/1988 surveys of adult dental health in the UK, whilst incorporating new conventions based on research findings and current epidemiological practice. In addition, criteria to assess clinical conditions that have emerged as significant in the last decade have also been introduced.

For the first time in one of the national Adult Dental Health Surveys, data will be entered directly on to a computer by the interviewer. For the purpose of early training, paper recording on printouts of the screen display will be used, but in the final stages of training and in the field all data will be recorded on a portable computer.

The criteria which follow should be studied in conjunction with the reproduction of the examination forms supplied. Each page of the forms shows several of the grids which the interviewer will complete on screen. The general and personal details will be entered by the interviewer before going into the household.

PROCEDURE BEFORE THE EXAMINATION

Medical Screening

Before the examination the dentist must ask the person some set questions about their medical history, specifically relating to any risk that the examination may pose. They will be asked about a history of rheumatic fever, endocarditis and valvular heart disease. They will also be asked about the presence of any artificial joints (usually hip or knee). Despite the extremely low risk of the examination, no risk is seen as acceptable in a survey of this sort, and those who respond positively to these questions will not undergo the periodontal examination. There is no reason at all why the rest of the examination should be a problem, as the gingivae will not be probed.

The precise wording of the questions is on a card which the interviewer will carry, but it is the examiner's responsibility to ask, and to pursue if necessary. A single code will be entered to record whether or not the periodontal component was omitted, and another to record whether or not the rest of the examination was completed in full. This will be at the end of the examination.

Equipment set-up and seating the participant

The participant should be seated in a comfortable chair which has good head support, and to which the examiner can get access. Individual examiner's preferences vary. Kitchens are sometimes difficult as the seats often have no head support. A comfortable chair in the sitting room is usually fine, but access and lighting can be a problem. Consideration needs to be given to the positioning of the "Daray" lamp, the availability of power points, and the convenience to the participants. The lamp can be clamped to an ironing board if necessary.

The instruments should be laid out on a clean tissue out of sight of the participant if possible, but allowing easy access. The light should be set up and adjusted. The Daray lamp should be set at the high power setting (II) and dark protective glasses placed on the subject.

Cross infection control

Each examiner will carry sufficient sets of sterile instruments to ensure that there are sterile instruments for every examination. Following the examination these will be placed in a sealed container for transport back to the examiners home clinic where the instruments will be autoclaved. Examiners will wear a clean pair of rubber gloves for the examination of each participant. These will be disposed of into a standard yellow bag with any tissues or wipes after the exam. This will be disposed of on return to the clinic along with normal clinical waste.

DIAGNOSTIC CRITERIA

Before the participant removes their dentures, the examiner may wish to look briefly in the mouth to assess the overall distribution of natural teeth and dentures. This may serve to put the participant at their ease before removing their prosthesis, but it is essential that the dentures are then removed for the rest of the examination. There is an initial box in the form which records the dental status (dentate in both arches, dentate in one arch, or edentate). This must be done with the dentures removed. This is completed at this stage.

The convention throughout is:
If in doubt - score low (i.e. "least disease").

1. Existence and state of coronal surfaces, and debris score

The first stage of the examination is to record the condition of the crowns of the teeth. As data are entered, the computer will automatically block out all missing teeth for the remaining **relevant** parts of the examination (namely the grids for roots, wear and periodontal condition), consequently there is no need to record which teeth are present or absent before starting. All the examiner has to decide is which tooth is being examined, data on spaces and replacement teeth will be recorded at a later stage.

Procedure

Using mirror and CPI probe the permanent teeth will be examined in the following order:
Upper right, upper left, lower left, lower right (i.e. clockwise as you look at the subject from in front).
The interviewer will probably prompt you with the tooth number as you move around the mouth. Before calling out the five surface codes, we would like you to call out a code to indicate whether or not there is any plaque (or supra-gingival calculus) on the tooth surface. The code to call is either "P" where there is plaque, or "C" (clean) where there is not.

Having called out the debris code for the tooth, and cleaned the surfaces of gross debris (if necessary), the surfaces should then be examined one at a time, distal first, then occlusal, mesial, buccal, lingual, and the codes called out clearly and unequivocally for each site. Each tooth, even the anterior teeth, has five surfaces because on the anterior teeth, the

incisal is considered a tooth surface equivalent to the occlusal surface on molars and premolars.

Most codes will be single codes, but multicoding (or sub-coding), where a second code is entered to qualify the first, is possible for some conditions. A tooth may be "5 sub-code X" or "5X". Clarity of calling is of the first importance if the examination is to be completed efficiently and accurately.

For this part of the examination the CPI type C probe is used. Being a ball-ended instrument, this means that it should not damage any incipient lesions. It should not be used for probing into fissures or early lesions, but it may be used for the following:
- *removing debris from around key areas if necessary.*
- *detecting and examining sealants*
- *placing into open crown margins or defects at the margin of restorations to estimate their dimension, but this should not be done with force.*

Codes and criteria

Debris
P = Any VISIBLE plaque (to naked eye, without running probe around)
C = Clean, no plaque visible to the naked eye
This is called out first, and then you should call the five surface codes for coronal condition below.

Coronal condition
M = Missing (and not a bridge pontic)
A = Adhesive bridge pontic
B = Conventional bridge pontic
Once these codes are entered for a single surface these teeth will be blocked through as missing for the remainder of the examination (except for contacts and spacing).

0 = Sound
> For this survey, the diagnostic threshold for caries follows the convention for dental epidemiology, only caries into dentine or thought to be into dentine is recorded as diseased, caries restricted to enamel is recorded as a sound surface. Note that there are three codes for caries (1,2 and 3) but that all of these represent caries into dentine. Where all surfaces are sound you may call "Q" and all 5 surfaces will be marked with zero.

1 = Visual caries
> The surface has decay present which is visible to the observer, but which is NOT obviously cavitated. This will usually manifest as shadowing under an occlusal surface or marginal ridge.

2 = Caries (cavitated)

> The surface has decay present which has caused the lesion to cavitate. Record '2' only if there is a cavity (but not "3", see below). In line with previous surveys, this also includes temporary dressings placed for the treatment of caries.

3 = Broken down/pulpal involvement evident or unequivocal

> Code 3 is for teeth which are so broken down that it is inconceivable that there is not pulp involvement and where restoration of the tooth would be very involved or impossible. This would usually be used for carious stumps or teeth so broken down that whole surfaces have been eliminated through caries and where more than two thirds of the marginal ridge has been destroyed. It should not be used for little bits of retained root left after extraction (which should be ignored at this stage), or for overdenture abutments (code 9). There must be active soft carious dentine. Although it is possible that on occasions pieces of restoration may be present on such a surface, by definition this is unlikely. In such cases, where a code 3 is to be used you should ignore them.

> Where a code 3 is entered it applies to all 5 surfaces of the tooth, even if there are pieces of restoration. This will be entered automatically for all 5 surfaces.

9 = Not possible to code

> Code 9 is used throughout the examination for occasions where you cannot make a reasonable judgement. It should be used **VERY** sparingly. In the case of coronal surfaces it represents circumstances where an entire surface is actually missing because it has fractured off or worn away, such that there is nothing that you can code. This is rare, if there is anything there you should score it. The most likely use for code 9 is for overdenture abutments. If a surface is missing because it has broken down through caries then 2 or 3 should be used. Code 9 should also be used for teeth which are partially erupted and where large parts of surfaces are obscured by flaps of mucosa. This is most likely to occur on lower third molars. Code 9 is used only for surfaces where more than half of the surface is covered.

The following codes indicate the presence of a restoration or sealant. **All** of the codes below must **always** be qualified by a second code which indicates the condition of the restoration or sealant

5 = Amalgam filling

6 = Intracoronal restoration, but not amalgam

> This will usually be composite or glass ionomer, but also includes inlays or onlays.

7 = Veneers, shims

> These are adhesive restorations. They are used simply to change the shape of a tooth or as adhesive retainers for resin bonded bridges. A shim is a thin metal restoration cemented onto a functional surface (such as the palatal surface of an upper anterior or a molar occlusal surface) to change it's shape. These are rare. A veneer is usually placed buccally to improve colour or shape, these are fairly common. The difference between them is not important, but neither is placed to treat caries. The key difference between code 7 and code 6 is that the restorations for code 6 are placed to treat caries, whilst those for 7 are stuck on to the surface to fulfil an aesthetic or occlusal need. Restorations placed on incisal edges of anterior teeth to repair fractures should also be coded 7, assuming that there is no question of them being placed to treat caries.

8 = Full crown

> This may be either permanent or temporary, and including full coverage bridge abutments for conventional bridges. It does not include ¾ crowns, these are coded '6' on the relevant surface. Temporary crowns are coded 8, but must be multicoded "Y" see below.

F = Sealants

> It is often impossible to be sure whether or not a sealant is a sealant alone or whether there is a restoration underneath. Where there is **clear** evidence of a sealant restoration (but **only** where there is clear evidence) this should be coded as 6 instead.

For codes 5,6,7,8 and F the computer will always need a second code to indicate the status of the restoration. The interviewer will not be able to move on unless you call a second code. The restoration may have visible caries but no cavitation associated with the restoration (similar to code 1 above), or there may be cavitation associated with caries at the margin (see code 2 above), it may have broken or been damaged but not carious, or it may be a perfectly sound restoration. There is a code to represent each of these and you must always use ONE of them with codes 5,6,7 and 8.

V = Recurrent caries (visual, no cavitation)

X = Recurrent caries (cavitation at the margin)

Y = Failed restoration, but not carious

> This may be a restoration which is chipped cracked or which has a margin into which a ball-ended probe tip will fit. Temporary crowns are included here.

Z = Sound

Priority

Data are collected on a surface by surface basis so the possibilities of having more than one code on a surface are limited. On rare occasions though there may be a restoration and completely unrelated caries. In these situations the caries code will ALWAYS take precedence, so if codes 2 and 5 or even 5X are possible, then code 2 should be entered. This is to ensure that new caries is never left unrecorded. Similarly if there is a filling which is fractured and carious, the caries code (X or V) is the one recorded as the multicode, not code Y, so that recurrent caries is always recorded unless there is new caries on the same surface.

Where there are two materials present on a surface, amalgam will take precedence.

Summary - coronal surfaces
- *move clockwise around the mouth*
- *the presence or absence of any plaque is called out first*
- *then five surface codes are called out (for D,M,O,B,L surfaces in that order)*
- *for each surface there will be a single code where there is no restoration or sealant*
- *where there is a restoration or sealant there are two codes, one to describe the restoration (e.g. amalgam, other material, crown, sealant etc) and a second code to indicate the condition*
- *on rare occasions where there is both new decay and **separate** restoration on the same surface, caries will always take priority. These situations will arise only rarely*

2. Condition of root surfaces

Procedure

Having completed the coronal surfaces the examiner should return to examine the roots in the same order as was used for those surfaces. There is no need for the recorder to mark out missing teeth, this will have been done automatically, but it is important that you keep the recorder orientated. You should call out which teeth you are on as you progress or at the very minimum you should indicate when the midline is reached. **On no account should you try to do the roots at the same time as the crowns.**

Diagnosis of root caries is different from that for coronal caries, and requires the use of a sharpened probe, because textural changes are at the heart of diagnosis. The examiner will now need to pick up the root probe. Note that this instrument is used for no other surface. The probe should be used on the surface of the roots to determine texture or detect cavitated defects. **Do not try to push the tip hard into dentine.** You will get some indication of the texture by dragging it across the surface, and gently feeling for any softness. Do this if there is any question of decay.

Anything exposed apical to the cemento-enamel junction is regarded as root surface. Each root may have four surfaces,

but in reality often only one or two will be exposed and in younger participants the number of exposed teeth will be rather low. However all four surfaces must be examined, to ensure complete coverage of the root surface.

Codes and criteria

Each root surface of every tooth should be examined and a single code for each tooth called using the codes below. Remember, if in doubt, score low (i.e. least disease)

N = No exposed root surface

0 = Exposed root surface present but no evidence of current or past disease
> Exposed root surface is any exposure of the root coronal to the gingival margin

2 = Caries on the root surface
> This is any caries which is believed to be active on the basis of texture. An active root lesion can be almost any colour from yellow or tan through to almost black. In some circumstances it can even be very difficult to tell caries from extrinsic staining. The texture is very important and the probe must be used to try to determine this. Anything which shows evidence of softening or frank cavitation should be coded as carious. **Shiny** dark areas are much less likely to be actively carious and more likely to be arrested, such areas should be coded as "4". Usually stained calculus and extrinsic staining will be fairly obvious, but if there is any doubt the texture is critical.

3 = Broken down tooth such that roots cannot be scored
> Code 3 is the same as 3 for coronal surfaces, it indicates that the root is present but is grossly carious and broken down.

4 = Hard, arrested decay
> The surface should be glossy and hard, despite being discoloured. There has been decay, but it is now arrested. See above for "2".

9 = Unscorable
> Code 9 should be used sparingly, and **only** if it is not clear whether or not there is any root exposure. This is most likely where there are very large deposits of calculus around lower incisors. If there is any visible root it should be coded with the appropriate letter. If there is no root surface exposed then a code 0 should be used. Only if the examiner suspects an exposed root surface, but cannot examine it should a code 9 be entered.

W = Worn to a depth of 2mm or more, but with no caries or restoration

The codes above are always used alone, the codes below are restoration codes and must always be used with a second code using exactly the same convention as for coronal surfaces. One of the codes V, X, Y, Z must be used to describe the condition of the root restorations. Situations where the

"V" code is required are expected to be very rare indeed.

.5 = Amalgam restoration (see note below)

6 = Filling or restoration, not amalgam (see note below)

Note:

Most restorations are either clearly crown or root restorations, but some restorations and lesions straddle the CEJ and these are difficult to call. Here the 3mm rule will apply. This goes as follows:

- If the restoration is clearly a coronal restoration which encroaches on to the root, it should ONLY be coded as a root restoration as well as a coronal restoration if it extends 3mm or more beyond the CEJ (or the estimated CEJ) and onto the root surface. The distal section of the CPI probe (above the ball end) can be used to measure this if necessary.

- If there is frank caries at the margin of the filling extending from the coronal onto the root surface then this will count as caries on the root, even where the restoration does not extend 3mm. In this case the condition of the coronal portion of the filling will be coded independently according to the condition of this part of the tooth.

- If a root restoration extends onto the crown, the same 3mm rule applies in reverse (i.e. there must be 3mm beyond the CEJ on to the crown to count as a coronal restoration), but any caries occurring on the coronal portion of a root restoration is recorded as coronal caries, whilst the root restoration is scored according to it's condition.

- Some lesions and some fillings are smaller, they straddle the CEJ and it is difficult to be sure whether they are primarily on the root or the crown and do not extend 3mm onto either. In this case they should be recorded as root as this is the more vulnerable surface if it is exposed.

- Artificial crowns cause a particular problem because it is often impossible to identify the CEJ. Where there is a crown and the CEJ is covered, the margin of the crown should be considered the same as the CEJ, unless the contour of the crown indicates where the CEJ lies in which case the extension of the crown beyond this can be measured. On the rare occasion where this extends 3mm or more on to the root surface, the surface should be recorded as filled.

Summary - root surfaces

- *root surfaces are examined in a separate single sweep of the mouth, examining the teeth in the same order as for crowns*
- *only a single code is entered for the whole root surface*
- *the codes are similar to those for crowns*
- *you must use the sharp probe to assess texture*
- *in younger patients the examination will usually be very easy and quick*

Tooth wear

Procedure

The assessment of tooth wear is a part-mouth examination.

The teeth should be inspected in good light, from the upper right canine to the upper left canine, and then left canine to right canine in the lower arch, just as for the previous parts of the examination. Each tooth should be assessed looking at each coronal surface (root surfaces have been recorded during the examination for roots). In order to provide comparable data with younger age groups from previous children's surveys, scores are recorded on three surfaces per tooth for the six upper teeth, the buccal, incisal and palatal. For the lower teeth, the worst surface score is the one recorded and this will almost always be the incisal score, but if buccal or lingual surfaces are worse, then this is recorded. In many cases there will be very heavily restored teeth or crowns, these cannot be scored, but are not missing and should be coded as unscorable. The computer programme will automatically mark off all missing teeth.

Remember the convention: If in doubt - score low.

Codes and criteria

Score	Surface	Criteria
0	All	Sound, Any wear is restricted to the enamel and does not extend into dentine
1	All	Loss of enamel just exposing dentine
2	B,L	Loss of enamel exposing dentine for more than an estimated one third of the individual surface area (B,L).
	Incisal	Loss of enamel and extensive loss of dentine, but not exposing secondary dentine or pulp. On incisal surfaces this will mean exposed dentine facets with a bucco-lingual dimension 2mm or greater at the widest point (see diagram)
3	B,L	**Complete** loss of enamel on a surface, pulp exposure, or exposure of secondary dentine where the pulp used to be. Frank pulp exposure is most unlikely.
	Incisal	Pulp exposure or exposure of secondary dentine
8	All	**Fractured tooth** - clear evidence of traumatic loss of tooth substance rather than wear.
9	All	**Unscorable.** >75% of surface obscured and no remaining incisal edge/tip which can be coded. If any incisal edge/tip is present and a score may be given, this should be done. All crowns and bridge abutments are given this code.

Notes:

1. Bridge pontics are coded as missing and will already be blocked out.

2. Code 2 is the most difficult one to judge. Use the CPI probe (shaded band) to measure the diameter of any exposed dentine facet if necessary.
3. Where wear is severe, it can often be contiguous from palatal onto incisal, such that it is difficult to distinguish the surfaces. In these instances, code both the same.
4. Frank pulpal exposure is very rare, but exposure of secondary dentine (where the pulp used to be), usually appearing as a small translucent area in the centre of a wide area of dentine exposure, is not uncommon in older people.

Summary - Tooth wear
● *only upper and lower anterior teeth are examined*
● *three upper surfaces and the worst lower surface of each of the teeth is recorded*
● *many teeth may be unscorable because of restorations*

Figure 1

The illustration below shows the measurement of a worn facet on a lower incisor. The CPI probe can be used to measure the bucco-lingual width of the **dentine** facet by using the shaded area on the instrument as a ruler. This part of the probe measures 2mm. In this case the exposed dentine is just less than 2mm so would be coded as 1.

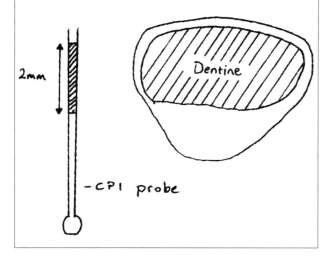

Occlusion - functional occlusal contacts

The assessment of occlusal contacts refers to occlusal contacts between **natural teeth and the pontics of fixed bridges only.** There are two parts to this short examination, first an assessment of the posterior (premolar and molar) regions, and then a simple count of the number of anterior tooth contacts.

Procedure
A contact is the same as an occlusal stop. For the purposes of this examination you should get the subject to close together normally on the back teeth (sometimes the phrase "clench your back teeth together" is the most effective) and then using

a mirror to hold back the cheek, look at the lower arch from the side and record the distribution of contacts. In a complete quadrant there will be 8 possible zones of contact in the posterior region (see diagram). Each of the premolars is a single zone, and each of the molars is about twice as wide, so we split them into two zones each.

We are interested in which zones the contacts are in, not the tooth involved - teeth can drift. You need not try to identify which tooth is doing what, just look at each zone in turn and work out whether or not there is a contact. For example, if a first premolar has been lost and the second premolar has moved forward, the mesial cusp of the first molar may have taken up the second premolar position, and the second premolar may have taken the first premolar position. However, although it is the 5 and 6 which are making the contacts, the contacts will be scored as being in the first and second premolar zone.

The scoring is quite easy if you **start at the front (distal to the canine) and work distally**, on the right first, then the left. Call out code 1 for contact present and 0 (zero) for no contact. Note that contact between a fixed bridge and a natural tooth, or even between two fixed bridge pontics, is considered as a contact. Obviously if there is NO lower tooth or bridge pontic in the zone you are looking at there cannot possibly be a contact.

Codes and criteria
Posterior functional contacts
0 = No posterior functional contact
1 = Posterior functional contact present
Notes:
● A posterior functional contact is classified as present where the contact forms a vertical occlusal stop. This is recorded according to the lower tooth (i.e. does the natural lower or bridge pontic contact with any natural upper or pontic), and is coded as a "1" even if the area of contact is small. In rare cases where there is contact but no occlusal stop (e.g. a scissors bite) a zero is recorded. Clearly there can be no contact if there is no lower tooth in the zone you are looking at
● In some cases it may be difficult to tell whether the teeth actually touch or not, you should assume that they do if you are in doubt.
● Where there are small spaces in the lower arch and you cannot decide whether you should consider it as a whole zone, count the space as a full "zone" if it is wider than a half a tooth, otherwise ignore it.

Anterior contacts
Anterior contacts are also recorded. To do this look at the six lower anterior teeth and count how many of them are in contact with natural uppers or fixed bridge pontics. The

recorder can enter any score in the range from 0 to 6. Most people have an overbite and it can be difficult to see whether there is contact, but in these cases assume there is. Where there is an anterior open bite, or where lower teeth are missing there clearly cannot be a contact.

Summary - contacts

- *it is the position of the contact which is critical, the tooth which makes it is irrelevant*
- *missing teeth are not marked out*
- *get the participant to close (or 'clench') together onto their back teeth*

- *start at the front (the position where the first premolar would be) and work back*
- *do the right first, then the left*
- *think of each side of the mouth as being split into 8 "zones" in the premolar and molar region, representing where each premolar or half molar tooth would normally be*
- *there are only two codes, '1' if there is a contact in that position, '0' if there is not (make sure to get them the right way around!)*

Figure 2:   Examples of scoring contacts

Right side: there are a couple of lower teeth present which do not make contact, and the two molars have drifted one "zone" forward into the distal half of the space where the first molar was. Remember the contacts are examined and called out starting at the front and working back (irrespective of the side concerned – follow the arrow), so the codes here would be......

0, 1, 0, 1, 1, 0, 0, 0 .... now check it and see if you agree

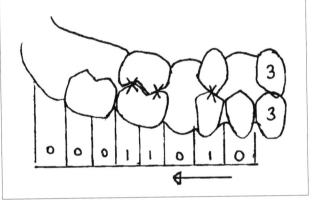

Left side: There has been wholesale loss of upper teeth, but one upper molar has drifted and tipped forward and makes a contact in about the fifth zone back (roughly where the mesial half of the second molar would be. Sometimes this position can be difficult to judge accurately. Whether the contact is actually in that position or one zone either side is not critical, what is important is that it is in the middle of the molar region. The codes are....

0, 0, 0, 0, 1, 0, 0, 0

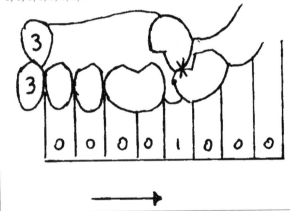

Left side: there has been a fair amount of drifting, which is not really relevant. Codes are called from front to back. The codes are.......

0, 0, 1, 1, 1, 0, 0, 0

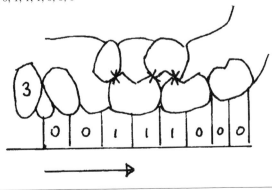

Right side: there are posterior teeth but they all miss each other. The upper 4 has slipped down into the lower premolar space, and although there may be contact between the lower molar and the upper premolar it is on the side of the tooth and does not constitute an occlusal "stop". These are called out as.......

0, 0, 0, 0, 0, 0, 0, 0

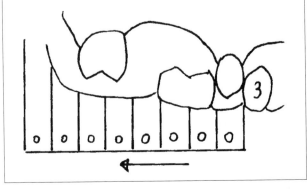

Figure 2: (continued)

Right side: This is a common situation where single upper and lower premolars have been removed and everything else has moved up one. Once again it does not matter that there are no second premolars, what matters is that there is a contact in that position. The codes are...

1, 1, 1, 1, 1, 1, 1, 0

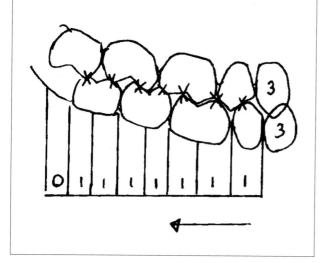

Spaces, aesthetics and dentures

The subject will have removed any dentures, but you may now need to look at them to help you decide on the correct codes. This examination is much easier to carry out from in front of the participant.

In this part of the examination you are looking for space in the anterior region, as far back as the second premolar zone to give some indication of aesthetics and the need for dentures and bridges. As with contacts, you are not recording which teeth are missing, that has been done already. For this reason it is again much easier if you start at the midline and work backwards but examine the quadrants in the same order as the rest of the exam (upper right, upper left, lower left and lower right). This way it is much more straightforward to assess the position of spaces as you can use the midline as a reference. As you look around you should look for spaces of half the width of the expected tooth at each zone. If there is a space present then call it out, the code depends on whether or not it is filled by an artificial tooth. If there is a natural tooth call it as "no space". Note that because teeth drift you may have a space at (for example) the upper second molar position even when that tooth is present (it may have drifted to a different position). What is important is that there is a space at that position, the teeth present are irrelevant. Your job is to map the spaces, you can completely ignore the tooth type. (see diagram)

Codes and criteria: spaces

Record for each tooth position the following codes:

N = No space (tooth present or space closed)
S = Space equal to, or more than, ½ the size of the tooth you would normally expect to be in that space
D = Space restored by a removable prosthesis
B = Space restored by a fixed bridge

Summary - spaces

- *once again, start in the midline and work out*
- *record the position of any spaces as far back as the second premolar space*
- *there are different codes for filled and unfilled spaces.*

Figure 3

A single example where there is one space in the first premolar position, but the first molar has come forward to the second premolar position eliminating the space.

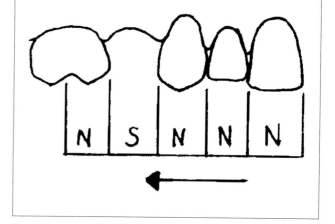

Codes and criteria: dentures

You will now have to hand any dentures the participant may have. The dentures, including full dentures opposed by natural teeth or partial dentures should be examined separately, upper and lower, for the following features.

Denture type
(recorded separately for upper and lower arches)

1 = Partial
2 = Full
3 = Complete overdenture
4 = Implant retained

Denture base material

1 = Metal, those dentures where the major connector is metal but not including those whose only metal component is clasps.
2 = Plastic, dentures whose major component is plastic.

Note:
Some people wear a lower free-end saddle denture with a small wrought metal bar linking the two saddles. If it does not have meshwork to retain the saddles and is not highly polished it is probably not cast. This kind of bar is about 2mm by 3mm and oval. This should not be recorded as a metal base. The denture in this case is really a plastic one with a small metal component. They are probably rare, but code as plastic.

Denture support
1 = Tooth borne, dentures with bounded saddles and rests
2 = Tissue borne, dentures without occlusal rests
3 = Tooth and tissue borne, dentures with rests and free-end saddles.

Denture status
0 = The denture is intact, not damaged
1 = The denture is in need of repair, for example, fractured, tooth missing or self-mended.

Summary - dentures
- *all dentures are examined*
- *the examination is self explanatory*

Periodontal condition

Remember to check that you have asked the screening 'medical' question at the outset before using the CPI 'C' periodontal probe in this section. Where there is a medical contra-indication, the periodontal examination will automatically be completely blocked out by the computer. Please also make sure that the probe you have is the "type C" probe which has marks at 8.5mm and 11.5mm as well as at 3.5mm and 5.5mm.

There are three parts to this examination, recording of pocket depths, loss of attachment and calculus.

Pocket depths and loss of attachment (LoA) will be recorded at two sites (mesial and distal) on each tooth, these two sites will be buccally on upper teeth and lingually on lower teeth. Gently insert the CPI probe into the sulcus distally on the tooth and observe the pocket depth and loss of attachment at which resistance is felt. This manoeuvre should not cause pain or blanching of the tissue, if it does, you are using too much pressure (as an indication of the force required when probing, place the probe below your fingernail, this should not be painful if the appropriate pressure is used). Reinsert the probe mesially on the tooth to obtain the readings for that surface. At each surface you need to record both the pocket depth and loss of attachment in that order. Having

called out the two distal codes, then call out the two mesial codes, then call out the single calculus score.

Start in the upper right and then work each site in sequence (distal then mesial), when you get to the midline call out "midline" and carry on in sequence, but now you are moving away from the midline, working distally. You will now be recording mesial then distal, the chart is set up to expect that. At each tooth call out pocket and LoA for the first site, pocket and LoA for the second and then the calculus score. The presence of calculus is called if it is visible or if it can be detected with the probe. You will thus be calling five codes for each tooth: probing depth, LoA, probing depth, LoA and finally calculus.

Codes and criteria: pocket depth and loss of attachment
The codes are the same for the two measures.
0 = Up to 3.5mm (first probe band)
1 = 4-5.5 mm (dark band)
2 = 6-8.5 mm (first area above the dark band)
3 = 9+ mm (second area above the dark band)
9 = Unscorable
Notes:
1. Pocketing is recorded from the gingival crest to the base of the pocket.
2. Loss of attachment is recorded from the base of the pocket to the cemento-enamel junction (CEJ). If this is damaged by a filling or restoration and there is no indication of where it should be then you should use the margin of the restoration. In most cases you can get an indication of where the CEJ should be, even where there are calculus deposits.
3. Code 9 should only be used if you cannot probe a pocket, either because of discomfort or because there is a physical barrier (e.g. a large shelf of calculus). In a few cases it may be necessary to use a code 9 where it is impossible to judge the position of the CEJ because of calculus.

Codes and criteria: calculus
Each surface, buccal on upper teeth, lingual on lowers should be examined for the presence of supra- or sub-gingival calculus, and a single code for each tooth is entered, using the following codes:

0 = No visible or detectable supra- or sub-gingival calculus
1 = Any supra- or sub-gingival calculus detectable with the probe or visible with the naked eye.
9 = Unscorable.

Summary - periodontal examination
- *the examination is only done if the medical history is clear according to the set questions*
- *the order of the examination is the same as for crowns (i.e. clockwise)*

- *there are 5 codes for each tooth, pocket then attachment loss for the distal, the same for the mesial and a score recording the presence of any calculus at the site*
- *the sites are mesial and distal, examining the buccal surfaces of upper teeth and the lingual surfaces of lowers*

Asterisks/comments

Examiners will be asked for any comments that they wish to make. If there are, they will be recorded on the computer, by the examiner. This has to be done in the household. The recorder will ask the dentist if they want to "finish off the record", and at this point you can type in the appropriate box any other findings. **The examiner should not dictate this.** If you wish to enter something into the record, you should remove your gloves and type it in yourself.

Note that these data are most unlikely to be analysed and reported, and you do not have to record anything. Only if there is pathology which you suspect is of a serious nature (e.g. suspected malignancy) are you obliged to enter any data. In such cases (which you are **very** unlikely to encounter) you must also complete a separate pro-forma which the recorder will carry. This should be done according to the detailed protocol described in the next section. Although you will probably never use it, please make sure that you are familiar with this protocol.

HANDLING PROFESSIONAL QUESTIONS AND REPORTING PATHOLOGY

In most circumstances the dental examiners do not make any comment about what they see during the examination. If the participant asks about their dental treatment need, or if questions related to the standard of previous dental care arise, the response will be that the survey is not designed to collect the sort of information on which a treatment can be planned, and that visiting a general dental practitioner is the best way of ensuring a thorough dental check-up. This is not only a way of deflecting potentially difficult questions, it is also absolutely true.

However, the interviewer is permitted to say, when recruiting participants, that as a dentist, you may be able to offer them some advice on the best way of looking after their mouth or teeth. If after the examination the subject wishes to know about their mouth you can give an indication of whether there is room for improvement in terms of the general oral hygiene/ cleanliness and/or a general statement along the lines of:

'The best way of getting information about any treatment you might need is by seeking advice from your dentist'.

If the participant does not have a dentist, you will have available a local contact telephone number in order for them to find a dentist.

If you are asked to comment on specific aspects of oral hygiene, we would suggest that you respond, if appropriate, by identifying areas for improvement but say that they will need more specific advice from a dentist or dental hygienist since there are many ways of achieving this. It is very important that you are not too prescriptive and that you adhere to general principles as there should be no scope for oral hygiene advice being given which conflicts with previous hygiene advice. You could preface this by saying:

'What I generally tell people is.............'

If you are asked to comment on specific aspects of past treatment, you need to say:

'This survey is limited and you need to see your (or a) dentist for specific advice and/or treatment'.

The only exception to this protocol is if the examining dentist notices a lesion which he /she considers may be serious and potentially life threatening (such as a suspected malignancy). Examiners are very unlikely to encounter such potentially serious pathology, the incidence of these lesions is very low, the examination is not a screening exercise and does not involve examination of the oral soft tissues (except the periodontium). However, it is possible that such a lesion may be noticed and, as the implications are serious, a protocol to deal with this eventuality is in place.

Protocol: reporting serious pathology

In the extremely unlikely event that such a lesion is noted, the examiner is obliged to follow a set protocol, which is designed to make sure that the participant's general medical practitioner is informed, whilst not causing the participant unnecessary worry.

The examiner should inform the participant using an appropriate form of words. As experienced clinicians, the examiners may wish to vary the approach or the tone they use to ensure good communication, but the basis of the wording should be the same in all cases. The dentist will usually want to introduce the subject, usually by asking whether or not the lesion causes any discomfort, and then state that "it is survey policy that a brief report of any ulcers or inflamed areas is passed to participant's family doctors. Would you have any objection to us doing that for you?" If they agree we ask them to sign a standard form agreeing to this, which records their doctor's name and details. If the participant says that they will arrange to see their doctor themselves then they should be encouraged to do this and it is left at that point. If the participant asks what the dentist thinks the lesion is, the dentist should answer honestly that they do not know, before re-iterating standard survey policy as above.

Once this is completed the dentist will leave the house before filling out a pro-forma recording the site and nature of the suspect lesion. This is sent immediately, along with the signed consent form, to one of the named survey contact consultants. The consultant will contact the doctor by letter with a copy of both the consent form and the dentist's record form as well as details of the nearest specialist unit where appropriate investigations can be undertaken.

It is most unlikely that any such lesions will be found, and it is also unlikely that, even those which are reported, will turn out to be serious. **It is the responsibility of the examiner not to alarm the participant unduly.**

Appendix C.2 Medical screening check

All questions below were answered before proceeding with the examination

1. Have you ever had Rheumatic fever or St Vitus Dance?

2. Do you have any artificial heart valves or a heart murmur?

3. Have you ever had any heart surgery?

4. Do you have any artificial joints, such as artificial hip or knee joints?

5. Have you ever had hepatitis or jaundice?

6. Do you have, or have you ever had any medical condition which has caused you a problem with dental treatment in the past?

Note: the responses to 5 and 6 are for the information of the examiner and a positive answer should not usually prevent the examination from proceeding. This is entirely at the discretion of the examiner.

Appendix C.3 The examination chart

Dentist..*Subject number*......................

ADULT DENTAL HEALTH 1998

PAPER FORM FOR DENTAL EXAMINATION

Area........Address.........Household.......Name..................................Person number............

	Yes	No
ASK ALL		
Medical history for periodontal examination?	1	2
Natural teeth in both arches?	1	2
IF NO:		
Natural teeth in upper arch only?	1	2
Natural teeth in lower arch only?	1	2

Dentist..Subject number.....................

TOOTH CONDITION:

Tooth Condition	PL	D	D M	O/I	O M	M	M M	B	B M	P	P M
UPPER RIGHT 8											
UPPER RIGHT 7											
UPPER RIGHT 6											
UPPER RIGHT 5											
UPPER RIGHT 4											
UPPER RIGHT 3											
UPPER RIGHT 2											
UPPER RIGHT 1											

Tooth Condition	PL	D	D M	O/I	O M	M	M M	B	B M	P	P M
UPPER LEFT 1											
UPPER LEFT 2											
UPPER LEFT 3											
UPPER LEFT 4											
UPPER LEFT 5											
UPPER LEFT 6											
UPPER LEFT 7											
UPPER LEFT 8											

Tooth Condition	PL	D	D M	O/I	O M	M	M M	B	B M	P	P M
** LOWER LEFT 8**											
** LOWER LEFT 7**											
** LOWER LEFT 6**											
** LOWER LEFT 5**											
** LOWER LEFT 4**											
** LOWER LEFT 3**											
** LOWER LEFT 2**											
** LOWER LEFT 1**											

Tooth Condition	PL	D	D M	O/I	O M	M	M M	B	B M	P	P M
** LOWER RIGHT 1**											
** LOWER RIGHT 2**											
** LOWER RIGHT 3**											
** LOWER RIGHT 4**											
** LOWER RIGHT 5**											
** LOWER RIGHT 6**											
** LOWER RIGHT 7**											
** LOWER RIGHT 8**											

Dentist..*Subject number*...................

ROOT CONDITION:

Root Condition	Root	Root M
UPPER RIGHT 8		
UPPER RIGHT 7		
UPPER RIGHT 6		
UPPER RIGHT 5		
UPPER RIGHT 4		
UPPER RIGHT 3		
UPPER RIGHT 2		
UPPER RIGHT 1		

Root Condition	Root	Root M
UPPER LEFT 1		
UPPER LEFT 2		
UPPER LEFT 3		
UPPER LEFT 4		
UPPER LEFT 5		
UPPER LEFT 6		
UPPER LEFT 7		
UPPER LEFT 8		

Root Condition	Root	Root M
LOWER LEFT 8		
LOWER LEFT 7		
LOWER LEFT 6		
LOWER LEFT 5		
LOWER LEFT 4		
LOWER LEFT 3		
LOWER LEFT 2		
LOWER LEFT 1		

Root Condition	Root	Root M
LOWER RIGHT 1		
LOWER RIGHT 2		
LOWER RIGHT 3		
LOWER RIGHT 4		
LOWER RIGHT 5		
LOWER RIGHT 6		
LOWER RIGHT 7		
LOWER RIGHT 8		

TOOTH WEAR

Tooth Wear	TWear B	TWear I	TWear L
UPPER RIGHT 3			
UPPER RIGHT 2			
UPPER RIGHT 1			
UPPER LEFT 1			
UPPER LEFT 2			
UPPER LEFT 3			

Tooth Wear	TWear
LOWER LEFT 3	
LOWER LEFT 2	
LOWER LEFT 1	
LOWER RIGHT 1	
LOWER RIGHT 2	
LOWER RIGHT 3	

Dentist..Subject number.....................

CONTACT

RIGHT	Contact		LEFT	Contact
** ZONE 1**			** ZONE 1**	
** ZONE 2**			** ZONE 2**	
** ZONE 3**			** ZONE 3**	
** ZONE 4**			** ZONE 4**	
** ZONE 5**			** ZONE 5**	
** ZONE 6**			** ZONE 6**	
** ZONE 7**			** ZONE 7**	
** ZONE 8**			** ZONE 8**	

What is the anterior occlusion total?   0 1 2 3 4 5 6

SPACES

Spaces	Space		Spaces	Space
** UPPER RIGHT 1**			** UPPER LEFT 1**	
** UPPER RIGHT 2**			** UPPER LEFT 2**	
** UPPER RIGHT 3**			** UPPER LEFT 3**	
** UPPER RIGHT 4**			** UPPER LEFT 4**	
** UPPER RIGHT 5**			** UPPER LEFT 5**	

Spaces	Space		Spaces	Space
** LOWER LEFT 1**			** LOWER RIGHT 1**	
** LOWER LEFT 2**			** LOWER RIGHT 2**	
** LOWER LEFT 3**			** LOWER RIGHT 3**	
** LOWER LEFT 4**			** LOWER RIGHT 4**	
** LOWER LEFT 5**			** LOWER RIGHT 5**	

Dentist..Subject number....................

DENTURES

	Yes	No		
Is there a denture present in the mouth?	1	2		

IF YES:

Is the denture upper, lower or both?	Upper	Lower	Both

IF UPPER OR BOTH

What is the upper denture type?	*Partial	Full Complete Implant
What is the upper denture base material?		Metal Plastic
What is the upper denture support?	*ToothBorne	TissueBorne BothBorne
What is the status of the upper denture?		Intact Repair

IF LOWER OR BOTH

What is the lower denture type?	*Partial	Full Complete Implant
What is the lower denture base material?		Metal Plastic
What is the lower denture support?	*ToothBorne	TissueBorne BothBorne
What is the status of the lower denture?		Intact Repair

*Complete	=	Complete Overdenture
Implant	=	Implant Retained
BothBorne	=	Both Tooth and Tissue Borne

Dentist..Subject number......................

PERIODONTAL CONDITION:

Periodontal Condition	DBDepth	DAttach	MBDepth	Mattach	Calculus
UPPER RIGHT 8					
UPPER RIGHT 7					
UPPER RIGHT 6					
UPPER RIGHT 5					
UPPER RIGHT 4					
UPPER RIGHT 3					
UPPER RIGHT 2					
UPPER RIGHT 1					

Periodontal Condition	DBDepth	DAttach	MBDepth	Mattach	Calculus
UPPER LEFT 1					
UPPER LEFT 2					
UPPER LEFT 3					
UPPER LEFT 4					
UPPER LEFT 5					
UPPER LEFT 6					
UPPER LEFT 7					
UPPER LEFT 8					

Periodontal Condition	DBDepth	DAttach	MBDepth	Mattach	Calculus
LOWER LEFT 8					
LOWER LEFT 7					
LOWER LEFT 6					
LOWER LEFT 5					
LOWER LEFT 4					
LOWER LEFT 3					
LOWER LEFT 2					
LOWER LEFT 1					

Periodontal Condition	DBDepth	DAttach	MBDepth	Mattach	Calculus
LOWER RIGHT 1					
LOWER RIGHT 2					
LOWER RIGHT 3					
LOWER RIGHT 4					
LOWER RIGHT 5					
LOWER RIGHT 6					
LOWER RIGHT 7					
LOWER RIGHT 8					

Appendix D Dental teams

Training Teams

University of Birmingham
Dr G Bradnock
Mr AJ Morris
Mrs DA White
University of Wales
Dr ET Treasure

University of Dundee
Professor NB Pitts
Dr CH Deery
Dr N Nuttall
Dr CM Pine
Mrs GVA Toppings

University of Newcastle-upon-Tyne
Dr JH Nunn
Dr JG Steele
University of Northern Ireland
Miss HMM Clarke

Pilot Examiners

Mr JR Francis
Ms G Jones
Mrs JM Kirk

Mr MCW Merrett
Mr MJ Prendergast
Mr D Trotter

Mrs DA White
Mr P Young

Dental Examiners

Mrs S Abayaratne
Miss CL Baglee
Mr GB Brown
Mr WH Challacombe
Mr DK Chersterton
Mr IS Chopra
Miss EHR Clarke
Mr AJ Clayton-Smith
Mr JA Clewett
Mrs EM Davies
Mr P Davies
Mr PW Dodds
Mr MA Donaldson
Mrs JT Duxbury
Mr D Evans
Mrs GL Fenn
Mr AC Gerrish
Ms LS Gilliat
Ms CD Gizzi
Miss SGM Gonsalves
Mrs D Gratrix
Mrs A Hallas
Mrs J Harlock
Mr IS Hay
Mr P Helliwell

Mr PB Higgins
Miss BMC Higgins
Mrs EL Jones
Dr DA Kavanagh
Mr PGL Laurie
Mr RH Lawson
Miss DLO Leverington
Miss EA Liptrot
Mrs PM Ludiman
Mrs AS Macmillan
Mrs EM Martin
Miss FA McAuley
Mrs RJ McCutcheon
Mrs RA McGonigle
Mrs GA McNaugher
Mr NP Monaghan
Mr AJ Morris
Mr J Morrison
Mr P Mortimer
Miss CJ Moss
Miss EM Moyes
Mrs CM Owen
Mr T Papadakis
Mr RM Parfitt
Mrs PM Pears

Ms NK Pearson
Mr PJ Pennington
Miss H Preston
Mrs GA Preston
Mr PN Rattenbury
Miss FG Reid
Miss EM Reilly
Mr FJM Sampson
Mr IC Shimeld
Mrs CJ Setchell
Miss CM Smart
Mr AJ Sprod
Mr DR Stevenson
Mr AJ Swan
Mrs A Sweeney
Mrs JA Waplington
Mrs DA White
Mrs KW Whittle
Mr CA Wilkinson
Mrs MG Williams
Mr PW Williams
Mrs I Wilson
Ms BA Wood
Mr DWM Young
Mr PH Zubkowski

Appendix E   List of authors by chapter

1.1 Background and methodology
Maureen Kelly

2.1 The loss of all natural teeth
Jimmy Steele

2.2 The number and distribution of teeth, and the functional dentition
Jimmy Steele

2.3 Replacement of missing teeth among dentate adults: dentures and bridges
Elizabeth Treasure
Jimmy Steele

3.1 The condition of natural teeth
June Nunn

3.2 Restorative treatment
Cynthia Pine
Jimmy Steele

3.3 The condition of supporting structures
John Morris
Jimmy Steele

4.1 Social and behavioural characteristics and oral health
Maureen Kelly
Elizabeth Treasure

7.4 Trends in total tooth loss
Jimmy Steele

5.2 Trends in the condition of natural teeth
Nigel Pitts

5.3 Trends in restorative treatment
Nigel Pitts

6.1 Usual reason for dental attendance
Nigel Nuttall

6.2 Opinions about dental visits
Maureen Kelly

6.3 Visiting the dentist
Deborah White
Nigel Nuttall

6.4 Treatment preferences of dentate adults
Deborah White

6.5 The impact of dental disease
Nigel Nuttall

6.6 Attitudes, expectations and experiences in relation to dentures
Nigel Nuttall

6.7 Dental hygiene behaviour
Nigel Nuttall

6.8 Other opinions about dental health and dentistry
Gillian Bradnock

6.9 Trends in dental attitudes, experience and behaviour
Gillian Bradnock

7.1 Dental health in England
Maureen Kelly

7.2 Dental health in Scotland
Nigel Nuttall

7.3 Dental health in Wales
Elizabeth Treasure

7.4 Dental health in Northern Ireland
John Morris

With additional contributions from research staff of the Social Survey Division of the Office for National Statistics; in particular, Eileen Goddard, Jan Gregory, Wendy Sykes and Alison Walker.

Appendix F Statistical and technical notes

F.1 Introduction

This appendix gives details of the sample design used in the 1998 Adult Dental Health Survey, the weighting procedures applied to the data and the sampling errors associated with the estimates shown in this report. It also describes the various statistical techniques and the measures of deprivation used in the analysis of the data.

F.2 The sample design

The sample design used in the 1998 survey was similar to that used for the 1988 survey. The sample size and design were determined by the need to provide data on the oral health of adults living in private households for the four constituent countries of the United Kingdom and to measure changes in oral health since 1988. In order to achieve this, proportionately larger samples were selected in Ireland, Scotland and Wales than in England. The sampling proportions in those three countries were increased by factors, in rounded terms, of six, four and three and a half respectively.

A multi-stage random probability sampling procedure was used, with the sample size at the final stage determined partly by consideration of the number of dental examinations that each examiner could reasonably be expected to cover both in terms of time and of cost. Different sampling procedures were used to obtain samples of addresses in Great Britain and Northern Ireland. In Great Britain, a sample of addresses was selected from the Postcode Address File (PAF) using a two-stage sample design. At the first stage, 76 postcode sectors in England, 32 postcode sectors in Scotland and 16 postcode sectors in Wales were selected by a systematic sampling method from a stratified list of postcode sectors. For this purpose all the postcode sectors were stratified first by Government Office Regions, then according to the proportion of heads of household in professional socio-economic groups (i.e. groups 1 to 5 and 13)[1] and finally by the proportion of households owning a car or van. The data on socio-economic groups and vehicle ownership were derived from the 1991 Census. The chance of selection of each postcode sector was proportional to the total number of delivery points in the sector. Forty addresses were then randomly selected from each sector. This gave a total of 4960 sampled addresses in Great Britain. In Northern Ireland a simple random sample of 580 of private addresses was selected from the Rating and Valuation list. In total, 5540 addresses were sampled for the survey.

The requirements of the survey were a sample of adults living in private households. Business addresses and institutions were excluded at the sample selection stage as far as possible by using the PAF small users' file as the sampling frame. Eleven per cent (599) of the selected addresses did not contain a private household and were excluded from the set sample of addresses. These ineligible addresses included demolished or permanently empty addresses, addresses used only on an occasional basis and business premises and institutions where there was no resident private household.

A small proportion (about 1%) of the sampled addresses contain more than one private household. Since each address in the two sampling frames was given only one chance of selection, additional procedures were carried out by interviewers at addresses found to contain more than one household in order to ensure that all households were given a chance of selection. When the sample address contained more than one private household, interviewers were asked to interview at all households up to a maximum of three. In the rare event that an address contained more than three households, the interviewer was instructed to list the households systematically and then three were chosen at random by reference to a selection table. In total, 43 extra households were identified in this way for inclusion in the survey, resulting in 4984 households to be approached by the interviewer.

Interviewers were instructed to seek interviews with all adults aged 16 and over living at each household. In total, 6204 interviews were carried out; at least one person was interviewed in 74% of all the households that were visited. If the respondent completed an interview and had some natural teeth they were eligible for a home dental examination; 3817 dental examinations were carried out representing a response rate of 72%. (See Table 1.1.1 for details of the response rates.)

F.3　Weighting procedures

Proportionately larger samples were drawn in Wales, Scotland and Northern Ireland than in England. The design produced a self-weighting sample for each of the constituent countries of the United Kingdom but necessitated re-weighting to produce representative figures for the United Kingdom as a whole.

In addition to the weighting required by the sample design, the data were also weighted to reduce the risk of non-response bias. Any survey that is based on several stages of data collection is liable to non-response at each of the stages. In this survey non-response occurred at the both the interview and examination stages. Non-response may bias the results of a survey if the characteristics of people who did not participate are different from those of people who did take part; the results based on data from the responding adults only would not be representative of the population as a whole. For example, if adults with natural teeth were more likely to respond to the survey than those without natural teeth, people with no natural teeth would be under-represented in the sample. If this occurred the results from the survey would give a higher proportion of dentate adults than existed in the population.

Weighting is used to try and reduce the risk of any non-response bias. One of the difficulties in trying to correct for non-response bias is identifying where and in what direction this bias may be found.

Weighting the interview data

Little or no information is known about the adults who were not contacted or refused to take part in the survey at the interview stage. Previous dental surveys have shown that many factors relating to dental health vary according to age, gender and region. It was therefore considered important that the age, gender and regional distributions of the responding sample reflected those of the population as a whole. These distributions of the sample of respondents in the Adult Dental Health (ADH) Survey were compared with the population distributions provided by the September to November 1998 Labour Force Survey (LFS). The LFS is a large sample survey carried out in the UK on a quarterly basis; the data are weighted and grossed to give estimates of the total UK population living in private households. The LFS estimates were taken as the most appropriate for comparison with the survey sample as they provide estimates of the age, sex and regional distribution of people living in private households in the UK.

The age and gender distributions for the ADH sample were weighted to the LFS distributions within the four countries within the UK, and to Standard Statistical Regions within England. The Standard Statistical Regions were used as they were the adminstrative regions which were closest geographically to the areas used in the analysis of the data for which population distributions were available. This weighting procedure incorporated both the adjustment for the oversampling in Wales, Scotland and Northern Ireland and the adjustment for possible non-response bias. The interview weights were calculated so that England retained the same sample size when weighted and unweighted; therefore, the weighted sample sizes for Wales, Scotland and Northern Ireland were much smaller than the unweighted sample sizes. Table F.1 shows the weighted and unweighted sample sizes and Table F.2 shows the weighting factors used for each the age and gender groups within each Region and country.

Tables F.1-2

As a whole, the sample of responding adults under-represented men. The sample also under-represented people aged 16 to 24 years and over-represented people aged 45 to 54 years and 65 to 74 years. Within each region, the age and sex distributions of the sample also varied when compared with the population distributions, although the subgroups that were under and over-represented differed in each region.

Weighting the examination data

All dentate adults who were interviewed were asked if they would have a home dental examination; 72% did so. To reduce the risk of bias it was possible to weight the data for the sample of people who had dental examinations to reflect the characteristics of the full sample of dentate adults who were interviewed.

During the interview a large amount of data were gathered from dentate adults on a variety of topics including demographic information; social background; and dental behaviour, attitudes and opinions. These data were analysed, using a program called CHAID[2] (Chi-squared Automatic Interaction Detector) to find which characteristics were most significant in distinguishing between those who took part in the dental examination and those who did not.

CHAID performs segmentational modelling, which divides a population into two or more distinct groups based on categories of the 'best' predictor of a dependent variable. It then splits each of these groups into smaller subgroups based on other predictor variables. The segments which CHAID derives are mutually exclusive and exhaustive; that is the segments do not overlap and each case is contained in exactly one segment. These segments can then be used as weighting classes.

The dependent variable for the analysis was the participation or non-participation in the dental examination. The analysis was carried out using weighted data (the interview weight) and therefore already contained an element of correction for any non-response bias in participation to the examination due to age, gender and region. Eight characteristics were found to be significant, giving a total of 14 segments or weighting classes (see Figure F.1). Social class of head of household was the best predictor of participation in the dental examination. Other predictors included characteristics relating to dental behaviour and opinions as well as socio-demographic information. Figure F.1 also shows how these characteristics were used by CHAID to produce the weighting classes. The weight for each class was calculated by dividing the overall response rate by the response rate for that class. The weighting classes together with the proportions in each class and the associated weights are shown in Table F.3.

Figure F.1, Table F.3

The overall weight for the examination data was created by multiplying the interview weight by the weight derived from the CHAID analysis.

F.4 The accuracy of survey results

Sources of error
Like all estimates based on samples, the results of the 1998 Adult Dental Health Survey are subject to variations and errors. The total error associated with any survey estimate is the difference between the estimate derived from the data collected and the true value for the population. The total error can be divided into two main types: random error and systematic error.

Random error
Random error occurs because survey estimates are based not on the whole population but only a sample of it. There may be chance variations between such a sample and the whole population. If a number of repeats of the same survey were carried out, this error could be expected to average to zero. The size of the sample and the sample design influence the magnitude of these variations due to sampling.

Systematic error
Systematic error is often referred to as bias. Bias can arise because the sampling frame is incomplete, because of variation in the way interviewers ask questions and record answers, because of variation in the way the dental examination was carried out, or because non-respondents to the survey have different characteristics to respondents. When designing this survey considerable effort was made to minimise systematic error; this included training interviewers to maximise response rates and ask questions in a standard way. The dental examiners were also trained to carry out the dental examination in a consistent way as possible (see Appendix B for further details on the training and calibration of the dental examiners). Nonetheless, some systematic error is likely to have remained, particularly from potential non-response bias, and the data were weighted to reduce any potential non-response bias.

Standard errors and design factors
Statistical theory enables estimates to made of how close the survey results are to the true population values for each characteristic. A statistical measure of the variation, the standard error, can be estimated from the value obtained for the sample, and provides a measure of the statistical precision of the survey estimate. This allows a confidence interval to be calculated around the sample estimate which gives an indication of the range in which the true population value is likely to fall. The confidence interval generally used in survey research is the 95% confidence interval; it comprises of approximately two (1.96) standard errors below the level of the estimate to two standard errors above the estimate. These calculations assess the errors associated with the sample design; they cannot take account of potential errors such as non-response bias or random error due to the misunderstanding of questions.

For results based on simple random samples, without clustering or stratification, the estimation of standard errors is straightforward. However, the sample design of the Adult Dental Health Survey was not a simple random sample and therefore a more complex calculation is needed which takes account of the stratification and clustering of the sample design is necessary[3]. Stratification tends to reduce the standard error, while clustering tends to increase it.

In a complex sample design, the size of the standard error depends on how the characteristic of interest is spread within and between the primary sampling units (PSUs) and strata. Different sample designs were used in Great Britain and Northern Ireland and therefore the countries had different PSUs; in Great Britain the strata were Government Office Regions and the PSUs were postcode sectors; whereas the data for Northern Ireland were treated as one strata and the PSUs were the individual sampled addresses.

Tables F.4 to F.28 show the standard error and 95% confidence intervals for selected survey estimates (calculated

Fig F.1   Weighting classes for examination data

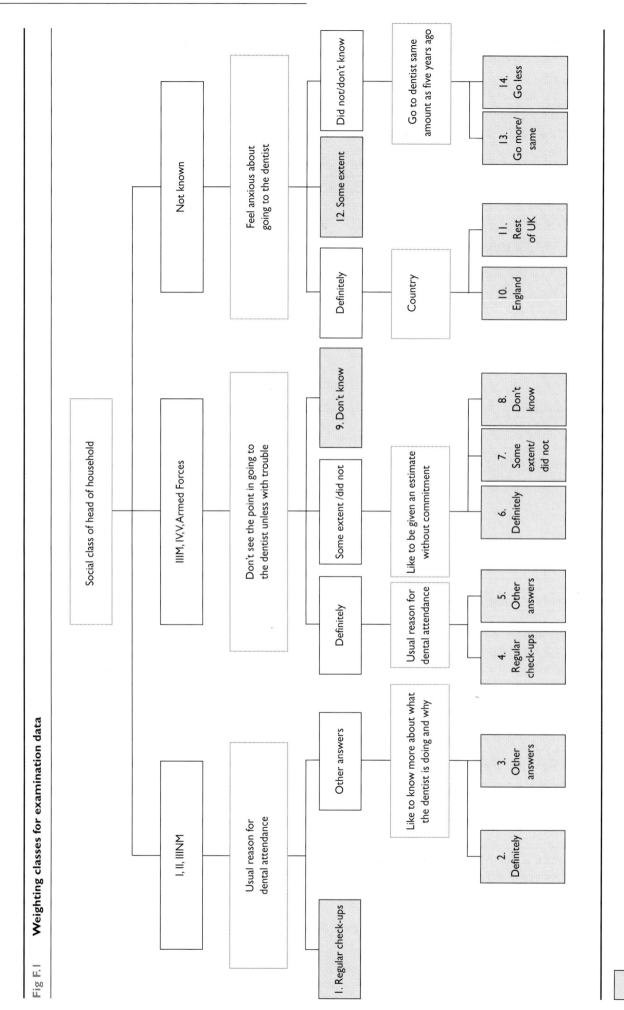

Final weighting class

using STATA, a statistical analysis software package). The tables do not cover all the topics discussed in the report but show a selection of estimates based on information from the both the interview and the dental examination. The tables also show the design factor, or deft; the ratio of the complex standard error to the standard error that who have resulted had the survey design been a simple random sample of the same size. This is often used to give a broad indication of the degree of clustering. The size of the design factor varies between survey variables reflecting the degree to which a characteristic is clustered within PSUs, or is distributed between strata. For a single variable the size of the factor also varies according to the size of the subgroup on which the estimate is based, and on the distribution of the subgroup between PSUs and strata. Design factors below 1.0 show that the complex sample design improved on the estimate that would have expected from a simple random sample, probably due to the benefits of stratification; design factors greater than 1.0 show less reliable estimates than might be gained from a simple random sample, due to the effects of clustering. Design factors of 1.2 or more were recorded for most of the estimates from both the dental examination and the interview, but few were greater than 2.0.

Tables F.4-28

The standard errors were generally lower for Northern Ireland than for other countries reflecting the different sampling strategies in Northern Ireland and Great Britain; in Northern Ireland clustering only occurred at address level, as all the adults within the address were interviewed. In Great Britain the addresses were also clustered by postal sector.

Where possible, the design factors for estimates from the 1998 survey were compared with those calculated for similar estimates from the 1978 survey. In general, the design factors for the estimates in 1998 were higher than those from 1978, except for the mean number of decayed and unsound teeth where substantially lower design factors were found in 1998 (1.39 in 1998 and 2.60 in 1978). The higher design factors found are probably due to the more clustered sample design used in 1998 and the different weighting procedures used in the two surveys.

Design factors can be used as multipliers when conducting significance tests which are based on the assumption of a simple random design, to adjust for the complex design of a particular sample. In this report, tests for significant differences between survey means generally used complex standard errors and the design factor was used as a multiplier in testing differences between proportions. In some cases the exact design factors were not calculated and conservative

design factors were estimated. The formulae used to test for significant differences were:

For proportions:

$$\sqrt{\frac{(p_1 q_1)}{n_1} + \frac{(p_2 q_2)}{n_2}} \times 1.96 \times deft$$

where p_1 and p_2 are the observed percentages for the two subsamples, q_1 and q_2 are respectively $(100-p_1)$ and $(100-p_2)$, and n_1 and n_2 are the (unweighted) subsample sizes. The value of 1.96 is used to test for significant differences at the 95% level ($p<0.05$).

For means:

$$\sqrt{se_1^2 + se_2^2} \times 1.96$$

where se_1 and se_2 are the complex standard errors of two means.

F.5   Statistical techniques

This section gives details of the statistical techniques used in the analysis of the data in Chapter 4.1. These were age-standardisation, logistic regression and multiple regression.

Age standardisation[4]

A strong relationship exists between age and many measures of dental health. It is therefore important to take age into account when investigating the relationship between dental health and other social characteristics such as social class or educational attainment, which also vary by age. One commonly-used method is the presentation of three-way tables, such as Table 3.1.7, which tabulates the proportion of dentate adults with 18 or more sound and untreated teeth by social class of head of household by a number of age groups. The resulting tables may, however, be difficult to interpret and can suffer from small cell sizes.

Alternatives are the use of modelling techniques, such as logistic or multiple regression which are discussed below, or age-standardisation, which presents the data in tabular format. The method of standardisation used in this report is that of indirect standardisation; this is considered more appropriate for survey data than the direct method of standardisation which is used in medical statistics. Indirect age standardisation involves applying age-specific rates for the whole population to the age distribution in the sub-group (for example the social class) of interest. The method does not make use of the rates observed for age groups within the subgroups which are likely to be based on small sample totals and to be affected by substantial sampling error.

The age standardised ratios shown in Chapter 4.1 (for example Table 4.1.3) were calculated by dividing the observed proportions of the particular measure of dental health by the expected proportion based on age alone and then multiplied by 100. An age-standardised ratio of more than 100 indicates a greater likelihood of having the aspect of dental health of interest than would be expected in that group on the basis of age distribution alone. Conversely, a ratio of less than 100 indicates that the members of the group were less likely to have the aspect of dental health of interest than would be expected given the age composition of the group. Since standardised ratios are calculated from survey data, they are subject to sampling error and a more precise assessment of their deviation from 100 involves the use of the standard error of the ratio in a convential test of statistical significance. To take Table 4.1.3 as an example, the standard error of the ratio for non-manual social classes (I, II, and IIINM) is 0.8; multiplying this by 1.96 (the 95% level of significance) gives 1.57. The difference between 105 (the standardised ratio for this group) and 100 is greater than the resulting figure, this difference is taken to be significant.

Logistic regression[5]

Logistic regresion is a multivariate statistical technique that predicts the outcome of a dependent variable which has only two possible outcomes, for example whether or not a dentate adult has any artificial crowns, from a set of independent variables. Variables with only two possible outcomes are also known as dichotomous or binary variables. Logistic regression was developed specifically for dichotomous variables and makes more appropriate assumptions about the underlying distributions and the range of possible proportions than the more familiar multiple regression method (see below).

Most of the tables with this report are based on crosstabulations. These tables show the proportion of people with a given aspect of dental health; an example would be the proportion of people aged 35 to 44 years who have artificial crowns. What such tables do not show, however, is how much other factors may interrelate with the independent variable; for example, how much the usual reason for attending the dentist interrelate with age to affect whether a person has artificial crowns. Logistic regression looks at how different independent variables interrelate by looking at the odds of the respondent having the given aspect of dental health for different combinations of the independent variables. Odds refers to the ratio of the probability that the

event will occur to the probability that the event will not occur. The odds can be converted into a probability (p) using the following formula:

$$p = \frac{odds}{1 + odds}$$

Logistic regression can therefore be used to predict the probability of a person having a particular aspect of dental health given a combination of characteristics, for example, it can be used to model the probability of a dentate adult having artificial crowns given their age, gender, social class of head of household and usual reason for dental attendance.

The logistic regression model can be written as:

$$Prob(event) = \frac{e^z}{1 + e^z}$$

where e is the base of the natural logarithms. Logistic regression actually models independent variables against the log odds (the natural logarithm of the odds) of an event because this forms a linear relationship:

$$Z = B_0 + B_1 X_1 + \dots \dots B_p X_p$$

where the X_1 to X_p are the independent variables, B_1 to B_p are the model parameters and B_0 is the baseline odds. The odds of having a particular aspect of dental health can then be calculated by multiplying the base line odds by the appropriate factors.

Before carrying out the logistic regression analysis, the data are crosstabulated to give an indication of which independent variables should be included in the regression. One of the categories is then defined as a reference category (with a value of 1); the reference category is usually the group least likely to display the aspect of dental health of interest. For each of the independent variables included in the regression a coefficient is produced which represents the factor by which the odds of a person having the aspect of dental health increases if the person has that characteristic. The odds produced by the regression are relative odds, that is they are relative to the reference category. Taking the above example where having artificial crowns is the dependent variable, the age group 16 to 24 years would be defined as the reference category as this is the group least likely to have artificial crowns and the odds given by the model would be relative to this; so it would be possible to show how much greater the odds were of a person aged 45 to 54 years having artificial crowns than of a person aged 16 to 24 years.

There are different methods of including independent variables in a logistic regression model. The method used in the analysis contained in this report was *forward stepwise selection* which is where the model starts off only containing the constant and then at each step the independent variable which is the most highly significant is added in. Variables are then examined and the coefficients which make the observed results 'most likely' are selected while the others are removed using either the Wald statistic or the Likelihood-Ratio test. The Likelihood-Ratio was test was used for this analysis.

The odds ratios produced by the regression are shown with 95% confidence intervals. If the confidence interval does not contain 1.00 then the odds ratio is likely to be significantly different from the reference category.

An additional statistic shown in the tables is the *Nagerkelke R^2 coefficient*. This measures the goodness of fit of the model. It indicates the proportion of the variation in the dependent variable that is accounted for by the regression. The closer this statistic is to 1 the better the fit of the model.

Multiple regression[5]

Multiple regression is a statistical technique that describes a relationship between one survey variable, the dependent variable, and several other variables, the independent variables. The dependent variable is a continuous variable, for example Table 4.1.4 shows the results of a regression where the dependent variable is the number of teeth of a dentate adult. The regression calculates a coefficient for each independent variable which quantifies the effect that a change in that variable will have on the dependent variable, leaving all other variables in the regression model unchanged. There are different methods of including independent variables in multiple regression modelling. The method used in the analysis contained in this report was the *enter method*, where all independent variables are entered into the model.

In the regression models used to analysis the data, all the independent variables used were categorical variables which indicate whether the respondent has a particular characteristic or not. Therefore the coefficients quantify the effect that having a particular characteristic will have on the dependent variable. The effect is measured in relative rather than absolute terms. For variables that have more than 2 categories, the coefficient for each category is estimated in comparison to the average effect of all the categories. For example, in Table 4.1.4, other things being equal, having an age of between 16 and 24 years increases the predicted number of teeth for an individual by 3.90 relative to the average effect of all six age groups. For variables with only two categories,

one category is chosen as a reference category. The coefficient of the other category is then estimated relative to that reference category. Again in Table 4.1.4, for gender, women is chosen as the reference category and the coefficient for men of 0.42 indicates that, other things being equal, one would expect men to have, on average, 0.42 more teeth than women.

In regression modelling, in order to determine whether there is a relationship between the dependent variable and the independent variables the coefficients for each category of the independent variables were tested to see whether or not they were significantly different from zero. For the above example, whether the coefficient for men (0.42) is significantly different from zero. A variable is judged to be statistically different from zero if it is significant at the 95% level or above. In the tables presenting results from multiple regression, characteristics that were significantly different at the 95%, 99% and 99.9% levels are indicated by one, two and three asterisks respectively. For example, in Table 4.1.4 the characteristic of being aged 16 to 24 years has three asterisks because the variable is significant at the 99.9% level. This means that if there were no relationship in the population between the number of teeth and being aged 16 to 24 years then there would be a probability of 0.1% or less that the regression would have estimated a coefficient that was so far away from zero.

An additional statistic shown in the tables is the *Adjusted R^2 coefficient*. This measures the goodness of fit of the model. It indicates the proportion of the variation in the dependent variable that is accounted for by the regression. The closer this statistic is to 1 the better the fit of the model.

F.6 Measures of deprivation based on area

This report has shown that the clinical measures of oral health (as measured during the survey) vary with respect to the social characteristics of the respondents, such as social class and educational attainment. An alternative method of investigating possible inequalities in oral health is to classify respondents according to the relative living standards of the area they lived in (sometimes known as geographical deprivation), rather than their individual characteristics. There are various measures of geographical deprivation available, although none of them are cover the UK as a whole. In Part 7 of this report, different measures of geographical deprivation were used in the analysis of the data for England and Scotland; the Jarman underprivileged area score and Carstairs and Morris index of deprivation respectively.

It should be noted that the measures of deprivation are based on geographical areas, rather than individual circumstances and that 'not all deprived people live in deprived wards, just as not everybody in a ward ranked as deprived are themselves deprived'[6].

Jarman underpriveliged area score

The data for England have been analysed using the Jarman underprivileged area score[7] (referred to as Jarman within the report). Jarman was not orginally constructed to measure deprivation but as a measure of General Practice workload. The Jarman score was derived to take account of geographic variations in the demand for primary care based on a survey of GPs who identified eight social factors among their patients that most affected their workload:

- Unemployment - percentage of economically active residents who were unemployed;
- Overcrowding - percentage of residents in overcrowded households (more than one person per room);
- Lone parents - percentage of residents in 'lone parent' households;
- Under 5s - percentage of residents aged under 5 years;
- Elderly living alone - percentage of elderly persons living alone;
- Ethnicity - percentage of households headed by a person born outside the United Kingdom;
- Social class - percentage of residents where head of household was unskilled (Social Class V); and
- Residential mobility - percentage of residents who changed their address in the previous year.

The combination of these social factors can be viewed as indicators of low living standards or a high need for services and therefore the Jarman scores can also be used as a measure of deprivation. A score for each electoral ward within England was calculated based on data from the 1991 Census relating to the eight social factors. The wards were ranked according to this score and then aggregated into five groups, each containing roughly a fifth of the total population. Thus, Area 1 contains the 20% of the population living in the most deprived wards and Area 5 contains the 20% living in the least deprived wards. The respondents from the survey were classified according to the score of the ward that they lived in.

Carstairs and Morris index of deprivation

Within Scotland, there has been increasing use of the Carstairs and Morris index of deprivation referred to as DEPCAT in analysing health data. This index classifies people on the basis of the postcode sector where they live and is based on four social factors within the postcode sectors:

- the level of overcrowding in households;
- male unemployment in the area;
- proportion of Social Classes IV or V in the area; and
- the proportion of people in private households with no car.

As with Jarman, a score is calculated based on these factors from data from the 1991 Census. The postcode sectors are then aggregated into seven categories on the basis of this score, which are referred to as DEPCAT areas; with area 1 containing the postcode sectors which were least deprived and area 7 containing the most deprived postcode sectors.

There are numerous examples of the relationship of the Carstairs and Morris deprivation index to health in Scotland. Examples of clinical measures of health associated with levels of deprivation are the incidence of mortality due to cerebrovascular disease, coronary heart disease and mental health problems of depression, suicide and anxiety.[9]

Notes and references

1. For futher details on socio-economic groups see OPCS, *Standard Occupation Classification, vol 3*, HMSO, 1991.
2. CHAID is a package within the statistical data analysis system SPSS.
3. For a full description of the method used to calculate standard errors for complex survey design, see Butcher B and Eliot D, *A Sampling Errors Manual*, HMSO, London, 1992
4. For a more detailed description of the indirect standardisation, see Foster K, The use of standardisation in survey analysis, *Survey Methodology Bulletin*, OPCS, vol. 33, 1993, pp. 19-27
5. For a more detailed description of logistic and multiple regression analysis, see *SPSS Advanced Statistics Users Guide*, SPSS Inc., 1990
6. Townsend P, Phillimore P and Beattie A, *Health and Deprivation: Inequality in the North*, Routledge, London, 1988
7. Jarman B, Identification of underprivileged areas, *British Medical Journal*, vol. 286, pp. 1705-1709
8. Carstairs V and Morris R, *Deprivation and health in Scotland*, Aberdeen University Press, Aberdeen, 1991.
9. McLaren G Bain M, *Deprivation and health in Scotland. Insights from NHS data*. Information & Statistics Division, Edinburgh, 1998.

Table F.1   Unweighted and weighted interview sample sizes by age, gender, Standard Statistical Region and country

	Standard Statistical Region			Country				United Kingdom
	Northern, North West, Yorkshire & Humberside	E Midlands, W Midlands, E Anglia	London, South East, South West	England	Scotland	Wales	Northern Ireland	
Unweighted sample								
Men								
16-24	53	44	75	172	52	68	39	331
25-34	91	64	131	286	60	79	52	477
35-44	94	84	137	315	49	99	57	520
45-54	90	84	124	298	82	89	51	520
55-64	66	46	86	198	59	80	42	379
65-74	73	47	73	193	44	65	38	340
75 and over	33	22	45	100	39	41	19	199
All	500	391	671	1562	385	521	298	2766
Women								
16-24	56	35	89	180	40	70	81	371
25-34	127	110	181	418	70	119	74	681
35-44	97	83	142	322	83	117	72	594
45-54	98	79	131	308	89	113	76	586
55-64	87	54	111	252	58	88	56	454
65-74	87	46	90	223	61	90	44	418
75 and over	63	39	69	171	44	86	33	334
All	615	446	813	1874	445	683	436	3438
Weighted sample								
Men								
16-24	73	57	109	239	15	26	10	289
25-34	98	80	164	342	18	35	11	406
35-44	92	74	152	318	17	33	10	379
45-54	85	70	133	287	17	29	8	342
55-64	65	52	97	214	14	23	6	257
65-74	50	41	76	167	11	17	5	201
75 and over	31	27	53	110	7	10	3	130
All	494	401	784	1679	100	172	53	2004
Women								
16-24	69	54	106	228	14	24	9	276
25-34	93	77	158	328	18	34	11	391
35-44	91	72	148	312	18	34	10	374
45-54	85	69	134	288	17	30	9	344
55-64	67	53	100	220	14	25	7	265
65-74	59	46	87	193	13	22	6	233
75 and over	54	44	90	188	12	18	5	223
All	519	415	823	1757	106	186	57	2106

Table F.2   Interview weighting factors

	Standard Statistical Region			Country		
	Northern, North West, Yorkshire & Humberside	E Midlands, W Midlands, E Anglia	London, South East, South West	Scotland	Wales	Northern Ireland
Men						
16-24	1.379	1.289	1.458	0.281	0.377	0.254
25-34	1.075	1.255	1.254	0.301	0.439	0.210
35-44	0.984	0.878	1.108	0.357	0.338	0.172
45-54	0.943	0.830	1.070	0.210	0.324	0.165
55-64	0.980	1.138	1.128	0.232	0.282	0.154
65-74	0.691	0.875	1.039	0.258	0.268	0.123
75 and over	0.934	1.213	1.171	0.183	0.241	0.142
Women						
16-24	1.236	1.529	1.186	0.340	0.348	0.115
25-34	0.735	0.703	0.872	0.251	0.288	0.151
35-44	0.943	0.870	1.044	0.213	0.289	0.143
45-54	0.863	0.875	1.027	0.196	0.262	0.113
55-64	0.769	0.975	0.901	0.244	0.280	0.123
65-74	0.682	1.008	0.970	0.213	0.239	0.133
75 and over	0.865	1.121	1.299	0.278	0.209	0.144

Table F.3   **Weighting classes for examination weights**

Dentate adults (interview data)

Weighting class†	Sample size††	Proportion of total sample	Response rate	Weight
1	1123	31	81	0.8958
2	278	8	79	0.9142
3	339	9	65	1.1098
4	72	2	76	0.9471
5	321	9	59	1.2353
6	595	17	78	0.9238
7	448	12	73	0.9852
8	95	3	58	1.2496
9	60	2	57	1.2767
10	53	1	19	3.8344
11	10	0	60	1.2058
12	66	2	41	1.7685
13	89	2	67	1.0732
14	42	1	48	1.5193
All	3591	100	72	

†   see Figure F.1 for description of weighting classes
†† Weighted sample sizes (by interview weight)

Table F.4   **Standard errors and 95% confidence intervals for dental status**

All adults

Base	Characteristic	% (p)	Unweighted sample size	Standard error of p	95% confidence intervals	Deft
All	Edentate	12.6	6204	0.69	11.3 - 14.0	1.64
	Natural teeth only	71.6	6204	0.97	69.7 - 73.5	1.69
	Natural teeth and dentures	15.8	6204	0.63	14.5 - 17.0	1.36
Age						
16-24	Edentate	0.1	702	0.06	0.0 - 0.2	0.66
	Natural teeth only	99.3	702	0.38	98.5 - 100	1.16
	Natural teeth and dentures	0.7	702	0.37	0.0 - 1.4	1.20
25-34	Edentate	0.5	1158	0.44	0.0 - 1.4	2.03
	Natural teeth only	96.8	1158	0.25	96.3 - 97.3	0.48
	Natural teeth and dentures	2.7	1158	0.52	1.7 - 3.7	1.09
35-44	Edentate	9.7	1114	0.36	9.0 - 10.4	0.40
	Natural teeth only	1.0	1114	1.00	0.0 - 2.9	3.42
	Natural teeth and dentures	89.3	1114	0.91	87.6 - 91.1	0.98
45-54	Edentate	6.5	1106	0.90	4.7 - 8.2	1.21
	Natural teeth only	75.2	1106	1.54	72.2 - 78.2	1.18
	Natural teeth and dentures	18.4	1106	1.21	16.0 - 20.7	1.04
55-64	Edentate	19.8	833	1.73	16.4 - 23.2	1.26
	Natural teeth only	45.8	833	1.99	41.9 - 49.7	1.15
	Natural teeth and dentures	34.3	833	2.01	30.4 - 38.3	1.22
65-74	Edentate	35.9	758	2.64	30.7 - 41.1	1.52
	Natural teeth only	29.4	758	2.33	24.8 - 34.0	1.41
	Natural teeth and dentures	34.7	758	2.44	29.9 - 39.5	1.41
75 and over	Edentate	57.5	533	2.63	52.3 - 62.6	1.23
	Natural teeth only	15.9	533	2.31	11.4 - 20.4	1.46
	Natural teeth and dentures	26.6	533	2.07	22.6 - 30.7	1.08
Gender						
Men	Edentate	10.4	2766	0.77	8.8 - 11.9	1.32
	Natural teeth only	73.5	2766	1.23	71.1 - 75.9	1.46
	Natural teeth and dentures	16.1	2766	0.81	14.6 - 17.7	1.16
Women	Edentate	14.8	3438	0.92	13.0 - 16.5	1.52
	Natural teeth only	69.8	3438	1.22	67.5 - 72.2	1.56
	Natural teeth and dentures	15.4	3438	0.77	13.9 - 16.9	1.24

Table F.5   **Standard errors and 95% confidence intervals for dental status**

All adults

Base	Characteristic	% (p)	Unweighted sample size	Standard error of p	95% confidence intervals	Deft
Country						
England	Edentate	11.9	3436	0.81	10.3 - 13.4	1.46
	Natural teeth only	73.0	3436	1.13	70.7 - 75.2	1.50
	Natural teeth and dentures	15.2	3436	0.73	13.7 - 16.6	1.19
Wales	Edentate	16.6	830	1.99	12.7 - 20.5	1.54
	Natural teeth only	66.0	830	1.50	63.0 - 68.9	0.91
	Natural teeth and dentures	17.4	830	2.69	12.2 - 22.7	2.04
Scotland	Edentate	17.6	1204	1.10	15.5 - 19.8	1.00
	Natural teeth only	62.5	1204	1.51	59.6 - 65.5	1.08
	Natural teeth and dentures	19.8	1204	1.50	16.9 - 22.8	1.31
Northern Ireland	Edentate	12.3	734	1.32	9.7 - 14.9	1.09
	Natural teeth only	69.7	734	1.84	66.1 - 73.3	1.09
	Natural teeth and dentures	18.0	734	1.47	15.1 - 20.9	1.04
English region						
North	Edentate	13.8	975	1.16	11.5 - 16.1	1.05
	Natural teeth only	69.1	975	1.42	66.3 - 71.9	0.96
	Natural teeth and dentures	17.2	975	1.16	14.9 - 19.4	0.96
Midlands	Edentate	14.6	856	1.78	11.1 - 18.1	1.48
	Natural teeth only	68.5	856	2.16	64.3 - 72.7	1.36
	Natural teeth and dentures	17.0	856	1.30	14.4 - 19.5	1.01
South	Edentate	9.6	1605	1.10	7.4 - 11.7	1.50
	Natural teeth only	77.1	1605	1.78	73.6 - 80.6	1.69
	Natural teeth and dentures	13.3	1605	1.19	11.0 - 15.6	1.40
Social class of head of household						
I, II, IIINM	Edentate	7.6	2732	0.63	6.4 - 8.8	1.24
	Natural teeth only	77.8	2732	1.36	75.2 - 80.5	1.70
	Natural teeth and dentures	14.6	2732	1.07	12.5 - 16.7	1.58
IIIM	Edentate	14.8	1729	1.06	12.8 - 16.9	1.23
	Natural teeth only	67.9	1729	1.15	65.6 - 70.1	1.03
	Natural teeth and dentures	17.3	1729	0.80	15.7 - 18.9	0.88
IV, V	Edentate	22.2	1221	1.55	19.2 - 25.2	1.30
	Natural teeth only	59.7	1221	1.26	57.3 - 62.2	0.89
	Natural teeth and dentures	18.1	1221	1.55	15.0 - 21.1	1.41

Table F.6   **Standard errors and 95% confidence intervals for usual reason for dental attendance**

Dentate adults (interview)

Base	Characteristic	% (p)	Unweighted sample size	Standard error of p	95% confidence intervals	Deft
All	Regular check-up	59.3	5281	1.03	57.3 - 61.3	1.52
	Occasional check-up	10.7	5281	0.57	9.6 - 11.8	1.35
	Only with trouble	30.0	5281	1.00	28.1 - 32.0	1.59
Age						
16-24	Regular check-up	48.2	701	2.32	43.6 - 52.7	1.23
	Occasional check-up	16.8	701	1.74	13.4 - 20.2	1.23
	Only with trouble	35.0	701	2.54	30.1 - 40.0	1.41
25-34	Regular check-up	53.2	1151	1.68	49.9 - 56.5	1.14
	Occasional check-up	12.0	1151	1.11	9.8 - 14.2	1.16
	Only with trouble	34.8	1151	1.64	31.6 - 38.0	1.17
35-44	Regular check-up	62.2	1099	1.92	58.4 - 66.0	1.31
	Occasional check-up	10.6	1099	1.11	8.4 - 12.8	1.20
	Only with trouble	27.2	1099	1.82	23.6 - 30.7	1.35
45-54	Regular check-up	63.9	1016	2.09	59.8 - 68.0	1.39
	Occasional check-up	10.7	1016	1.13	8.5 - 12.9	1.17
	Only with trouble	25.4	1016	1.86	21.8 - 29.1	1.36
55-64	Regular check-up	67.5	645	2.40	62.8 - 72.2	1.30
	Occasional check-up	5.8	645	1.00	3.9 - 7.8	1.08
	Only with trouble	26.7	645	2.34	22.1 - 31.2	1.34
65 and over	Regular check-up	65.2	669	2.13	61.0 - 69.3	1.15
	Occasional check-up	5.2	669	1.08	3.1 - 7.3	1.26
	Only with trouble	29.6	669	1.92	25.9 - 33.4	1.08
Gender						
Men	Regular check-up	52.4	2406	1.38	49.7 - 55.1	1.36
	Occasional check-up	10.7	2406	0.64	9.4 - 11.9	1.02
	Only with trouble	36.9	2406	1.50	34.0 - 39.9	1.53
Women	Regular check-up	66.1	2875	1.19	63.8 - 68.5	1.35
	Occasional check-up	10.7	2875	0.69	9.4 - 12.1	1.20
	Only with trouble	23.1	2875	1.15	20.9 - 25.4	1.46

Table F.7   **Standard errors and 95% confidence intervals for usual reason for dental attendance**

Dentate adults (interview)

Base	Characteristic	% (p)	Unweighted sample size	Standard error of p	95% confidence intervals	Deft
Country						
England	Regular check-up	60.0	3010	1.19	57.6 - 62.3	1.33
	Occasional check-up	10.4	3010	0.66	9.1 - 11.7	1.18
	Only with trouble	29.6	3010	1.17	27.3 - 31.9	1.41
Wales	Regular check-up	59.1	682	2.97	53.3 - 64.9	1.57
	Occasional check-up	11.6	682	1.76	8.1 - 15.0	1.43
	Only with trouble	29.3	682	2.30	24.8 - 33.8	1.32
Scotland	Regular check-up	54.8	953	2.04	50.8 - 58.8	1.26
	Occasional check-up	12.2	953	1.33	9.6 - 14.8	1.25
	Only with trouble	32.9	953	1.62	29.8 - 36.1	1.06
Northern Ireland	Regular check-up	51.2	636	2.16	47.0 - 55.5	1.09
	Occasional check-up	13.6	636	1.47	10.7 - 16.5	1.08
	Only with trouble	35.1	636	2.19	30.9 - 39.4	1.15
English region						
North	Regular check-up	57.8	823	2.57	52.8 - 62.8	1.49
	Occasional check-up	11.6	823	1.68	8.3 - 14.9	1.50
	Only with trouble	30.6	823	2.78	25.2 - 36.1	1.73
Midlands	Regular check-up	62.9	731	1.97	59.1 - 66.8	1.10
	Occasional check-up	6.9	731	0.67	5.6 - 8.3	0.71
	Only with trouble	30.1	731	2.22	25.8 - 34.5	1.31
South	Regular check-up	59.7	1456	1.68	56.4 - 63.0	1.31
	Occasional check-up	11.4	1456	0.98	9.5 - 13.3	1.17
	Only with trouble	28.9	1456	1.49	26.0 - 31.8	1.25
Social class of head of household						
I, II, IIINM	Regular check-up	64.6	1926	1.58	61.5 - 67.7	1.45
	Occasional check-up	11.8	1926	0.86	10.1 - 13.5	1.18
	Only with trouble	23.6	1926	1.38	20.9 - 26.3	1.43
IIIM	Regular check-up	57.0	1025	1.95	53.2 - 60.9	1.26
	Occasional check-up	8.5	1025	0.91	6.7 - 10.2	1.04
	Only with trouble	34.5	1025	1.93	30.7 - 38.3	1.30
IV,V	Regular check-up	48.6	625	2.49	43.8 - 53.5	1.24
	Occasional check-up	94.8	625	1.38	92.1 - 97.6	1.56
	Only with trouble	41.9	625	2.27	37.4 - 46.3	1.15

Table F.8   **Standard errors and 95% confidence intervals for type of dental service**

Dentate adults (interview)

Base	Characteristic	% (p)	Unweighted sample size	Standard error of p	95% confidence intervals	Deft
All	NHS	76.2	5281	1.17	73.9 - 78.5	1.99
	Private	18.3	5281	1.00	16.4 - 20.3	1.88
	Other	5.4	5281	0.56	4.3 - 6.5	1.78
Age						
16-24	NHS	85.6	701	1.58	82.5 - 88.7	1.19
	Private	10.8	701	1.33	8.2 - 13.5	1.14
	Other	3.6	701	0.81	2.0 - 5.2	1.16
25-34	NHS	75.8	1151	1.30	73.3 - 78.3	1.03
	Private	18.9	1151	1.18	16.6 - 21.3	1.02
	Other	5.3	1151	0.65	4.0 - 6.5	0.99
35-44	NHS	74.6	1099	2.12	70.5 - 78.8	1.61
	Private	19.2	1099	1.88	15.5 - 22.9	1.58
	Other	6.2	1099	1.03	4.2 - 8.2	1.42
45-54	NHS	72.9	1016	2.30	68.3 - 77.4	1.65
	Private	21.6	1016	2.07	17.6 - 25.7	1.60
	Other	5.5	1016	1.03	3.5 - 7.5	1.43
55-64	NHS	72.9	645	2.57	67.8 - 77.9	1.47
	Private	21.0	645	2.34	16.4 - 25.6	1.46
	Other	6.1	645	1.53	3.1 - 9.1	1.62
65 and over	NHS	75.8	669	2.76	70.4 - 81.2	1.66
	Private	18.0	669	2.59	13.0 - 23.1	1.74
	Other	6.2	669	1.30	3.6 - 8.7	1.40
Gender						
Men	NHS	75.3	2406	1.25	72.9 - 77.7	1.42
	Private	18.8	2406	1.09	16.7 - 21.0	1.37
	Other	5.9	2406	0.65	4.6 - 7.2	1.35
Women	NHS	77.1	2875	1.37	74.5 - 79.8	1.74
	Private	17.9	2875	1.16	15.6 - 20.1	1.62
	Other	5.0	2875	0.62	3.8 - 6.2	1.53

Table F.9   Standard errors and 95% confidence intervals for type of dental service

Dentate adults (interview)

Base	Characteristic	% (p)	Unweighted sample size	Standard error of p	95% confidence intervals	Deft
Country						
England	NHS	75.4	3010	1.36	72.7 - 78.0	1.73
	Private	19.2	3010	1.17	16.9 - 21.5	1.64
	Other	5.4	3010	0.64	4.2 - 6.7	1.55
Wales	NHS	77.2	682	3.87	69.6 - 84.7	2.41
	Private	16.5	682	2.96	10.7 - 22.3	2.08
	Other	6.3	682	1.91	2.6 - 10.1	2.04
Scotland	NHS	81.5	953	2.01	77.5 - 85.4	1.59
	Private	13.6	953	1.39	10.9 - 16.3	1.25
	Other	4.9	953	0.96	3.1 - 6.8	1.37
Northern	NHS	84.8	636	1.74	81.4 - 88.2	1.22
Ireland	Private	10.2	636	1.43	7.4 - 13.0	1.19
	Other	5.0	636	0.99	3.0 - 6.9	1.14
English region						
North	NHS	86.8	823	2.23	82.4 - 91.2	1.89
	Private	9.2	823	1.45	6.4 - 12.1	1.43
	Other	3.9	823	1.19	1.6 - 6.3	1.75
Midlands	NHS	77.7	731	2.49	72.8 - 82.6	1.61
	Private	18.0	731	2.02	14.0 - 21.9	1.42
	Other	4.3	731	1.00	2.4 - 6.3	1.33
South	NHS	68.8	1456	2.24	64.4 - 73.2	1.84
	Private	24.5	1456	2.13	20.4 - 28.7	1.89
	Other	6.7	1456	0.94	4.8 - 8.5	1.43
Social class of head of household						
I, II, IIINM	NHS	73.2	2483	1.79	69.6 - 76.7	2.02
	Private	21.2	2483	1.62	18.0 - 24.4	1.97
	Other	5.6	2483	0.74	4.2 - 7.1	1.59
IIIM	NHS	79.6	1431	1.62	76.5 - 82.8	1.52
	Private	15.7	1431	1.35	13.1 - 18.4	1.40
	Other	4.6	1431	0.81	3.1 - 6.2	1.46
IV, V	NHS	80.9	915	1.46	78.0 - 83.8	1.12
	Private	14.3	915	1.68	11.0 - 17.6	1.46
	Other	4.8	915	0.86	3.2 - 6.5	1.20
Usual reason for dental attendance						
Regular	NHS	75.3	3121	1.71	72.0 - 78.7	2.22
check-up	Private	18.9	3121	1.50	16.0 - 21.8	2.14
	Other	5.8	3121	0.89	4.0 - 7.5	2.14
Occasional	NHS	76.7	575	2.26	72.3 - 81.2	1.28
check-up	Private	19.7	575	1.78	16.2 - 23.2	1.08
	Other	3.6	575	1.07	1.5 - 5.7	1.37
Only with	NHS	77.8	1572	1.43	75.0 - 80.6	1.36
trouble	Private	16.9	1572	1.11	14.7 - 19.1	1.18
	Other	5.3	1572	0.78	3.8 - 6.9	1.38

Table F.10   Standard errors and 95% confidence intervals for time since last visit

Dentate adults (interview)

Base	Characteristic	% (p)	Unweighted sample size	Standard error of p	95% confidence intervals	Deft
All	Up to 1 year	70.8	5281	1.04	68.8 - 72.8	1.66
	Over 1 year	7.5	5281	0.43	6.6 - 8.3	1.20
	Over 2 years	11.5	5281	0.62	10.3 - 12.7	1.42
	Over 5 years	5.6	5281	0.46	4.7 - 6.5	1.45
	Over 10 years	2.9	5281	0.23	2.4 - 3.3	0.99
	Never been	1.8	5281	0.20	1.4 - 2.1	1.10
Age						
16-24	Up to 1 year	65.1	701	2.46	60.3 - 69.9	1.36
	Over 1 year	10.8	701	1.59	7.7 - 13.9	1.35
	Over 2 years	16.5	701	1.77	13.0 - 19.9	1.26
	Over 5 years	6.0	701	1.12	3.8 - 8.2	1.25
	Over 10 years	3.3	701	0.52	2.2 - 4.3	0.78
	Never been	0.3	701	0.26	-0.3 - 0.8	1.37
25-34	Up to 1 year	66.7	1151	1.68	63.4 - 70.0	1.21
	Over 1 year	8.5	1151	0.84	6.9 - 10.2	1.02
	Over 2 years	15.3	1151	1.21	12.9 - 17.7	1.14
	Over 5 years	6.6	1151	0.96	4.7 - 8.5	1.32
	Over 10 years	2.7	1151	0.52	1.7 - 3.7	1.08
	Never been	0.2	1151	0.16	-0.1 - 0.5	1.25
35-44	Up to 1 year	73.1	1099	1.68	69.8 - 76.4	1.26
	Over 1 year	7.2	1099	0.84	5.5 - 8.8	1.08
	Over 2 years	9.9	1099	0.95	8.1 - 11.8	1.05
	Over 5 years	5.3	1099	0.86	3.6 - 7.0	1.27
	Over 10 years	2.6	1099	0.56	1.5 - 3.7	1.16
	Never been	1.9	1099	0.43	1.0 - 2.7	1.05
45-54	Up to 1 year	75.1	1016	1.67	71.8 - 78.4	1.23
	Over 1 year	5.8	1016	0.76	4.3 - 7.3	1.04
	Over 2 years	9.1	1016	0.87	7.4 - 10.8	0.96
	Over 5 years	5.1	1016	0.82	3.5 - 6.7	1.19
	Over 10 years	3.9	1016	0.76	2.4 - 5.3	1.25
	Never been	1.1	1016	0.33	0.4 - 1.8	1.03
55-64	Up to 1 year	75.4	645	2.26	70.9 - 79.8	1.33
	Over 1 year	5.0	645	1.02	3.0 - 7.0	1.19
	Over 2 years	7.6	645	1.07	5.5 - 9.7	1.03
	Over 5 years	5.6	645	0.93	3.7 - 7.4	1.03
	Over 10 years	3.6	645	0.95	1.7 - 5.4	1.29
	Never been	2.8	645	0.80	1.3 - 4.4	1.21
65 and over	Up to 1 year	70.8	669	2.15	66.6 - 75.1	1.22
	Over 1 year	6.6	669	1.18	4.3 - 8.9	1.22
	Over 2 years	8.2	669	1.23	5.8 - 10.6	1.16
	Over 5 years	4.5	669	1.04	2.5 - 6.6	1.29
	Over 10 years	3.5	669	0.79	1.9 - 5.0	1.11
	Never been	6.3	669	1.21	4.0 - 8.7	1.28

Table F.11   **Standard errors and 95% confidence intervals for time since last visit**

Dentate adults (interview)

Base	Characteristic	% (p)	Unweighted sample size	Standard error of p	95% confidence intervals	Deft
Country						
England	Up to 1 year	70.8	3010	1.21	68.4 - 73.1	1.46
	Over 1 year	7.3	3010	0.51	6.3 - 8.3	1.07
	Over 2 years	11.5	3010	0.72	10.1 - 12.9	1.24
	Over 5 years	57.1	3010	0.54	56.1 - 58.2	0.60
	Over 10 years	3.0	3010	0.26	2.5 - 3.5	0.84
	Never been	1.7	3010	0.23	1.3 - 2.2	0.96
Wales	Up to 1 year	71.5	682	2.60	66.4 - 76.6	1.50
	Over 1 year	8.1	682	1.02	6.1 - 10.1	0.98
	Over 2 years	10.4	682	1.44	7.6 - 13.3	1.23
	Over 5 years	5.3	682	0.55	4.2 - 6.4	0.64
	Over 10 years	2.2	682	0.71	0.8 - 3.6	1.27
	Never been	2.5	682	0.52	1.5 - 3.5	0.87
Scotland	Up to 1 year	71.0	953	1.47	68.1 - 73.9	1.00
	Over 1 year	8.1	953	0.61	6.9 - 9.3	0.69
	Over 2 years	12.6	953	1.18	10.3 - 14.9	1.09
	Over 5 years	4.2	953	0.63	2.9 - 5.4	0.97
	Over 10 years	2.2	953	0.43	1.4 - 3.1	0.90
	Never been	1.9	953	0.49	0.9 - 2.8	1.11
Northern Ireland	Up to 1 year	69.4	636	2.07	65.3 - 73.4	1.13
	Over 1 year	9.8	636	1.37	7.1 - 12.5	1.16
	Over 2 years	11.1	636	1.42	8.3 - 13.9	1.14
	Over 5 years	6.8	636	1.08	4.7 - 9.0	1.08
	Over 10 years	1.5	636	0.47	0.6 - 2.4	0.98
	Never been	1.5	636	0.46	0.6 - 2.4	0.97
English region						
North	Up to 1 year	70.8	823	1.64	67.6 - 74.0	1.04
	Over 1 year	7.8	823	0.96	5.9 - 9.6	1.03
	Over 2 years	12.1	823	1.48	9.2 - 15.1	1.30
	Over 5 years	4.9	823	1.02	2.9 - 6.9	1.36
	Over 10 years	2.6	823	0.59	1.5 - 3.8	1.05
	Never been	1.8	823	0.35	1.1 - 2.5	0.76
Midlands	Up to 1 year	70.3	731	2.48	65.4 - 75.1	1.46
	Over 1 year	5.5	731	0.97	3.6 - 7.4	1.15
	Over 2 years	13.0	731	1.64	9.8 - 16.2	1.32
	Over 5 years	5.4	731	0.94	3.6 - 7.3	1.12
	Over 10 years	3.0	731	0.66	1.7 - 4.3	1.06
	Never been	2.8	731	0.73	1.4 - 4.3	1.19
South	Up to 1 year	71.0	1456	1.81	67.4 - 74.5	1.52
	Over 1 year	7.9	1456	0.75	6.4 - 9.3	1.06
	Over 2 years	10.5	1456	0.96	8.6 - 12.4	1.20
	Over 5 years	6.3	1456	0.78	4.7 - 7.8	1.23
	Over 10 years	3.2	1456	0.38	2.5 - 4.0	0.82
	Never been	1.2	1456	0.27	0.6 - 1.7	0.96

Table F.12   **Standard errors and 95% confidence intervals for time since last visit**

Dentate adults (interview)

Base	Characteristic	% (p)	Unweighted sample size	Standard error of p	95% confidence intervals	Deft
Gender						
Men	Up to 1 year	65.0	2406	1.56	62.0 - 68.1	1.60
	Over 1 year	7.6	2406	0.60	6.4 - 8.8	1.12
	Over 2 years	13.7	2406	0.92	11.9 - 15.4	1.31
	Over 5 years	7.4	2406	0.79	5.9 - 9.0	1.48
	Over 10 years	3.7	2406	0.38	3.0 - 4.5	0.98
	Never been	2.6	2406	0.33	1.9 - 3.2	1.04
Women	Up to 1 year	76.6	2875	0.96	74.7 - 78.4	1.21
	Over 1 year	7.3	2875	0.63	6.1 - 8.6	1.30
	Over 2 years	9.4	2875	0.62	8.2 - 10.6	1.15
	Over 5 years	3.8	2875	0.36	3.1 - 4.5	1.02
	Over 10 years	2.0	2875	0.37	1.3 - 2.7	1.42
	Never been	0.9	2875	0.17	0.6 - 1.3	0.91
Social class of head of household						
I, II, IIINM	Up to 1 year	75.4	2483	1.43	72.6 - 78.2	1.65
	Over 1 year	7.5	2483	0.65	6.3 - 8.8	1.23
	Over 2 years	9.1	2483	0.65	7.8 - 10.4	1.13
	Over 5 years	4.9	2483	0.52	3.9 - 5.9	1.19
	Over 10 years	2.0	2483	0.33	1.3 - 2.6	1.19
	Never been	1.2	2483	0.22	0.7 - 1.6	1.02
IIIM	Up to 1 year	67.8	1431	1.98	63.9 - 71.7	1.60
	Over 1 year	6.8	1431	0.64	5.6 - 8.1	0.96
	Over 2 years	14.3	1431	1.31	11.7 - 16.9	1.41
	Over 5 years	5.5	1431	0.79	4.0 - 7.1	1.31
	Over 10 years	3.6	1431	0.47	2.7 - 4.5	0.94
	Never been	2.0	1431	0.53	0.9 - 3.0	1.46
IV, V	Up to 1 year	62.8	915	2.03	58.8 - 66.8	1.27
	Over 1 year	83.0	915	1.39	80.2 - 85.7	1.11
	Over 2 years	13.6	915	1.43	10.7 - 16.4	1.26
	Over 5 years	7.8	915	1.21	5.4 - 10.2	1.37
	Over 10 years	4.5	915	0.86	2.8 - 6.2	1.25
	Never been	3.1	915	0.64	1.8 - 4.3	1.12
Usual reason for dental attendance						
Regular check-up	Up to 1 year	97.2	3121	0.41	96.4 - 98.0	1.38
	Over 1 year	2.2	3121	0.32	1.6 - 2.8	1.21
	Over 2 years	0.5	3121	0.15	0.2 - 0.8	1.17
	Over 5 years	0.1	3121	0.07	0.0 - 0.2	1.26
	Over 10 years†	0.0	3121	-		
	Never been†	0.0	3121	-		
Occasional check-up	Up to 1 year	58.8	575	2.61	53.7 - 63.9	1.27
	Over 1 year	24.3	575	1.67	21.0 - 27.6	0.93
	Over 2 years	14.2	575	1.98	10.3 - 18.1	1.36
	Over 5 years	2.7	575	0.74	1.2 - 4.1	1.10
	Over 10 years†	0.0	575	-		
	Never been†	0.0	575	-		
Only with trouble	Up to 1 year	23.3	1572	1.47	20.5 - 26.2	1.38
	Over 1 year	12.0	1572	0.98	10.0 - 13.9	1.20
	Over 2 years	32.4	1572	1.51	29.4 - 35.4	1.28
	Over 5 years	17.4	1572	1.25	15.0 - 19.9	1.31
	Over 10 years	9.4	1572	0.74	7.9 - 10.9	1.01
	Never been	5.5	1572	0.65	4.2 - 6.8	1.12

† No cases observed

Table F.13   **Standard errors and 95% confidence intervals for frequency of teeth cleaning**

Dentate adults (interview)

Base	Characteristic	% (p)	Unweighted sample size	Standard error of p	95% confidence intervals	Deft
All	Twice or more	73.9	5281	0.75	72.5 - 75.4	1.24
	Once	21.8	5281	0.74	20.3 - 23.2	1.31
	Less than once	3.8	5281	0.32	3.2 - 4.5	1.22
	Never	0.5	5281	0.08	0.3 - 0.6	0.87
Age						
16-24	Twice or more	78.7	701	1.82	75.1 - 82.3	1.18
	Once	17.5	701	1.90	13.7 - 21.2	1.33
	Less than once	3.7	701	0.95	1.8 - 5.5	1.34
	Never	0.2	701	0.10	0.0 - 0.3	0.65
25-34	Twice or more	76.8	1151	1.59	73.7 - 79.9	1.27
	Once	19.7	1151	1.40	16.9 - 22.4	1.19
	Less than once	3.4	1151	0.74	1.9 - 4.8	1.39
	Never	0.2	1151	0.10	0.0 - 0.4	0.80
35-44	Twice or more	75.2	1099	1.43	72.4 - 78.0	1.10
	Once	20.3	1099	1.33	17.7 - 22.9	1.10
	Less than once	4.1	1099	0.64	2.8 - 5.3	1.07
	Never	0.4	1099	0.22	0.0 - 0.9	1.13
45-54	Twice or more	71.4	1016	1.77	68.0 - 74.9	1.25
	Once	24.0	1016	1.65	20.8 - 27.2	1.23
	Less than once	4.1	1016	0.86	2.4 - 5.8	1.38
	Never	0.5	1016	0.25	0.0 - 1.0	1.14
55-64	Twice or more	71.0	645	2.24	66.6 - 75.4	1.25
	Once	24.5	645	2.16	20.3 - 28.7	1.27
	Less than once	4.1	645	0.95	2.3 - 6.0	1.22
	Never	0.4	645	0.28	0.0 - 0.9	1.17
65 and over	Twice or more	66.8	669	2.14	62.6 - 71.0	1.17
	Once	27.8	669	2.08	23.7 - 31.9	1.20
	Less than once	4.0	699	0.77	2.4 - 5.5	1.05
	Never	1.4	699	0.65	0.1 - 2.7	1.45
Gender						
Men	Twice or more	64.4	2406	1.05	62.4 - 66.5	1.08
	Once	28.4	2406	0.98	26.5 - 30.3	1.07
	Less than once	6.3	2406	0.53	5.2 - 7.3	1.07
	Never	0.9	2406	0.15	0.6 - 1.2	0.81
Women	Twice or more	83.4	2875	0.88	81.7 - 85.2	1.26
	Once	15.1	2875	0.90	13.3 - 16.9	1.35
	Less than once	1.4	2875	0.25	0.9 - 1.9	1.14
	Never	0.0	2875	0.04	0.0 - 0.1	0.98

Table F.14   **Standard errors and 95% confidence intervals for frequency of teeth cleaning**

Dentate adults (interview)

Base	Characteristic	% (p)	Unweighted sample size	Standard error of p	95% confidence intervals		Deft
Country							
England	Twice or more	74.3	3010	0.86	72.6 -	76.0	1.08
	Once	21.7	3010	0.87	20.0 -	23.4	1.15
	Less than once	3.5	3010	0.37	2.8 -	4.3	1.11
	Never	0.4	3010	0.09	0.2 -	0.6	0.74
Wales	Twice or more	74.5	682	1.91	70.8 -	78.3	1.14
	Once	20.9	682	1.77	17.4 -	24.4	1.14
	Less than once	4.3	682	0.51	3.3 -	5.3	0.66
	Never	0.3	682	0.20	0.0 -	0.7	0.99
Scotland	Twice or more	70.4	953	1.56	67.3 -	73.5	1.05
	Once	22.5	953	1.22	20.1 -	24.9	0.90
	Less than once	6.1	953	0.69	4.7 -	7.4	0.89
	Never	1.0	953	0.38	0.3 -	1.8	1.18
Northern Ireland	Twice or more	71.7	636	2.06	67.7 -	75.8	1.15
	Once	21.8	636	1.92	18.1 -	25.6	1.17
	Less than once	5.8	636	1.09	3.7 -	8.0	1.17
	Never	0.6	636	0.30	0.0 -	1.2	0.99
English region							
North	Twice or more	72.5	823	2.04	68.5 -	76.5	1.31
	Once	23.5	823	1.65	20.3 -	26.7	1.11
	Less than once	3.7	823	0.42	2.8 -	4.5	0.64
	Never	0.3	823	0.21	0.0 -	0.7	1.00
Midlands	Twice or more	71.8	731	1.78	68.3 -	75.3	1.07
	Once	23.8	731	1.79	20.3 -	27.3	1.14
	Less than once	3.2	731	0.65	1.9 -	4.5	1.00
	Never	1.2	731	0.27	0.7 -	1.7	0.66
South	Twice or more	76.3	1456	1.15	74.1 -	78.6	1.03
	Once	19.9	1456	1.30	17.4 -	22.5	1.24
	Less than once	3.6	1456	0.66	2.3 -	4.9	1.35
	Never	0.1	1456	0.07	0.0 -	0.2	1.01
Social class of head of household							
I, II, IIINM	Twice or more	78.4	2483	1.14	76.1 -	80.6	1.38
	Once	19.8	2483	1.07	17.7 -	21.9	1.33
	Less than once	1.6	2483	0.29	1.0 -	2.2	1.16
	Never	0.2	2483	0.10	0.0 -	0.4	1.09
IIIM	Twice or more	69.1	1431	1.50	66.1 -	72.0	1.23
	Once	24.5	1431	1.32	21.9 -	27.1	1.16
	Less than once	6.1	1431	0.80	4.5 -	7.6	1.26
	Never	0.3	1431	0.17	0.0 -	0.7	1.13
IV, V	Twice or more	65.5	915	2.10	61.4 -	69.6	1.34
	Once	25.8	915	1.90	22.1 -	29.5	1.31
	Less than once	7.5	915	1.06	5.4 -	9.5	1.22
	Never	1.2	915	0.31	0.6 -	1.8	0.86
Usual reason for dental attendance							
Regular check-up	Twice or more	80.0	3121	0.92	78.2 -	81.8	1.28
	Once	17.9	3121	0.83	16.3 -	19.6	1.22
	Less than once	2.0	3121	0.37	1.3 -	2.8	1.46
	Never	0.1	3121	0.06	0.0 -	0.2	1.04
Occasional check-up	Twice or more	77.8	575	2.03	73.8 -	81.8	1.17
	Once	20.3	575	2.03	16.3 -	24.3	1.21
	Less than once	1.9	575	0.51	0.9 -	2.9	0.91
	Never†	0.0	575	-			
Only with trouble	Twice or more	60.9	1572	1.59	57.7 -	64.0	1.29
	Once	29.8	1572	1.60	26.7 -	33.0	1.38
	Less than once	8.1	1572	0.90	6.3 -	9.8	1.31
	Never	1.2	1572	0.24	0.8 -	1.7	0.87

† No observed cases

Table F.15   **Standard errors and 95% confidence intervals for proportion with 21 or more teeth**

Dentate adults (exam data)

Base	% (p)	Unweighted sample size	Standard error of p	95% confidence intervals	Deft
All	82.6	3817	0.80	81.1 - 84.2	1.30
Age					
16-24	99.6	491	0.32	99.0 - 100.2	1.11
25-34	98.4	854	0.35	97.7 - 99.1	0.83
35-44	94.0	781	0.92	92.3 - 95.8	1.08
45-54	81.7	746	1.57	78.7 - 84.8	1.11
55-64	56.8	461	2.61	51.7 - 61.9	1.13
65-74	38.3	484	2.80	32.8 - 43.8	1.27
Gender					
Men	82.5	1745	0.87	80.8 - 84.2	0.95
Women	82.7	2072	1.10	80.6 - 84.9	1.33
Country					
England	83.3	2186	0.93	81.5 - 85.1	1.16
Wales	81.2	502	2.11	77.1 - 85.4	1.21
Scotland	77.5	668	1.55	74.5 - 80.6	0.96
Northern Ireland	80.6	461	1.99	76.7 - 84.5	1.08
English region					
North	82.1	617	1.61	78.9 - 85.3	1.04
Midlands	79.7	495	1.94	75.9 - 83.5	1.07
South	85.4	1074	1.39	82.6 - 88.1	1.29
Social class of head of household					
I, II, IIINM	86.4	1926	1.18	84.1 - 88.7	1.51
IIIM	79.4	1025	1.20	77.0 - 81.7	0.95
IV,V	75.5	625	1.81	71.9 - 79.0	1.05
Usual reason for dental attendance					
Regular check-up	84.3	2400	1.13	82.0 - 86.5	1.52
Occasional check-up	88.3	408	1.78	84.9 - 91.8	1.12
Only with trouble	77.0	1003	1.42	74.2 - 79.8	1.07

Table F.16   **Standard errors and 95% confidence intervals for proportion with 18 or more sound and untreated teeth (1998 criteria)**

Dentate adults (exam data)

Base	% (p)	Unweighted sample size	Standard error of p	95% confidence intervals	Deft
All	39.9	3817	0.98	38.0 - 41.8	1.23
Age					
16-24	89.3	491	1.71	86.0 - 92.7	1.23
25-34	65.0	854	1.94	61.2 - 68.8	1.19
35-44	36.7	781	1.93	32.9 - 40.5	1.12
45-54	15.1	746	1.64	11.9 - 18.3	1.25
55-64	4.7	461	0.98	2.7 - 6.6	0.99
65-74	5.0	484	0.85	3.4 - 6.7	0.85
Gender					
Men	42.2	1745	1.19	39.9 - 44.5	1.00
Women	37.6	2072	1.42	34.8 - 40.4	1.33
Country					
England	41.4	2186	1.14	39.1 - 43.6	1.08
Wales	36.9	502	1.98	33.0 - 40.8	0.92
Scotland	29.2	668	1.48	26.3 - 32.1	0.84
Northern Ireland	31.8	461	2.49	26.9 - 36.7	1.15
English region					
North	40.9	617	2.61	35.8 - 46.0	1.32
Midlands	42.0	495	1.77	38.5 - 45.4	0.80
South	41.4	1074	1.57	38.3 - 44.4	1.04
Social class of head of household					
I, II, IIINM	39.5	1926	1.62	36.4 - 42.7	1.45
IIIM	37.4	1025	1.78	33.9 - 40.9	1.18
IV,V	40.2	625	2.24	35.8 - 44.6	1.14
Usual reason for dental attendance					
Regular check-up	35.1	2400	1.29	32.5 - 37.6	1.32
Occasional check-up	52.2	408	3.00	46.3 - 58.1	1.21
Only with trouble	45.1	1003	1.61	41.9 - 48.2	1.03

Table F.17   **Standard errors and 95% confidence intervals for proportion with 12 or more restored otherwise sound teeth (1998 criteria)**

Dentate adults (exam data)

Base	% (p)	Unweighted sample size	Standard error of p	95% confidence intervals	Deft
All	26.7	3817	0.85	25.0 - 28.4	1.19
Age					
16-24	1.6	491	0.50	0.6 - 2.6	0.89
25-34	16.4	854	1.52	13.4 - 19.4	1.20
35-44	35.6	781	1.85	32.0 - 39.2	1.08
45-54	46.4	746	2.56	41.4 - 51.4	1.40
55-64	35.7	461	2.93	29.9 - 41.4	1.31
65 and over	25.6	484	2.53	20.7 - 30.6	1.27
Gender					
Men	23.5	1745	1.06	21.4 - 25.6	1.04
Women	29.9	2072	1.24	27.5 - 32.4	1.24
Country					
England	26.0	2186	0.99	24.1 - 28.0	1.05
Wales	25.8	502	1.07	23.7 - 27.9	0.55
Scotland	32.0	668	2.05	28.0 - 36.0	1.13
Northern Ireland	33.1	461	2.43	28.4 - 37.9	1.11
English region					
North	21.5	617	2.18	17.2 - 25.8	1.32
Midlands	24.1	495	2.42	19.4 - 28.9	1.26
South	29.1	1074	1.14	26.8 - 31.3	0.83
Social class of head of household					
I, II, IIINM	32.1	1926	1.61	28.9 - 35.2	1.52
IIIM	23.3	1025	1.56	20.2 - 26.4	1.18
IV,V	16.8	625	1.51	13.8 - 19.7	1.01
Usual reason for dental attendance					
Regular check-up	34.0	2400	1.24	31.6 - 36.4	1.28
Occasional check-up	23.5	408	2.34	18.9 - 28.1	1.11
Only with trouble	12.7	1003	1.26	10.2 - 15.1	1.20

Table F.18   **Standard errors and 95% confidence intervals for proportion with 1 or more decayed or unsound teeth (1998 criteria)**

Dentate adults (exam data)

Base	% (p)	Unweighted sample size	Standard error of p	95% confidence intervals	Deft
All	54.8	3817	1.26	52.3 - 57.3	1.57
Age					
16-24	51.0	491	2.64	45.8 - 56.1	1.17
25-34	59.9	854	2.60	54.8 - 65.0	1.55
35-44	51.1	781	2.28	46.6 - 55.5	1.27
45-54	56.6	746	2.34	52.1 - 61.2	1.29
55-64	54.4	461	2.78	49.0 - 59.9	1.20
65 and over	53.8	484	2.69	48.6 - 59.1	1.19
Gender					
Men	57.7	1745	1.55	54.6 - 60.7	1.31
Women	51.8	2072	1.70	48.5 - 55.2	1.55
Country					
England	54.7	2186	1.46	51.8 - 57.5	1.37
Wales	47.9	502	3.60	40.9 - 55.0	1.61
Scotland	58.1	668	2.97	52.3 - 63.9	1.55
Northern Ireland	59.7	461	2.76	54.3 - 65.1	1.21
English region					
North	64.7	617	2.82	59.2 - 70.2	1.46
Midlands	52.0	495	3.58	45.0 - 59.0	1.59
South	50.8	1074	1.90	47.1 - 54.5	1.24
Social class of head of household					
I, II, IIINM	49.9	1926	1.82	46.3 - 53.4	1.60
IIIM	57.4	1025	1.96	53.6 - 61.2	1.27
IV,V	62.5	625	2.12	58.4 - 66.6	1.09
Usual reason for dental attendance					
Regular check-up	48.4	2400	1.60	45.3 - 51.5	1.57
Occasional check-up	56.1	408	3.54	49.1 - 63.0	1.44
Only with trouble	67.4	1003	1.57	64.3 - 70.5	1.06

Table F.19   **Standard errors and 95% confidence intervals for proportion with 18 or more sound and untreated teeth (1988 criteria)**

Dentate adults (exam data)

Base	% (p)	Unweighted sample size	Standard error of p	95% confidence intervals	Deft
All	41.7	3817	0.95	39.8 - 43.5	1.19
Age					
16-24	92.3	491	1.33	89.7 - 94.9	1.10
25-34	68.4	854	1.77	64.9 - 71.9	1.11
35-44	38.3	781	2.17	34.0 - 42.5	1.25
45-54	16.3	746	1.70	13.0 - 19.6	1.25
55-64	4.7	461	0.98	2.8 - 6.6	0.99
65 and over	5.0	484	0.85	3.4 - 6.7	0.85
Gender					
Men	44.0	1745	1.20	41.7 - 46.4	1.01
Women	39.3	2072	1.30	36.8 - 41.9	1.21
Country					
England	42.9	2186	1.10	40.7 - 45.0	1.04
Wales	38.8	502	1.99	34.9 - 42.7	0.91
Scotland	33.5	668	1.58	30.4 - 36.6	0.86
Northern Ireland	34.1	461	2.54	29.2 - 39.1	1.15
English region					
North	44.7	617	2.19	40.4 - 49.0	1.09
Midlands	43.0	495	2.22	38.6 - 47.3	1.00
South	41.9	1074	1.55	38.9 - 45.0	1.03
Social class of head of household					
I, II, IIINM	41.1	1926	1.59	38.0 - 44.2	1.42
IIIM	39.7	1025	1.95	35.9 - 43.5	1.27
IV,V	42.2	625	2.27	37.7 - 46.7	1.15
Usual reason for dental attendance					
Regular check-up	36.2	2400	1.28	33.7 - 38.7	1.30
Occasional check-up	54.4	408	2.92	48.7 - 60.1	1.18
Only with trouble	48.1	1003	1.60	44.9 - 51.2	1.01

Table F.20   **Standard errors and 95% confidence intervals for proportion with 12 or more restored otherwise sound teeth (1988 criteria)**

Dentate adults (exam data)

Base	% (p)	Unweighted sample size	Standard error of p	95% confidence intervals	Deft
All	28.0	3817	0.87	26.3 - 29.7	1.20
Age					
16-24	1.8	491	0.52	0.8 - 2.8	0.87
25-34	18.0	854	1.53	15.0 - 21.0	1.17
35-44	38.0	781	1.94	34.2 - 41.8	1.12
45-54	48.0	746	2.55	43.0 - 53.0	1.40
55-64	36.7	461	2.94	31.0 - 42.5	1.31
65 and over	26.1	484	2.54	21.1 - 31.1	1.27
Gender					
Men	25.2	1745	1.08	23.1 - 27.4	1.04
Women	30.9	2072	1.17	28.6 - 33.2	1.15
Country					
England	27.3	2186	1.01	25.3 - 29.2	1.06
Wales	27.4	502	1.27	25.0 - 29.9	0.64
Scotland	34.2	668	2.13	30.0 - 38.4	1.16
Northern Ireland	35.0	461	2.46	30.2 - 39.8	1.10
English region					
North	23.9	617	2.32	19.4 - 28.5	1.35
Midlands	24.7	495	2.50	19.8 - 29.6	1.29
South	30.0	1074	1.14	27.7 - 32.2	0.82
Social class of head of household					
I, II, IIINM	33.8	1926	1.64	30.6 - 37.0	1.52
IIIM	24.8	1025	1.51	21.9 - 27.8	1.12
IV,V	17.4	625	1.55	14.4 - 20.4	1.02
Usual reason for dental attendance					
Regular check-up	35.2	2400	1.22	32.8 - 37.6	1.25
Occasional check-up	25.5	408	2.25	21.1 - 29.9	1.04
Only with trouble	14.2	1003	1.40	11.5 - 17.0	1.27

Table F.21   **Standard errors and 95% confidence intervals for proportion with 1 or more decayed or unsound teeth (1988 criteria)**

Dentate adults (exam data)

Base	% (p)	Unweighted sample size	Standard error of p	95% confidence intervals	Deft
All	41.8	3817	1.28	39.3 - 44.3	1.60
Age					
16-24	30.8	491	2.78	25.4 - 36.3	1.33
25-34	41.7	854	2.39	37.0 - 46.4	1.42
35-44	39.9	781	2.16	35.7 - 44.2	1.23
45-54	45.7	746	2.65	40.5 - 50.9	1.45
55-64	47.6	461	2.90	41.9 - 53.3	1.25
65 and over	48.4	484	2.87	42.7 - 54.0	1.26
Gender					
Men	45.2	1745	1.49	42.3 - 48.1	1.25
Women	38.4	2072	1.68	35.1 - 41.7	1.58
Country					
England	42.4	2186	1.49	39.5 - 45.3	1.41
Wales	34.1	502	2.00	30.2 - 38.0	0.94
Scotland	41.3	668	2.58	36.3 - 46.4	1.35
Northern Ireland	40.8	461	2.74	35.5 - 46.2	1.19
English region					
North	46.3	617	2.56	41.3 - 51.3	1.28
Midlands	37.8	495	2.50	32.9 - 42.7	1.14
South	42.3	1074	2.35	37.7 - 46.9	1.56
Social class of head of household					
I, II, IIINM	37.1	1926	1.70	33.8 - 40.5	1.55
IIIM	45.3	1025	1.94	41.5 - 49.1	1.25
IV,V	48.0	625	2.52	43.1 - 53.0	1.26
Usual reason for dental attendance					
Regular check-up	35.8	2400	1.51	32.8 - 38.7	1.54
Occasional check-up	37.9	408	3.16	31.7 - 44.1	1.31
Only with trouble	55.8	1003	1.77	52.3 - 59.2	1.13

Table F.22   Standard errors and 95% confidence intervals for mean number of teeth (1998 criteria)

Dentate adults (exam data)

Base	Mean (p)	Unweighted sample size	Standard error of p	95% confidence intervals	Deft
All	24.8	3817	0.15	24.5 - 25.1	1.53
Age					
16-24	27.9	491	0.15	27.6 - 28.2	1.72
25-34	28.1	854	0.95	27.9 - 28.3	1.00
35-44	29.7	781	0.16	26.4 - 27.0	1.06
45-54	24.0	746	0.22	23.6 - 24.5	1.08
55-64	19.9	461	0.40	19.1 - 20.7	1.21
65 and over	17.3	484	0.48	16.4 - 18.3	1.42
Gender					
Men	25.0	1745	0.17	24.6 - 25.3	1.20
Women	24.6	2072	0.19	24.2 - 25.0	1.39
Country					
England	24.9	2186	0.17	24.6 - 25.3	1.64
Wales	24.2	502	0.41	23.4 - 25.0	0.89
Scotland	23.8	668	0.26	23.3 - 24.3	0.69
Northern Ireland	24.4	461	0.32	23.8 - 25.1	0.53
English region					
North	24.6	617	0.23	24.1 - 25.1	1.14
Midlands	24.6	495	0.43	23.7 - 25.4	1.74
South	25.3	1074	0.26	24.7 - 25.8	1.86
Social class of head of household					
I, II, IIINM	25.4	1926	0.21	25.0 - 25.8	1.65
IIIM	24.2	1025	0.20	23.8 - 24.6	0.99
IV,V	23.6	625	0.31	23.0 - 24.2	1.05
Usual reason for dental attendance					
Regular check-up	24.9	2400	0.19	24.5 - 25.3	1.63
Occasional check-up	26.3	408	0.28	25.7 - 26.8	1.17
Only with trouble	24.0	1003	0.26	23.4 - 24.5	1.19

Table F.23   **Standard errors and 95% confidence intervals for mean number of sound and untreated teeth (1998 criteria)**

Dentate adults (exam data)

Base	Mean (p)	Unweighted sample size	Standard error of p	95% confidence intervals		Deft
All	15.3	3817	0.17	15.0 -	15.7	1.45
Age						
16-24	23.4	491	0.28	22.8 -	23.9	1.47
25-34	19.1	854	0.24	18.6 -	19.6	2.24
35-44	15.4	781	0.24	14.9 -	15.9	1.11
45-54	11.7	746	0.17	11.4 -	12.1	0.88
55-64	9.6	461	0.25	9.1 -	10.1	1.15
65 and over	8.5	484	0.27	8.0 -	9.1	1.17
Gender						
Men	15.7	1745	0.19	15.4 -	16.1	1.14
Women	14.9	2072	0.21	14.5 -	15.3	1.29
Country						
England	15.6	2186	0.20	15.2 -	16.0	1.55
Wales	15.1	502	0.22	14.6 -	15.5	0.43
Scotland	13.6	668	0.20	13.2 -	14.0	0.51
Northern Ireland	13.9	461	0.38	13.2 -	14.7	0.57
English region						
North	15.4	617	0.34	14.7 -	16.0	1.37
Midlands	15.6	495	0.32	14.9 -	16.2	1.14
South	15.7	1074	0.33	15.0 -	16.3	1.82
Social class of head of household						
I, II, IIINM	15.4	1926	0.27	14.9 -	16.0	1.60
IIIM	14.8	1025	0.24	14.3 -	15.2	1.11
IV,V	15.3	625	0.36	14.6 -	16.0	2.24
Usual reason for dental attendance						
Regular check-up	14.6	2400	0.21	14.2 -	15.0	1.44
Occasional check-up	17.5	408	0.48	16.6 -	18.5	1.36
Only with trouble	16.0	1003	0.26	15.4 -	16.5	1.14

Table F.24   **Standard errors and 95% confidence intervals for mean number of restored otherwise teeth (1998 criteria)**

Dentate adults (exam data)

Base	Mean (p)	Unweighted sample size	Standard error of p	95% confidence intervals	Deft
All	7.9	3817	0.12	7.7 - 8.1	1.33
Age					
16-24	2.8	491	0.13	2.5 - 3.0	1.08
25-34	7.1	854	0.20	6.7 - 7.5	1.31
35-44	9.8	781	0.21	9.4 - 10.3	1.17
45-54	10.8	746	0.25	10.3 - 11.3	1.27
55-64	8.9	461	0.35	8.2 - 9.6	1.32
65 and over	7.5	484	0.38	6.8 - 8.3	1.46
Gender					
Men	7.4	1745	0.13	7.2 - 7.7	1.05
Women	8.4	2072	0.17	8.1 - 8.7	1.34
Country					
England	7.8	2186	0.13	7.6 - 8.1	1.41
Wales	7.9	502	0.22	7.5 - 8.3	0.54
Scotland	8.4	668	0.32	7.7 - 9.0	0.97
Northern Ireland	9.0	461	0.30	8.4 - 9.6	0.55
English region					
North	7.3	617	0.21	6.9 - 7.7	1.18
Midlands	7.5	495	0.35	6.8 - 8.2	1.78
South	8.2	1074	0.18	7.9 - 8.6	1.35
Social class of head of household					
I, II, IIINM	8.7	1926	0.19	8.3 - 9.1	1.45
IIIM	7.7	1025	0.19	7.3 - 8.1	1.22
IV,V	6.3	625	0.21	5.9 - 6.7	1.09
Usual reason for dental attendance					
Regular check-up	9.1	2400	0.15	8.8 - 9.4	1.36
Occasional check-up	7.3	408	0.30	6.7 - 7.9	1.14
Only with trouble	5.6	1003	0.18	5.3 - 6.0	1.22

Table F.25 **Standard errors and 95% confidence intervals for mean number of decayed or unsound teeth (1998 criteria)**

Dentate adults (exam data)

Base	Mean (p)	Unweighted sample size	Standard error of p	95% confidence intervals		Deft
All	1.5	3817	0.68	1.4 -	1.6	1.83
Age						
16-24	1.6	491	0.19	1.3 -	2.0	1.61
25-34	1.8	854	0.13	1.5 -	2.0	1.57
35-44	1.4	781	0.93	1.2 -	1.6	1.19
45-54	1.4	746	0.11	1.2 -	1.7	1.48
55-64	1.3	461	0.11	1.1 -	1.5	1.20
65 and over	1.2	484	0.11	1.0 -	1.5	0.33
Gender						
Men	1.7	1745	0.09	1.6 -	1.9	1.52
Women	1.3	2072	0.74	1.1 -	1.4	1.67
Country						
England	1.5	2186	0.08	1.3 -	1.6	1.93
Wales	1.2	502	0.14	0.9 -	1.5	0.92
Scotland	1.8	668	0.21	0.4 -	2.2	1.42
Northern Ireland	1.5	461	0.12	1.3 -	1.7	0.61
English region						
North	1.9	617	0.20	1.5 -	2.3	2.58
Midlands	1.4	495	0.19	1.1 -	1.8	2.02
South	1.3	1074	0.08	1.2 -	1.5	1.48
Social class of head of household						
I, II, IIINM	1.2	1926	0.83	1.1 -	1.4	1.81
IIIM	1.7	1025	0.10	1.5 -	1.9	1.32
IV,V	1.9	625	0.15	1.6 -	2.2	1.48
Usual reason for dental attendance						
Regular check-up	1.1	2400	0.06	1.0 -	1.2	1.58
Occasional check-up	1.4	408	0.17	1.0 -	1.7	1.67
Only with trouble	2.3	1003	0.13	2.1 -	2.6	1.44

Table F.26   **Standard errors and 95% confidence intervals for mean number of sound and untreated teeth (1988 criteria)**

Dentate adults (exam data)

Base	Mean (p)	Unweighted sample size	Standard error of p	95% confidence intervals	Deft
All	15.7	3817	0.17	15.4 - 16.0	1.38
Age					
16-24	24.0	491	0.27	23.5 - 24.5	1.49
25-34	19.7	854	0.23	19.2 - 20.1	1.18
35-44	15.7	781	0.24	15.2 - 16.2	1.11
45-54	11.9	746	0.02	11.6 - 12.2	0.89
55-64	9.8	461	0.26	9.3 - 10.3	1.17
65 and over	8.6	484	0.28	8.1 - 9.2	1.19
Gender					
Men	16.1	1745	0.19	15.8 - 16.5	1.12
Women	15.2	2072	0.20	14.8 - 15.6	1.22
Country					
England	15.9	2186	0.19	15.5 - 16.3	1.48
Wales	15.4	502	0.25	14.9 - 15.9	0.48
Scotland	14.1	668	0.23	13.6 - 14.5	0.55
Northern Ireland	14.3	461	0.39	13.6 - 15.1	0.57
English region					
North	15.9	617	0.26	15.4 - 16.4	1.01
Midlands	16.0	495	0.39	15.2 - 16.7	1.35
South	15.8	1074	0.32	15.2 - 16.5	1.76
Social class of head of household					
I, II, IIINM	15.7	1926	0.27	15.2 - 16.3	1.57
IIIM	15.1	1025	0.24	15.6 - 15.6	1.12
IV,V	15.7	625	0.36	15.0 - 16.4	1.20
Usual reason for dental attendance					
Regular check-up	14.9	2400	0.21	14.4 - 15.3	1.42
Occasional check-up	18.0	408	0.46	17.1 - 18.9	1.29
Only with trouble	16.5	1003	0.26	16.0 - 17.0	1.11

Table F.27   **Standard errors and 95% confidence intervals for mean number of restored (otherwise sound) teeth (1988 criteria)**

Dentate adults (exam data)

Base	Mean (p)	Unweighted sample size	Standard error of p	95% confidence intervals		Deft
All	8.1	3817	0.12	7.9 -	8.3	1.35
Age						
16-24	2.9	491	0.15	2.6 -	3.2	1.16
25-34	7.4	854	0.21	7.0 -	7.8	1.36
35-44	10.1	781	0.21	9.7 -	10.5	1.18
45-54	11.1	746	0.24	10.6 -	11.5	1.24
55-64	9.0	461	0.35	8.3 -	9.7	1.32
65 and over	7.7	484	0.39	6.9 -	8.5	1.49
Gender						
Men	7.6	1745	0.13	7.4 -	7.9	1.08
Women	8.6	2072	0.17	8.3 -	8.9	1.32
Country						
England	8.0	2186	0.14	7.7 -	8.3	1.44
Wales	8.1	502	0.23	7.6 -	8.5	0.57
Scotland	8.8	668	0.29	8.2 -	9.4	0.89
Northern Ireland	9.3	461	0.29	8.7 -	9.9	0.54
English region						
North	7.6	617	0.22	7.21 -	8.1	1.2
Midlands	7.7	495	0.35	7.01 -	8.4	1.7
South	8.3	1074	0.19	7.96 -	8.7	1.4
Social class of head of household						
I, II, IIINM	8.9	1926	0.19	8.5 -	9.3	1.45
IIIM	7.9	1025	0.19	7.5 -	8.3	1.21
IV,V	6.6	625	0.21	6.2 -	7.0	1.09
Usual reason for dental attendance						
Regular check-up	9.3	2400	0.15	9.0 -	9.6	1.35
Occasional check-up	7.5	408	0.31	6.9 -	8.1	1.15
Only with trouble	5.8	1003	0.18	5.5 -	6.2	1.22

Table F.28   **Standard errors and 95% confidence intervals for mean number of decayed or unsound teeth (1988 criteria)**

Dentate adults (exam data)

Base	Mean (p)	Unweighted sample size	Standard error of p	95% confidence intervals		Deft
All	1.0	3817	0.40	0.9 -	1.0	1.39
Age						
16-24	0.8	491	0.14	0.6 -	1.1	1.59
25-34	1.0	854	0.07	0.8 -	1.1	1.29
35-44	0.9	781	0.07	0.7 -	1.0	1.21
45-54	1.0	746	0.10	0.8 -	1.2	1.55
55-64	1.0	461	0.09	0.8 -	1.2	1.18
65 and over	1.0	484	0.10	0.8 -	1.2	1.46
Gender						
Men	1.1	1745	0.06	1.0 -	1.2	1.20
Women	0.8	2072	0.05	0.7 -	0.9	1.52
Country						
England	1.0	2186	0.45	0.9 -	1.1	1.44
Wales	0.7	502	0.61	0.5 -	0.8	0.62
Scotland	0.9	668	0.12	0.7 -	1.2	1.26
Northern Ireland	0.8	461	0.07	0.7 -	0.9	0.55
English region						
North	1.0	617	0.08	0.8 -	1.1	1.56
Midlands	0.9	495	0.93	0.7 -	1.1	1.27
South	1.0	1074	0.74	0.9 -	1.2	1.62
Social class of head of household						
I, II, IIINM	0.7	1926	0.05	0.6 -	0.8	1.62
IIIM	1.1	1025	0.69	1.0 -	1.3	1.10
IV,V	1.2	625	0.98	1.0 -	1.4	1.19
Usual reason for dental attendance						
Regular check-up	0.7	2400	0.43	0.6 -	0.8	1.59
Occasional check-up	0.8	408	0.09	0.6 -	0.9	1.23
Only with trouble	1.6	1003	0.08	1.4 -	1.7	1.05

Appendix G   Base Tables

Notes on bases

The following notes refer to bases presented in this appendix and in the main body of the report

1998: all bases are unweighted. This appendix also gives values for the weighted bases.

1988: bases for the total number of people interviewed and examined are unweighted; those for the four countries are weighted for non-response only; those for all other sub-groups are weighted for non-response and to compensate for the proportionately larger samples drawn in Scotland, Wales and Northern Ireland.

1978: bases for the total number of people interviewed and examined in the UK and for sub-groups by country are unweighted; those for all other sub-groups are weighted to compensate for the proportionately larger samples drawn in Scotland and Wales.

1972 Scotland: bases are weighted to compensate for the proportionately larger samples drawn in the Highlands and Islands.

1968 England and Wales: all bases are unweighted

Table G1   **Unweighted bases for all adults, 1998**

All adults   *United Kingdom*

	16-24	25-34	35-44	45-54	55-64	65-74	75 and over	All
All	702	1158	1114	1106	833	758	533	6204
Gender								
Men	331	477	520	520	379	340	199	2766
Women	371	681	594	586	454	418	334	3438
Country								
England	352	704	637	606	450	416	271	3436
Wales	92	130	132	171	117	105	83	830
Scotland	138	198	216	202	168	155	127	1204
Northern Ireland	120	126	129	127	98	82	52	734
English region								
North	102	193	163	165	133	140	79	975
Midlands	73	172	172	168	103	97	71	856
South	177	339	302	273	214	179	121	1605
Social class								
of head of household								
I,II,IIINM	279	521	530	499	345	338	220	2732
IIIM	176	339	305	322	249	220	118	1729
IV,V	133	205	181	192	196	170	144	1221

Table G2   **Unweighted bases for all adults, 1998**

All adults

	England	Wales	Scotland	Northern Ireland	United Kingdom
All	3436	830	1204	734	6204
Gender					
Men	1562	385	521	298	2766
Women	1874	445	683	436	3438
Social class of					
head of household					
I, II, IIINM	1611	343	517	261	2732
IIIM	948	266	307	208	1729
IV,V	628	165	271	157	1221

Table G3   **Unweighted bases for dentate adults (interview data), 1998**

Dentate adults (interview)   *United Kingdom*

	16-24	25-34	35-44	45-54	55-64	65-74	75 and over	All
All	701	1151	1099	1016	645	456	213	5281
Gender								
Men	331	472	513	481	296	224	89	2406
Women	370	679	586	535	349	232	124	2875
Country								
England	352	701	633	572	367	269	116	3010
Wales	92	129	131	148	88	64	30	682
Scotland	137	197	208	176	112	77	46	953
Northern Ireland	120	124	127	120	78	46	21	636
English region								
North	102	191	163	156	100	79	32	823
Midlands	73	171	170	150	76	62	29	731
South	177	339	300	266	191	128	55	1456
Social class **of head of household**								
I,II,IIINM	279	520	526	480	316	240	122	2483
IIIM	176	336	300	287	170	125	37	1431
IV,V	132	203	176	166	124	77	37	915
Usual reason for **dental attendance**								
Regular check-up	330	625	684	644	429	294	115	3121
Occasional check-up	132	145	115	108	37	25	13	575
Only with trouble	238	379	298	262	178	134	83	1572

Table G4   Unweighted bases for dentate adults (examination data), 1998

Dentate adults (exam)   *United Kingdom*

	16-24	25-34	35-44	45-54	55-64	65-74	75 and over	All
All	491	854	781	746	461	344	140	3817
Gender								
Men	227	339	364	356	216	178	65	1745
Women	264	515	417	390	245	166	75	2072
Country								
England	246	517	447	423	259	211	83	2186
Wales	69	90	96	109	73	46	19	502
Scotland	80	157	142	132	75	53	29	668
Northern Ireland	96	90	96	82	54	34	9	461
English region								
North	82	146	126	121	67	56	19	617
Midlands	42	125	114	101	47	46	20	495
South	122	246	207	201	145	109	44	1074
Social class								
of head of household								
I,II,IIINM	215	402	395	394	247	191	82	1926
IIIM	116	254	214	204	115	96	26	1025
IV,V	90	148	121	107	85	49	25	625
Usual reason for								
dental attendance								
Regular check-up	241	479	515	516	325	239	85	2400
Occasional check-up	95	102	81	81	24	18	7	408
Only with trouble	155	271	184	148	112	86	47	1003

Table G5   **Weighted bases for all adults, 1998**

All adults   *United Kingdom*

	16-24	25-34	35-44	45-54	55-64	65-74	75 and over	All
All	565	798	753	686	522	434	353	4110
Gender								
Men	289	406	379	342	257	201	130	2004
Women	276	391	374	344	265	233	223	2106
Country								
England	467	671	630	576	434	360	298	3436
Wales	28	36	35	35	28	24	19	205
Scotland	50	69	67	58	47	39	28	359
Northern Ireland	19	22	20	17	13	11	7	110
English region								
North	133	169	157	149	115	96	70	889
Midlands	102	156	153	144	104	86	77	822
South	233	346	320	283	215	178	150	1725
Social class								
of head of household								
I,II,IIINM	246	375	374	327	220	201	140	1883
IIIM	144	229	202	200	153	130	84	1143
IV,V	102	135	116	107	124	91	98	772

Underlined bases are weighted

Table G6   **Weighted bases for dentate adults (interview data), 1998**

Dentate adults (interview)   *United Kingdom*

	16-24	25-34	35-44	45-54	55-64	65-74	75 and over	All
All	565	793	745	641	419	278	150	3592
Gender								
Men	289	402	376	318	210	139	62	1796
Women	275	391	369	323	209	139	88	1795
Country								
England	467	668	626	544	356	238	130	3029
Wales	28	35	35	30	21	15	7	171
Scotland	50	69	65	51	31	20	10	295
Northern Ireland	19	22	20	16	11	6	3	96
English region								
North	133	167	157	140	86	54	29	766
Midlands	102	154	151	129	77	56	33	703
South	233	346	318	275	192	127	68	1560
Social class of head of household								
I,II,IIINM	246	373	372	317	205	147	79	1740
IIIM	144	227	199	181	111	78	32	973
IV,V	101	134	113	96	81	46	29	600
Usual reason for dental attendance								
Regular check-up	271	420	463	409	282	188	89	2123
Occasional check-up	94	95	79	69	24	13	9	384
Only with trouble	197	275	202	163	111	75	51	1075

Underlined bases are weighted

Table G7   **Weighted bases for dentate adults (examination data), 1998**

Dentate adults (exam)   United Kingdom

	16-24	25-34	35-44	45-54	55-64	65-74	75 and over	All
All	407	586	523	472	292	214	104	2599
Gender								
Men	215	288	262	235	151	110	49	1309
Women	192	298	261	236	142	104	55	1290
Country								
England	336	491	437	402	247	186	92	2190
Wales	23	25	26	21	17	11	4	127
Scotland	31	54	45	37	21	14	7	208
Northern Ireland	17	16	16	11	7	5	1	73
English region								
North	113	126	119	107	55	38	18	575
Midlands	58	115	98	85	47	45	23	470
South	166	250	219	209	146	103	52	1145
Social class								
of head of household								
I,II,IIINM	184	265	256	242	146	112	55	1259
IIIM	91	172	147	130	77	63	25	706
IV,V	78	102	82	64	56	31	21	434
Usual reason for								
dental attendance								
Regular check-up	197	307	331	311	197	149	67	1560
Occasional check-up	69	71	57	55	16	9	6	283
Only with trouble	141	206	135	104	79	55	30	751

Underlined bases are weighted

Table G8   **Bases for all adults from previous surveys in the series**

All adults

	United Kingdom		England		Wales		Scotland			Northern Ireland[†]
	1978	1988	1978	1988	1978	1988	1972	1978	1988	1988
All	5967	6825	3833	3751	580	807	2717	1420	1582	685
Age										
16-24	729	834	605	599	72	104	417	244	248	117
25-34	860	821	709	644	110	134	374	272	297	165
35-44	785	781	647	678	104	154	405	236	272	106
45-54	728	618	598	533	94	136	394	228	233	93
55-64	711	600	584	752	99	117	339	213	233	94
65 and over	821	882	684	741	101	162	360	227	299	110
Gender										
Men	2176	2179	1799	1823	273	108	1070	662	193	56
Women	2463	2357	2034	1966	307	118	1220	758	214	59
Social class of head of household										
I, II, IIINM	1658	2076	1382	1770	202	90	615	502	171	46
IIIM	1561	1433	1294	1184	211	81	869	489	133	34
IV,V	976	860	801	697	114	44	602	321	88	30

	England and Wales			North		Midlands		South	
	1968	1978	1998	1978	1988	1978	1988	1978	1988
All	2932	4075	4556	1105	1042	965	917	1763	1790
Age									
16-24	395	635	703	171	191	146	173	288	326
25-34	515	755	778	218	184	168	166	197	335
35-44	550	690	832	187	181	161	156	299	320
45-54	475	637	667	167	146	152	117	279	252
55-64	494	625	650	173	137	153	130	258	234
65 and over	503	728	914	188	207	182	175	314	359
Gender									
Men	1382	1912	1930	531	501	467	448	800	875
Women	1550	2163	2084	574	545	498	469	963	952
Social class of head of household									
I, II, IIINM	950	1466	1860	341	454	292	332	749	984
IIIM	1074	1382	1266	559	344	381	338	529	503
IV,V	672	848	741	273	212	227	223	301	261

† For Northern Ireland bases for 1979 see the Northern Ireland report (see Chapter 1.1 for reference)
Underlined bases are weighted.

Table G9 **Bases for dentate adults (interview data), 1978-88**

Dentate adults (interview)

	United Kingdom	
	1978	1988
All	4082	5280
Age		
16-24	<u>726</u>	<u>818</u>
25-34	<u>825</u>	<u>800</u>
35-44	<u>684</u>	<u>745</u>
45-54	<u>499</u>	<u>508</u>
55 and over	<u>524</u>	<u>661</u>
Gender		
Men	<u>1624</u>	<u>1824</u>
Women	<u>1639</u>	<u>1758</u>
Social class **of head of household**		
"I,II,IIINM"	<u>1300</u>	<u>1786</u>
IIIM	<u>1108</u>	<u>1090</u>
"IV,V"	<u>605</u>	<u>586</u>
Usual reason for **dental attendance**		
Regular check-up	<u>1404</u>	<u>1764</u>
Occasional check-up	<u>444</u>	<u>513</u>
Only with trouble	<u>1392</u>	<u>1258</u>

Underlined bases are weighted

Table G11 **Bases for dentate adults (examination data) for United Kingdom 1978-88 and England and Wales 1968-88**

Dentate adults (exam)

	United Kingdom		England and Wales		
	1978	1988	1968	1978	1988
All	<u>2784</u>	4331	1694	<u>2486</u>	<u>2649</u>
Age					
16-24	<u>646</u>	<u>706</u>	371	<u>567</u>	<u>622</u>
25-34	<u>742</u>	<u>678</u>	445	<u>641</u>	<u>599</u>
35-44	<u>589</u>	<u>618</u>	396	<u>528</u>	<u>550</u>
45-54	<u>419</u>	<u>417</u>	257	<u>383</u>	<u>379</u>
55 and over	<u>403</u>	<u>552</u>	225	<u>367</u>	<u>499</u>
Gender					
Male	<u>1398</u>	<u>1540</u>	858	<u>1244</u>	<u>1375</u>
Female	<u>1386</u>	<u>1430</u>	836	<u>1242</u>	<u>1274</u>
Social class of **head of household**					
"I, II, IIINM"	<u>1136</u>	<u>1524</u>	n/a	n/a	<u>1374</u>
IIIM	<u>950</u>	<u>890</u>	n/a	n/a	<u>787</u>
"IV,V"	<u>494</u>	<u>468</u>	n/a	n/a	<u>407</u>
Usual reason for **dental attendance**					
Regular check-up	<u>1270</u>	<u>1504</u>	680	<u>1148</u>	<u>1347</u>
Only with trouble	<u>1113</u>	<u>999</u>	817	<u>983</u>	<u>881</u>

Underlined bases are weighted

Table G10 **Bases for dentate adults (interview data) for England and Wales, 1968-88 and Scotland, 1972-88**

Dentate adults (interview)

	England and Wales			Scotland		
	1968	1978	1988	1972	1978	1988
All	1854	<u>2912</u>	<u>3143</u>	<u>1170</u>	872	1145
Age						
16-24	391	<u>634</u>	<u>717</u>	<u>382</u>	239	n/a
25-34	480	<u>729</u>	<u>706</u>	<u>297</u>	246	n/a
35-44	429	<u>610</u>	<u>665</u>	<u>234</u>	173	n/a
45-54	282	<u>454</u>	<u>457</u>	<u>152</u>	105	n/a
55 and over	272	<u>481</u>	<u>598</u>	<u>105</u>	109	n/a

Underlined bases are weighted.

Table G12 **Bases for dentate adults (examination data) usual reason for dental attendance by age, 1988**

Dentate adults (exam)   *United Kingdom*

	Age					
	16-24	25-34	35-44	45-54	55-64	65 and over
Usual reason for dental attendance						
Regular check-up	323	327	364	229	154	108
Only with trouble	220	234	171	137	125	112

Underlined bases are weighted

Table G13 **Bases for dentate adults (examination data) dental status by age, 1988**

Dentate adults (exam)   *United Kingdom*

Age	Dental status		All
	Natural teeth and dentures	Natural teeth only	
16-34	80	1304	1384
35-54	266	769	1035
55 and over	339	213	552

Underlined bases are weighted

Appendix H Adult Dental Health Questionnaire

ADULT DENTAL HEALTH QUESTIONNAIRE

Unless otherwise specified, questions were asked of the group identified at the beginning of each block.

HOUSEHOLD COMPOSITION

All household members

Sex
    (1) Male
    (2) Female

Birth
    What is your date of birth?
    FOR DAY NOT GIVEN....ENTER 15 FOR DAY
    FOR MONTH NOT GIVEN....ENTER 6 FOR MONTH

ASK IF: (Birth = DONTKNOW) OR (Birth = REFUSAL)

AgeIf
    What was your age last birthday?
    98 or more = CODE 97
    0..97

Age
    Age for whole sample, from Birth and AgeIf
    0..120

ASK IF: Age >= 16

MarStat
    ASK OR RECORD CODE FIRST THAT APPLIES
    Are you :
    (1) single, that is, never married
    (2) married and living with your husband/wife
    (3) married and separated from your husband/wife
    (4) divorced
    (5) or widowed?

ASK IF: Age >= 16
AND: DVHSIZE > 1
AND: MarStat <> MarrLiv

LiveWith
    ASK OR RECORD
    May I just check, are you living with someone in the household as a couple?
    (1) Yes
    (2) No
    (3) SPONTANEOUS ONLY - same sex couple

ASK IF: Age >= 16 AND: NOT (DVHSIZE = 1)

Hhldr
    In whose name is the accommodation owned or rented?
    ASK OR RECORD
(1)   This person alone
(3)   This person jointly
(5)   NOT owner/renter

ASK IF: Age >= 16

Natural
    (May I just check) Have you still got some of your natural teeth or have you lost them all?
    (1) Got some
    (2) Lost them all

DVMarDF
    De facto marital status
    (1) Married
    (2) Cohabiting
    (3) Single
    (4) Widowed
    (5) Divorced
    (6) Separated
    (7) Same sex couple

HEAD OF HOUSEHOLD

HoHnum
    ENTER PERSON NUMBER OF HOH.
    1..10

ASK IF: HoHnum = RESPONSE
AND: (QTHComp.QHComp[HoHnum].MarStat = MarrLiv)
OR (QTHComp.QHComp[HoHnum].LiveWith = Yes)
{HOH is married or cohabiting}

HoHprtnr
    THE HoH IS (DMNAMES[HoHnum])
    ENTER THE PERSON NUMBER OF DMNAMES[HoHNum]'s SPOUSE/PARTNER
    NO SPOUSE/PARTNER = 11
    1..11

INTERVIEW

All adults

ISwitch
    THIS IS WHERE YOU START RECORDING ANSWERS FOR INDIVIDUALS. DO YOU WANT TO RECORD ANSWERS FOR NOW OR LATER?

    (1) Yes, now
    (2) Later
    (3) or is there no interview with this person?

ASK IF: ISwitch = YesNow

PersProx
    INTERVIEWER: IS THE INTERVIEW ABOUT BEING GIVEN:
    (1) In person
    (2) or by someone else?

ASK IF: PersProx = ByProxy

ProxyNum
    ENTER PERSON NUMBER OF PERSON GIVING THE INFORMATION
    1..10

NATURAL TEETH

All adults with natural teeth

NatNum
    Adults can have up to 32 natural teeth but over time some people lose some of them.
    How many natural teeth have you got, is it..
    RUNNING PROMPT
    (1) fewer than 10
    (2) between 10 and 19
    (3) or do you have 20 or more natural teeth?

Denture   Do you have (require) a denture, even if you don't wear it?
(1)   Yes
(2)   No

Filling

Do you have any fillings?
(1)   Yes
(2)   No

ASK IF: Filling = Yes

NumFill   How many filled teeth have you got (now) is it...
RUNNING PROMPT
(1)   fewer than 10
(2)   between 10 and 19
(3)   or do you have 20 or more fillings?

ORAL HEALTH IMPACT PROFILE

All adults with natural teeth

RespSC   Just for the next few questions, if you are willing, I would like you to take the computer and put in your answers by yourself. The computer will show you how to tell it what your answers are. I can also help if necessary.
CODE HOW SELF-COMPLETION QUESTIONS ARE ANSWERED, COMING BACK HERE AND RE-CODING IF NECESSARY
(1)   Self-completion accepted and completed
(2)   Self-completion started, but interviewer was asked to take over
(3)   All questions completed by interviewer
(4)   All self-completion questions refused

ASK IF: RespSC = SCAccept

Practice   This is the first time I have used a computer.
(1)   Yes
(2)   No
(3)   Don't want to answer

ASK IF: ((RespSC = SCAccept) OR (RespSC = SCstart)) OR (RespSC = SCHelp)

Words   In the last 12 months, that is, since {DATE} ...
have you had trouble PRONOUNCING ANY WORDS because of problems with your teeth, mouth or dentures?
(1)   ....never
(2)   ....hardly ever
(3)   ....occasionally
(4)   ....fairly often
(5)   .... very often

ASK IF: ((RespSC = SCAccept) OR (RespSC = SCstart)) OR (RespSC = SCHelp)

Taste   In the last 12 months, that is, since {DATE} ...
have you felt that your SENSE OF TASTE has worsened because of problems with your teeth, mouth or dentures?
(1)   ....never

(2)   ....hardly ever
(3)   ....occasionally
(4)   ....fairly often
(5)   .... very often

ASK IF: ((RespSC = SCAccept) OR (RespSC = SCstart)) OR (RespSC = SCHelp)

Aching   In the last 12 months, that is, since {DATE} ...
have you had PAINFUL ACHING in your mouth?
(1)   ....never
(2)   ....hardly ever
(3)   ....occasionally
(4)   ....fairly often
(5)   .... very often

ASK IF: ((RespSC = SCAccept) OR (RespSC = SCstart)) OR (RespSC = SCHelp)

Foods   In the last 12 months, that is, since {DATE} ...
have you found it UNCOMFORTABLE TO EAT ANY FOODS because of problems with your teeth, mouth or dentures?
(1)   ....never
(2)   ....hardly ever
(3)   ....occasionally
(4)   ....fairly often
(5)   .... very often

ASK IF: ((RespSC = SCAccept) OR (RespSC = SCstart)) OR (RespSC = SCHelp)

Self   In the last 12 months, that is, since {DATE} ...
have you been SELF-CONSCIOUS because of your teeth, mouth or dentures?
(1)   ....never
(2)   ....hardly ever
(3)   ....occasionally
(4)   ....fairly often
(5)   .... very often

ASK IF: ((RespSC = SCAccept) OR (RespSC = SCstart)) OR (RespSC = SCHelp)

Tense   In the last 12 months, that is, since {DATE} ...
have you FELT TENSE because of problems with your teeth, mouth or dentures?
(1)   ....never
(2)   ....hardly ever
(3)   ....occasionally
(4)   ....fairly often
(5)   .... very often

ASK IF: ((RespSC = SCAccept) OR (RespSC = SCstart)) OR (RespSC = SCHelp)

Diet   In the last 12 months, that is, since {DATE} ...
has your DIET BEEN UNSATISFACTORY because of problems with your teeth, mouth or dentures?
(1)   ....never
(2)   ....hardly ever
(3)   ....occasionally
(4)   ....fairly often
(5)   .... very often

ASK IF: ((RespSC = SCAccept) OR (RespSC = SCstart)) OR (RespSC = SCHelp)

Meals   In the last 12 months, that is, since {DATE} ... have you had to INTERRUPT MEALS because of problems with your teeth, mouth or dentures?
- (1) ....never
- (2) ....hardly ever
- (3) ....occasionally
- (4) ....fairly often
- (5) .... very often

ASK IF: ((RespSC = SCAccept) OR (RespSC = SCstart)) OR (RespSC = SCHelp)

Relax   In the last 12 months, that is, since {DATE} ... have you found it DIFFICULT TO RELAX because of problems with your teeth, mouth or dentures?
- (1) ....never
- (2) ....hardly ever
- (3) ....occasionally
- (4) ....fairly often
- (5) .... very often

ASK IF: ((RespSC = SCAccept) OR (RespSC = SCstart)) OR (RespSC = SCHelp)

Embarass   In the last 12 months, that is, since {DATE} ... have you been a bit EMBARRASSED because of problems with your teeth, mouth or dentures?
- (1) ....never
- (2) ....hardly ever
- (3) ....occasionally
- (4) ....fairly often
- (5) .... very often

ASK IF: ((RespSC = SCAccept) OR (RespSC = SCstart)) OR (RespSC = SCHelp)

Irritabl   In the last 12 months, that is, since {DATE} ... have you been a bit IRRITABLE WITH OTHER PEOPLE because of problems with your teeth, mouth or dentures?
- (1) ....never
- (2) ....hardly ever
- (3) ....occasionally
- (4) ....fairly often
- (5) .... very often

ASK IF: ((RespSC = SCAccept) OR (RespSC = SCstart)) OR (RespSC = SCHelp)

Jobs   In the last 12 months, that is, since {DATE} ... have you had DIFFICULTY DOING YOUR USUAL JOBS because of problems with your teeth, mouth or dentures?
- (1) ....never
- (2) ....hardly ever
- (3) ....occasionally
- (4) ....fairly often
- (5) .... very often

ASK IF: ((RespSC = SCAccept) OR (RespSC = SCstart)) OR (RespSC = SCHelp)

Less   In the last 12 months, that is, since {DATE} ... have you felt that life in general was LESS SATISFYING because of problems with your teeth, mouth or dentures?
- (1) ....never
- (2) ....hardly ever
- (3) ....occasionally
- (4) ....fairly often
- (5) .... very often

ASK IF: ((RespSC = SCAccept) OR (RespSC = SCstart)) OR (RespSC = SCHelp)

Functin   In the last 12 months, that is, since {DATE} ... have you been TOTALLY UNABLE TO FUNCTION because of problems with your teeth, mouth or dentures?
- (1) ....never
- (2) ....hardly ever
- (3) ....occasionally
- (4) ....fairly often
- (5) .... very often

PRESENT STATE OF TEETH

Adults with natural teeth

Treat   If you went to the dentist tomorrow do you think you would need any treatment or not?
- (1) Need treament
- (2) Not
- (3) Don't know

DENTURES

Adults with no natural teeth or adults with natural teeth and dentures

(Natural = No) OR (Denture = Yes)

ASK IF: Natural = Yes

Upper   Do you have a denture in your upper jaw?
- (1) Yes
- (2) No

ASK IF: Upper = Yes

UFull   Is the denture in your upper jaw full or partial?
- (1) Full
- (2) Partial

ULong   (You said earlier that you have no natural teeth. May I just check)
How long ago did you get your present top plate?
PROMPT AS NECESSARY
- (1) Less than a year
- (2) 1 year less than 2 years
- (3) 2 years less than 5 years
- (4) 5 years less than 10 years
- (5) 10 years less than 20 years

(6) 20 yrs or more

(7) Can't remember/don't know

ASK IF: Natural = No

UNHSPriv   Did you get your top plate through the National Health Service or did you get it privately?

(1) N.H.S.

(2) Privately

(3) Before N.H.S.

(4) Can't remember, don't know

ASK IF: Natural = Yes

Lower   Do you have a denture in your lower jaw?

(1) Yes

(2) No

ASK IF: Lower = Yes

LFull   Is the denture in your lower jaw full or partial?

(1) Full

(2) Partial

ASK IF: (Natural = No) OR (Lower = Yes)

LLong   How long ago did you get your present bottom plate?

PROMPT AS NECESSARY

(1) Less than a year

(2) 1 year less than 2 years

(3) 2 years less than 5 years

(4) 5 years less than 10 years

(5) 10 years less than 20 years

(6) 20 yrs or more

(7) Can't remember/don't know

ASK IF: Natural = No

LNHSPriv   Did you get your bottom plate through the National Health Service or did you get it privately?

(1) N.H.S.

(2) Privately

(3) Before N.H.S.

(4) Can't remember, don't know

ASK IF: (Lower = Yes)) OR (Natural = No)

LWorn   Have you worn your bottom plate at all in the last 4 weeks?

(1) Worn

(2) Not worn

ASK IF: LWorn = Yes

LNight   Do you usually keep your bottom plate in at night?

(1) In at night

(2) Not

ASK IF: LWorn = Yes

LDay   ATTITUDE TO DENTURES

Do you wear your bottom plate from the time you get up to when you go to bed?

(1) All daytime

(2) Not

ASK IF: LDay = No

LOut   Do you usually wear your bottom plate when you go out?

(1) Yes

(2) No

ASK IF: LDay = No

LEat   Do you usually wear your bottom plate when you are eating?

(1) Yes

(2) No

ASK IF: LDay = No

LHouse   Do you usually wear your bottom plate when you are about the house?

(1) Yes

(2) No

ASK IF: (Upper = Yes) OR (Natural = No)

UWorn   Have you worn your top plate at all in the last 4 weeks?

(1) Worn

(2) Not worn

ASK IF: UWorn = Yes

UNight   Do you usually keep your top plate in at night?

(1) In at night

(2) Not

ASK IF: UWorn = Yes

UDay   Do you wear your top plate from the time you get up to when you go to bed?

(1) All daytime

(2) Not

ASK IF: UDay = No

UOut   Do you usually wear your top plate when you go out?

(1) Yes

(2) No

ASK IF: UDay = No

UEat   Do you usually wear your top plate when you are eating?

(1) Yes

(2) No

ASK IF: UDay = No

UHouse   Do you usually wear your top plate when you are about the house?

(1) Yes

(2) No

SpTroub   Do (would) you have any trouble speaking clearly with your dentures?

(1) Yes

(2) No

ASK IF: SpTroub = Yes

WSpTroub   [*]What kind of trouble do (would) you have with speaking?
SET [4] OF
(1) Slurs/alters speech
(2) Loose/slips while talking
(3) Teeth rattle
(4) Teeth whistle
(5) Some other kind of trouble

ASK IF: Oth IN WSpTroub

WSpTrouO   INTERVIEWER - RECORD OTHER KINDS OF TROUBLE
STRING[100]

EaTTroub   Do (would) you have any discomfort or trouble with your dentures when you are eating or drinking?
(1) Yes
(2) No

ASK IF: EaTTroub = Yes

WEtTroub   [*]What other kind of discomfort or trouble do (would) you have with eating or drinking?
SET [4] OF
(1) Food gets stuck underneath plate
(2) Slips/loose while eating
(3) Uncomfortable/sore gums/plate rubs
(4) Get ulcers
(5) Can't chew/bite well
(6) Denture worn down
(7) Problems because of medical disorder
(8) Some other kind of trouble

ASK IF: Oth IN WEtTroub

WEtTrouO   INTERVIEWER - RECORD OTHER KINDS OF TROUBLE
STRING[100]

Probs   Do (would) you have any other problems with your dentures?
(1) Yes
(2) No

ASK IF: Probs = Yes

WProbs   [*]What other problems do (would) you have?
CODE ALL THAT APPLY
SET [4] OF
(1) Dentures move, are loose, don't meet, don't fit properly
(2) uncomfortable/sore gums/plate rubs
(3) get ulcers
(4) can't chew/bite well
(5) denture is worn down
(6) get food under plate
(7) individual has a medical disorder which causes problems with dentures
(8) other

ASK IF: Other IN WProbs

WWprobs   INTERVIEWER - RECORD OTHER PROBLEMS.
STRING[100]

ASK IF: DVDProb = 1

Dentist   Are you planning to visit the dentist to see about your dentures?
(1) Planning to visit
(2) Not

APPEARANCE OF TEETH

All adults

Appear   (Thinking about both your natural teeth and your dentures) In general, how do you feel about the appearance of your teeth (and/or dentures), are you satisfied or not satisfied with the way they look?
(1) Satisfied
(2) Not satisfied

ASK IF: Natural = Yes
AND: Appear = unsatisfied

NSatis   [*]What is it about the way your teeth or dentures look that makes you not satisfied?
CODE ALL THAT APPLY
SET [4] OF
(1) Crooked/slanting/protruding/irregular teeth
(2) gaps/space between teeth/teeth missing
(3) size of teeth/shape of teeth
(4) broken/chipped teeth
(5) colour of teeth - stained teeth/need cleaning
(6) fillings/colour of fillings
(7) decayed teeth/bad teeth/teeth going black
(8) They need filling
(9) any other

ASK IF: Other IN NSatis

NNSatis   INTERVIEWER - RECORD OTHER REASONS
STRING[200]

ASK IF:: Natural = Yes

BAche   If you went to the dentist with an aching back tooth would you prefer the dentist to take it out or fill it (supposing it could be filled)?
(1) Take it out
(2) Fill it

ASK IF:: Natural = Yes

FCrown   If the dentist said a front tooth would have to be extracted (taken out) or crowned, what would you prefer?
(1) Extracted
(2) Crowned

ASK IF: Natural = Yes

BCrown   If the dentist said a back tooth would have to be extracted (taken out) or crowned, what would you prefer?
(1) Extracted
(2) Crowned

ASK IF: Natural = Yes

BMiss   If you had several missing teeth at the back would you prefer to have a partial denture or manage without?
(1) Back partial denture
(2) Manage without

TOOTH CARE AND ATTENDING THE DENTIST

Adults with natural teeth

Cleaner   Now I'd like to talk a little about cleaning your teeth. How often do you clean your teeth nowadays?
(1) Never
(2) Less than once a day
(3) Once a day
(4) Twice a day
(5) More than twice a day
(6) other

ASK IF: Cleaner = other

CleanerO   INTERVIEWER - RECORD HOW OFTEN RESPONDENT CLEANS THEIR TEETH
STRING[100]

ASK IF: (((Cleaner = once)
OR (Cleaner = less))
OR (Cleaner = twice))
OR (Cleaner = more)

TimDay   At what time of day do you clean your teeth?
IF 'MORNING & DOESNT HAVE BREAKFAST'
CODE 8
CODE ALL THAT APPLY
SET [9] OF
(1) Before breakfast
(2) After breakfast
(3) Midday
(4) Tea time
(5) After evening meal
(6) When going out
(7) Last thing at night
(8) In morning (but no breakfast)
(9) Other times

ASK IF: other IN TimDay

TimDayO   INTERVIEWER - RECORD OTHER TIMES
STRING[120]

ASK IF: (((Cleaner = once)
OR (Cleaner = less))

OR (Cleaner = twice))
OR (Cleaner = more)

TPaste   Nowadays there are more things available in chemist shops to help with dental hygiene. Do you use anything other than an ordinary toothbrush and toothpaste for dental hygiene purposes?
(1) Yes
(2) No

ASK IF: AND: TPaste = Yes

SpPaste   What do you use?
CODE ALL THAT APPLY
SET [9] OF
(1) Dental floss
(2) Interdens/toothpicks/woodsticks
(3) Mouthwash
(4) Interspace brush
(5) Electric toothbrush
(6) Dental disclosing tablets
(7) dental chewing gum
(8) Sensodyne or smokers' toothpaste
(9) Something else

ASK IF: other IN SpPaste

SpPasteO   INTERVIEWER
PLEASE SPECIFY OTHER THINGS USED
STRING[50]

Brush   Has a dentist or any of the dental staff demonstrated to you how to clean your teeth?
(1) Demonstrated
(2) not

Care   Has a dentist or any of the dental staff given you advice on caring for your gums?
(1) Given gum advice
(2) not

Work   Some people go to a dentist near to their work. Can I just check, are you working at the moment?
(1) Yes
(2) No

ASK IF: Work = Yes

TOff   Do you usually take time off work when you go to the dentist?
(1) Yes
(2) No

ASK IF: TOff = Yes

Time   How much work time does a dental visit usually take?
(in hours)
(1) Under 1 hour
(2) 1 hour but less than 2
(3) 2 hours but less than 3
(4) 3 hours but less than 4
(5) or 4 hours or more?

ASK IF: Work = No

Study   Are you studying at the moment?
         (1)   Yes
         (2)   No

ASK IF: (Work = Yes) OR (Study = Yes)

Near   Is the dental practice (you went to last time) nearer to your home or to your work/college?
         (1)   Nearer to home
         (2)   Nearer to work or college
         (3)   The same

Far   How far is the dental practice from home/work/college? (whichever is nearest)
         (1)   Up to half a mile
         (2)   More than half up to 1 mile
         (3)   More than 1 up to 2 miles
         (4)   More than 2 up to 5 miles
         (5)   More than 5 up to 10 miles
         (6)   More than 10 up to 20 miles
         (7)   More than 20 up to 30 miles
         (8)   More than 30 miles
         (9)   Other

ASK IF: DVHSIZE > 1

AllSame   Do all people in this household (that is ...[NAMES] ...) go to the same dental practice as you or not?
         (1)   All go to the same practice
         (2)   Do not

Regular   In general do you go to the dentist for...
         RUNNING PROMPT
         (1)   a regular check up
         (2)   an occasional check up
         (3)   or only when you're having trouble with your teeth?

Attend   Would you say that nowadays you go to the dentist more often, about the same or less often than you did 5 years ago?
         (1)   more often
         (2)   about the same
         (3)   less often

ASK IF: Attend = more

MOften   [*]What has made you go more often?
         CODE ALL THAT APPLY
         SET [4] OF
         (1)   Have (more) trouble with teeth(or gums)/neglected teeth so need more treatment
         (2)   more dentally aware/want to keep my teeth
         (3)   less frightened of dentist/treatment now
         (4)   have good dentist now
         (5)   more incentive to go
         (6)   Other

ASK IF: Other IN MOften

MMOften   INTERVIEWER - RECORD OTHER REASONS
         STRING[200]

ASK IF: Attend = Less

LOften   [*]What has made you go less often?
         CODE ALL THAT APPLY
         SET [9] OF
         (1)   Have no/less trouble with my teeth
         (2)   have fewer teeth left
         (3)   can't be bothered/got out of the habit/don't know/no reason
         (4)   scared/afraid of dentists/scared of injections
         (5)   lack of time/would go if surgery opened at different hours
         (6)   difficult to get to dentist/make the journey
         (7)   too expensive
         (8)   don't have a regular dentist/dentist retired
         (9)   other

ASK IF: Other IN LOften

LLOften   INTERVIEWER - RECORD OTHER REASONS
         STRING[200]

OPINIONS ABOUT VISITING THE DENTIST

Adults with natural teeth

Intros   INTERVIEWER -Introduce
         In the past we've talked to some people who don't like going to the dentist. We asked them what they didn't like and what changes they would suggest. Now I'd like to ask you whether you feel the same or not about the things they said.
         INTERVIEWER - SHUFFLE CARDS AND LAY OUT BASE CARDS
         Would you read these cards and then place them on the base card that is closest to your view, that is whether you definitely feel like that, feel like that to some extent or don't feel like that?
         INTERVIEWER - PRESS <ENTER> TO CONTINUE
         STRING[1]

Defo   INTERVIEWER - PLEASE COLLECT ALL THE CARDS LAID ON TOP OF THE STATEMENT 'DEFINITELY FEEL LIKE THAT'
         AND KEY IN THE NUMBERS ON THE BACKS OF THE CARDS.
         THEN PRESS <ENTER>
         REMEMBER TO KEEP THESE CARDS SEPERATE FOR LATER USE
         SET [11] OF
         (1)   I always feel anxious about going to the dentist
         (2)   I'm nervous of some kinds of dental treatment
         (3)   I don't see any point in visiting the dentist unless I need to
         (4)   The worst part of going to the dentist is the waiting
         (5)   I'd like to be able to drop in at the dentist without an appointment
         (6)   I'd like to know more about what the dentist is going to do and why
         (7)   I find NHS dental treatment expensive
         (8)   I'd like to be able to pay my dental treatment by instalments
         (9)   I'd like to be given an estimate without commitment

(10) It will cost me less in the long run if I only go to the dentist when I have trouble with my teeth
(11) I don't want fancy (intricate) dental treatment
(12) No Cards

Xtent INTERVIEWER - PLEASE COLLECT ALL THE CARDS LAID ON TOP OF THE STATEMENT 'TO SOME EXTENT' AND KEY IN THE NUMBERS ON THE BACK OF THE CARDS.
THEN PRESS <ENTER>
SET [11] OF
(1) I always feel anxious about going to the dentist
(2) I'm nervous of some kinds of dental treatment
(3) I don't see any point in visiting the dentist unless I need to
(4) The worst part of going to the dentist is the waiting
(5) I'd like to be able to drop in at the dentist without an appointment
(6) I'd like to know more about what the dentist is going to do and why
(7) I find NHS dental treatment expensive
(8) I'd like to be able to pay my dental treatment by instalments
(9) I'd like to be given an estimate without commitment
(10) It will cost me less in the long run if I only go to the dentist when I have trouble with my teeth
(11) I don't want fancy (intricate) dental treatment
(12) No Cards

NotFeel INTERVIEWER - PLEASE COLLECT ALL THE CARDS LAID ON TOP OF THE STATEMENT 'DON'T FEEL LIKE THAT' AND KEY IN THE NUMBERS ON THE BACK OF THE CARDS.
THEN PRESS <ENTER>
SET [11] OF
(1) I always feel anxious about going to the dentist
(2) I'm nervous of some kinds of dental treatment
(3) I don't see any point in visiting the dentist unless I need to
(4) The worst part of going to the dentist is the waiting
(5) I'd like to be able to drop in at the dentist without an appointment
(6) I'd like to know more about what the dentist is going to do and why
(7) I find NHS dental treatment expensive
(8) I'd like to be able to pay my dental treatment by instalments
(9) I'd like to be given an estimate without commitment
(10) It will cost me less in the long run if I only go to the dentist when I have trouble with my teeth
(11) I don't want fancy (intricate) dental treatment
(12) No Cards

DKnow INTERVIEWER - PLEASE COLLECT ALL THE CARDS LAID ON TOP OF THE STATEMENT 'DON'T KNOW' AND KEY IN THE NUMBERS ON THE BACK OF THE CARDS.
THEN PRESS <ENTER>
SET [11] OF
(1) I always feel anxious about going to the dentist
(2) I'm nervous of some kinds of dental treatment

(3) I don't see any point in visiting the dentist unless I need to
(4) The worst part of going to the dentist is the waiting
(5) I'd like to be able to drop in at the dentist without an appointment
(6) I'd like to know more about what the dentist is going to do and why
(7) I find NHS dental treatment expensive
(8) I'd like to be able to pay my dental treatment by instalments
(9) I'd like to be given an estimate without commitment
(10) It will cost me less in the long run if I only go to the dentist when I have trouble with my teeth
(11) I don't want fancy (intricate) dental treatment
(12) No Cards

Rank INTERVIEWER - PLEASE COLLECT ALL THE CARDS FROM THE STATEMENT 'DEFINITELY FEEL LIKE THAT' AND KEY IN THE NUMBERS ON THE BACK OF THE CARDS IN THE RESPONDENT'S RANKING ORDER - STARTING WITH THE MOST IMPORTANT
THEN PRESS <ENTER>
SET [8] OF
(1) I always feel anxious about going to the dentist
(2) I'm nervous of some kinds of dental treatment
(3) I don't see any point in visiting the dentist unless I need to
(4) The worst part of going to the dentist is the waiting
(5) I'd like to be able to drop in at the dentist without an appointment
(6) I'd like to know more about what the dentist is going to do and why
(7) I find NHS dental treatment expensive
(8) I'd like to be able to pay my dental treatment by instalments
(9) I'd like to be given an estimate without commitment
(10) It will cost me less in the long run if I only go to the dentist when I have trouble with my teeth
(11) I don't want fancy (intricate) dental treatment
(12) No Cards

HISTORY OF TOOTH LOSS

Adults with no natural teeth

Loss I'd like to talk to you about when you lost the last of your natural teeth.
How many years ago did you lose the last of your natural teeth?
PROMPT AS NECESSARY
(1) Up to 5 years ago
(2) Over 5 up to 10 years ago
(3) Over 10 up to 15 years ago
(4) Over 15 up to 20 years ago
(5) Over 20 up to 30 years ago
(6) Over 30 years ago
(7) Can't remember/Don't know

Ageloss   How old were you when you lost the last of your natural teeth?
ENTER YEARS
1..97

Numloss   When you lost the last of your own teeth how many teeth were there to come out altogether?
IN WHOLE COURSE OF TREATMENT
PROMPT AS NECESSARY
(1)   1 to 11
(2)   12 to 20
(3)   21 or more

Whyloss   [*]Why did you lose the last of your natural teeth, was it because...
CODE ALL THAT APPLY
RUNNING PROMPT
SET [3] OF
(1)   the teeth were decayed
(2)   the gums were bad
(3)   or was it for some other reason?

ASK IF: other IN Whyloss

WhyLossO   INTERVIEWER - RECORD OTHER REASONS
SET [4] OF
(4)   Decided (dentist advised) to remove last few good teeth
(5)   Partial dentures caused problems
(6)   Illness/accident
(7)   Teeth could not be filled/restored
(8)   Toothache (all other mentions)
(9)   Other reasons

ASK IF: Oth IN WhyLossO

WhyLosOO   INTERVIEWER RECORD OTHER REASONS
STRING[50]

Upset   [*]Did you find losing the last of your natural teeth and having full dentures..
RUNNING PROMPT
(1)   very upsetting
(2)   a little upsetting
(3)   or not at all upsetting?
(4)   never had full dentures

ASK IF: ((Upset = very)
OR (Upset = little))
OR (Upset = ornot)

Suggest   Did you suggest to the dentist that the last of your natural teeth should come out or did the dentist suggest it to you?
(1)   Respondent suggested it
(2)   Dentist suggested it
(3)   Teeth fell out of own accord
(4)   Doctor/nurse/other medical person suggested it
(5)   Hospital suggested it
(6)   Other

ASK IF: Suggest = other

SuggestO   INTERVIEWER - RECORD OTHER REASONS
STRING[50]

Regular2   When you had your own (natural) teeth did you go to the dentist for..
RUNNING PROMPT
(1)   a regular check-up
(2)   an occasional check-up
(3)   or only when you were having trouble with your teeth?

Denbef   Some people have some false teeth (partial dentures) before they lose all of their own teeth. When you lost the last of your own teeth did you previously have any dentures?
(1)   Previously had dentures
(2)   Did not

ASK IF: Denbef = prev

Agefalse   How old were you when you first had some false teeth?
ENTER YEARS
IF NEVER HAD FALSE TEETH, CODE 1
1..97

ASK IF:: Denbef = prev

Whereden   Just before you lost the last of your own teeth did you have a denture in your upper jaw?
(1)   Yes
(2)   No

ASK IF: Denbef = prev

Lower2   Did you have a denture in your lower jaw?
(1)   Yes
(2)   No

ASK IF: Whereden = Yes

UFull2   Was the denture in your upper jaw full or partial?
(1)   Full
(2)   Partial

ASK IF: Lower2 = Yes

LFull2   Was the denture in your lower jaw full or partial?
(1)   Full
(2)   Partial

ASK IF: (Loss = nto5)
OR (Loss = n5to10n)

Prbs   What kind of dental problems did you have in the five years before you lost the last of your natural teeth?
SET [4] OF
(1)   Gum decay/teeth loose
(2)   Toothache/abcess
(3)   Broken/decaying teeth
(4)   No real problems (incl cosmetic)
(5)   Other

ASK IF: (Loss = nto5) OR (Loss = n5to10n)

Expectls   [*]Did you expect to lose your teeth around then or were you surprised to lose them at that age?
(1)   Expected to lose
(2)   Surprised at that age
(3)   Other

ASK IF: Expectls = other

ExpectLO INTERVIEWER - RECORD OTHER REASONS
STRING[200]

ASK IF: (Loss = nto5) OR (Loss = n5to10n)

Used When you first had full dentures about how long did it take you to get used to having them?
(1) No time at all
(2) Up to a week
(3) Over a week, up to a month
(4) Over a month, up to 3 months
(5) Never had FULL dentures
(6) Never got used to them
(7) Other

ASK IF: Used = other

UsedO INTERVIEWER - RECORD HOW LONG IT TOOK TO GET USED TO THE DENTURES
STRING[200]

ASK IF: AND: Used <> NevFull

Food Since you've had full dentures have you had to change the kind of food you eat?
(1) Had to change
(2) Not

ASK IF: Food = Had

Caneat What can you eat now that you couldn't eat before?
(1) Nothing
(2) Other answers

ASK IF: Caneat = Oth

CaneatO INTERVIEWER - RECORD ANSWERS TO PREVIOUS QUESTION (What can you eat now that you couldn't eat before?)
STRING[200]

ASK IF: Food = Had

Canteat What could you eat before that you can't eat now?
(1) Nothing
(2) Other answers

ASK IF: Canteat = Oth

CanteatO INTERVIEWER - RECORD ANSWERS TO PREVIOUS QUESTION (What could you eat before that you can't eat now?)
STRING[200]

ASK IF: Used <> NevFull

Advice [*]If you knew someone who thought that they might soon have the rest of their teeth out and full dentures fitted what advice would you give them?
SET [4] OF
(1) If they are bad (don't hesitate) have them out
(2) Keep natural teeth as long as possible/Dentures should only be a last resort
(3) See/get advice from a good dentist

(4) Wait a while between having natural teeth out and the dentures put in
(5) Should have new dentures put in as soon as possible after having natural teeth out
(6) After dentures are fitted, keep them in/persevere with them
(7) It's nothing to worry about
(8) Wouldn't presume to give advice/not qualified to give advice
(9) Other

ASK IF: Other IN Advice

AdviceO INTERVIEWER - RECORD OTHER ADVICE
STRING[200]

TREATMENT HISTORY

All adults

Xray I'd like to talk next about the kind of dental treatment you've had over the whole of your lifetime right from the first time you went to the dentist, including when you were a child.
Have you ever had an x-ray of your teeth?
(1) Yes
(2) No
(3) Don't know

ASK IF: (Natural = No) OR (Filling = No)

EverFil Have you ever had any fillings?
(1) Yes
(2) No
(3) Don't know

ASK IF: (Filling = Yes) OR (EverFil = Yes)

InjGum Have you ever had an injection in the gum for a filling?
(1) Yes
(2) No
(3) Don't know

ASK IF: (Filling = Yes) OR (EverFil = Yes)

InjArm Have you ever had an injection in the arm for a filling?
(1) Yes
(2) No
(3) Don't know

Wisdom Have you ever had any wisdom teeth removed?
(1) Yes
(2) No
(3) Don't know

ASK IF: Wisdom = Yes

Remove Were your wisdom teeth removed in hospital or at the dentist's?
(1) Hospital
(2) Dentist
(3) Both

ASK IF: Wisdom = Yes

Where   Were the wisdom teeth that were removed..
CODE ALL THAT APPLY
RUNNING PROMPT
SET [4] OF
(1)   fully through the gums
(2)   part way through the gum
(3)   or still underneath the gum?
(4)   Don't know

Extract   Have you ever had any other teeth extracted (taken out)?
(1)   Yes
(2)   No
(3)   Don't know

ASK IF: Extract = Yes

ExArm   Have you ever had an injection in the arm for an extraction?
(1)   Yes
(2)   No
(3)   Don't know

ASK IF: Extract = Yes

ExGum   Have you ever had an injection in the gum for an extraction?
(1)   Yes
(2)   No
(3)   Don't know

ASK IF: Extract = Yes

ExGas   Have you ever had gas for extractions?
(1)   Yes
(2)   No
(3)   Don't know

Fluoride   Have you ever had fluoride treatment or fissure sealants?
(1)   Yes
(2)   No
(3)   Don't know

Brace   Have you ever had a brace to straighten your teeth?
(1)   Yes
(2)   No
(3)   Don't know

AbTreat   Have you ever had an abcess treated?
(1)   Yes
(2)   No
(3)   Don't know

Nerve   Have you ever had a nerve removed?
(1)   Yes
(2)   No
(3)   Don't know

TCrown   Have you ever had a tooth crowned?
(1)   Yes
(2)   No
(3)   Don't know

Bridge   Have you ever had a dental bridge?
(1)   Yes
(2)   No
(3)   Don't know

Polish   Have you ever had a scale and polish by the dentist?
(1)   Yes
(2)   No
(3)   Don't know

Hygiene   Have you ever had treatment from a dental hygienist?
(1)   Yes
(2)   No
(3)   Don't know

ASK IF: Natural = No

Part   Have you ever had a partial denture fitted?
(1)   Yes
(2)   No

ASK IF: Part = Yes

Rep   Have you ever had a partial denture repaired?
(1)   Yes
(2)   No

ASK IF: Natural = No

Full   Have you ever had a full denture fitted?
(1)   Yes
(2)   No

ASK IF: Full = Yes

FRep   Have you ever had a full denture repaired?
(1)   Yes
(2)   No

ASK IF: (Natural = Yes)
AND (Denture = No)

Denfit   Have you ever had a denture fitted?
(1)   Yes
(2)   No
(3)   Don't know

ASK IF: (Denfit = Yes)
OR (Denture = Yes)

Denrep   Have you ever had a denture repaired?
(1)   Yes
(2)   No
(3)   Don't know

ASK IF: (Denfit = Yes)
OR (Denture = Yes)

Hlpeat   Was your first denture mainly for the sake of appearance or mainly to help you eat?
(1)   For appearance
(2)   For eating
(3)   Both

ASK IF: Denture = No
AND: Denfit = No

Partup   [*]You said you have never had a denture, but when people lose some of their natural teeth they may need a denture.
Do you find the thought of having a partial denture to replace some of your teeth..
RUNNING PROMPT
(1) very upsetting
(2) a little upsetting
(3) or not at all upsetting?

ASK IF: Denture = No
AND: Denfit = No

Partfive   [*]During the next five years some people will have a partial denture for the first time. Do you think you are likely or unlikely to have a partial denture within the next five years?
(1) Likely to
(2) Unlikely to
(3) Don't know

ASK IF: Denture = No

Fullup   [*]Many people eventually have all their natural teeth out and have full dentures. Do you find the thought of losing all your own teeth and having full dentures..
RUNNING PROMPT
(1) very upsetting
(2) a little upsetting
(3) or not at all upsetting?

ASK IF:: Denture = No

Fullfive   [*]During the next five years some people will have full dentures for the first time. Do you think you are likely or unlikely to have full dentures within the next five years?
(1) Likely to
(2) Unlikely to
(3) Don't know

ASK IF: Fullfive = Unlikely

Keepall   [*]Do you think that at sometime, you will have full dentures or do you think you will always keep some of your natural teeth?
(1) Full dentures sometime
(2) Always keep natural teeth
(3) Don't know

Othtreat   Have you ever had any other kind of dental treatment?
(1) Yes
(2) No
(3) Don't know

ASK IF: Othtreat = Yes

WOThTr   What other treatment have you had?
CODE ALL THAT APPLY
SET [3] OF
(1) Gum treatment
(2) Root treatment
(3) Some other kind of treatment

ASK IF: Oth IN WOThTr

OthtreatO   INTERVIEWER - RECORD OTHER TREATMENT
STRING[100]

HISTORY OF TOOTH LOSS

Have natural teeth and wear dentures or have worn dentures in the past

(Denture = Yes) OR (Denfit = Yes)

DenAge   How old were you when you first had a denture?
1..97

DenFirst   Was your first denture for the upper jaw, the lower jaw or for both jaws?
(1) For the upper jaw
(2) For the lower jaw
(3) For both jaws

NatLoss   Since you had your first denture how many more of your natural teeth have you lost (had out)?
0..30

ATTENDANCE AT DENTIST

All adults

ASK IF: (Natural = Yes)
AND (Regular = reg)

YReg   Has there ever been a time in your life when you have not been for a regular dental check up?
(1) Has not always been regularly
(2) Always been regularly

ASK IF: YReg = No

Whyno   [*]Why did you not go regularly at that time?
CODE ALL THAT APPLY
SET [4] OF
(1) Scared/afraid of dentist/don't like the thought of going to the dentist
(2) have no trouble with teeth
(3) lack of time/would go if surgery opened at different hours
(4) apathy/laziness/don't bother/don't think about it/ no reason
(5) afraid the dentist will find trouble or will cause damage
(6) not worth it now
(7) difficult to get an appointment/can't get an appointment for even a check-up
(8) difficult to get to the dentist/make the journey
(9) Too expensive
(10) other

ASK IF: Other IN Whyno

WWhyno   INTERVIEWER - RECORD OTHER REASONS
STRING[120]

ASK IF: ((Natural = Yes)
AND (Regular = occ))
OR ((Natural = Yes)
AND (Regular = only))

NReg   Has there ever been any time in your life when you have been for regular dental checks?
      (1)   Used to go for regular checks
      (2)   Never been for regular checks

ASK IF: NReg = Yes

WhyYes   [*]Why did you go regularly at that time?
      CODE ALL THAT APPLY
      SET [4] OF
      (1)   Had no choice
      (2)   wanted to look after my teeth
      (3)   dentist sent reminder card
      (4)   had more time/fewer commitments
      (5)   needed/wanted/was receiving treatment
      (6)   treatment was free
      (7)   was taking the family
      (8)   other

ASK IF: AND: Other IN WhyYes

WWhyYes   INTERVIEWER - RECORD OTHER REASONS
      STRING[200]

ASK IF: AND: NReg = No

WhyNev   [*]Why is it that you have never been for regular checks?
      STRING[200]

NumCheck How many times have you been to the dentist in the last five years purely for a check up?
      FIVE YEARS AGO = DMD5YR
      0..97

NumTroub How many times have you been to the dentist in the last five years because you've had trouble?
      FIVE YEARS AGO = DMD5YR
      0..97

ASK IF: (NumCheck > 0) OR (NumTroub > 0)

YearDen   Have you been to the dentist in the last 12 months, that is since {DATE}?
      (1)   Yes
      (2)   No

ASK IF: YearDen = Yes

Treat2   (Can I just check) are you in the middle of a course of treatment or not?
      (1)   In the middle of treatment
      (2)   Not

ASK IF: Treat2 = mid

Thisone   INTERVIEWER - IN THE FOLLOWING QUESTIONS REFER TO THIS COURSE OF TREATMENT.
      PRESS <ENTER> TO CONTINUE
      STRING[1]

ASK IF: YearDen = Yes

LastDen   Have you been to the dentist in the last 6 months, that is since last DMDL6MTH?
      (1)   Yes
      (2)   No

ASK IF: (YearDen = No)
OR ((NumCheck = 0)
AND (NumTroub = 0))

HowDen   About how long ago was your last visit to the dentist?
      PROMPT AS NECESSARY
      (1)   More than 1 up to 2 years ago
      (2)   More than 2 up to 3 years ago
      (3)   More than 3 up to 5 years ago
      (4)   More than 5 up to 10 years ago
      (5)   More than 10 up to 20 years ago
      (6)   More than 20 years ago
      (7)   Never

ASK IF: (LastDen = Yes)
OR (HowDen < never)

Checkup   The last time you went to the dentist what made you go? Was it because you were having some trouble with your teeth or for a check-up or for some other reason?
      (1)   Trouble with teeth
      (2)   For a check-up
      (3)   Other

ASK IF: Checkup = other

Checkup1 INTERVIEWER - RECORD OTHER REASONS
      SET [4] OF
      (1)   To have tooth out
      (2)   New dentures/plate fitted
      (3)   Dentures/plate repaired
      (4)   Dentures/plate adjusted
      (5)   To have root/tooth stump removed
      (6)   Other trouble with dentures
      (7)   Other trouble with natural teeth
      (8)   Check-up for dentures

ASK IF:
(HowDen < never) or (HowDen = DNA)

DVisits   When people go to the dentist they sometimes have to make more than one visit (for a course of treatment) When you last went to the dentist how many visits did you make?
      PROMPT AS NECESSARY
      (1)   One visit
      (2)   Two visits
      (3)   Three visits
      (4)   Four visits
      (5)   Five visits or more

ASK IF:
(HowDen < never) or (HowDen = DNA)

DenDone   {introductory question - not on data file}
      (Can I just check) in the visit(s) you made to the dentist what did you have done? Did you have...

ASK IF:
(HowDen < never) or (HowDen = DNA)

ChekUp ...a check-up (examination)?
    (1) Yes
    (2) No

ASK IF:
(HowDen < never) or (HowDen = DNA)

TeethOut ...teeth taken out?
    (1) Yes
    (2) No

ASK IF:
(HowDen < never) or (HowDen = DNA)

XRays ...X-rays taken?
    (1) Yes
    (2) No

ASK IF:
(HowDen < never) or (HowDen = DNA)

Impress ...impressions taken?
    (1) Yes
    (2) No

ASK IF:
(HowDen < never) or (HowDen = DNA)

DenRepr ...dentures repaired
    (1) Yes
    (2) No

ASK IF:
(HowDen < never) or (HowDen = DNA)

DenFitt ...new dentures fitted?
    (1) Yes
    (2) No

ASK IF: (HowDen < never) or (HowDen = DNA)

Abcess ...treatment for an abcess?
    (1) Yes
    (2) No

ASK IF:
(HowDen < never) or (HowDen = DNA)
AND: Natural = Yes

TeethFil ...teeth filled?
    (1) Yes
    (2) No

ASK IF:
(HowDen < never) or (HowDen = DNA)
AND: Natural = Yes

CrownFit ...crowns (re)fitted?
    (1) Yes
    (2) No

ASK IF:
(HowDen < never) or (HowDen = DNA)
AND: Natural = Yes

TeethSca ...teeth scaled (scraped, cleaned) and polished?
    (1) Yes
    (2) No

ASK IF:
(HowDen < never) or (HowDen = DNA)

DenDoneO ...some other treatment?
    (1) Yes
    (2) No

ASK IF: DenDoneO = Yes

DenDonOO INTERVIEWER - RECORD ANY OTHER
TREATMENT.
SET [3] OF
    (2) Dentures removed and mouth checked
    (3) Gum treatment
    (4) Some other treatment

ASK IF: (HowDen < never) or (HowDen = DNA)

NHS Was your treatment under the NHS, was it private or was it something else?
DO NOT PROMPT
    (1) National Health Service
    (2) Private
    (3) N.H.S. and private
    (4) School/Community dental service
    (5) Armed forces
    (6) Dental hospital (hospital)
    (7) Dentist at your workplace
    (8) Through insurance
    (9) With a dental plan
    (10) Something else?

ASK IF: NHS = other

NHSO INTERVIEWER - RECORD OTHER SOURCE OF
TREATMENT.
STRING[120]

ASK IF:
(HowDen < never) or (HowDen = DNA)
AND: Natural = Yes

Cost How much did the treatment cost you?
PENCE GO AFTER THE DECIMAL PLACE eg 25.45
0.00..9997.00

ASK IF: Cost > 0

Install Did you pay through installments, either monthly or annually?
    (1) Yes, monthly
    (2) Yes, annually
    (3) Yes, at other intervals
    (4) No

ASK IF: Cost > 0

CExpect — Did the treatment cost more than you expected, about what you expected or less than you expected?
   (1)   More than expected
   (2)   About what expected
   (3)   Less than expected
   (4)   Other expectation
   (5)   Did not know what to expect

ASK IF: CExpect = other

CExpectO — INTERVIEWER - RECORD OTHER
EXPECTATIONS
STRING[100]

ASK IF:
(HowDen < never) or (HowDen = DNA)
AND: Natural = Yes

Free — If you had dental treatment under the NHS in the next 4 weeks do you think you would be entitled to free treatment or would you have to pay something towards the cost?
   (1)   Free
   (2)   Have to pay
   (3)   Don't know

ASK IF:
(HowDen < never) or (HowDen = DNA)
AND: Natural = Yes

First — Thinking about the dental practice you went to last time had you been there before or was that your first time at that practice?
   (1)   Been before
   (2)   First time

ASK IF: AND: First = bef

NumAtend — For about how many years have you been going to that dental practice?
   (1)   Less than a year
   (2)   One year less than two
   (3)   Two years less than five
   (4)   5 years or more
   (5)   Don't know

ASK IF:
(HowDen < never) or (HowDen = DNA)
AND: Natural = Yes

HChoose — How did you choose that particular dental practice?
DO NOT PROMPT CODE ALL THAT APPLY
SET [4] OF
   (1)   Family dentist
   (2)   Recommended
   (3)   Nearest
   (4)   Can't remember
   (6)   No choice
   (7)   Emergency dentist
   (8)   By chance
   (9)   Lady dentist
   (10) Other

ASK IF: AND: other IN HChoose

HChooseO — INTERVIEWER - RECORD OTHER REASONS
STRING[120]

ASK IF:
(HowDen < never) or (HowDen = DNA)
AND: Natural = Yes

Next — Will you go to the practice again next time?
   (1)   Yes
   (2)   No

ASK IF: AND: Next = No

WChange — [*]Why will you change your dental practice next time?
CODE ALL THAT APPLY
SET [4] OF
   (1)   No longer convenient
   (2)   dentist retired/moved/died
   (3)   informant not satisfied with dental treatment
   (4)   it's difficult to get an appointment
   (5)   Informant never goes regularly to the dentist
   (6)   last visit was extraordinary, not the informants usual dentist
   (7)   heard of better dentist
   (8)   no longer in system
   (9)   Other

ASK IF: AND: Other IN WChange

WWChange — INTERVIEWER - RECORD OTHER REASONS
STRING[200]

Comment — We have asked you a lot about dental health and dentistry. Is there anything you would like to say that we haven't asked you about?
   (1)   Yes
   (2)   No

ASK IF: Comment = Yes

CommentO — [*]What would you like to add?
CODE ALL THAT APPLY
SET [10] OF
   (1)   No NHS dentist available
   (2)   Dislike drift from NHS
   (3)   Treatment should be free
   (4)   Costs too much (no mention of NHS/free)
   (5)   Can't get appointment
   (6)   Dentist over-loaded
   (7)   Satisfied
   (8)   Better than in past
   (9)   Frightened of dentist
   (10) Other

ASK IF: Other IN CommentO

CommntOO — INTERVIEWER
RECORD OTHER REASONS
STRING[200]

ILO EMPLOYMENT STATUS

All adults

Wrking — Did you do any paid work in the 7 days ending Sunday the [DATE], either as an employee or as self-employed?
    (1) Yes
    (2) No

ASK IF: (Age < 63)
OR ((Age < 65)
AND (Sex = Male))

SchemeET — Were you on a government scheme for employment training?
    (1) Yes
    (2) No

ASK IF: AND: Wrking = No
AND: (LILO1 = 1)
OR (SchemeET = No)

JbAway — Did you have a job or business that you were away from?.....
    (1) Yes
    (2) No
    (3) Waiting to take up a new job/business already obtained

ASK IF: (JbAway = No)
OR (JbAway = Waiting)

OwnBus — Did you do any unpaid work in that week for any business that you own?
    (1) Yes
    (2) No

ASK IF: OwnBus = No

RelBus — ...or that a relative owns?....
    (1) Yes
    (2) No

ASK IF: AND: RelBus = No

Looked — Thinking of the 4 weeks ending Sunday the [DATE], were you looking for any kind of paid work or government training scheme at any time in those 4 weeks?
    (1) Yes
    (2) No
    (3) Waiting to take up a new job or business already obtained

ASK IF: AND: ((Looked = Y)
OR (Looked = Wait))
OR (JbAway = Waiting)

StartJ — If a job or a place on a government scheme had been available in the week ending Sunday the [DATE], would you have been able to start within 2 weeks?
    (1) Yes
    (2) No

ASK IF: AND: (Looked = N)
OR (StartJ = No)

YInAct — What was the main reason you did not seek any work in the last 4 weeks/would not be able to start in the next 2 weeks?
    (1) Student
    (2) Looking after the family/home
    (3) Temporarily sick or injured
    (4) Long-term sick or disabled
    (5) Retired from paid work
    (6) None of these

LAST JOB

If economically inactive or unemployed

(DVILO3a = EcInAct) OR (DVILO3a = Unemp)

Everwk — Have you ever had a paid job, apart from casual or holiday work?
    (1) Yes
    (2) No

ASK IF: Everwk = Yes

DtJbL — When did you leave your last PAID job?
FOR DAY NOT GIVEN....ENTER 15 FOR DAY
FOR MONTH NOT GIVEN....ENTER 6 FOR MONTH

DVJb12ML — DV for unemployed/inactive but has worked in last 12 months
    (1) Worked in last 12 months
    (2) NOT worked in last 12 months

JOB DETAILS

Adults currently working or have worked in the past

(DVILO3a = InEmp) OR (EverWk = Yes)

IndD {not on data file)
    CURRENT OR LAST JOB
    What did the firm/organisation you worked for mainly make or do (at the place where you worked)?
    DESCRIBE FULLY - PROBE MANUFACTURING or PROCESSING or DISTRIBUTING ETC. AND MAIN GOODS PRODUCED, MATERIALS USED, WHOLESALE or RETAIL ETC.
    STRING[80]

OccT {not on data file)
    JOBTITLE CURRENT OR LAST JOB
    What was your (main) job (LMainJb3 [DATE])?
    STRING[30]

OccD {not on data file)
    CURRENT OR LAST JOB
    What did you mainly do in your job?
    CHECK SPECIAL QUALIFICATIONS/TRAINING NEEDED TO DO THE JOB
    STRING[80]

Stat — Were you working as an employee or were you self-employed ?
(1) Employee
(2) Self-employed

ASK IF: Stat = Emp

Manage — Did you have any managerial duties, or were you supervising any other employees?
ASK OR RECORD
(1) Manager
(2) Foreman/supervisor
(3) Not manager/supervisor

ASK IF: Stat = Emp

EmpNo — How many employees were there at the place where you worked?
(1) 1-24
(2) 25 or more

ASK IF: Stat = SelfEmp

Solo — Were you working on your own or did you have employees?
(1) on own/with partner(s) but no employees
(2) with employees

ASK IF: Solo = WithEmp

SENo — How many people did you employ at the place where you worked?
(1) 1-24
(2) 25 or more

OEmpstat — Observed employment status
1..8

FtPtWk — In your (main) job were you working:
(1) full time
(2) or part time?

EDUCATIONAL ATTAINMENT

All adults

EdAttn1 — Do you have any educational qualifications for which you received a certificate? ....
(1) Yes
(2) No

EdAttn2 — Do you have any professional, vocational or other work-related qualifications for which you received a certificate? ....
(1) Yes
(2) No

ASK IF: (EdAttn1 = Yes)
OR (EdAttn2 = Yes)

EdAttn3 — Was your highest qualification:
(1) at degree-level or above
(2) or another kind of qualification?

EdYears — ASK OR RECORD
How old were you when you finished your continuous full-time education
STILL IN = 96
NEVER HAD = 97
5..97

OTHER CLASSIFICATION QUESTIONS

Head of household only

Car — A lot of people own cars these days.
(Can I just Check) is there a car or van normally available for use by you or any members of your household?
(1) Yes
(2) No

longlive — How long have you lived in this part of the country, that is within 50 miles of here....
RUNNING PROMPT
(1) ...all your life...
(2) 10 years or more...
(3) or less than 10 years?

CHECK FOR INCOME QUESTIONS

Head of household or partner

AskInc — THE NEXT SET OF QUESTIONS ARE ON HOUSEHOLD INCOME. THE INFORMATION SHOULD BE COLLECTED FROM EITHER THE HEAD OF HOUSEHOLD OR THEIR PARTNER. At least one person in the household should be coded Yes at THIS question (and stay coded Yes here, even if you have to record REFUSED at the income questions themselves).
IS THE RESPONDENT THE PERSON YOU WANT TO COLLECT THE HOUSEHOLD'S INCOME DATA FROM?
(1) Yes
(2) No

INCOME

Head of household or partner

((DVHSIZE=1) OR ((LDM1=QHoH.HoHnum) AND (QHoH.HoHprtnr = 11))) OR (AskInc = Yes)

SrcInc — SHOW CARD T
This card shows various possible sources of income. Can you please tell me which kinds of income you (HoH) (and spouse/partner) receive?
CODE ALL THAT APPLY
SET [9] OF
(1) Earnings from employment or self-employment
(2) Pension from former employer
(3) State Pension
(4) Child Benefit
(5) Income Support

(6) Other state benefits
(7) Interest from savings etc.
(8) Other kinds of regular allowance from outside the household
(9) Other sources e.g. rent
(10) No source of income

ASK IF: AND: (Earn IN SrcInc)
OR (Pens IN SrcInc)
OR (ChldBn IN SrcInc)
OR (Mob IN SrcInc)
OR (IS IN SrcInc)
OR (Ben IN SrcInc)
OR (Intrst IN SrcInc)
OR (OthReg IN SrcInc)
OR (Other IN SrcInc)

Gross   (I've just been asking you about where you both get your income from but can I first ask about HOH's income). Will you please look at this card and tell me which group represents HOH's total income from all these sources before deductions for income tax, National Insurance etc. SHOW CARD U AND EXPLAIN
ENTER BAND NUMBER
1..32
ASK I: Gross = 32

Gross3   Could you please look at the next card and give me DMNAMES[QHoH.HoHnum]'s total income as an annual amount from this card?
SHOW CARD V
ENTER BAND NUMBER
1..60

ASK IF: HoHprtnr < 11
AND: Gross <> REFUSAL

Spinc   Does [HOH's partner] have any separate income of her own?
(1) Yes
(2) No

ASK IF: Spinc = Yes

SGross   Which group represents [HOH partner]'s total income from all these sources before deductions for income tax, National Insurance etc.
SHOW CARD U AND EXPLAIN
ENTER BAND NUMBER
1..32

ASK IF: SGross = 32

SGross3   Could you please look at the next card and give me DMNAMES[QHoH.HoHprtnr]'s total income as an annual amount from this card?
SHOW CARD V
ENTER BAND NUMBER
1..60

ASK IF: QHoH.HoHprtnr < 11
AND: (Gross = DONTKNOW)
OR (SGross = DONTKNOW)

JntInc   Would it be possible for you to tell me which group represents the total income of [HOH] and [HOH's partner] taken together - before any deductions?
SHOW CARD U
ENTER BAND NUMBER
1..32

ASK IF: AND: JntInc = 32

Gross5   Could you please look at the next card and give me that total income taken together as an annual amount from this card?
SHOW CARD V
ENTER BAND NUMBER
1..60

JWeekGr   THIS IS CALCULATED BY THE COMPUTER. YOU DO NOT NEED TO ENTER ANYTHING HERE. PRESS ENTER TO CONTINUE.
(IF YOU WANT TO CHANGE IT USE THE UP ARROWS TO GO BACK AND CHANGE THE INCOME CODE(S)).
0..999999

ASK IF: ((NSrc IN SrcInc)
OR (JWeekGr = RESPONSE))
AND ((QTHComp.NumAdult > 2)
OR ((QTHComp.NumAdult = 2)
AND (QHoH.HoHprtnr = 11)))

IfHSrc   Can I just check, does anyone else in the household have a source of income?
(1) Yes
(2) No

ASK IF: AND: IfHSrc = Yes

HGross   (And now) thinking of the income of the household as a Whole, Which of the groups on this card represents the total income of the Whole household before deductions for income tax, National Insurance etc.
SHOW CARD U
ENTER BAND NUMBER
1..32

ASK IF: HGross = 32

HGross2   Could you please look at the next card and give me that as an annual amount from this card?
SHOW CARD V
ENTER BAND NUMBER
1..60

HWeekGr   THIS IS CALCULATED BY THE COMPUTER. YOU DO NOT NEED TO ENTER ANYTHING HERE. PRESS ENTER TO CONTINUE.
(IF YOU WANT TO CHANGE IT USE THE UP ARROWS TO GO BACK AND CHANGE HGROSS/HGROSS2)
0..9999